Advances in 21st Century Human Settlements

Series editor

Bharat Dahiya, Chulalongkorn University, Bangkok, Thailand

More information about this series at http://www.springer.com/series/13196

T.M. Vinod Kumar
Editor

E-Democracy for Smart Cities

Springer

Editor
T.M. Vinod Kumar
School of Planning and Architecture, New Delhi (SPA-D)
Calicut, Kerala
India

ISSN 2198-2546 ISSN 2198-2554 (electronic)
Advances in 21st Century Human Settlements
ISBN 978-981-10-4034-4 ISBN 978-981-10-4035-1 (eBook)
DOI 10.1007/978-981-10-4035-1

Library of Congress Control Number: 2017932771

Printed on acid-free paper

This Springer imprint is published by Springer Nature
The registered company is Springer Nature Singapore Pte Ltd.
The registered company address is: 152 Beach Road, #21-01/04 Gateway East, Singapore 189721, Singapore

Contents

Editor and Contributors

About the Editor

Professor T.M. Vinod Kumar had 46 years of experience in Urban Planning, as teacher, researcher, adviser and consultant and worked in India, China, Bhutan, Nepal, Malaysia, Indonesia and Hawaii USA. He was Dean of Studies, Head of the Department of Urban Planning, Head Centre for Systems Studies and Analysis, Centre for GIS and Remote Sensing, and Centre for Urban Studies of School of Planning and Architecture, New Delhi; Visiting Professor National Institute of Technology, Calicut, and Institute of Technology Bandung, Indonesia and Professional Associates, East West Resources Systems Institute Hawaii, Fellow Centre for the Study of Developing Societies, Delhi, Project Manager in Council for Social Development, New Delhi, Regional Program Coordinator at International Centre for Integrated Mountain Development (ICIMOD) and Planner-Engineer the Ford Foundation. He is author of many books and journal articles. He edited recently "Geographic Information System for Smart Cities" (2014), "E-Governance for Smart Cities" (2015: Springer), and "Smart Economy in Smart Cities" (2016: Springer).

Contributors

Dr. Femi Ola Aiyegbajeje has a Ph.D. in Geography, a Master of Science (M.Sc) degree in Geography and a Bachelor of Science (B.Sc) degree in Geography from the University of Ibadan, Ibadan Nigeria. Dr. Aiyegbajeje is currently with the Distance Learning Centre, University of Ibadan. His research interest is in Human Geography with specialization in Transport and Communication Geography, Urban and Regional Development.

Dr. Prabh Bedi has Ph.D. from School of Planning and Architecture, New Delhi. She completed her Post Baccalaureate Certificate in Geographic Information Systems from Pennsylvania State University, USA. She graduated from School of Planning and Architecture, New Delhi (India) with Master of Planning (Regional Planning). She completed her Master of Arts (Geography) and Bachelor of Arts (Honours-Geography) from Panjab University, India. Having 20 years in experience in planning and geospatial technologies, she's the founding director of Nipun Planners and Infrastructure Pvt. Ltd., India and visiting faculty at School of Planning and Architecture, New Delhi and other planning institutes around Delhi. Her research interests include Information Systems, geospatial technologies, NUIS, Smart Cities, demography, hydrology and climate change.

Prof. Jignesh G. Bhatt received B.E. (Instrumentation and Control Engineering), Gujarat University, India, 1997 and M.Tech. (Electrical Engineering, Specialization: Measurement and Instrumentation) from IIT Roorkee, India, 2010. After acquiring industrial experience on managerial positions during 1997–1999, since 2000, he has been serving Dharmsinh Desai University (DDU), Nadiad, India as Sr. Assistant Professor (Instrumentation and Control Engineering Department). He served as Co-ordinator and Principal Investigator for Govt. Sponsored Research, Extension and Community Development Projects, PG Thesis Supervisor, Invited Expert-Session Chair in Conferences and Journal Reviewer-Editorial Board Member. His research interests include Automation, e-Democracy, e-Governance, e-Learning, Instrumentation and Control, Smart City, Smart Grid, Solar City, Solar Energy and Wireless Sensor Networks.

P. Bimal holds a master's degree in City Planning from Indian Institute of Technology Kharagpur, and bachelors' degree in architecture from Kerala University. He had been working at JUSCO, Jamshedpur, before joining as assistant professor at Department of Architecture, NIT Calicut. He is currently pursuing Ph.D. in urban modelling. His research interests include, Urban modelling, Transportation, GIS and Remote Sensing, Urban design, Parametric modelling in architecture and architectural visualisations. He has authored chapters in previous books in this series.

Ms. Krista Busa is an Associate of Marketing and Business Development for UDP International. She holds a Bachelor's Degree in The Engineering of Regional Development and Urban Economics from Riga Technical University, Lativa, and earned a Bachelor's in Real Estate Economics while on exchange at Aalto University School of Engineering, Finland. Her main areas of focus are Economic development, Community Engagement, ICT based planning, and Marketing. She has worked with several public-sector projects on local and regional development levels while working for the European Institute of Public Administration in Barcelona (Spain) in 2015. She has also gained experience in marketing, fundraising and business development while working as a Corporate Relations Responsible on several international projects for European Students of Industrial Engineering and Management (ESTIEM) and working as a Marketing Manager for several private companies.

Dr. Krupesh A. Chauhan is Associate professor and Section Head in Urban Planning Section of CED, SVNIT, Surat. He got his master in WR Engineering and Town and Country Planning from South Gujarat University. He got his Ph.D. in Civil Engineering from SGU. He has field experience of 13 years and teaching experience of 19 years. His areas of interest area Traffic and Transportation planning, Housing, Infrastructure planning and management etc. He has published more than 20 International Journal papers and more than 40 conferences attended. He has guided 2 Ph.D. scholars and 6 ongoing.

Ms. Ho Yan Claudia Chow is a Research Assistant at the Institute for Sustainable Urbanisation and Planning Assistant at UDP International. She holds a Bachelor's Degree from the University of California, San Diego in Urban Planning & Studies. Her town planning experience includes planning applications; adaptive re-use in historic buildings; public engagement; and planning studies and research. Key projects with her involvement in Hong Kong include the Kai Tak Sports Park; a Dedicated Quarantine Facility in Lei Yue Mun Park; a temporary eating place in the New Territories. Key planning research and studies with her participation include the Smart Cultural Triangle Precinct Conservation Plan for Central district and the E-Engagement in Hong Kong planning processes. During her practice at UDP International, she is responsible for assisting town planning related projects and research. Her research interests include: Town Planning, Active Transportation (i.e. creating great mobility choices; and promoting sustainable growth); and innovative community outreach.

Prof. R.D. Desai got his bachelor of architecture (B. Arch) from M.S. University, Baroda. Prof. R.D. Desai got his Post Graduate in Town and Country Planning with "Problems of Fringe Areas of Cities" as specialization from School of Planning and Architecture, New Delhi. He got his L. L. B. degree from South Gujarat University, Surat. He has served as Advisor in Surat Urban Development Authority for more than two years. He has served as Director of Planning in Surat Municipal Corporation for over

Thirteen years with duties like town planning schemes, Town Development etc.

He is founder chairman of

1. Surat Centre of Indian Institute of Architecture, Gujarat chapter
2. Surat Centre of Gujarat Regional Chapter of Institute of Town Planners, India
3. Sanskruti Cultural Organization arranging cultural programs and dramas
4. Trustee, Shri Sogayo Kelavani Mandal conducting primary to higher secondary schools for urban poor-Shri D.K. Bhatkar School at Althan, Surat.

Currently he is President of Surat Centre, Gujarat Regional Chapter of ITPI and Hon. Director of GCPIA. His areas of expertise are Town Planning Schemes, Development Plan, City Fringe area Development etc.

Dr. Chen Feng received the M.A.Sc. and Ph.D. degrees from the Department of Electrical and Computer Engineering, University of Toronto, Canada, in 2009 and 2014, respectively. From 2014 to 2015, he was a Postdoctoral Fellow with Boston University, USA, and EPFL, Switzerland. He joined the School of Engineering, University of British Columbia, Kelowna, Canada, in July 2015, where he is currently an Assistant Professor.

His research interests include data networks, coding theory, information theory, and network coding. Dr. Feng was a recipient of the prestigious NSERC Postdoctoral Fellowship in 2014. He was recognized by the IEEE Transactions on Communications (TCOM) as an Exemplary Reviewer in 2015.

Dr. Sujata S. Govada is an award winning qualified urban designer, certified town planner and a registered architect in India, with over thirty-five years of diverse international experience in Hong Kong, PRC, India, Philippines and the US working on design and planning projects. She is the Founding Director of the Institute for Sustainable Urbanisation and the Founding and Managing Director of UDP International, a boutique global practice. Her expertise and research interests include sustainable urbanization, smart city development, transit and pedestrian oriented development, harbourfront planning, new town development, urban renewal, heritage conservation, affordable housing and community engagement. She is Adjunct Associate Professor at the School of Architecture, Chinese University of Hong Kong and previously taught in the urban planning and urban design programs at The University of Hong Kong. She is the Global Trustee of ULI and the Founding Vice President of the Hong Kong Institute of Urban Design and the Past President of the American Institute of Hong Kong.

Mrs. Kshama Gupta is currently working as Scientist in Indian Institute of Remote Sensing (Indian Space research Organization), Dehradun. She had completed her M.Tech. in Urban Planning from School of Planning and Architecture, New Delhi, India in 2001 after the completion of Bachelors in Architecture from Malviya National Institute of Technology, Jaipur, India. Since then she is working as researcher in the field of remote sensing and GIS applications for urban management and made contributions in many national level projects and research areas. She has more than 50 publications to her name as research papers in international/national journals/conferences and ISRO technical reports. Her research interest includes smart planning, urban climate and micro climate, urban green spaces and 3D modeling of urban areas.

Dr. Omkar K. Jani is currently the Principal Research Scientist (Solar Energy) at Gujarat Energy Research and Management Institute (GERMI). His primary goals include advisory for various solar-related government and private initiatives, developing and undertaking fundamental and applied research, and capacity building. Dr. Jani is a member of the State Advisory Committee of the Gujarat Electricity Regulatory Commission (GERC). He is also an advisor to several clean-tech companies and NGOs. Dr. Jani received his Ph.D. in Electrical Engineering from Georgia Institute of Technology with specialization in Solar Photovoltaic Science and Engineering, and his post-doctoral fellowship from the Institute of Energy Conversion, Delaware, USA.

Yury Kabanov is a lecturer at the Department of Applied Political Science, the National Research University Higher School of Economics (St. Petersburg). His research interests include comparative authoritarianism, political regime dynamics, electronic participation and Internet politics. He has published several articles in Russian and English on the development of e-government/open government/open data in Russia and comparative e-participation development in non-democracies.

Ashmita Karmakar A graduate in Architecture, Masters in Town and Regional Planning and Ph.D. in Environmental Planning from School of Planning and Architecture, New Delhi, India. Ashmita has been working in the field of Urban Planning since year 2002. She has worked with various Consulting companies, Semi Government companies and Government Authorities in the UAE from 2007 focusing on developing concept plan to final master plan proposal, feasibility studies, impact assessment studies and formulation of Development control guidelines and regulations.

Apart from the professional field, she is actively participating in academics by mentoring young planners of the region and contributing articles for journals and international publications.

Dr. Harish Chandra Karnatak is currently working as a Scientist and Head, Geoweb Services, IT and Distance Learning Department at Indian Institute of Remote Sensing (IIRS), ISRO Dehradun, India. He received his Ph.D. Degree in Computer Science with specialization in Geoinformatics. Dr. Karnatak has made significant contributions in various national level projects and outreach programmes on applications of space technology for natural resources and disaster management in India. His area of specialization includes web/Internet based GIS, spatial DBMS, online geoprocessing and analysis. Currently he is leading a team of scientists and technologists for development of online digital content and capacity building programmes on space technology and its applications under IIRS outreach programme. He has published more than 60 peer reviewed research papers in various scientific journals, seminars and conferences. Dr. Karnatak is also a recipient of various professional awards in the area of remote sensing and geospatial technologies.

Mikhail Karyagin is a Ph.D. student at the National Research University Higher School of Economics (St. Petersburg). He is Master of Arts in Political Science and currently holds the position of a leading expert at the Infometer Project Center. His scientific interests are the open data research, ICTs and politics, political regimes and elections. He has carried out and published several articles on the topics in Russian and English.

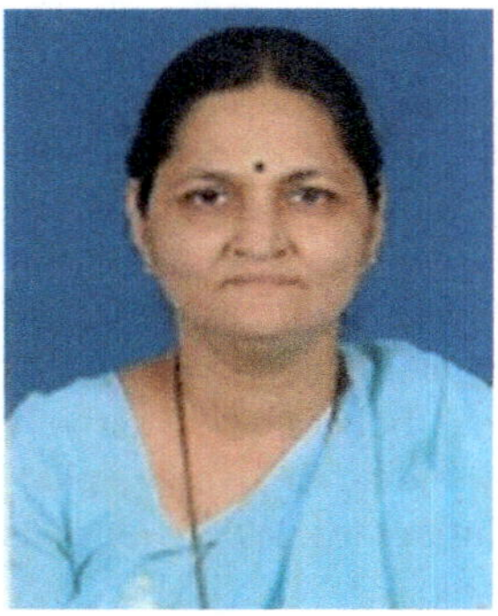

Dr. (Mrs.) Anjali K. Khambete is Associate professor of Civil Engineering Department, SVNIT, and Surat. She got her Bachelors' Degree in Civil engineering and masters in Environment engineering from SVNIT, Surat. She did her Ph.D. in "Integrated Efficiency based performance evaluation of wastewater treatment plant using fuzzy logic approach" from SVNIT, Surat. She has teaching experience of over 32 years. She has published 7 books and many papers in national and international journal and reviewer for various journals. She has guided 4 Ph.D. students and 38 M.Tech. Students throughout her career. She is active in consultancy work as auditor regarding environment engineering with government agencies and member in audit committee at Gujarat Pollution Control Board, Gandhinagar. She is acting as nominee director at Ms/Vapi Green Enviro Ltd. at Vapi and Sarigam CETP (SCT), Sarigam.

Professor Ashok Kumar, Ph.D., is working as a Professor in the School of Planning and Architecture, New Delhi since 1993. His research interests include Inclusive City Planning, Spatial Justice, and Planning Epistemologies. He has published nearly 100 papers in national and international refereed journals, written and edited 8 books, and published 18 book chapters. He is the Editor of the *ITPI Journal* since 2002. A co-authored chapter is slated for publication in *The Routledge Handbook of Planning Theory* in early 2017. Currently, Prof. Kumar is working on a co-authored book manuscript on "City Planning in India" under contract with Routledge, which is expected to be published in 2017.

Pramod Kumar is Head, Urban and Regional Studies Department, IIRS, Dehradun, India. He is also the Group Head of Programme Planning and Evaluation Group of IIRS. He is an alumnus of IIT, Kharagpur, India and joined Indian Space Research Organization in 1991. Earlier, he has worked as Assistant Engineer at CES, New Delhi. He has been involved in more than 50 national level/ technology demonstration and research projects using geospatial data and techniques to evolve solutions for natural resources management and brought out technical reports and research publications. He has published more than 40 papers in journals and conference proceedings and many technical reports. He is the recipient of ISRO Team Excellence Awards for two projects. At present, he has research interests in urban water utilities and urban hydrology.

Dr. Zheng Liu received the Doctorate degree in engineering from Kyoto University, Kyoto, Japan, in 2000. He joined and served the National Research Council (NRC) Canada, Ottawa, ON, Canada from 2001–2012. He was working for the Toyota Technological Institute, Nagoya, Japan from 2012–2015 and is currently with UBC Okanagan. His research interests include image and data fusion, computer vision, pattern recognition, sensors and sensor networks, IoT, structural health monitoring, and non-destructive inspection and evaluation. Dr. Liu is senior member of IEEE and serves on the editorial board for a number of peer-reviewed journals of IEEE, Elsevier, and Springer.

Dr. J.E.M. Macwan did his B.Tech in Civil Engineering from South Gujarat University, Surat. In the same university he got his Master degree in civil engineering with specialization in "Evaluation of … Bio-kinetic Constants of Industrial wastes". He got his Ph.D. from South Gujarat University with specialization in "Fuzzy Simulation analysis of Residential Location behavior for Metropolitan". He has total teaching experience of 33 years in S V National Institute of Technology, Surat. Currently he is Head of Civil Engineering Department. He is Author of over 23 International and 11 National Journal papers. He guided more than 60 Post Graduates in various fields of civil engineering and 4 Ph.D. Scholars. His areas of interest are Smart Mass Transportation, Smart Housing, Affordable Housing, Water and Waste water etc.

Prof. Mahavir holds a degree in Architecture followed by a Master's degree in Urban and Regional Planning. He subsequently obtained a PG Diploma in Remote Sensing from ITC, Netherlands. He was also awarded PhD jointly by the ITC and Utrecht University, Netherlands. He has been engaged in teaching, practice and research in the fields of Urban and Regional Planning and GeoSpatial information for over 30 years. Prof. Mahavir has authored and edited several books. He has published over 40 papers, nationally and internationally. He is currently Professor and Editor of the SPACE Journal at School of Planning and Architecture, New Delhi, India.

Dr. Nezih Mrad received the Ph.D. degree in engineering from Pennsylvania State University in 1995, and two Masters degrees, in applied mathematics and in mechanical engineering, from the University of South Carolina, Columbia, in 1987 and 1988, respectively.

He is currently a Defence Scientist with the Defence R&D Canada (DRDC), Department of National Defence, Ottawa, ON, Canada. He is also an Adjunct Professor with Carleton University and the University of Sherbrooke. He has intensive experience in advanced sensing, wireless sensors and sensor networks, IoT, energy harvesting, and

RFID sensors. He conducts and manages R&D activities in CBM, SHM, PHM, micro- and nanotechnologies, and autonomic logistics. He has authored or co-authored over 250 publications.

Keziah Mwang'a is an urban planner and currently a Ph.D. candidate at the Gran Sasso Science Institute, Italy. She holds a BA in Urban and Regional Planning from the University of Nairobi and a Master of City and Regional Planning from the University of California Berkeley, USA. Her research and career interests are centered on the role of political governance, power relations and planning education in shaping urban development policy and practice. She has worked both in Kenya and USA with governments, non-profit organizations and research institutions working to promote equitable development.

Dennis Mwaniki is the Associate Director for Urban Planning and GIS at GORA Corp. Before joining GORA Corp in 2014, he worked for 3.5 years as a GIS and research consultant at the Global Urban Observatory (GUO) and the Best Practices Section of UN-Habitat. In GUO, he developed and implemented the methodology for calculating the amount of land allocated to streets for over 100 cities. He contributed to the State of the World's Cities report: Prosperity of Cities in 2012-13; and to the publication "streets as public spaces and drivers of urban prosperity. He has also worked as a GIS and urban planning consultant with the Slum Dwellers International—Kenya; the Center for Urban Research and Innovations at the University of Nairobi; and various private companies based in Nairobi. He holds a Masters in Environmental Planning and Management and a Bachelors degree in Urban and Regional Planning. He is a member of the Global Human Settlements Working Group of the Group on Earth Observations (GHS WG) and was a co-author in the publication "Smart Economy in Smart Cities".

Baraka Mwau is an Urban Planner and Researcher, currently a consultant at UN-Habitat and Associate Researcher: Centre for Urban Research and Innovations, University of Nairobi. He has previously worked with Shack/Slum Dwellers International and Affiliate Organizations in Kenya, South Africa, Tanzania and Malawi as an urban planner and as a community planning studio facilitator for projects done in collaboration with Association of African Planning Schools. Mwau's research interests are: urbanization and urban development in sub-Saharan Africa, urban informality, participatory planning and development.

Femi Olokesusi has a Doctorate Degree in Geography from the University of Ibadan, Oyo State, Nigeria with specialisation in Environmental Planning and Management. A fellow of the Nigerian Institute of Town Planning, he retired as a Research Professor from the Nigerian Institute of Social and Economic Research (NISER) in Ibadan, Nigeria. He held several positions NISER including Acting Director-General/Chief Executive Officer. A 1994/95 Fulbright African Senior Scholar, he has conducted several studies for many national and international agencies such as UNDP, UNCTAD, and UN-Habitat. He served as a resource person to draft Nigeria Vision 20:2020 and the National Infrastructure Master Plan. He was a Land Use/Urban Planning consultant for UN-Habitat in 2014/2015 on the preparation of City Structure Plans for Lokoja and Dekina in Nigeria. He is currently a member of the Independent Advisory Group of the World Bank-assisted Ibadan Urban Flood Management Project and a Professor of Urban and Regional Planning at the College of Social and Management Sciences, Afe Babalola University, Ado-Ekiti, Ekiti State, Nigeria.

Romanus Opiyo is a lecturer at the Department of Urban and Regional Planning and an associate researcher with Centre for Urban Research and Innovations, University of Nairobi, Kenya. He has ten years University teaching and research supervision experience, and more than fifteen years undertaking research work. His research experience and interests are on urban security, transportation and general land use planning, climate change and institutional governance and livelihoods for which he has published widely. Dr. Opiyo has also consulted at various capacities with the Government of Kenya, UNEP, UN-HABITAT, UNDP, UNCRD, World Bank, DFID, IDRC, SIDA, among other national and international NGOs. He is a member of the Kenya Institute of Planners (KIP), the Town and County Planners Association of Kenya (TCPAK) and Environmental Institute of Kenya (EIK). He is also a member of the Kenya Urban Sector Reference Group which is supporting the Council of Governors Urban Development Committee, a Member of Kenya Civil Society Network for NMT and Road Safety (CIVNET) and a Member of Kenya Transport Research Network. Dr. Opiyo was a co-author in the publication "Smart Economy in Smart Cities". He holds a Ph.D. and Masters in Urban and Regional Planning and a Bachelors degree in Social Sciences.

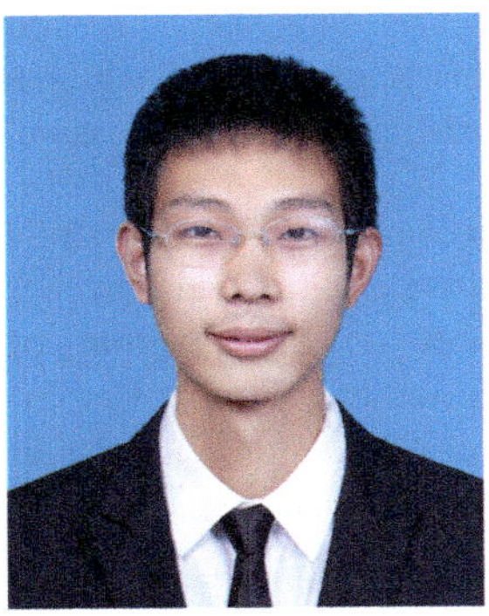

Renming Qi received the B.Eng. degree from University of Science and Technology of China in 2016 and is now pursuing an M.A.Sc. degree at the School of Engineering, University of British Columbia, Kelowna, Canada, starting from September 2016. His current research interests include information theory and distributed systems, with a particular focus on blockchain techniques.

Aruna Sri Reddi has completed her Ph.D. from the Department of Architecture and Regional Planning, Indian Institute of Technology, Kharagpur. She has dual Master's degree in Urban and Regional Planning from India and USA. Aruna has more than five years of experience in the area of Transportation planning and GIS from USA. She is a trained architect and worked in India, Belarus, and Russia in various architectural projects and has ten years of experience. She started her career in teaching from National Institute of Technology, Calicut. Currently she is working as associate professor at Pillai College of Architecture, Maharashtra.

Mr. Timothy Rodgers is a Research Associate at the Institute for Sustainable Urbanisation and UDP International. He holds a Bachelor's Degree from McGill University in Economics and Geography (Urban Systems) and a Master's Degree from the University of Hong Kong in Transport Policy and Planning. His experience at UDP International includes research and report writing on current issues including Smart Cities and Affordable Housing, assisting in the development of a Smart City Framework with a focus on People, Place and Planet as core elements. He was the team lead for a research chapter contributing to a collaborative international research book publication on the topic of "Smart Economy in Smart Cities", and has benchmarked affordable housing policies and strategies of major global cities to aid Hong Kong's affordable housing discussion.

Viacheslav Romanov has graduated from the Urals State Law Academy Justice Institute (lawyer/barrister specialization). Presently he holds the position of Director for Analytics in the Infometer Project Center. His area of interest includes data analysis and visualization, as well as the impact of ICTs on everyday life.

Ummer Sahib is a City Planner with 30 years of experience in geospatial technology and development of innovative solutions for City Planning and Management. He has been responsible for process automation in executing large GIS, LIS and Urban mapping projects in Asia, Middle East, Europe and Africa. Mr. Sahib is currently the Executive Director of Informap Technology Center in Sharjah, UAE heading Future Technologies and application development.

Starting his career in GIS software development project for Department of Electronics, Govt. of India, Mr. Sahib's expertise lies in GIS, Open source web mapping applications, navigation data production, Geo-analytics and perfecting a virtual addressing system called GRL (Geo Reference Locator) for the UAE. Since 2005, Mr. Sahib is the driving force behind his company's creative innovation on GPS and NFC technology solutions. He is passionate on integrating IoT, Big Data and cloud technologies for establishing Urban Operating System to realize the true meaning of a SMART CITY.

Mr. Sahib holds Post Graduate Diploma (M.Tech) in City Planning (1985) from School of Planning and Architecture, CEPT Ahmedabad, India. He is also the Associate member of the Institute of Town Planners, ITPI India, since 1986.

He has co-authored book on **E-Governance for Smart Cities** published by Springer in January 2015. He was a panel member in the 4th GCC Municipalities & smart cities conference and speaker at the GITEX 2014 Smart Cities conference held in Dubai. He has published several papers on GIS/LIS and accident management and is currently leading a team on designing and developing Mobile Apps for Smart Cities.

Dr. A. Senthil Kumar received M.Sc. (Engg.) and Ph.D. from the Indian Institute of Science, Bangalore in the field of image processing in 1985 and 1990, respectively. He worked as consultant and post-doctoral fellow at the Raman Research Institute, Bangalore for six months before joining ISRO in 1991. Since then, he has served in Indian Remote Sensing programmes in various capacities. He has been involved in sensor characterization, radiometric data processing, image restoration, data fusion techniques, and artificial neural networks. As an outcome of his research work, he has published more than 60 technical papers in international journals and conferences and ISRO technical reports in the areas of image processing and remote sensing. He has received prestigious Prof. Satish Dhawan Award conferred by the Indian Society of Remote Sensing apart from ISRO Team Awards for his contribution to Chandrayaan-1 and Cartosat-1 missions.

He is currently serving as the Director of Indian Institute of Remote Sensing, Dehradun and UN affiliated Centre for Space Science and Technology Education in Asia and the Pacific. Besides, he is the President of ISPRS Technical Commission V on Education and Outreach and Vice Chair of CEOS Working Group on Capacity Building and Data Democracy.

Swapnil Shrivastav I am a 23-year old entrepreneur working in the domain of clean-tech, water and Internet of Things (IoT). I graduated as an architect from National Institute of Technology Calicut in 2016. Design thinking and sustainable solutions are my strong focus areas. I have previously worked on projects involving rain-water harvesting, grey-water treatment systems, architectural design and UI/UX. At my startup, Uravu Labs, we are working on technologies which involve generating drinking water from air and are on a mission to make water accessible to everyone!

Asfa Siddiqui is currently working as Scientist in Indian Institute of Remote Sensing (Indian Space research Organization) since 2014. She did her Bachelors in Architecture from Govt. College of Architecture, Lucknow and Masters in Urban Planning from School of Planning and Architecture, New Delhi. She has also worked in NIT Calicut prior to joining ISRO. Her research interests are in areas of smart energy, smart environment, thermal remote sensing and hyperspectral remote sensing.

Dr. Neha Goel Tripathi is an environmental planner with over 15 years of experience in Environmental Management and Architecture. She is presently working as Assistant Professor in School of Planning and Architecture, New Delhi. She did her post graduation from School of Planning and Architecture, New Delhi in Master of Planning, specialization in Environmental Planning (2003) and was awarded the Gold medal for overall performance. She is Associate member, Institute of Town Planners, New Delhi, Member, Council of Architecture, New Delhi and Member, Indian Building Council, New Delhi. National Environmental Science Academy. Experience of more than 10 years in the field of environmental planning and architecture and worked on projects like Zonal Development Plan for Eco sensitive Area, Mount Abu, Indian Institute of Management, Indore, Manali Redevelopment Plan etc. She has published articles on field of Ecological foortprint, Solar Zoning, Climate change etc. Awarded the first prize from the Hon'ble Minister for Housing & Urban Poverty Alleviation, Kumari Selja for paper presentation, Climate Change and Indian Cities Perspectives presented for World Habitat Day 2011 organized by HUDCO, New Delhi.

Partha Tripathy has a Bachelors in Engineering from SVNIT, Surat, Post Graduation in Project Management from CEPT, Ahmedabad and a MA in Regulatory Studies from TERI University, New Delhi. Partha is pursuing his Ph.D. in Urban Planning from SVNIT, Surat working on Smartness Index for Indian Cities. He has published 3 research papers in International Conferences on E-governance, Mobility in Smart Cities and financing of Smart Cities projects. He has nearly 16 years of experience in Project Development in the PPP projects and has worked in State Government PSUs in Gujarat as well as Odisha. Partha is a certified PPP trainer under the National PPP Capacity Building Programme (NPCBP) of the Department of Economic Affairs (DEA), Ministry of Finance, Govt. of India. Partha is currently working as a consultant with International Finance Corporation.

Mr. Frank Wong is a versatile and multi-disciplinary professional in the development industry specializing in cross-disciplinary project management and town planning. Throughout his career, he has acquired a variety of project management experience including cultural district development, new town development, and design and implementation of sustainable greening framework. His town planning experience includes planning studies, statutory planning, public engagement, development consultancy, urban design and master planning, strategic planning and cross-boundary facilities planning. During his practice at UDP International, he is responsible for business development as well as providing advice to clients on various aspects ranging from real estate development (incl. preliminary valuation, retail and market studies for commercial property, strategic advice on acquisition), town planning and urban design.

Chapter 1
State of the Art of E-Democracy for Smart Cities

T.M. Vinod Kumar

Abstract A well-functioning smart city require a vibrant e-democracy. The theoretical foundation for "E-Democracy for Smart Cities" is presented here. This chapter in detail studies the relationship of smart cities and e-democracy. The basic democratic principles are stated and then the culture of democracy that evolved through many centuries were studied. The present constitutional provision and ancient Greek and Indian models of democracy were discussed and democracy in panchayat system were also studied. Advantages and disadvantages of democracy were discussed. E-democracy was defined, E-democratic models investigated and e-democracy typologies were postulated as well as ten models of e-democracy. Finally, several tools used in e-democracy were enumerated. The chapter concluded with a vision of e-democracy in large urban agglomeration.

Keywords Smart cities · Democracy · Culture of democracy · Ancient Indian and Greece practice of democracy · Citizen participation · Democracy and panchayat system · Advantages and disadvantages of democracy · E-democracy defined · E-democracy and smart cities · E-democracy and smart economic development · E-democracy typologies · Ten models of e-democracy · Tools of e-democracy

1.1 Smart Cities

Smart city is a knowledge-based city that develops extraordinary capabilities to be self-aware, how it functions 24 h and 7 days a week and communicate, selectively, in real time knowledge to citizen end users for satisfactory way of life with easy public delivery of services, comfortable mobility, conserve energy, environment and other natural resources, and create energetic face to face communities and a vibrant urban economy even at a time there are National economic downturns [1].

T.M. Vinod Kumar (✉)
National Institute of Technology, Calicut, Kerala, India
e-mail: tmvinod@gmail.com

T.M. Vinod Kumar (ed.), *E-Democracy for Smart Cities*, Advances in 21st Century Human Settlements, DOI 10.1007/978-981-10-4035-1_1

This self-awareness of smart cities is derived out of successfully modelling through ICT technology means, (such as for example geospatial and sensors technologies) a real-world smart city at every instance to a cyberspace Smart city model that can be used to intervene in real world smart city. Simplistically, it can be a website which can be used to intervene automatically in the real-world cities using big data cloud and apps by intelligent entities like Internet of Things. Using a ubiquitous network of sensors, the smart city is associated with the physical city. A cloud computing platform handles the massive data storage, computation, analysis, and decision-making process, and conducts automated control based on the results of those analyses and decisions. At the level of the smart city, the digital and physical cities can be linked by the Internet of Things, thereby forming an integrated cyber-physical space. Smart Water supply or intelligent water is a good example of this modelling. Therefore, a simple formula, we can denote the smart city as follows: Smart City = digital city in cyberspace + Internet of Things + cloud computing. Using radio frequency identification (RFID), infrared sensors, global positioning systems (GPS), laser scanners, and other information-sensing devices, the Internet of Things connects things and the Internet based on protocols and conducts information exchange and communication to achieve intelligent identification, positioning, tracking, monitoring, and management. Cloud computing is an Internet-based computing model for public participation. Its computing resources (including computing power, storage capacity, interactivity, etc.) are dynamic, scalable, virtualized, and provided as a service. Cloud computing is a model of Internet-based computing, an extension of distributed computing and grid computing. Cloud computing can rapidly process the huge amounts of data produced by the smart cities and simultaneously service millions of users using apps.

A smart city is characterised by six components such as Smart People, Smart Economy, Smart Mobility, Smart Environment, Smart Living and Smart Government [1–3]. The ability of any city to transform and progress towards these six components of smart cities from the existing conditions in an integrated and comprehensive manner is the ability to transform the city to smart city. Prime movers for this capacity to transform cities, are Smart People and Smart Economy. This proposition contradicts the highly simplified approach of many Governments in developing countries world over who considers Smart Cities are there because of investment decisions made for deployment of smart ICT enabled infrastructure technologies such as Internet of Things and other smart city infrastructure which can be purchased from Multi-National Companies. This is far from the truth. The smart cities shall be considered as centres of emerging third industrial revolution [4, 5] where sharing economy develops and move towards marginal cost of production nearing to zero, triggered by a smart economy which is sustainable and resilient. A Smart city has the capacity to grow more and more smart people to multiply their opportunities for engaging in smart economy. The emergent needs of the smart economy and smart people fix the configuration of smart mobility, smart environment and resulting smart living. There can be many variations of these components as per the ecology. Here the technologies used is evolving over time and what is used some three years before will be totally obsolete afterwards and technologies are developed in these cities by smart people and not imported from multinationals. Without these two prime components of this smart city system, it is impossible to transform a city to smart city and smart city creation cannot take place by simply investing in smart infrastructure and smart ICT enabled technologies in a city.

'Smart City Economy', requires the following attributes.

1. A Smart City understands its economic DNA.
2. A Smart City is driven by innovation and supported by universities that focus on cutting-edge research, not only for science, industry and business but also for cultural heritage, architecture, planning, development and the like.
3. A Smart City highly values creativity and welcomes new ideas.
4. A Smart City has enlightened entrepreneurial leadership.
5. A Smart City offers its citizens diverse economic opportunities.
6. A Smart City knows that all economics works at the local level.
7. A Smart City is prepared for the challenges posed by and opportunities of economic globalisation.
8. A Smart City experiments to support and promotes sharing economy.
9. A Smart City thinks locally, acts regionally and competes globally.
10. A Smart City makes strategic investments on its strategic assets.
11. A Smart City develops and supports compelling national brand/s.
12. A Smart City insists on balanced and sustainable economic development (growth).
13. A Smart City is a destination that people want to visit (tourism).
14. A Smart City is nationally competitive on selected and significant factors.
15. A Smart City is resourceful, making the most of its assets while finding solutions to problems.
16. A Smart City excels in productivity.
17. A Smart City has the high flexibility of labour market.
18. A Smart City welcomes human resources that enhance its wealth.
19. A Smart City's inhabitants strive for sustainable natural resource management and understand that without this its economy will not function indefinitely [3].

These six components of smart cities are characterised by two important omissions. There is no explicit mention of smart technologies in these six components that can create smart cities which are generally ICT enabled and culturally bound. Also, there is no explicit mention about e-democracy. These two are implicit in these six components. Smart Technologies and e-democracy is an integral part of all these six components. A 24 h 7 days uninterrupted water supply which we may call smart water supply in smart city standards can only be possible if we embrace an ICT and SCADA (Supervisory Control And Data Acquisition)-enabled city water supply well known to engineers' world over. Technology development in all six components of smart cities is available off the shelf, however with the short life cycle and uncertain feasibilities.

1.1.1 Smart City as an Ecological and Cultural System and E-Democracy

The smart city is an ecological and cultural system. Because of the unique ecology of Smart Cities, a generalised model of the cyber city in cyberspace is not possible. So also, a generalised solution for a smart city is not workable. The spatial and ecological structure of built space forming land use and transportation system as also varies and other ecological elements of the city. However, these ecological elements are ideal candidates for dealing with the smart city conceptualisation and for cyberspace interventions at real time using Internet of Things.

The smart city is also having unique cultural characteristics. There are innumerable factors that influence the culture of a people in a smart city. The culture of the city is determined by the regularity of behaviour of inhabitants of smart cities. These factors determine the cultural attributes of smart cities. These cultural factors also have a bearing on smart city functioning and will influence the functioning and managing of smart cities in 24 h and 7 days a week functioning. Also, there is a cultural element that influences the deployment of smart technologies. While water supply can be efficiently provided for by smart technology, water use pattern in a smart city which aims for water conservation is culturally conditioned and require cultural adaptation to be water conserving. In any country where there is open air, defecation is culturally prevalent then the use pattern of built toilet must be by cultural transformation if it all it is to be used.

While Smart city as Ecosystem can be managed in cyberspace-modelling by Internet of Things, this is not sufficient to manage the smart city as socio- economic and cultural system. There are two additional systems along with Internet of Things are required. They are Internet of People or e-democracy and Internet of Governance or E-Governance.

Smart cities are the meeting place of heterogeneous socio-cultural groups, and cultures which demand heterogeneous use pattern of urban space and urban infrastructure. This can only be ensured by a powerful e-democracy system which we call Internet of People. Our experience shows that Government alone cannot manage and sort out all problems of smart cities and leads the city to the path of prosperity and wellbeing. For example, there is no history in any city where potholes are repaired the next day it is sighted. To achieve this, an e-democracy system for road repairs and drainage correction is required using smartphone apps.

Now Internet of People is not the legitimate party to address a variety of actions Government can do such as punishing for crime, collecting taxes, conducting a judicial enquiry and so on. These actions are based on legislations that protect the interest of all sections of society framed after great deliberation and scrutiny. They are implemented by judiciary and bureaucracy and people are not entitled to these powers.

Hence a right mixture of Internet of Things, Internet of Government and Internet of People are to be in place in a city for the effective functioning of smart cities.

Efficiency, transparency and accountability of bureaucracy charged with implementing Government policy go up if there is a proper mixture of Internet of Things, Internet of Governance and Internet of People and every action is visible to all on an internet.

In ecological management, energy conservation is paramount important. Many ecological issues can be easily solved by the deployment of Internet of Things. For example, energy conservation in street lighting can be affected by Internet of Things in such a way street lights brightens up only when a moving object approaches near the service area of street lights and otherwise it is very dim or even off. E-democracy can be used for reporting for replacement of bulbs but the operation of energy conservation in street lighting can only be executed by Interne to Things.

It may be noted that Internet of Governance is most expensive than Interne of Things and least Expensive is the Internet to People. For example, there are cases where the major part of revenue collected goes to the salary of Municipal staff leaving an insignificant part of revenue for city development. In such a situation, there is no value addition to the city by Municipality, a constitutional body. Further administrative activities as such will not bring about any value addition to the urban economy and hence the role of administration in economic development is debatable and often found not contributing to smart city economic development. Internet of Governance shall reduce the administrative cost and further reduction of cost can be achieved by a city by a proper combination of functions allocated to Internet of People, Internet of Things and Internet of Governance. This is a decision which Jointly Government and People of the city must make.

However, e-democracy cannot be purchased off the shelf from a technology provider unlike Internet of Things but developed within the Nation, State and Cities and within the framework of Constitutions which changes from country to country. The cultural and social system has a greater influence in framing the constitution which is also the basis of democracy.

This book address one important component of Smart Cities called e-democracy that is all pervading in all the six components of smart cities and I would like to use the term Internet of People or Internet of Democracy alternatively for e-democracy and Internet of Governance for E-Governance. The effective combination of Internet of Governance [4] and Internet of Democracy is the foundation of the development of any smart cities. Constitution of India provides for the active role of Democracy in the spatial-social-economic development of cities and villages. The seventy-fourth and seventy-third constitutional amendment of India states the role of participatory democracy in Governance which includes identification of issues, policy making, planning and implementation. The Constitution also spells out the limits of power of Government and People in a democracy based on various rights enshrined in Indian constitutions, all of which are protected by the constitution. Therefore, there are definite legal boundaries between Internet of People and Internet of Governance enshrined in the constitution and one cannot overpower and subjugate the other.

The foundation of the smart city is built based on the ability of a smart city to convert the total population to Smart People irrespective of their income, caste, class, religion, and educational attainment. So, the smart people are all equal like the citizen of India. There can be many means such as continuous functional education, distant education, ICT deployment, and broadband accessibility and so on are means of continuing distant education. The minimum requirement for a Smart People is that they should be the citizen who can vote and who can even be Prime Minister or President of this country and other constitutional functionaries. For this, Indian constitutions do not give any prescription to limit the population of the citizen about to religion, sex, caste, educational attainment and so on but the age limit is prescribed. The additional requirement is that citizen is willing to learn and can be part of continuous learning that makes him part of the ever changing and dynamic world, with changing economic opportunities, requiring changing capability. 'Collectives of Commons' of the third industrial revolution is only possible in that way. The 'Smart People', the fundamental building block of a Smart City System, require many crucial attributes as given,

(a) Smart people excel in skills what they do professionally and otherwise.
(b) Smart people have a high Human Development Index.
(c) A Smart City integrates its universities and colleges into all aspects of city life and bring them closer to every household.
(d) It attracts high human capital, for example, knowledge workers from any part of the world.
(e) A Smart City maintains high Graduate Enrolment Ratio and has ever increasing the pool of people with high level of qualifications and expertise.
(f) Its inhabitants opt for lifelong learning and use e-learning models.
(g) People in a Smart City are highly flexible and resilient to the changing circumstances.
(h) Smart City inhabitants excel in creativity and find unique solutions to challenging issues.
(i) Smart people are cosmopolitan, open-minded, and hold a multicultural perspective.
(j) Smart people maintain a healthy lifestyle [3].

Many documentations available regarding the functioning of parliament deliberation of ancient Republics whether it is in Greece or India was most dignified deliberation unlike what we see in Indian parliament where deliberation was obstructed by slogan shouting and parliamentarian moving to the well where the speaker is seated calling for adjournment of parliament. There was even fist fighting and destruction of property in parliament. In e-democracy, E-Parliament and E-Town Hall such things are not possible since it is in a virtual world where furniture of parliament cannot be destroyed or deliberation of E-Parliament cannot be obstructed and there is no adjournment of Mr Speaker. What you can do in this virtual world deliberately, discuss, and e-vote your point of view with dignity. Thereby e-democracy creates a parliament of the whole city more dignified and away from uncivilised actions in parliament.

1.2 Democracy

Democracy, or democratic government, is defined by the Oxford English Dictionary as "a system of government in which all the people of a state or polity ... are involved in making decisions about its affairs, typically by voting to elect representatives to a parliament or similar assembly". Democracy is further defined as (a) "government by the people; especially: rule of the majority" (b) "a government in which the supreme power is vested in the people and exercised by them directly or indirectly through a system of representation usually involving periodically held free elections". When we look democracy from top down, these definition looks alright, but when we look at democracy from ward level and citizen, the lowest constitutional level in India to multi-level constitutional bodies upwards, such as Municipality, Metropolitan Planning Committee Area, District State legislative assemblies and National Parliament, there can be questions about the two definitions given above especially regarding representative democracy. Ward committee meetings are generally attended by all people and debate the development issues. Ward member will not have greater say than people of the ward in such deliberations. Why not this model is replicated in multi levels using ICT. In such legislative debates, there can be not only these elected members but also all people if not those who could attend the deliberations using the broadband internet. Under participatory democracy, Indian Constitution is in full agreement with this type of deliberations at ward level. The centre of the debate is always the development issue of the ward and not those at a higher level of constitutional bodies and often overlook multi-level development issues. The only requirement is revised rulebooks of Parliament and Assemblies procedures must be rewritten to incorporate e-democracy so that people can voice their point of view at multi-level higher than ward on multi-level sector issues starting from ward level.

If this is agreeable why not the same approach of face to face discussion involving all people be applied at the higher level of constitutional bodies such as Municipal Level, Metropolitan regional level, District Level, State Level and National Level above ward level. The answer is it is difficult just because of the size of the population involved and so the representative democracy is feasible and hence election to these constitutional bodies. These bodies shall have quick response reaction by interested population by ICT for all policy and plan decision through what we call Internet of Democracy. The recent development of ICT makes it easy for even 1.25 billion plus people of India can react to policy decisions using technologies. Hence, much-debated land legislation for land acquisition in Indian Parliament can be amended after eVoting of 1.25 billion populations if systems and procedures are well formulated for eVoting which is within the capability of ICT technologies. This will give out most feasible legislative option for the country or state overlooking the vested interest of political parties in the parliament.

Political scientist Larry Diamond postulates that, democracy consists of four key elements: (a) A political system for choosing and replacing the government through free and fair elections; (b) The active participation of the people, as citizens, in politics and civic life; (c) Protection of the human rights of all citizens, and (d) A rule of law, in which the laws and procedures apply equally to all citizens [6].

The term democracy originates from the Greek δημοκρατία (dēmokratía) "rule of the people" [7] which was found from δῆμος (dêmos) "people" and κράτος (krátos) "power" or "rule", in the 5th century BC to denote the political systems then existing in Greek city-states, notably Athens; the term is an antonym to ἀριστοκρατία (aristokratía) "rule of an elite". While theoretically, these definitions are in opposition, in practice the distinction has been blurred historically [8]. The political system of Classical Athens, for example, granted democratic citizenship to an elite class of free men and excluded slaves and women from political participation which is not what Indian democracy follows. In virtually all democratic governments throughout ancient and modern history, democratic citizenship consisted of an elite class until full enfranchisement was won for all adult citizens in most modern democracies through the suffrage movements of the 19th and 20th centuries. Prior to Greek, Lichhavis in Eastern India practised a system of democracy and Sangha of Buddhism also practised democracy. So, democracy can be applied to factories, schools, hospitals and commercial establishments. This is important for a smart city which is a collective of commons and for the smart economic development of cities.

Democracy contrasts with forms of government where power is either held by an individual, as in an absolute monarchy, or where power is held by a small number of individuals, as in an oligarchy. Nevertheless, these oppositions, inherited from Greek philosophy [9], are now ambiguous because contemporary governments have mixed democratic, oligarchic, and monarchic elements. Karl Popper defined democracy in contrast to dictatorship or tyranny, thus focusing on opportunities for the people to control their leaders and to oust them without the need for a revolution the dream of every Socialist and Marxist.

No consensus exists on how to define democracy, but legal equality, political freedom, and the rule of law have been identified as important characteristics [10, 11]. These principles are reflected in all eligible citizens being equal before the law and having equal access to legislative processes. For example, in a representative democracy, every vote has equal weight, no unreasonable restrictions can apply to anyone seeking to become a representative, and the freedom of its eligible citizens is secured by legitimised rights and liberties which are typically protected by a constitution [12, 13].

One theory holds that democracy requires three fundamental principles: (1) upward control, i.e. sovereignty residing at the lowest levels of authority, (2) political equality, and (3) social norms by which individuals and institutions only consider acceptable acts that reflect the first two principles of upward control and political equality [14].

The term "democracy" is sometimes used as shorthand for liberal democracy, which is a variant of representative democracy that may include elements such as political pluralism; equality before the law; the right to petition elected officials for redress of grievances; due process; civil liberties; human rights; and elements of civil society outside the government. Roger Scruton argues that democracy alone cannot provide personal and political freedom unless the institutions of civil society are also present [15].

In some countries, notably in the United Kingdom which originated the Westminster system, the dominant principle is that of "parliamentary sovereignty", while maintaining judicial independence [15, 16] which is also followed in Indian Democracy. Democracy practised in India borrows many characteristics of British Democracy. In the United States, separation of powers is often cited as a central attribute. In India, parliamentary sovereignty is subject to the Constitution of India which includes judicial review [17]. Other uses of "democracy" include that of direct democracy. Though the term "democracy" is typically used in the context of a political state, the principles also are applicable to private organisations such as business and industries, as stated. In the third industrial revolution, democracy shall be practised by collective commons having the same interest.

Majority rule is often listed as a characteristic of democracy. Hence, democracy allows for political minorities to be oppressed by the "tyranny of the majority" in the absence of legal protections of individual or group rights. In Indian Democracy if majoritarian goes against the rights enshrined in the constitution the majority legislation is invalid. An essential part of an "ideal" representative democracy is competitive elections that are fair both substantively and procedurally. Freedom of political expression, freedom of speech, and freedom of the press are considered important to ensure that voters are well informed enabling them to vote per their own interests [18, 19]. The freedom is all subjected to rights given in the constitutions to all.

It has also been suggested that a basic feature of democracy is the capacity of all voters to participate freely and fully in the life of their society [20]. With its emphasis on notions of social contract and the collective will of all the voters, democracy can also be characterised as a form of political collectivism because it is defined as a form of government in which all eligible citizens have an equal say in law-making [20].

While representative democracy is sometimes equated with the republican form of government, the term "republic" classically has encompassed both democracies and aristocracies [21]. Many democracies are constitutional monarchies, such as the United Kingdom.

The term "democracy" first appeared in ancient Greek political and philosophical thought in the city-state of Athens [22, 23]. Led by Cleisthenes, Athenians established what is generally held as the first democracy in 508–507 BC. Cleisthenes is referred to as "the father of Athenian democracy" [24].

Athenian democracy took the form of a direct democracy, and it had two distinguishing features: the random selection of ordinary citizens to fill the few existing government administrative and judicial offices [25], and a legislative assembly consisting of all eligible Athenian citizens [26]. All eligible citizens could speak and vote in the assembly, which set the laws of the city-state. However, as stated Athenian citizenship excluded women, slaves, foreigners, and males under 20 years old. Of the estimated 200,000–400,000 inhabitants of Athens, there were between 30,000 and 60,000 citizens. The exclusion of large parts of the population from the citizen body is closely related to the ancient understanding of citizenship. In those time, the benefit of citizenship was tied to the obligation to fight war campaigns.

Athenian democracy was not only direct in the sense that decisions were made by the assembled people, but also the most direct in the sense that the people through the assembly and courts of law controlled the entire political process and a large proportion of citizens were involved constantly in the public business [27]. Even though the rights of the individual were not secured by the Athenian constitution in the modern sense (the ancient Greeks had no word for "rights") [28], the Athenians enjoyed their liberties not in opposition to the government but by living in a city that was not colony to another power [29].

In some countries for example Bhutan, a king would voluntarily relinquish his authority, by establishing a constitution, empowering a legislature, expanding voting rights, and extending civil liberties and even restricting the King. This perhaps may deter revolution in some other country, but an unlikely situation in Bhutan and encourages the private sector to invest (and, thereby, to expand the pie that the government can tax). An increase in human capital that raises in democracy may result in the people's ability to resist oppression but also raises the ruler's benefits from subjugating them.

Similarly, a rise in urbanisation and transformation of cities to smart cities makes it easier for people to meet and communicate—which presumably makes them harder to suppress—but also makes it easier for an autocrat to monitor and control activities using the same ICT.

The data for a large panel of countries confirm the Lipset/Aristotle hypothesis [30, 31], which says that a higher standard of living promotes democracy [32]. This relation shows up when democracy is represented by electoral rights or civil liberties and when the standard of living is measured by per capita GDP, primary school attainment, the gap between male and female, and the importance of the middle class. Democracy does not relate significantly to school attainment at the secondary and higher levels. For a given standard of living, democracy tends to fall with urbanisation and a greater reliance on natural resources but has little relation to country size. The apparently strong relation of democracy to colonial heritage mostly disappears when the measures of standard of living are held constant. Similarly, the allowance for the standard of living weakens the interplay between democracy and religious affiliation.

1.2.1 Democracy and Indian Panchayat System

Panchayat system that is practised in India and many parts of the world for many centuries was a perfect democratic system. All people of the village sat together to discuss the problem they are faced with. The problem can be a household where people get sick continuously or problem of sharing water for agriculture or issues arising out of sharing common properties like village land. The cause and remedies were discussed and an action plan formulated which all villagers follows jointly without resorting to the bureaucracy to implement. As the size of the population of habitat became bigger and bigger, it was found impractical to discuss such issues in a face to face manner in a situation where high-speed broadband was not there for video conferencing. They found the best way out is to accept representative democracy where one person is selected based on some representation ratio for example one state legislative assembly seat for every 200,000 populations to represent the constituency and discuss the problem. Then an administrative system was developed to see the implementation is attempted with adequate financial resources from taxing and accepted administrative procedures. There was no provision to evaluate bureaucrats performance in 24 h and 7 days a week unlike in a Panchayat with a face to face community able to oversee the outcome of joint decision making. Soon, there was a general dissatisfaction with this system. Many time democracies were subverted to mere non-accountable and irresponsible bureaucracy with no proper accountability and unsatisfactory outcomes and no immediately actionable interventions by people to replace habitual corrupt and erring politicians and bureaucrats. In an Athenian democracy, instances are there that this was easily done and so also in ancient Indian panchayats. In India, many members of parliament had criminal records and were easily prone to all corrupt practices of a criminal that violate Indian Penal Codes. There was no strict control on them. No wonder such parliamentarians disrupted for passing of legislation likes Goods and Services Tax for certain narrow political considerations not necessarily related to the wellbeing of the country. There was no way to monitor the functioning of both administration and represented elected members in 24 h and 7 days a week. This made them often corrupt. Barring few discussions that appear in the National TV and News Papers on corruption based on which some remedial action precipitated, most instances of corruption and inefficiency in the execution of legislative orders escape public scrutiny. No effort could weed out corrupt politician and administrators immediately since several laws protect the corrupt. For example, Peoples Republic of China could identify just more than 50,000 corrupt politician and administrators last two years each making this number 100,000 plus and action are being taken to punish them as reported in media. This can be a lower number or a tip of the iceberg of corruption in China. A similar attempt is also made to weed out these undesirable actors of democracy in many Nations like India, and so on. Global experience shows that socialist and communist countries which depend on bureaucracies to implemented their regimental goals are most corrupt.

Cities are growing bigger day by day with the addition of millions of populations. The major issue is to find ways and means to reinvent well tested and ancient panchayat system in a practical way in very large metropolises. This Panchayat system in large size cities can continue as face to face community to analyse their issues of habitat and find a joint solution and attempt to implement the solutions jointly in an ICT enriched with high-speed broadband cities using Internet of Things, Internet of Democracy/People and Internet of Governance.

1.2.2 Culture of Democracy

The culture of democracy is one which is something natural to people and passed over to the present generation from times immemorial from the earliest known history. The configuration and layout of Mohan Je Daro cities in India show these cities cannot function without a democracy since it was not a king's city but that of people's city. The Sumerian city-states are believed to have had some form of Democratic setup initially. They became monarchies over time.

Evidence of a Democratic system of government in India is originally found in the Vedas which is oldest written document available in India. There is distinctive evidence from Rig Veda, which mentions a thriving republican form of Government in India. Therefore, after independence from colonial rulers, India naturally went back to declare India as a Republic, the cultural root of India. Here I quote shlokas from Rig Veda which were to be sung in unison at the beginning of the republican assembly in several centuries back in Indian republics

> We pray for a spirit of unity; may we discuss and resolve all issues amicably, may we reflect on all matters (of state) without rancour, may we distribute all resources (of the state) to all stakeholders equitably, may we accept our share with humility—[33].

In ancient Indian written history, one of the earliest instances of democracy in a civilisation was found in many republics in ancient India, which were established sometime before the 6th century BC, and prior to the birth of Gautama Buddha. These republics were known as *Maha Janapadas*, and among these states, Vaishali (in the present Bihar, India) was considered as the world's first republic. The democratic Sangha, Gana, and Panchayat systems were used in some of these republics; the Panchayat from 6th century BC and the republican system is still used today in Indian villages and cities as a part of village culture.

Later during the time of Alexander, the Great in the 4th century BC, the Greeks wrote about the Sabarcae and Samba star states in what is now Pakistan and Afghanistan, whose "*form of government was democratic and not a monarchy*" according to Greek scholars at the time. Another example was Gopala's rise to power by democratic election in Bengal, which was documented by the Tibetan historian Taranath. Republican and Panchayat system was spread in all parts of India North, South, East, and West from where the Indian democratic culture emerged.

The following conversation with Gautama Buddha and Ananda describe how Democracy functioned in Vaishali.

The Buddha's commitment to republicanism (the ideal republican virtues) was a strong one if we are to believe the Mahaparinibbanasuttanta, among the oldest of Buddhist texts [34]. As is common in the Buddhist scriptures, a precept is illustrated by a story. Here Ajatasastru, the King of Maghada, wishes to destroy the Vajjian confederacy (of the Licchavis of Vaishali) and sends a minister, Vassakara the Brahman, to the Buddha to ask his advice. Will his attack be a success? Rather than answer directly, the Buddha speaks to Ananda, his closest disciples:

> "Have you heard, Ananda, that the Vajjians hold full and frequent public assemblies?"
>
> "Lord, so I have heard," replied he.
>
> "So long, Ananda," re-joined the Blessed One, "as the Vajjians hold these full and frequent public assemblies; so long may they be expected not to decline, but to prosper...

In a series of rhetorical questions to Ananda, the Buddha outlines other requirements for Vajjian prosperity:

> So long, Ananda, as the Vajjians meet together in concord, and rise in concord, and carry out their undertakings in concord...so long as they enact nothing not already established, abrogate nothing that has been already enacted, and act in accordance with the ancient institutions of the Vajjians as established in former days...so long as they honour and esteem and revere and support the Vajjian elders, and hold it a point of duty to hearken to their words...so long as no women or girls belonging to their clans are detained among them by force or abduction...so long as they honour and esteem and revere and support the Vajjian shrines in town or country, and allow not the proper offerings and rites, as formerly given and performed, to fall into desuetude...so long as the rightful protection, defence, and support shall be fully provided for the Arahats among them, so that Arahats from a distance may enter the realm, and the Arahats therein may live at ease so long may the Vajjians be expected not to decline, but to prosper.

Then the Blessed One addressed Vassakara the Brahman, and said,

> When I was once staying, O Brahman, at Vesali at the Sarandada Temple, I taught the Vajjians these conditions of welfare; and so long as those conditions shall continue to exist among the Vajjians, so long as the Vajjians shall be well instructed in those conditions, so long may we expect them not to decline, but to prosper.

The comment of the king's ambassador underlines the point of this advice: "So, Gotama, the Vajjians cannot be overcome by the king of Magadha; that is, not in battle, without diplomacy or breaking up their alliance" [35].

The same story tells us that once the king's envoy had departed, the Buddha and Ananda went to meet the assembly of monks. Buddha told the monks that they too must observe seven conditions if they were to prosper: Full and frequent assemblies, concord, preserving and not abrogating established institutions, honouring elders, falling "not under the influence of that craving which, springing up within them, would give rise to renewed existence," delighting in a life of solitude, and training "*their minds that good and holy men shall come to them, and those who have come shall dwell at ease*".

It may be emphasised here that the Buddha saw the virtues necessary for a righteous and prosperous community, whether secular or monastic, as being much the same. Foremost among those virtues was the holding of "*full and frequent assemblies.*" In this, the Buddha spoke not only for himself, and not only out of his personal view of justice and virtue. He based himself on what may be called the democratic tradition in ancient Indian politics democratic in that it argued for a wide rather than narrow distribution of political rights, and government by discussion rather than by command and submission [36]. The Pali Canon gives us our earliest, and perhaps our best, detailed look at Indian republicanism, its workings, and its political philosophy. About no other republics do we know as much as we do about the Buddhist Sangha and the Licchavis in the time of Buddha even though we do know that republics survived and were a significant factor until perhaps the fourth century A.D., a period of over 800 years? Scattered inscriptions, a great number of coins, and the occasional notice in Greek sources, the Jatakas or other Indian literature give us a few facts, but any history of Indian republicanism is necessarily a rather schematic one.

In Ancient Greece, present before the world, the speaker's platform in the Pnyx, the meeting ground of the assembly where all the great political struggles of Athens were fought out during the "Golden Age". Here Athenian statesmen stood to speak, such as Pericles and Aristides in the 5th century and Demosthenes and Aeschines in the 4th, along with countless humbler citizens as well. In the background high on the Acropolis is the Parthenon, the temple of Athena, the city's protective goddess, looking down upon their deliberations.

The culture of democracy was well studied by Gautama Buddha and adopted it in codifying the Sangha he built. Many pillars of democracy are discussed by Gautama Buddha as follows.

1. Buddha spoke about the equality of all human beings (long before Abraham Lincoln on the USA) the existing oldest democracy, and that classes and castes are artificial barriers erected by society. This means all people in a smart city is equal. The only classification of human beings, per the Buddha, is based on the quality of their moral conduct which is valid today when a corrupt politician, businessman, and administrators are not respected in Indian Society.
2. Buddha encouraged the spirit of social -co-operation and active participation in society. This spirit is actively promoted in the political process of modern societies. The third industrial revolution that may take place in Smart Cities requires "collectives of commons" which can only be achieved through e-democracy.
3. Since no one was appointed as the Buddha's successor, the members of the Order were to be guided by the Dhamma and Vinaya, or in short, the Rule of Law. Until today very member of the Sangha is to abide by the Rule of Law which governs and guides their conduct. In a democracy, successive leaders are selected by votes and not based on the qualifications. In India, dynastic politics still exists in many political parties, some hundred-year-old.
4. Buddha encouraged the spirit of consultation and the democratic process. This is shown within the community of the Order in which all members have the right to decide on matters of general concern. When a serious question arose demanding attention, the issues were put before the monks and discussed in a manner like the democratic parliamentary system used today. This self-governing procedure may come as a surprise to many to learn that in the assemblies of Buddhists in India 2500 years and more ago are to be found the culture of the parliamentary practice of the present day. A special officer like 'Mr. Speaker' was appointed to preserve the dignity of the Parliamentary Chief Whip, was also appointed to see if the quorum was secured. Matters were put forward in the form of a motion which was open to discussion. In some cases,

it was done once, in others three times, thus anticipating the practice of Parliament in requiring that a bill is read a third time before it becomes law. If the discussion showed a difference of opinion, it was to be settled by the vote of the majority through balloting.

The Buddhist approach to political power is the moralization and the responsible use of public power. The Buddha preached non-violence and peace as a universal message. He did not approve of violence or the destruction of life and declared that there is no such thing as a 'just' war. He taught: '*The victor breeds hatred, the defeated lives in misery. He who renounces both victory and defeat is happy and peaceful.*'

In the Jataka, the Buddha had given to rules for Good Government, known as '*Dasa Raja Dharma*'. These ten rules can be applied even today by any government which wishes to rule the country peacefully. The rules are as follows:

1. be liberal and avoid selfishness,
2. maintain a high moral character,
3. be prepared to sacrifice one's own pleasure for the well-being of the subjects,
4. be honest and maintain absolute integrity,
5. be kind and gentle,
6. lead a simple life for the subjects to emulate,
7. be free from hatred of any kind,
8. exercise non-violence,
9. practice patience, and
10. respect public opinion to promote peace and harmony.

Regarding the behaviour of rulers, He further advised:

- A good ruler should act impartially and should not be biased and discriminate between one group of subjects against another.
- A good king should not harbour any form of hatred against any of his subjects.
- A good ruler should show no fear whatsoever in the enforcement of the law if it is justifiable.
- A good ruler must possess a clear understanding of the law to be enforced. It should not be enforced just because the ruler has the authority to enforce the law. It must be done in a reasonable manner and with common sense—(Cakkavatti Sihananda Sutta).

1.2.3 Advantages and Disadvantages of Democracy

Democracy is a type of political system that requires a popular vote (representative election) to take place to elect the leader of the country and other officials. The leaders are chosen by the people. Many countries in the world, operate under a democratic form of government. While democracy is noted as one of the most efficient government systems ever, it is, however, not without any downsides.

Gautama Buddha spoke about the prosperity of Vaishali that goes hand in hand with democratic practice in earlier para. Empirical studies [29, 30] shows there is the impact of economic development on a country's propensity when democracy is experienced. A common view since Lipset's [31] research is that prosperity stimulates democracy; this idea is often called the Lipset hypothesis [32]. Lipset credits the idea to Aristotle: "From Aristotle down to the present, men have argued that only in a wealthy society in which relatively few citizens lived in real poverty could a situation exist in which the mass of the population could intelligently participate in politics and could develop the self-restraint necessary to avoid succumbing to the appeals of irresponsible demagogues" (For a statement of Aristotle's views, see Aristotle [1932, book 6].). Theoretical models of the effect of economic conditions on the extent of democracy are not well developed. Lipset et al. [32] emphasised increased education and an enlarged middle class as key elements, and he also stressed the thinking of Tocqueville's (1835) idea that private organisations and institutions are important as checks on centralised government power. This point has been extended by Putnam (1993), who argues that the propensity for civic activity is the key underpinning of good government in the regions of Italy. For Huber et al. (1993, pp. 74–75), the crucial concept is that capitalist development lowers the power of the landlord class and raises the power and ability to organise the working and middle classes.

1.2.3.1 Advantages of Democracy

1. It protects the interest of citizens.

Citizens in a democratic country are given the right to vote on political, social and economic issues, particularly the representatives they want to oversee making major decisions, as the Members of Parliament and legislative assembly. This can greatly protect the people from anything they would disagree to occur.

2. It prevents monopoly of authority.

Since the government is bound by an election term where parties compete to regain authority, democracy prevents monopoly of the ruling authority. The elected ruling party would make sure their policies will work for the people as they will not be able to remain in power after their term with bad records—they will not be re-elected.

3. It promotes equality.

Generally, democracy is based on the rule of equality, which means that all people are equal as far as the law is concerned. Every person has the right to experience and enjoy equal political, social and economic rights, and the state is not allowed to discriminate him on the standard of sex, class, caste, religion and income.

4. It makes for a responsible and stable administration.

When there are elected and fixed representatives, a more responsible government is formed. Thus, democracy can be efficient, firm and stable. Its administration is ruled and conducted with a sense of dedication, and people under this system discuss matters and problems thoroughly to come up with sensible decisions.

5. It brings a feeling of obligation towards the citizens.

The ruling authorities owe their success to elections by the citizens, so they would feel grateful to and socially responsible for them. This can serve as their motivating factor to work for the citizens, for they have the right of choosing their government.

6. It imparts political education to the people.

One argument in favour of democracy is that it can serve as a training school for citizens—they are driven to take part in state affairs. During elections, political parties propose their programs and policies in support of their candidates through public meetings, demonstrations, television, radio, posters and speeches by their leaders to win public favour. These can impact political consciousness among the people.

7. It helps make good citizens.

Democracy aims to create the ideal environment that is conducive to personality improvement, character cultivation, and good habits. As per the experts, this political system seems to function as the first school for good citizenship, where individuals can learn about their rights and duties from birth to the time of death.

8. It allows a little chance of revolution.

Since this system is based upon the public will, there will be little to no chance of public revolt. Elected representatives conduct state affairs with public support, and if they do not work efficiently or do not meet the public's expectations, they will probably not do well during the next elections. Democracy or other popular governments often function with consensus; thus, the question of revolution would not arise.

9. It promotes change.

This political system can promote changes in the government without having to resort to any form of violence. It tries to make citizens feel great and even provides them with a good sense of participation and involvement.

1.2.3.2 Disadvantages of Democracy

1. It might allow misuse of public funds and time. Nothing can be done if the parliament does not legislate by obstructing the proceedings by opposition parliamentarians as can be seen in the Indian parliament as routine

adjournments. Democratic governments can lead to wasted time and resources, considering that it takes a huge amount of time on formulating laws and requires a lot of money to be spent during elections. It is also highly possible that the country will be ruled by incompetent and irresponsible leaders who will just spend public funds for their own tours and recreation.

2. It instigates corruption.

Those who are elected to power might resort to unethical means for personal interests and engage in corrupt practices. During their tenure in office, they might take advantage of authority for personal gains, putting the interests of the masses in the backseat.

3. It risks the wrong choice of public servants.

Not all individuals under a democratic government are aware of the political and social circumstances in their country. In a voting system, majority wins, and there is no distinction between the votes cast by the literate and the illiterate; informed and uninformed member. People may favour a candidate based on other factors other than pure and required capability. Taking these things into consideration, the elected official may not always be the perfect person for the seat, leading to erroneous decisions.

4. It allows not exercising the right to vote.

Sadly, in some democratic countries, people fail to exercise their right to vote. Perhaps, they are reluctant to do it or are just less aware of the impact of their votes. Or, perhaps they do not see it as a privilege and take the process less seriously.

5. It may put more emphasis on quantity, rather than quality.

In terms of providing services—it tends to put more emphasis on quantity, rather than quality. Also, considering that the system might be governed by irresponsible and incompetent leaders, equality might be in question for only the rich and famous might be prioritised more than the poor. Since money is required for contesting election, poor cannot be a contesting election.

6. It can take long to make decisions.

Because it takes long to make decisions, it will also take long to implement them. Unlike in a monarchy where one person is making decisions that are implemented quickly, democracy requires majority voting in implementation, thus, it is relatively less prompt in taking actions.

7. It may involve immoral practices during elections.

To lure the masses, election campaigns might involve immoral practices, where candidates would use muscle power to draw most votes, even trying to tarnish their opponents' reputation. Money and power may be abused to influence the people to disregard opposing parties.

By weighing their advantages and disadvantages, of a democracy, we can conclude that the democracy is best for the people.

1.3 Nature of E-Democracy in Cities

E-democracy in a city is the support and enhancement of democracy, democratic institutions and democratic processes by means of ICT. E-democracy concerns all sectors of democracy, all democratic institutions, and all levels of government. E-democracy cannot be isolated from traditional democratic processes. It is additional, complementary to, and interlinked with traditional democratic processes, to widen the choices available to the public for taking part in political processes [37].

1. E-democracy does not in itself affect the constitutional duties and other duties and responsibilities of decision makers; rather, it can provide them with additional benefits.
2. E-democracy requires information, dialogue, communication, deliberation and the ongoing creation of open cyberspaces where citizens can get together to uphold their civic interests.
3. E-democracy is an integral part of the information society, which has ushered in a range of traditional and innovative tools that can usefully be applied in democratic processes and institutions.
4. E-democracy flourishes best where there are the political will and leadership to make it work effectively by introducing the structural changes needed to take account of the opinions expressed. The incorporation of ICT into democratic processes usually requires structural changes and procedural reform.

E-democracy in smart cities is a mean for disseminating more of city-based political information and for enhancing communication and participation, as well as hopefully in the long run for the transformation of the political debate and the political culture of a city. Participants in the field of e-democracy in cities include civil society (organised and non-organized), Non-Governmental Organisations, journalists, the administration, and politicians from ward level to national level and —the citizen irrespective of their caste, class, income, religion, gender or age.

E-democracy as related to smart city development should be defined broadly as given above, since computers and telecommunications, particularly tied to the Internet and web, are connected to nearly all aspects of politics and governance in any city. From paving roads or filling in potholes in the road, to electing politicians at the ward, municipal or above levels, electronic media are reshaping access to what people know, who they communicate with, and what they need to know to get things done. In these ways, technical change can enable democratic patterns of communicative power.

E-democracy covers those arrangements by which electronic communications are used by those with power and the citizens they serve to interact with each other to inform and modify the way that power is used. E-democracy is NOT about paying speeding fines over the Internet (that is e-government); it IS about consulting on whether the speed limit on a stretch of road should be raised, lowered or left as it is or whether the road shall be converted to cycling path and walkway and debar cars and buses to use it. Thereby, the spatial structure of the smart city is reshaped by people using electronic means using past experiences based on data. It may, one day, be used as a way of empowering citizens in the process of making major national decisions when the power of e-democracy expand from city level to national level.

E-democracy is a deliberate way citizen act to facilitate greater and most convenient participation in government using digital or electronic means. These initiatives can include e-forums, e-town hall meetings (virtual and not real), e-consultations, e-referenda, e-voting, e-rule-making, and other forms of e-participation, and any form of 'digital engagement'.

1.4 Definition of E-Democracy

Democracy is defined by Webster's dictionary as 'a government in which the supreme power is vested in the people and exercised by them directly or indirectly through a system of representation.' Electronic democracy is simply the use of technology tools to facilitate democratic activities. "E-democracy consists of all electronic means of communication that enable/empower citizens in their efforts to hold rulers/politicians accountable for their actions in the public realm. Depending on the aspect of democracy being promoted, e-democracy can employ different techniques: (1) for increasing the transparency of the political process; (2) for enhancing the direct involvement and participation of citizens, and (3) improving the quality of opinion formation by opening new spaces of information and deliberation" [38]. The Internet and e-democracy present one way to positively redefine democratic processes and reinvigorate the relationship between citizens and their elected representatives.

There are many different definitions of e-democracy. A recent study for the European Parliament adopted a definition focusing on accountability as an objective of e-democracy. "E-democracy consists of all electronic means of communication that enable/empower citizens in their efforts to hold rulers/politicians accountable for their actions in the public realm.

Other definitions focus on the procedures and tools by which the enhancement of democracy shall be achieved. E-democracy is concerned with "the use of information and communication technologies to engage citizens, support the democratic decision-making processes and strengthen representative democracy. The principal ICT mechanism is the internet accessed through an increasing variety of channels, including PCs, both in the home and in public locations, mobile phones, and interactive digital TV. The democratic decision-making process can be divided into two main categories: one addressing the electoral process, including e-voting, and the other addressing citizen e-participation in democratic decision-making." Following this segmentation into e-voting and e-participation differentiates e-participation into

- "Information (eEnabling)—a one-way relation in which government produces and delivers information for use by citizens.
- Consultation (eEngaging)—a two-way relationship in which citizens provide feedback to government, based on the prior definition of the government of the issue on which citizens' views are being sought.
- Active participation (eEmpowerment)—a relation based on a partnership with government, in which citizens actively engage in the policy-making process. It acknowledges a role for citizens in proposing policy options and shaping the policy dialogue, although the responsibility for the final decision or policy formulation rests with the government.

"E-democracy consists of all electronic means of communication that enable/empower citizens in their efforts to hold rulers/politicians accountable for their actions in the public realm. Depending on the aspect of democracy being promoted, e-democracy can employ different techniques: (1) for increasing the transparency of the political process; (2) for enhancing the direct involvement and participation of citizens, and (3) improving the-the efficiency [39].

1.5 E-Democracy

E-democracy is Democracy plus ICT or specifically Internet. E-democracy, therefore, can be called Internet of People (IOP). The resulting solution, therefore, is supplementing representative democracy at multi-levels, reducing proliferating, expensive and largely ineffective bureaucracy to e-democracy and leave implementation directed by empowered people instead of bureaucracy. It has been demonstrated e-budgeting by people do not depend on bureaucracy and implementation can be electronically monitored and reported in real time on the internet accessible to all concerned. These practical and workable processes supplement and complement representative democracy and bureaucracy.

"Dick Morris, the one-time strategic adviser to Bill Clinton, claimed in his 1999 book, "www.Vote.com", that Jefferson's 'utopian vision of a democracy based on town meetings and direct popular participation is about to become a reality then.' Although such plebiscitary visions persist, they have been supplemented in recent years by two other trends. Firstly, there have been several experiments in using the internet to make representative democracy more effective. These range from local initiatives, such as municipal planning consultations in Germany and Sweden, to national projects, such as the Estonian Government's TOM portal where citizens can propose new laws, and a series of online consultations in the UK Parliament. Secondly, there has been the emergence of policies for e-democracy, most notably in the UK where both Parliament and the Government has outlined a set of policy principles for e-democracy, and several e-democracy projects have been publicly funded. In his direction, but there has not been much progress in India focusing on e-democracy.

Today, all OECD member countries recognise new ICTs to be powerful tools for enhancing citizen engagement in public policy-making. Despite the limited experience to date, some initial lessons for online citizen engagement in policy-making are emerging: Technology is an enabler, not the solution. Integration with traditional, 'offline' tools for access to information, consultation and public participation in policy-making is needed to make the most of the ICTs. The online provision of information is an essential precondition for engagement, but quantity does not mean quality. Active promotion and competent moderation are key to effective online consultations. The barriers to greater online citizen engagement in policy-making are cultural, organisational and constitutional not technological. Overcoming these challenges will require greater efforts to raise awareness and capacity both within governments and among citizens.

E-democracy for Smart Cities is based on the belief that Smart People in a city can transform the urban economy to smart economy, urban environment to smart environment, urban mobility to smart mobility, urban governance to smart governance and urban living to smart living [1–3]. The infrastructure requirement is ICT accessible to all through ownership of smartphones, or computer services can be rented at nominal cost from a place which is within walking distance and above nearly full e-literacy. The only additional requirement is continuing education and empowering people to smart people for articulating their views on all matter of cities that relates to their living, governance, environment, and personal mobility and see it is heard. There are many aspects of empowerment of citizen to smart people and e-democracy is one and the most important.

1.5.1 E-Democracy for Smart Cities Versus E-Democracy in Smart Cities

If authors of this book are alive and writing another revised volume of this book in 2117 AD, probably we may use the book title "IN" and not use "FOR" for that book. By 2116 AD there may be experiences accumulated on e-democracy for Smart cities in some smart cities to be studied, evaluated and reapplied in many cultural environments. There can be a big portfolio of Internet of People apps which can be adapted to many cities. There can also be trillions of Internet of Things and apps for Internet of Governance which can be available as open sources. Probably the way we plan and execute plan will change and will be based on Internet of Things, Internet of People and Internet of Governance through a judicious mixture of all. As of today, however, the number of smart cities being born per year globally is very small [3] and the e-democracy experiences accumulated is also insignificant.

The current program of 100 smart cities in India is not giving any stress on e-democracy there and there may not be much to learn from that experience for e-democracy development in the year 2116.

We are now in Research and Development stage to design some prototypes and most authors in this book are trying to answer what e-democracy to be designed suited for Smart Cities in different geographic regions under differing constitutions. Indeed, we are much ahead of time in this book and it may give unexpected solutions. Chapter 3 shows a new way. This is not the way Town Planners world over prepare execution plan for Outline Development Plans. The interesting part is everyone knows the usefulness of e-democracy for Smart Cities and otherwise all authors who agreed to write this book will not join me. This fully justifies our title of the book.

I find Smart Cities are the cities of Third Industrial Revolution what Rifkin [4, 5] talks about and we do not have examples of third industrial revolution which the European Union is making a blueprint for action and China is also taking some baby steps. The third industrial revolution is highly sustainable ecologically, economically and socially since Industrial production, mobility, marketing, services and all will be in near zero marginal cost. The third industrial revolution requires the collaboration of commons and e-democracy provides for that.

1.6 E-Democracy Typologies of Initiative in City Level

One of the most important developments to occur with the emergence of the Internet and other new ICTs are networks of online communities. These online communities are the founding blocks of e-democracy. The networked world allows the flow of knowledge and ideas in ways not known before succinctly summarises the characteristics of networked interaction, of which the Internet is a prime example, as follows [40]:

- "distributed intelligence in place of monopolies of information;
- flatter structure, as there is less need for layers of knowledge keepers and cultivators;
- simultaneous action as against vertical hierarchy of administrative action known for very slow responses;
- an exponential increase in actors and participants as against few administrators;
- a structure organised around flexible nodes of information and exchange replacing the conventional hierarchies;
- three-dimensional structure in the sense that a network consists of an almost infinite number of "layers" of other networks;
- non-differentiation in potential influence between organisations and individuals;
- interactivity, as information flows in all directions and is constantly exchanged, not simply channelled."

While this development has created the potential for non-hierarchical activities that could lead to more distributed power from government—and give the government wider opportunities to communicate and interact with the citizen—it does not necessarily mean this is happening.

E-democracy enhances transparency, participation, and deliberation continuously at a city level from the earlier model of representative democracy in which citizen vote once in five years and then forget direct proactive involvement, only to wake up from sleep after five years to vote again like zombies. In e-democracy, ICT enables citizens to monitor their elected representatives in all his daily performance continuously and elected members gets an opportunity instantly broadcast their message of their achievements, and create their own image. ICTs bring the potential for much farther reaching citizen consultation and other, novel means of participating in the city governing the process. These activities achieve transparency and broadly falls into two categories. The first type simply provides a central source for information about the government and its activities. The second type of e-democracy initiative that works to increase transparency is that which uses ICTs and especially Web 2.0 technology to make massive text-based, audio, and even visual records of Municipal government activity available on citizens' computer screens. Often the information made available in this way can be pre-sorted in terms of topic or author to present the citizen with exactly that information which he or she desires. These two major types of initiatives effectively summarise the larger dimension of transparency-enhancing e-democracy initiatives. This e-democracy initiative uses of ICTs that promote citizen participation in the governing process. There are three major subtypes to this dimension. The first type is a direct evolution of the voting process that allows citizens to vote online on all major issues. The second, and by far most common, the subtype is that of "citizen feedback" or "e-consultation." This type of e-democracy initiative can be as simple as offering a form like a google form for citizens to fill out online or as well-organized as a question and answer chat sessions with mayors or could even include a system that allows citizens to offer advice via SMS. The third type offers citizens the option of participating in tradition political processes online. For instance, a system that allows citizens to give online feedback on Municipal budget allocation or to launch a ballot initiative online transform how citizens can participate in political processes, making it much easier. While the latter two subtypes offer citizens the chance to communicate with and give feedback to their representatives in new and potentially simpler ways, the former offers a new way to participate in selecting those representatives in the first place. In both cases, however, ICTs are beginning to make participation in modern democracies, the level of which has been decreasing over the last generations, as easy as the click of a button. The result of this is not only the potential for higher participation in processes. The third type of e-democracy initiative is that which attempts to recreate a sense of the public sphere online using opportunities given by Facebook and Twitter. In general, these initiatives offer citizens of the local region or municipality the option of participating in a discussion forum that is intended to be a space to raise issues of local importance. In some cases, these discussion forums are simply a place for discussion, supplanting the literal public space which is becoming less and less important in an era dominated by computerised communication and telecommuting. However, these discussion forums serve as the ingredients with which legislation is formed as the discussion threads are often officially monitored by legislators and administrators to gain ideas. Overall, those projects that fit this third archetype of e-democracy initiative harness the interactive capacity of ICTs to allow citizens to talk to each other via the new medium that is replacing physical public space. In addition to online, text-based forums, some of the most exciting initiatives make available the option of more interactive forums. The critical departure from the web forum is the real-time element. This could be an online chat session with a municipal leader. A more innovative example is the provision of a video Question and Answer session with an administrative leader where citizens can create a dialogue directly with a politician. Some of these have been already practised in many States in India.

There are many types of e-democracy initiative oriented toward increasing transparency. The first type of e-technique used is simply the addition of a personalization element to a web page of Mayor of Municipality who address the citizen from time to time. It allows citizens to filter all the information provided by their governments.

The second type is an official blog. In all the cases identified, this blog is kept by the chief administrator of a municipality. The mayor's blog posts are interspersed with occasional posts from other members of his administration, who share important events or decisions they have made. In this way, the municipality's administration and the way it works become more open to the people. To be able to track these change in zonal legislation required a trip to the physical place where that legislation was being considered. With the advent of ICTs, many legislatures have begun to facilitate this process. It is not becoming quite common to find that legislative websites at both the State and the municipal level are equipped with information not only about every bill that is currently being debated but about all bills either decided upon or still on the table for any legislative session.

To be able to monitor the activity of the legislative branch from any computer is quite a large step forward in the effort to reinvigorate open vigorous discussion in a democratic context. However, even this provision of information appears rudimentary in comparison to the most overwhelming e-democracy. Webcasting harnesses the full audio-visual power of communication technology. A webcast is a stream of data that a host sends to a user which relies upon the presence of a digital media player to be turned into audio or visual playback on the user-end. In other words, a webcast is a soundtrack or a video sent to a PC tool.

1.7 Smart City and E-Democracy

E-democracy in a smart city is nothing else but a virtual face-to-face democracy practised in ancient Licchavi or Athens discussed in earlier para using multimedia. This is possible in a smart city with a high level endowment of broadband and ICT infrastructure even if the city population size is in several million unlike ancient Athens or Licchavi. E-democracy is the technological adaptation of ancient face-to-face democratic tradition in an ICT-enriched smart city. It is the use of information and communication technologies and strategies by 'democratic sectors' within the political processes of local communities, states/regions, nations and on the global stage. Democratic actors and sectors in this context include, in order of importance, citizens/voters, political organisations, the media, elected officials and governments. E-democracy, like democracy in its proactive form, is a direct democracy which is the primary requirement of smart cities. E-democracy often refers to technological adjuncts to a smart city republic, i.e. the use of information and communication technologies and strategies in political and governance processes. In few cases, the word 'e-democracy' is used to refer to anything political that involves the Internet. It may be also called Internet of Democracy. E-democracy is concerned with the use of information and communication technologies to engage citizens in supporting the democratic decision- making processes and strengthen representative democracy. E-democracy is said to aim for more active citizen participation by using the Internet, mobile communications, and other technologies in today's representative democracy, as well as through more participatory or direct forms of citizen involvement in addressing public challenges.

1.8 Strategic Approach of E-Democracy in Selected Countries

1. E-democracy as part of the Information Society Strategy (Estonia)

Estonia—The Estonian Information Society Strategy 2013–14 (approved November 2006) sets out objectives and defines fields of action and measures for the development of the information society in a long-term perspective.

2. E-Government strategy with some elements of e-democracy and e-Inclusion Strategy (Hungary, Latvia, Slovakia, Slovenia)

Hungary—There is no specific strategy for e-democracy, but there is a strategy for e-Public Administration which covers some elements of e-democracy and an e-Inclusion Strategy.

Latvia—Declaration of the Intended Activities of the Cabinet of Ministers (available in English) Guidelines for Government Communication Policy 2008–2013, (available only in Latvian); Programme of Consolidation of Civil Society 2008–2012 (available only in Latvian).

Slovakia—The e-Government Strategy of the Slovak republic Slovenia—The strategies are outlined in the SEP2010-11 e-Government Strategy of the Republic of Slovenia for the period 2006 to 2010. In SEP 2010, e-democracy represents one of the several strategic orientations for the Republic of Slovenia in the development of e-Government.

3. Special policy document for e-democracy, Whitepapers (Austria)

Austria—Finished: Standards of Public Participation (offline- and online participation; adopted by the Austrian Council of Ministers on 2 July 2008); Whitepaper EDEM (principles for e-democracy and e-participation); Austrian e-democracy Strategy; Overview of national e-democracy and e-participation projects (to be updated); In process: e-democracy and e-participation Tools; e-participation Prototype Processes; e-participation Guidelines.

E-participation is a necessary component, or even, more precisely, a prerequisite of e-democracy. It refers to the means of ICT-supported participation in processes concerning administration, policy making, decision making, service delivery, information provision, consultation, deliberation, etc.

E-voting enables people to vote from anywhere using their identity card and a smart-card reader to select their favoured candidate from lists posted on the relevant website.

4. Regional or local strategies (Italy, Spain, Switzerland)

Italy—Among the most important initiatives, the initiatives of the Tuscany Region have highlighted: The Regional Law on Participation (2006) and the 2007–2013 regional planning initiative, in which many ICT projects to support e-participation are included.

Spain—In the Spanish highly decentralised model of government, the regions have extensive legislative powers. In exercising these powers, the Valencia Regional Government has adopted a Citizen Participation Regional Act of Valencia (Ley de la Comunidad Valenciana 11/2008), which includes an electronic channel among the means of participation that citizens have at their disposal. The regional government of the Canary Islands is undertaking a similar regional Act for parliamentary approval. These regional legislative frameworks define specific policies enforced within each geographic area.

Switzerland—National e-Government strategy. This strategy was developed in close cooperation with representatives of the cantons and the municipalities, under the direction of the FSUIT. The strategy constitutes the basis for the Confederation, the cantons, and the municipalities to orient their efforts towards common goals.

5. Strategies based on the CoE Recommendations, still in process, envisaged for 2010—(Denmark, Sweden)

The main objectives of this document include:

1. Increasing transparency and traceability
2. Improving e-Participation and communication
3. Developing new cooperation models
4. More efficient participation projects
5. Promoting social networks (web 2.0).

In general terms, the administration models can be divided into two types:

- Centralized model—Austria, Slovenia, Switzerland, Slovenia
- Decentralized model—Norway; Estonia, Latvia, Denmark, Spain.

1.8.1 Models of E-Democracy for Smart Cities

Historically, the city-state of Licchavi, Vaishali and Athens were the first well-recorded and widely known democratic state that engaged in direct democracy. Nearly 200 years that democracy thrived in Athens and it changed through several evolutions. At the height of their practices of democracy, 6000–8000 citizens of Athens would gather almost weekly to form a consensus on legislative issues which looks like an ideal democracy. In fact, this was not a modern inclusive democracy as we understand it today, as vast numbers of Athenians were excluded. Those not participating were the slaves as well as members of families who were considered immigrants (an immigrant family could be one that arrived in Athens as much as four or more generations earlier. Women were also excluded, thus eliminating over half the population from the process. However, the democratic practice in ancient Athens is an early, though flawed because non-inclusive, an example of the practice of direct democracy through the reaching of consensus amongst the participants. Ancient Indonesian village community decides based on discussion and consensus where no one is excluded. This is a model for smart city e-democracy which requires the full potential of democracy to bring out the best in smart cities. In fact, there are about ten models of democracy for the smart city to choose from which are briefly outlined below. Held's [37] ideal forms of democracy contain four historical models: Classical Athenian, Republicanism, Liberal, and Direct Marxism and four contemporary models: Competitive Elitist, Pluralism, Legal, and Participatory.

1.8.2 Liberal E-Democracy

Liberal democracy as we practice in India, in terms of its structural characteristics, is as follows:

1. Pluralistic competition among parties and individuals for all positions of government power;
2. Participation among equal citizens in the selection of parties and representatives through free, fair and periodic elections; and civil and political liberties to speak, publish, assemble and organize, as necessary conditions, to ensure effective competition and participation.

1.8.3 Ten Models of E-Democracy

Ten models of democracy with few subtle variations are given below.

1. The participatory democracy model emphasizes the importance of participation. Equal rights can only be obtained in a participatory society which encourages a sense of political efficacy [37, 39]. Related models are the deliberative model [40, 41] and the strong model [42]. This model emphasizes the role of public discussions, the importance of citizen participation, and the existence of a well-functioning public sphere.
2. In accordance with the direct democracy model, network-based groups and individuals take over the role of traditional institutions [43, 44]. The quick Democracy model allows that the majority should directly influence decision making although the representative model of decision-making be necessary in certain situations. The direct democracy model represents a radical alternative to the representative models of democracy which is technically feasible in large ICT-enriched smart cities. The idea of citizens participating directly in political decision making originates from the classical ideas of democracy and participation in the state, focusing on equal rights to rule and be ruled in turn by the collective of free male citizens. In smart cities, ICT plays a critical role in implementations where the Internet no longer represents a supplement to traditional communication channels, but instead a crucial precondition for democracy [43]. A direct e-democracy initiative requires communication technology to support coordination among a great number of decision makers, i.e. citizens, possibly geographically scattered, with diverse interests and backgrounds. The direct democracy model emphasizes the importance of bringing power to the public. Decisions are justified by consensus (communism) or by frequent elections giving mandates to elected members of government (socialism).
3. The liberal democracy model is characterized by a representative government. Citizens are involved as voters as well as elected representatives at the multi-level such as ward, municipality, district, state and nation level, and as participators in the public debate.
4. Competitive elitism is represented in Held's model [37] on modern democracy. Experts represent (or claim to represent) different interests in society and act in

policy networks. Political experts are heavily involved in the formation of policies and the definition of public services.

5. The pluralist democracy model exhibits, a competitive electoral system, consisting of a minimum of two parties or two major coalitions, which may include regional parties along with national parties, secures the government through electoral majority and political liberty. In a multi-party system like that practiced in India, two parties get transformed into two coalitions. This political system divides power through checks and balances into three branches of government: the legislative, the executive, and the Judiciary, in addition to the operational administration. In the legalise democracy model, the majority principle protects individuals from random governance. Effective political leadership is underpinned by liberal principles, characterized by minimal state intervention with civil society and the general public's private life. The participatory democracy model emphasizes that equal rights are achieved through a participatory process where skilled members of society are profoundly involved in the governing process.
6. Partisan democracy initiatives are characterized by citizen-initiated participation and implicit citizen intervention in the decision-making process. Active citizens participate in the political debate, but not through traditional channels or solely through representatives. Information technology seeks to obtain visibility for alternative political expressions and criticism without interruptions from the political elite [44–54]. Unrestricted discussions set the agenda. Examples include use of independent online communities discussing politics [54], chat room discussions [45], Usenet discussions [55], and blogging [56]. The opportunity to be heard and to meet an audience can be considered important, even when the audience is scarce or absent. New voices in the political arena and empowered citizens expressing alternative ideologies might fortify the importance of partisan democracy solutions, even when the connection to the prevailing decision-making processes remains implicit or absent. The review showed a gap between the reported cases and the theoretical discussions on various democracy models. Under the partisan e-democracy model, the main challenge is to create a movement which would involve the audience and facilitate the evolution of emerging arguments. Partisan e-democracy implementations have the greatest influence in situations where alternative arenas for expressing political viewpoints are absent or difficult to access. Citizen participation can be increased both by addressing a common objective and by addressing the existence of contradictory views in the political debate and cultivating such contradictions to crystallize the standpoints of the participants. Any implementation of partisan e-democracy should facilitate free and independent debate.
7. The government-based agenda for decision making and implicit citizen participation in the decision-making process outside elections characterizes liberal democracy. This category includes several previously defined democracy models and concepts. Liberal democracy, in general, is characterized by a representative government, where citizens from the electorate, giving mandates

to representatives at the local level but also participating in the public debate. The purpose of politics is to reconcile conflicting interests, and politicians are responsible for mediating these conflicts as they occur, through negotiation [40]. In a liberal democracy, the majority rule protects individuals from random government. Effective political leadership is underpinned by liberal principles such as minimum state intervention in civil society and respect for individual privacy [57, 58].

8. The ideal of deliberative democracy connects citizens more explicitly and directly to decision-making processes [39] and emphasizes the role of open discussions in a well-functioning public sphere [37]. Politicians and citizens share an interest in dialogue and discourse leading to the formation of political opinion. Still, as it is a form of representative democracy, the input, and cooperation of citizens and politicians constitute the legalization of display of power. In relation to deliberative democracy, several ideas and concepts have been suggested. The concepts of participative, protective, and developmental democracies emphasize the achievement of equal rights and a balance of power that can only be accomplished in a participatory society. The importance of citizen participation and involvement are emphasized further in the concepts of neo-republican and plebiscitary democracy [57]. In smart cities, information technologies are developed with the purpose of increasing citizen participation and involvement in political decision making beyond casting their vote in elections or participating in electoral campaigns.
9. The consumer model of democracy focuses on the role of citizens as consumers of public services while accepting well-established features of democracy such as parliamentary institutions, elections, and parties. The main democratic value is the legal right to services, and the model seeks to re-focus democracy around the effective provision of these services emphasising value for money. Right to service being legislated in India is an expression of consumer model of democracy. Elitist democracy also accepts traditional institutional features as the basis for democracy. The general public's main personal interests are assumed to lie outside 'high' politics, rather on services and economic prosperity. This lack of interest leads to the fact that political decision-making is the responsibility of an elite of specialists and experts, whose task is to be mediators in conflicting interests and claims. Experts representing (or claiming to represent) differing interests in society act in extended policy networks, forming the elite which is effectively involved in the formation of consensus, the articulation of policy and the defining of services.
10. Neo-republican democracy focuses on the quality of citizen's participation and involvement. Citizens are expected to be active, especially at the micro- and local level, their citizenship contributing both to the emancipation of the political sphere and to their own moral, social, and intellectual development. The model is rooted in radical assumptions of common social rights and responsibilities, where the revitalizing of civic spirit is a central objective.

The cyber-democratic model represents the most radical change to traditional democratic institutions and features. Here, traditional institutions lose power and influence in favour of network-based groups forging identities with alienated individuals defying barriers caused by society, and the nation states, cultural background and class. This virtual society of networks is dependent on the use of the self-organizing Internet communities. In this post-modern vision of extended pluralism escaping the bounds of cultural hegemony, yet struggling to retain social cohesion and collective political will, ICT is no longer a supplement to traditional communication channels, but a condition for e-democracy in smart cities.

1.8.4 Smart City E-Democracy Framework

The concept of e-democracy refers to the use of information and communication technology (ICT) in political debates and decision-making processes, complementing or contrasting traditional means of communication, such as face-to-face interaction or one-way mass media. Initiatives for e-democracy from the local municipality level [58] to international programs [59], for instance, e-Europe 2005, are addressing a belief that the new ICT may increase democratization. Held's framework forms the background for a body of theoretical work which relates ICT to forms of political organizations, suggesting models of e-democracy. Van Dijk [57] abandons the obviously less relevant historical models, and analyses the role of ICT in relation to Held's four modern democracy models: pluralist, participatory, legalise and competitive democracy. He considers the models to be primarily characterized by:

1. Whether the main goal of democracy is opinion forming or decision making, and
2. Whether the primary means of democracy is the use of elected representatives or direct voting by the people.

A related, but the simpler framework of four e-democracy models is introduced by Bellamy [57]. Whereas Van Dijk discusses the role of ICT as a supplement to traditional communication mechanisms, Bellamy adds a post-Internet democracy model (Cyberdemocracy), where the Internet no longer represents a supplement to traditional communication channels but emerges as a crucial pre-condition for democracy. Bellamy claims that her four models (consumer, demo-elitist, neo-republican, and cyberdemocratic) serve as logically coherent constructs abstracted from specific social settings or from competing for political values, seeking to 'ground electronic democracy' in a set of rival discourses connecting democratic values to technological change.

1.9 Practical Point of View

From the practical viewpoint, ten models of democracy indicate that e-democracy implementations need to be holistically considered from the viewpoint of the political and decision-making context as a part of the larger issue of democratisation in society, not as a target for development. Unless new communication practices are made official, let alone recognised, by the key stakeholders such as politicians and citizens, the development may have little impact. Traditional communication channels and e-democracy will continue to exist side by side. So far, an incredibly high number of e-democracy applications have been left without an explicit connection to the traditional democratic communication and decision-making processes.

Practitioners should be also aware that the different stakeholders of e-democracy may have different ideas and ideals of democracy, which may affect the use of certain e-democracy applications. Considering the framework, these differences could be considered by the practitioners who may want to choose to promote a certain kind of e-democracy. Based on the suggested framework, any implementation of e-democracy can be specific about the actual democracy model pursued in an initiative.

Full implementation of ICT provides for more effective participatory e-democracy than that exists today anywhere globally. There are many examples of e-democracy emerging in different countries. It generally triggers community action in urban space based on mass mobilisation. There is an organisation in few countries promoting e-democracy. There are attempts to spread open source e-democracy software. In Kerala (India), this participatory democratic mobilisation was utilised to achieve eradication of illiteracy and achieving implementation of low population growth rate without using any coercive policy measures or punitive administrative interventions which are the traditional tools of legacy democracy. This was also done without e-democracy with great effort, expenditure, and mass mobilisation. These were based on people's movement which is logically an integral part of smart city democracy. Now Kerala is about to reach zero population growth rates by 2031 without using policy measures China has adopted. If e-democracy is practised it could have been accomplished with far lesser effort and money and result would have been much faster. Now there is well tested, since the 1990s, a participatory annual planning system in Kerala also called as people's planning movement (Janakiya Asuthranam) yielding results for participatory democracy. No doubt there are policy changes in participatory planning when government changes which often dilutes decentralisation. Community action for the spatial development of locality can be achieved by better ICT implementation. Spatially, urban compaction and urban design of public realm can be executed as part of annual planning through participatory e-democracy. This will show much better result than the bureaucratic implementation of legislations which is, in general, are not accepted by people.

Imagine how easy citizens' life will be if democratically elected ward committee member using ICT can be the only community focal point or as an interface of central, state and municipal government, acting as a local representative for all public service issues of government. Citizen need not meet officers in many departments whose service tradition is deplorably very bad although citizen pays for their salaries by taxes. If e-democracy is practised, this could have been accomplished with far lesser effort and money.

Indian democracy has been strengthened with 73rd and 74th constitutional amendment and has empowered panchayat and municipal system of democracy as the third tier after State Assembly and National Parliament. Unlike state legislative assembly and National Parliament, panchayat and municipality are more progressive in working out participatory democracy in India. They have mandatory 50% female representative as elected persons, unlike the national and state assemblies/parliaments. Participatory democracy and governance are what this constitutional amendment has introduced and it is working but can be strengthened considerably with e-democracy by opening it up to the virtual face-to-face community in a million-plus city.

Governance system thus formed needs considerable refinement. Like cities, governance system evolves out of accumulated experiences as a goal seeking automatic system. They must evolve in the direction of e-democracy. Today city is for the people and by the people but has grown into a very complex large entity like urban agglomeration, unlike ancient Mohenjo-Daro cities, Licchavi cities or Greek cities where strong democratic tradition is known to be of existence and were small settlements. When our cities get transformed to smart cities as the next phase of development, e-democracy is the logical extension of the present democracy. In e-democracy the size of settlement does not matter with the application of broadband, the Internet, and ICT.

Today in Indian democracy, 500 or so members of parliament in India frame legislations on a variety of topics related to urban and rural living without consulting the opinion of 1.25 billion diverse people in diverse and heterogeneous circumstances at the grass roots who are expected to benefit or suffer out of this legislation. Can these 500 people have all the wisdom of 1.25 billion people? Do they have the qualification of all knowledgeable and expert representative of 1.25 billion populations? E-democracy facilitates answers to this question and effect such consultation and participation at least cost.

Although the city is largely a high-density living, city dwellers exist as isolated atomised individual entities in an urban society characterised by urban anonymity. In a participatory democracy, they must come together, articulate their thinking and shape the future of our cities. It is only possible through e-democracy. This can be more successful in smart cities, hopefully, to make urban living more enjoyable.

1.9.1 From E-Democracy to E-Governance or Internet of Governance for Smart Cities

Electronic governance means governments online to deliver their services and programs, to provide government information, and to interact with the citizen. E-governance differs from e-government. E-government constitutes the way government institutions use ICT to apply public administration principles and conducts the business of government, using new tools to enhance the delivery of existing services [60]. E-governance includes the vision, strategies, planning, leadership and resources needed to carry this out. Included within the concept of e-governance is e-democracy, which deals with how the citizen interacts with government or influences the legislative process. It seeks to engage the citizen with governments and legislatures using the ICT.

There are two main avenues of citizen participation in e-governance: online voting and online consultation. However, there are other forms of participatory democracy, such as input into policy decisions, town meetings, online public hearings, and organisation of like-minded individuals and groups through global networking and the building of online communities.

The United Nations' Public Economics and Public Administration division and the American Society for Public Administration conducted a global survey [61], which broadly defined e-government as: Including the use of all information and communications technologies from outdated fax machines to yester years wireless palm pilots or modern mobile smartphones to facilitate the daily administration of government. However, like e-commerce, the popular interpretation of e-government is one that defines it exclusively as an Internet-driven activity that improves citizen access to government information, services and expertise to ensure citizen participation in, and satisfaction with the governing process. Although it speaks of e-government, the UN definition captures the essential role of e-governance and the evolving, new relationship of government with the citizen. E-governance results in the formation of new relationships with non-governmental organisations, citizen's groups, unions, associations, volunteer groups, business, along with citizens and other levels of government. One aspect of the changing nature of this relationship is the capacity that citizens now must access a wide range of services and information from all manner of organisations.

The UN's five guiding principles on e-government objectives are [61]:

1. Building services around citizens' choices
2. Making government and its services more accessible
3. Social inclusion
4. Providing information responsibly
5. Using IT and human resources effectively and efficiently.

The UN's eight issues on the content of current e-government are as follows:

1. Are governments doing enough to maximise the use of online services?
2. Coordination of e-government initiatives among the country's different levels of government
3. Implementing technology options, i.e. leapfrogging with wireless communications
4. Legislative and policy-making environments
5. Policy initiatives that governments are taking
6. Greater citizen participation in the policy-making environment (e-democracy)
7. Addressing the digital divide
8. E-government participation: users and non-users of e-government; measuring e-government performance.

The OECD's (Organisation for Economic Co-operation and Development) ten guiding principles on e-government are [62]:

1. Commitment: proclaim feasible goals and provide financial support
2. Rights: assure access, privacy and confidentiality to all users
3. Clarity: adopt measurement standards for electronic service delivery
4. Time: provide long-term time frames; avoid artificial deadlines
5. Objectivity: set criteria for network performance and user satisfaction
6. Resources: hire skilled personnel to design, implement and operate facilities
7. Co-ordination: use common look and feel, and document content control
8. Accountability: be accountable for electronic service quality and quantity
9. Evaluation: conduct annual e-government audits and performance reviews
10. Active citizenship: encourage active use; incorporate suggested improvements.

The approaches to delivering local e-government fell into three broad categories:

1. E-services—securing and providing government services by electronic means
2. E-governance—linking up citizens, stakeholders and elected representatives to participate in the governance of communities by electronic means (including e-democracy)
3. E-knowledge—developing the skills and the ICT infrastructure to exploit knowledge for competitive advantage.

In 2001, the US National Electronic Commerce Coordinating Council (NECCC) put forward definitions for e-government portals based on five levels, i.e. the five ways in which the web will develop [63]:

The first-level portal provides information or services easily, hiding organisational complexity and showing government as the citizen wants to see it.
The second-level portal offers online transactions such as vehicle registration, business licensing, tax filing and bill payment.
The third-level portal lets people jump from one service to the next without having to authenticate themselves again.

The fourth-level portal draws out data needed for a transaction from all available government sources.
The fifth—and highest—level portal adds value and allows people to interact with government on their own terms, providing aggregated and customised information and services in subject areas corresponding to the citizen's own circumstance.

Singapore's e-Citizen portal is considered by many to be closest to the level-5 portal. Their government website can integrate information, determine the speed of a person's connection when a request comes in, identify who the person is making the request and then do an automatic search which profiles the person. Students can come online to register for university, someone else can see a birth certificate, another be directed to make an appointment with a doctor or request a health service. People can apply for a TV license, change their address, or receive information on a variety of government services, such as renting or buying flats (the government owns 80% of all flats in Singapore) [61].

1.10 Citizen Participation in Smart City

These past experiences clearly demonstrate that simple ICT applications can have widespread impacts when the appropriate needs of the community in which they are to be applied are addressed. Doing this for jurisdictions with large national populations is a more difficult challenge unlike in smart city scale. However, what is clear from a review of many surveys and studies conducted is that most the e-government applications, as well as discussion of their successes and failures, revolve around the effectiveness of ICTs in reducing costs and increasing the efficiency of government service delivery programs. Even where there is recognition of the need for greater citizen participation if all aspects of e-governance are to be achieved, there is a noticeable lack of direction or programs for implementation of this dimension.

It is also evident that the movement towards citizen participation is limited in larger jurisdictions, though there may be more engagement of citizens in local jurisdictions that are closer to the public like in the smart city. Nevertheless, the danger in the tendency to merge e-governance and e-government and to use the terms interchangeably has the effect of discounting e-democracy—and true citizen e-participation—from the e-governance equation, where it then may be given an even lower priority by the government.

Electronic democracy suggests that the rise of ICTs offers some hope in that they may be used to facilitate a greater participation of the citizen by providing:

1. Access to more government information and interactivity using online government services, thereby contributing to the creation of a more informed citizenry.

2. A forum for the free exchange of ideas and the ability to share informed debate on issues of the day.
3. Input by citizens, through online consultations, into the decision-making process of government on those issues that directly affect them.

1.11 Relationship of E-Democracy with E-Governance

Clift [64, 65] describes e-democracy as referring to "how the Internet can be used to enhance our democratic processes and provide increased opportunities for individuals and communities to interact with government and for the government to seek input from the community." Characteristics of the Internet which he claims facilitate e-democracy are those which provide an opportunity to participate in debates as they happen. This e-participation is less limited by geography, disability or community networks, and facilitates the access to information and provision of input by individuals and groups who previously had not been included in these debates.

One of the proponents of e-government in Canada, the Centre for Collaborative Government, has dispensed with the term 'e-governance'. It uses e-government to encompass all such electronic activities and programs, with e-democracy included as a 'growing' part of e-government. Terms such as 'digital government' and 'digital voice' have also come into use. E-democracy is treated more because of e-government rather than an equal part of the equation. The Centre's emphasis on the use of ICTs by the government and elected officials often overshadows the real difficulties involved in online citizen engagement, which is presented as if it were merely an extension of more traditional consultation methods. E-government presents a real transformation in democratic governance, including design, decision-making, and service delivery capabilities. E-governance refers to new processes of coordination made possible or even necessary by the advent of technology—and the spreading of online activities. Thus, e-government refers to an IT-led reconfiguration of governance. Key elements found to enhance the democratic experience are:

1. Multiple channels for dialogue and interaction;
2. Partnership between government and a civil society group; and how knowledge, power, and purpose are redistributed considering new technological realities;
3. Terms of engagement made clear at the start—government shares in the risk but can reach out to some not normally reached. If the process is to be credible, it must be reciprocal;
4. Even though not every suggested solution could be implemented, active listening is important, giving citizens the feeling their opinions are valuable;
5. Fostering of civic literacy, with abundant information and background materials provided, and time for deliberation;

6. Transparent dialogue where answers of others could be accessed (if permission to publish was given); and
7. Active moderation by the research group per civil rules (which were published on the site) so that the discussion could not be hijacked nor turn into a free-for-all.

The two main thrusts of their democratic reform agenda are e-participation and e-voting.

E-voting is the use of ICT to facilitate participation in elections or other ballots under statutory control. This comprises e-voting in:

- elections;
- referendums and
- private ballots under statutory control.

This track also includes online registration of voters and online application to be an absentee voter.

E-participation—the use of ICT to open new channels for participation in the democratic process between elections. This comprises e-participation of citizens in:

- government's policy process;
- the processes of policy-making, law-making and scrutiny by elected representatives;
- the processes of policy formulation in political parties; and other civil society organisations.

While there have been several e-consultations, as well as an experiment with e-voting in few countries, the Office of the e-Envoy has been disappointed with the utilisation of these electronic methods both by government departments and by the citizens. British MP Edward Leigh, Chairman of the House of Commons Public Accounts Committee, reported that "Where take-up figures are available they often tell a depressing tale of low usage" (BBC News, December 13, 2002). With the aim of intensifying the engagement of its citizens, the city council of Kalix in Sweden conducted an experiment in online deliberation. It enabled the citizens to have online discussions with local politicians and each other and to give their opinions on the renovation of the town centre. Over a two-week period, citizens could participate through the Internet, as well as through traditional means of communication such as the telephone, post, and fax. Most (86%) of those participating chose to use the Internet. The citizens were also able to vote on the issue online. To ensure that only those entitled to do so could vote and that they voted only once, the council set up a password-protected website and issued a password to registered voters. For citizens without computer access, the council arranged for PCs to be made available at schools and libraries. Approximately 1200 of the 15,000 inhabitants participated, 72% of whom reported the experience as a valuable democratic initiative.

1.12 Tools for E-Democracy

The most commonly used e-tools are listed below;

1. e-discussion
2. e-initiative
3. e-petitions
4. e-consultations
5. e-feedback
6. e-complaints
7. e-polls
8. e-voting
9. e-campaigning
10. e-budgeting
11. e-consulate/e-embassy
12. webcasts
13. e-meeting
14. e-democracy games
15. e-awards.

Some of these tools are described and tabulated below

1. Web applications for getting feedback from citizens on policies and supporting citizen participation in planning procedures.
2. e-discussions (discussion forums monitored by government bodies).
3. Webcasts (recordings of meetings transmitted over the internet that allows people to watch and listen to events such as Parliamentary debates or Council Committees).

Some of the Models of the democratic process to ensure interaction between government and society are given below in terms of Models, basic functions, Concepts and period of emergence.

Model	Basic foundations	Concepts	Period
Legitimate democracy	Rule of law and the priority of a free market society. Constitutional State (Anglo-American political tradition), with a clear division of powers	New public management, E-Government	1980–1990s
Participatory democracy	Maintaining an open institutional system. Minimising unaccountable bureaucratic power in public life	E-Governance, E-Participation Government 2.0	2000s
Deliberative democracy	The mutual validity of political decisions. Developing civic education programs. Subsidies for advisory bodies and practices, as well as relevant associations	E-democracy, E-Participation Open Government	2005–2010s

(continued)

(continued)

Model	Basic foundations	Concepts	Period
Democratic autonomy	Full use of deliberative democratic mechanisms and procedures. The central and local administrative offices are built with the inclusion of citizens direct participation	E-Participation Government 3.0	2010 and further

Electronic formulation of components of e-democracy is given below.

eParticipation chat rooms	Web applications where a chat session takes place in real time especially launched for eParticipation purposes
eParticipation discussion forum/board	Web applications for online discussion where users with common interests can exchange open messages on specific eParticipation issues, pick a topic, see a "thread" of messages, reply and post their own message
Decision-making games	These typically allow users to view and interact with animations that describe, illustrate or simulate relevant aspects of an issue; herewith the specific scope of policy decision-making
Virtual communities	Web applications in which users with a shared interest can meet in virtual space to communicate and build relationships; the shared interest being within participation contexts
ePanels	Web applications where a 'recruited' set, as opposed to a self-selected set, of participants, give their views on a variety of issues at specific intervals over a period
ePetitioning	Web applications that host online petitions and allow citizens to sign in for a petition by adding their name and address online
eDeliberative polling	Web applications which combine deliberation in small group discussions with random sampling to facilitate public engagement on specific issues
eConsultation	Web applications designed for consultations which allow a stakeholder to provide information on an issue and others to answer specific questions and/or submit open comments
eVoting	Remote internet enabled voting or voting via mobile phone, providing a secure environment for casting a vote and tallying of the votes
Suggestion tools for (formal) planning proceed	Web applications supporting participation in formal planning procedures where citizens' comments are expected to official documents within a restricted period
Webcasts	Real-time recordings of meetings transmitted over the internet
Podcasts	Publishing multimedia files (audio and video) over the Internet where the content can be downloaded automatically using software capable of reading RSS feeds

(continued)

(continued)

eParticipation chat rooms	Web applications where a chat session takes place in real time especially launched for eParticipation purposes
Wikis	Web applications that allow users to add and edit content collectively
Blogs	Frequently modified web pages that look like a diary as dated entries are listed in reverse chronological order
Quick polls	Web-based instant survey
Surveys	Web-based, self-administered questionnaires, where the website shows a list of questions which users answer and submit their responses online
GIS-tools	Web applications that enable the users to have a look at maps underlying planning issues and to use them in various ways
Search Engines	Web applications to support users find and retrieve relevant information typically using keyword searching
Alert services	One-way communication alerts to inform people about a news item or an event, e.g. email Alerts and RSS Feeds
Online newsletters	One-way communication tools to inform a general audience or a pre-registered audience of specific news items and events
Frequent asked questions (FAQ)	A 'tree' of questions and answers that can be searched using keywords or by inputting a question or statement
Web portals	Web sites providing a gateway to a set of specific information and applications
Groupware tools	Tool environment to support computer-based group works

1.13 Conclusion

Although it is desirable to convert as many cities in India to smart cities, it is not simply possible. There should be potential for smart cities in states and union territories. Smart people and smart communities are primary requirements to trigger the development of smart cities. They work within the framework of e-democracy, e-governance, and e-government. They are there wherever there is high HDI. It is only smart people who can create smart communities and smart cities. Evidence of emerging urban economy is indicated by the high urban content of the states. The Large impact of the smart economy can be there only in million-plus urban agglomeration. When one combines all these requirements, not all 53 million-plus cities in India that appeared in 2011 census in India can be potential candidates for smart cities. Even if few of this million-plus agglomeration or cities are shortlisted as candidates, they can never be smart cities unless they are having an appropriate institutional framework which can shape them to smart cities such as e-democracy, e-government, and e-governance. Seventy-fourth constitutional amendments provide for the metropolitan planning committees which need to be legislated by the respective states. Triple Helix Model is the only dynamic framework that can shape smart cities. It presupposes that members of this constitutional body, the metropolitan committee shall have representatives of democratically elected representatives such as mayor, chairman/president municipality, and panchayat in the metropolitan region. Members should be there from industrial sector or its representative bodies such as the chamber of commerce and industries. There shall be academicians such as vice chancellors, directors of higher educational institutions, directors of public research institutions, and eminent professors in metropolitan committee. This committee shall have an implementing authority with senior administrator heading implementing body whose main function is to implement the decisions of the metropolitan committee. The administrator shall be assisted by three wings namely metropolitan planning wing with highest level of capability for metropolitan regional development planning and capability to manage the modern centre for metropolitan geographic information system using e-democracy for e-government and e-governance; an engineering wing with senior engineer managing it and a finance wing whose main capability shall be the mobilisation of large-scale finances from national and international sources for implementing large-scale infrastructure and other development tasks in addition to traditional account keeping. All these wings shall have provision to employ the highest level of expert consultants to accomplish tasks which normally they cannot perform at market rate.

A global alliance of smart cities shall be formed where Indian smart cities shall be members to exchange experience and share software development. Metropolitan planning committee shall have 20-year spatial and infrastructure development plan, 5-year metropolitan social and economic development plans in tune with state and national five-year plans and annual plans for implementation of the plan as per annual budget. A share of annual budget shall come from state government, and another central government and metropolitan planning committee shall have their own financial resources from entering commercial developments such as land development and leasing of land for industries, development of infrastructure to potential entrepreneurs or running major transport networks. Institute of Town Planners India can have a positive role in formulating model metropolitan planning committee legislation and model organisation framework for the smart city metropolitan committee and its implementing authority. This will considerably help the state government to formulate legislations for the creation of metropolitan planning committee. India's experience in the direction of the metropolitan planning committee is very meagre or one can say almost non-existent.

There are specialised hardware and software required for the functioning of smart cities and smart communities. Many of these technologies are not implemented in Indian cities. Most of them aim at resource conserving metropolitan functioning of smart cities whether it is of energy or water use or on transportation. Outright purchase of technology and licence for local manufacture and commercial production is one approach for large-scale deployment of these technologies in emerging Indian smart cities. Another approach is collaborative research and joint production. Still another approach is to develop appropriate smart city technologies in Indian research centres based on specific needs. There is a need to appraise globally available smart city technologies and formulate appropriate Indian standards. Indian Standard Institution is competent to develop such standards, and Ministry of Science and Technology can conduct a global smart city urban technologies assessment.

There are no smart city policies for India or states of India. There is a need to conduct basic research leading towards the formulation of policies. This policy formulation can be from the study of other countries such as for example European community who is taking many steps in this direction. Another approach is to evaluate the existing smart cities to learn the experience and transfer it to Indian condition.

There are few experiences such as e-governance and e-commerce available in India. There are also resources conserving technologies being developed in India which can be used in smart cities. These needs to be studied and expanded in scope.

Smart cities open to an information or knowledge society. It is likely that information society may trespass in the privacy of a citizen. These needs to be studied and appropriate norms need to be applied to smart cities 1. E-democracy for Smart Cities.

This book "E-Democracy for Smart Cities" is organized into five sections and within each section there are one or more chapters.

The first section introduction is meant to lay out the dominant thinking on Democracy and e-democracy mainly derived from literature surveys and experiences across the world to evolve an operational e-democracy model for Smart Cities.

The second session presents a series of case studies on the state of the Art of E-Democracies in emerging smart cities across the world. Unlike E-Governance, e-democracy is not widely practiced. Therefore, the nascent state of e-democracy in all these cities will be evident. These chapters also point out the potentials.

The third section deals with a series of case studies on different domains in e-democracy. These domain studies show there has been progress in selected sectors of domain studies in e-democracy while the full practice of e-democracy is nowhere to be found in any cities.

The fourth section discusses issues and the emerging tools of e-democracy.

The fifth and final section combines all conclusions derived out of all chapters of the book in one chapter.

References

1. Vinod Kumar TM et al (ed) (2014) Geographic information system for smart cities. Copal Publishing Group, New Delhi
2. Vinod Kumar TM (ed) (2015) E-governance for smart cities. Springer, Singapore
3. Vinod Kumar TM (ed) (2016) Smart economy in smart cities. Springer, Singapore
4. Rifkin J (2013) The third industrial revolution: how lateral power is transforming energy, the economy, and the world. Kindle Edition

5. Rifkin J (2015) The zero marginal cost society: the internet of things, the collaborative commons, and the eclipse of capitalism. Kindle Edition
6. Diamond L, Lecture at Hilla University for Humanistic Studies 21 Jan 2004: What is democracy?; Diamond L, Morlino L (2016) The quality of democracy. In: Diamond L (ed) Search of democracy. Routledge, London. ISBN 978-0-415-78128-2
7. δημοκρατία in Henry George Liddell, Robert Scott, A Greek-English Lexicon, at Perseus
8. Wilson NG (2006) Encyclopedia of ancient Greece. Routledge, New York. p 511. ISBN 0-415-97334-1
9. Barker E (1906) The political thought of Plato and Aristotle, Chapter VII, Section 2. G.P. Putnam's Sons
10. Staff writer (22 Aug 2007) Liberty and justice for some. The Economist (Economist Group)
11. O'Donnell G (2005) Why the rule of law matters? In: Diamond L, Morlino L (eds) Assessing the quality of democracy. Johns Hopkins University Press, Baltimore, pp 3–17. ISBN 9780801882876. Preview
12. Dahl RA, Shapiro I, Cheibub JA (2003) The democracy sourcebook. MIT Press, Cambridge Details. ISBN 9780262541473
13. Hénaff M, Strong TB (2001) Public space and democracy. University of Minnesota Press, Minneapolis. ISBN 9780816633883
14. Kimber R (1989) On democracy. Scand Polit Stud 12(3):201, 199–219. doi:10.1111/j.1467-9477.1989.tb00090.x. Full text
15. Scruton R (9 Aug 2013) A point of view: is democracy overrated? BBC News (BBC)
16. Kopstein J, Lichbach M, Hanson SE (eds) (2014) Comparative politics: interests, identities, and institutions in a changing global order (4, revised edn). Cambridge University Press, Cambridge, pp 37–39. ISBN 1139991388
17. Parliamentary sovereignty. UK Parliament. Retrieved 18 Aug 2014; Independence. Courts and Tribunals Judiciary. Retrieved 9 Nov 2014
18. Daily Express News (2 Aug 2013) All-party meet vows to uphold Parliament supremacy. The New Indian Express (Express Publications (Madurai) Limited). Retrieved 18 Aug 2013
19. Barak A (2006) Protecting the constitution and democracy. In: Barak A (ed) The judge in a democracy. Princeton University Press, Princeton, New Jersey, p 27. ISBN 9780691120171. Preview
20. Kelsen H (1955) Foundations of democracy. Ethics, special issue: Part 2: foundations of democracy (Chicago J) 66(1):1–101. doi:10.1086/291036.JSTOR2378551
21. Nussbaum M (2000) Women and human development: the capabilities approach. Cambridge University Press, Cambridge New York. ISBN 9780521003858
22. Snyder R, Samuels D (2006) Devaluing the vote in Latin America. In: Diamond L, Plattner MF (eds) Electoral systems and democracy. Johns Hopkins University Press, Baltimore, p 168. ISBN 9780801884757
23. Montesquieu, Spirit of the laws, Book II, Chap. 2–3
24. Everdell WR (2000) [1983] The end of kings: a history of republics and republicans, 2nd edn. University of Chicago Press, Chicago. ISBN 9780226224824
25. Dunn J (1994) Democracy: the unfinished journey 508 BC–1993 AD, 91994. Oxford University Press, Oxford. ISBN 0-19-827934-5. Raaflaub, Ober & Wallace 2007
26. Po-chia Hsia R, Hunt L, Martin TR, Rosenwein BH, Smith BG (2007) The making of the West, peoples and cultures, a concise history, vol I: To 1740. Bedford/St. Martin's, Boston and New York, p 44
27. Grinin LE (2004) The early state, its alternatives and analogues. 'Uchitel' Publishing House
28. Lape S (2009) Reproducing Athens: Menander's comedy, democratic culture, and the Hellenistic city. Princeton University Press, Princeton, p 4. ISBN 1400825911
29. Huber E, Rueschemeyer D, Stephens JD (1993) The impact of economic development on democracy. J Econ Pers 7:71–85
30. Aristotle (1932) Politics (Translated by H Rackham). Harvard University Press, Cambridge
31. Lipset SM (1959) Some social requisites of democracy: economic development and political legitimacy. Am Polit Sci Rev 53(01):69–105

32. Lipset SM, Seong K-R, Torres JC (1993) A comparative analysis of the social requisites of democracy. Int Soc Sci J 45(2):155–175
33. The Mahaparinibbanasuttanta is the story of the "great decease of the Buddha" and as such includes both colourful anecdotes and important last minute instructions to his followers
34. The Pali Canon uses both the term Vajji (Vriji in Sanskrit) and Licchavi to designate a republican polity based at Vesali. Scholars believe that the Licchavi were the people who lived at Vesali, while Vajji was the name of a confederation that they headed. For a detailed discussion, see Sharma
35. http://faculty.nipissingu.ca/muhlberger/HISTDEM/INDIADEM.HTM 8/16
36. In this sense, R.C Majumdar was right in calling the Buddha "an apostle of democracy;"
37. Held D (1996) Models of democracy. Blackwell, Oxford
38. Corporate Life, p 219. Contra, Drekmeier, Kingship and community in early India, p 113 (Alexander H. Trechsel and Fernando Mendez, The European Union and E-Voting (Routledge, London, 2005), p 5 (Italics added)
39. Pateman C (1970) Participation and democratic theory. Cambridge University Press, Cambridge
40. Eriksen EO, Weigård J (1999) Kommunikativ handling of deliberative democratic: Jürgen Habermas' teori om politikk og samfunn. Fagbokforlaget, Bergen
41. Gimmler A (2001) Deliberative democracy, the public sphere and the internet. Philos Soc Criticism 27(4):21–39
42. Barber BR (1984) Strong democracy: participatory politics for a new age. University of California Press, Berkeley
43. Bellamy C (2000) Modelling electronic democracy, towards democratic discourses for an information age. In: Hoff J, Horrocks I, Tops P (eds) Democratic governance and new technology, technologically mediated innovations in political practice in Western Europe. Routledge, London, pp 33–54
44. Lynne E (2004) Direkte demokrati. Pax, Oslo
45. Fung A (2002) One city, two systems: democracy in an electronic chat room in Hong Kong. Javnost-The Publ 9(2):77–93
46. Hurwitz R (1999) Who needs politics? Who needs people? The ironies of democracy in cyberspace. Contemp Sociol 28:655–661
47. Moon JY, Yang S (2003) The internet as an agent of political change: the case of "Rohsamo" in the South Korean Presidential Campaign of 2002. In: Twenty-fourth international conference on information systems
48. Olsson T, Sandstrom H, Dahlgren P (2003) An information society for everyone? Gaz: Int J Commun Stud 65(4–5):347–363
49. Paolillo JC, Heald D (2002) Democratic participation in the discursive management of Usenet. In: Proceedings of the 35th annual Hawaii international conference on system sciences, vol 35, pp 1040–1049
50. Papacharissi Z (2004) Democracy online: civility, politeness, and the democratic potential of online political discussion groups. New Media Soc 6(2):259–283
51. Rodan G (1998) The internet and political control in Singapore. Polit Sci Q 113(1):63–89
52. Schneider SM (1996) Creating a democratic public sphere through political discussion—a case study of abortion conversation on the internet. Soc Sci Comput Rev 14(4):373–393
53. Stromer-Galley J (2002) New voices in the public sphere: a comparative analysis of interpersonal and online political talk. Javnost-The Publ 9(2):23–41
54. Tsaliki L (2002) Online forums and the enlargement of public space: research findings from a European project. Javnost-The Publ 9(2):95–112
55. Hill KA, Hughes JE (1998) Cyberpolitics: citizen activism in the age of the internet. Rowman & Littlefield, Lanham, MD
56. Griffiths M (2004) e-Citizens: blogging as democratic practice. Eur J E-Gov 2(3):155–166
57. Van Dijk J (2000) Models of democracy and concepts of communication. In: Hacker KL, Van Dijk J (eds) Digital democracy, issues of theory and practice. Sage, London

58. Grönlund Å (2003) Emerging electronic infrastructures—exploring democratic components. Soc Sci Comput Rev 21(1):55–72
59. Commission E (2002) eEurope 2005: an information society for all. http://europa.eu.int/information_society/eeurope/2002/news_library/documents/eeurope2005/eeurope2005_en.pdf, current 10 June 2004
60. Okot-Uma RWO (2000) Electronic governance: re-inventing good governance. Commonwealth Secretariat, London
61. United Nations (2002) A global survey of e-government 2002. www.nuasurveys.org online newsletter, 24 June 2002: 2
62. Holmes D (2001) E-gov: e-business strategies for government. Nicholas Brealey Publishing, London
63. Clift S (2001) E-governance to e-democracy: progress in Australia and New Zealand. Online http://www.electronicgov.net
64. Clift S (2002) The future of e-democracy. Democracy Online Newswire
65. http://www.e-democracy.org

Part I
E-Democracy State of the Art City Studies

Chapter 2
E-Democracy for Smart City Lagos

Femi Olokesusi and Femi Ola Aiyegbajeje

Abstract Information and Communication Technologies (ICTs) have strengthened the democratic processes of most democracies in the world. The adoption of E-democracy is to encourage good governance and to properly monitor the democratic processes for the benefits of Lagos residents. Lagos has the fastest and most lucrative ICT market in Nigeria and Africa, yet in spite of this obvious and significant progress, little is known about its E-democracy adoption. This paper provides a comprehensive review of E-democracy in order to assess the levels of E-democracy implementation, evaluate critical success factors of E-democracy implementation and discusses the successes and failures of E-democracy. The analysis of the impact of E-democracy on service delivery in Lagos is also provided. It notes that Lagos is facing a number of challenges in the introduction of E-democracy. Given the importance of the successful implementation of electronic democratic services and from a practical perspective, the paper suggests that government should take a positive position towards the factors which will bring about effective and efficient E-democracy in Nigeria, particularly in Lagos. Lagos State Government needs to widen access to ICT services and guarantee enabling environment for attracting the right level of investments. The paper concludes that there is a lot of hope on the potential of E-democracy to transform the internal efficiency of government and the relationship of government with citizens.

Keywords E-democracy · Smart city Lagos · Information and communications technology · Lagos

F. Olokesusi
College of Social and Management Sciences, Afe Babalola University,
Ado-Ekiti, Ekiti State, Nigeria
e-mail: femioloke@yahoo.com

F.O. Aiyegbajeje (✉)
Department of Geography, Faculty of the Social Sciences,
University of Ibadan, Ibadan, Nigeria
e-mail: femidavid2002@yahoo.com

T.M. Vinod Kumar (ed.), *E-Democracy for Smart Cities*, Advances in 21st Century Human Settlements, DOI 10.1007/978-981-10-4035-1_2

2.1 Introduction

Democracy is a system of governance that establishes a social contract between the citizens and the representatives [1]. This therefore necessitated prompt responsive communication between the people and the government. Democracy as a form of governance stretches beyond campaign, voting and other electoral practices as its basic components. It also involves interaction among participants in a democratic setup. Democracy actually entails the governing of a people by the people but with the use of electronic (Information and Communication Technology), democracy becomes e-democracy. The 'people' in this context, refers to the elected people and the governed or voters resident in the state. E-democracy is the use of information technologies (such as Wide Area Networks (WAN), the Internet, and mobile computing) that have the ability to transform relations with citizens, businesses, and other forms of human endeavours. These technologies can serve a variety of different ends: better delivery of government services to citizens, improved interactions with business and industry, citizen empowerment through access to information, or more efficient government management. By putting a suffix "e" in front of democracy means nothing more than using information technology tools to facilitate, improve and ultimately extend the exercise of democracy [1].

E-democracy is positioned as a tool in enhancing and making the democratic process more accessible; ensuring citizen participation in public policy decision-making. E-democracy allows a broader influence in policy outcomes as more individuals involved could yield smarter policies; increasing transparency and accountability, and keeping the government closer to the consent of the governed, increasing its political legitimacy, especially considering Lagos state where residents continuously accused the state government of lack of transparency and bad governance which constitute hindrances to the development of democracy and democratic principles in the state. In this regard, representative democracy has failed and hence has led to frustration and reduced trust in democracy but the adoption of electronic based democratic practices in Lagos state has re-enacted a ray of hope. The growth of ICT does not only focus its attention on the internet but also spreads its wings to cover telecommunication, whereby people are connected and can communicate to themselves through cell phones, home personal computers and other medium. From the available statistics, the Internet facility has been discovered to be the most used medium of communication than the personal computers and the telephone put together [2]. For instance, the use of *Facebook* and *Twitter* in the last general elections of 2011 and 2015 in Lagos Nigeria was very impressive. It created a necessary platform for the contestants and the electorates to interact.

E-democracy is an attempt aimed at developing digital citizenship (Netizen) through the use of information and communication technology (ICT) to create personal contact, dialogue and consultation among participants in democracy [3]. E-democracy offers a level playing ground and bridge communication gap among administrators, citizens, associations and public and private entities across the various tiers of government. E-democracy, presents a new form of democratic practice carried out through information and communication technology. One essential elements of E-democracy is the relevance of residents' participation in decision making. According to Azeta et al. [4], E-democracy is an innovation still in its infant stage; and is still subject to much debate and activity within government, civic oriented groups and societies around the world. Through the internet and social media, government and its officials communicate with the residents more effectively, and also aid the communication between citizens and their fellow citizens to discuss political and governmental issues which could be a resourceful contribution to the improvement of governance and most democracy. Citizen-participation in governance is at an increasing level in the nation owing to the involvement of ICT in governance. E-democracy does not just stop at the levels of the citizens, but goes as far as fostering the communication among administrators, associations, public and private entities among the various tiers of governments.

ICT has the potential to engage people in all areas of the political process such as the generation of information, enhanced deliberation among citizens, and most of all enhance participation in decision making [3]. According to Policy and Legal Advocacy Centre [5], the 2011 elections in Nigeria witnessed a remarkable use of the social media as a tool for political communication. However, in the 2015 election in Lagos, the social media became more potent and largely deployed to execute the elections in Lagos. The social media pass information freely because it is unregulated. Information pertaining to elections and campaign flew around on social media applications such as *WhatsApp, BBM, blogs, Facebook, and Twitter.* ICT is also one of the best means of bridging the communication gaps between the people and the government. Through the internet, it is now possible for the government to communicate with the residents easily via social media, and also aids the communication between residents and their fellow residents of the state to discuss political and governmental issues which could be a resourceful contribution to the improvement of governance and most democracy.

One of the most embraced aspects of ICT in Nigeria is the mobile telephone technology. A new publication *sub-Saharan Africa Mobility Report* by Ericson says that Nigeria is the largest and fastest telecommunication market in Africa with about US$38 billion investment and active mobile subscription of over 150 million as at September 2016 [6]. Further, with respect to subscription to mobile telephone services, the Nigeria Bureau of Statistics (NBS) report of the first quarter of 2016 puts Lagos State as having the highest active mobile telephone and Internet subscribers in the country [7]. Nigeria recorded about 148.74 million active voice and internet subscribers in the first quarter (Q1) of 2016. In its quarterly report on Nigerian Telecommunications in the first quarter of 2016, the National Bureau of Statistics (NBS) revealed that the report represents the total number of active voice and internet subscribers in the 36 states of the country including the FCT at the end of the first quarter 2016 disaggregated by States. The report showed that the number of active voice subscriptions in the country stood at 148.74 million at the end of Q1 2016.

According to the NBS, out of the 36 States in Nigeria, Lagos State accounted for the largest share of active voice subscribers with 19.04 million (12.8%) of the total, followed by Ogun State with 8.53 million subscribers (5.7%), Kano State with 7.81 million (5.25%), Oyo State with 7.53 million subscribers (5.06%), then FCT and Rivers State with 6.03 million (4.05%) and 5.84 million (3.93%) respectively [7]. It is observed that all of the top five States with respect to active voice subscriptions had total subscribers in excess of their official population number as provided by the Nigerian Population Commission, except for Kano and Rivers, whose official population figures exceeded the number of active voice subscribers at the end of Q1 2016. On the other hand, the following States of the Federation; Bayelsa (1.11 million), Yobe (1.40 million), Ekiti (1.42 million) and Ebonyi (1.43 million) had the smallest number of active subscriber as of Q1 2016 [7].

Against this backdrop, the objectives of this study are to appraise E-democracy practice in Lagos city by assessing the levels of E-democracy implementation, critical success factors of e-democracy implementation, and analyse the impact of E-democracy on service delivery in Lagos state. Lagos is one of the 36 federating units of Nigeria and not a fully autonomous state because of the several provisions of the Constitution of the Federal Republic of Nigeria which does not give the Federating States absolute powers to operate as independent units of the Federation as obtained in developed democracy such as the United State of America.

2.1.1 *Lagos in Brief*

Lagos is one of the 36 federating units in Nigeria. It is located in the coastal part of the south-western part of the country Nigeria (see Fig. 2.1). Lagos is the largest city in Nigeria and arguably the most populous city in Africa. Lagos is often referred to as the economic capital of Nigeria. Lagos city is a huge metropolis with streaming population in the south-western archipelagos of the Nigerian coastline which includes the mainland and the surrounding Islands. Lagos is also the former capital city of Nigeria; it has since been replaced by Abuja. Abuja city officially gained its

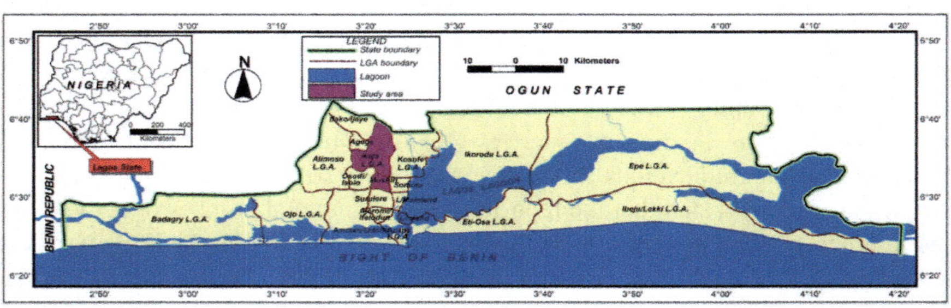

Fig. 2.1 Political map of Lagos State, Nigeria. *Source* [20]

status as the capital city of Nigeria on 12 December 1991; although the decision to move the federal capital had been made in Decree No. 6 of 1976.

The leading air and sea ports in Nigeria are situated in Lagos. The Lagos Sea Port is operated by the Nigerian Port Authority and it is divided into three main sections: Lagos port, Apapa Port and Tin Can Port, all located on the Gulf of Guinea. Oil and petroleum products provide 20% of GDP and about 95% of foreign exchange earnings in Nigeria. Lagos is also Nigeria's commercial centre as majority of the country's largest banks and financial institutions are located here. More than half of Nigeria's industrial capacity is located in Lagos's mainland suburbs, particularly in the Apapa, Matori, Ilupeju and Ikeja industrial estates. A wide range of manufactured goods are produced in the city.

2.2 E-Democracy Implementation and Achievements in Lagos-Nigeria

Basically, E-democracy represents the use of electronic communications technologies such as the Internet to enhance and advance the democratic process of a republic. In this part of the world, there exists a low attitude of participation of the citizens in governance, but with the advent of electronic communication technology, the level of awareness and participation has tremendously increased among the citizenry. The benefits of having an E-Democracy system are numerous: It reduces the success of rigging during elections. It is time saving because it reduced some undue electoral processes. It allows for easy communication between the citizens and their elected candidates. And it allows for participatory governance.

The E-democracy implementation assesses the citizens' E-participation (E-information, E-consultation, E-decision-making), E-campaign, E-voting, E-forum, and E-monitoring of projects. However, in Lagos city, only E-information, E-campaign and E-monitoring of projects have been successfully utilized. E-information involves dissemination of information about government ministries, departments and agencies while E-campaign involves the use of social media to canvass for votes. Government also uses it to disseminate vital information to the residents of the city. Electronic-monitoring of projects involve the use of the citizens residing where government projects are being executed to report to the government via a dedicated social media platform such as *Facebook, Twitter*, and various dedicated websites) created by government for the purpose of monitoring projects. Another important aspect of E-monitoring of projects is the use of electronic facilities to monitor the traffic situation in the metropolis. This has largely help motorists to plan their route effectively. To some extent, these techniques have been successful but not without some shortcomings such as irregular update of information, disseminating of hate campaigns against political opponents as evidenced in the 2015 elections in Lagos.

2.2.1 E-Participation

E-participation comprises of a participatory processes including E-information, E-consultation, E-decision-making. *E-information:* This assesses the state government's websites and portals to determine if governments are providing the basic information that serves as the foundation for citizen participation. Lagos state government has about twenty-eight (28) website addresses for the state ministries [8] and eighty-seven (87) official websites for the state departments and agencies [9]. Although these websites are not interactive, but provide vital information about government activities in the respected ministries, departments and agencies (MDAs). The major problems associated with this platform are poor or no feedback, not providing updated information on the website, and in some cases, the websites are not functioning. Appendix 1 shows few examples of the website addresses of the ministries in Lagos State Government while Appendix 2 similarly indicates the website addresses of the State Government's Departments and Agencies.

2.2.2 E-Campaign

E-Campaign is the utilisation of various interactive platforms on electronic communication devices by electoral candidates as promotional campaign strategies to attract more support from voters. It also provides the political parties a platform not only to disseminate their ideologies and manifestoes to the voters but also to get feedback. E-campaign of various forms includes *Facebook, Twitter, WhatsApp, E-mail, Blog, Youtube, Friendster, Myspace, SMS, Forum* and others. E-campaign in Lagos is fast gaining ground in the political space since 2007 to date. This is largely because of the wide acceptance of ICT and coupled with high number of mobile telephone users among the citizens. The use of ICT in political campaigns was highly deployed in the last previous elections of 2011 and particularly 2015 in Lagos State because most of the social media platforms were accessed via mobile telephone.

Adoption of the Internet usage as one of the promotional political campaign strategies is widespread in the developed countries. For instance, in the 1997 and 2001 UK general elections, it was asserted that the internet plays an important campaign role [10]. Similarly, the 2012 and 2016 US Presidential elections are examples in justifying the transformed use of Internet through the development of websites and email [11]. The adoption of this technology was also experienced in the 2015 governorship and state house of assembly elections held in Lagos State and some of other States of the Nigerian Federation.

The Internet technology adoption took a similar trend in the Nigeria political scene, especially in the 2011 and 2015 general elections. The application of e-campaign strategies in the Nigeria's 2011 and 2015 elections in Lagos further raised the level of awareness among voters and citizens giving the high level of mobile telephone subscription in Lagos State (12.8%) which is the highest in Nigeria [7]. During the 2011 and 2015 general elections in Lagos, the E-campaign strategy was adopted by all electoral candidates across party lines in all the elections.

The social media was hugely deployed with more than 134 million users of mobile phones and about 70 million Nigerians on the internet, according to statistics from the Nigerian Communication Commission. It therefore explains the reason social media played an influential role in electioneering campaign in Lagos in the 2015 elections. The most accessible new media consisting of *Facebook, Twitter, Youtube, Instagram, Google Plus, WhatsApp and LinkedIn* among many others have become a veritable tool for interacting and mobilizing citizens towards active participation in the political process and democratic projects. Their flexibility, accessibility and affordability have made them attractive to the youths who are the most active users and the largest voting bloc.

2.2.3 E-Voting

The goal of any voting system is to establish the intent of the voter and transfer that intent to the vote counter. The efficiency of the voting method and the accuracy of the vote counter are the crucial determinants of the ability and capacity of the system to correctly determine the wish of the voters [12]. Electronic voting (E-voting) is the use of ICTs in the conduct of elections. This idea was adopted in Nigeria during the 2015 elections to address the major problems associated with elections. These problems include missing names of some registered voters, intimidation and disenfranchisement of voters, multiple and under aged voting, snatching or destruction of ballot boxes, miscomputation and falsification of results [13, 14]. The only aspect of the E-voting that is still practicable in Nigeria is the use of Card Reader Machine (CRM). The introduction of card reader machine and permanent voters card (PVC) by the Independent National Electoral Commission (INEC) was aimed at stemming the tides of the problems listed earlier. This card reader machine is dedicated to read the Permanent Voters Card (PVC) issued to eligible voters. The PVC is screened by the card reader machine in order to check for fraudulent practices among the voters.

Unfortunately, as beautiful and workable this initiative is, there were still some shortcomings and challenges identified in the 2015 general elections in Nigeria.These include the inability of card reader to identify some PVCs, lack of back-up batteries to power the card readers, poor electricity, people's attitude towards the technology, poor technical know-how among the INEC staff and ad hoc staff among others.

For the success of implementing the e-voting in Lagos and by extension Nigeria, the provision for the adoption of E-voting must be inserted in the constitution of the country to make it legal for conducting future elections. Also there should be a concerted effort in reviving the power and ICT sectors for quality service delivery.

The National Orientation Agency (NOA) needs to embark on a serious campaign and advocacy on behavioural change and technology acceptance of e-democracy.

Despite the foregoing shortcomings, E-voting still retains a measure of attraction for the electoral process in Nigeria. It is believed that E-voting can facilitate the processes of free and fair elections in Nigeria. The proposition on the introduction of E-voting in the electoral process of Nigeria subsequently reached the Nigerian Senate again, as component of a new Electoral Act to drive the 2015 general elections but was rejected by the Nigerian Senate April 2014.

2.2.4 E-Poll

Provides available functionalities for the creation of opinion polls for users to cast their opinion on trending issues. This is a method of assessing and evaluating government's progress and public opinion on a particular issue of interest. It is also used in determining the acceptance of electoral candidates before the election. In Nigeria, this is mainly done by both print and non-print media houses e.g. Channels Television, African Independent Television, Punch Newspaper, Guardian Newspaper etc. Sometimes, the use of mobile telephone is deployed and this involves the use of mobile telephone to send short messages (sms) to dedicated members of the public for or against an opinion.

2.3 Strengthening E-Democracy in Nigeria

Whereas, there is evidence of the diffusion of E-democracy in Lagos, E-democracy is yet to be fully deployed in the democratic space of Nigeria especially in the area of E-voting despite its immense importance. A critical area where E-democracy needs to be seriously strengthened includes provision of virile legal frameworks, because the existing legal frameworks only allows for a physical and manual voting where a voter is expected to cast his vote by thumb-printing a ballot paper and then drop the ballot paper in a ballot box provided by INEC at the polling booth. Also, the electricity provision must be enhanced and with a serious advocacy on behavioural change and technology acceptance.

All this efforts will go a long way to entrench E-democracy in Lagos and Nigeria at large. As suggested, before adopting any form of electronic voting method, a critical appraisal of such a method must be undertaken [12]. Such chosen system must as a matter of fundamental necessity possesses a certain basic attributes that will recommend it. These include among others: the system must increase and guarantee voter confidence and secrecy; must be voter and user friendly; offer the best in terms of reliability, usability and recountability; and must be secure.

2.3.1 *Success of E-Democracy in Lagos*

E-democracy has attained some level of successes in making Lagos a smart city. As earlier mentioned, one of the areas of success of E-democracy in Lagos State are in the area of E-participation, E-campaign, e-monitoring of projects, E-poll. Key among these successes is seen in the E-monitoring of projects and one major area this has contributed is the rapid growth of infrastructural facilities. This was made possible because projects are being monitored via the social media where citizens can update government officials in charge of the projects latest development without these officials being on the site. Also, by monitoring the traffic situation in the city has a way of helping to create an enabling environment for businesses to thrive because we all understand the negative impact that a chaotic traffic environment can do to businesses.

To better understand the successes and achievements what E-democracy has brought to infrastructure development in Lagos metropolis through the adoption of E-monitoring of projects concept, few critical infrastructures are critically reviewed below to buttress this fact.

2.3.1.1 Transportation

The transportation sector is a key factor for the economic development of any society because of the important roles it plays in the movement of people, services and information. For government to achieve this, E-democracy has helped the movement of about 7 million passengers per day [15]. The scope of E-democracy in this regard include the introduction of E-ticketing for Bus Rapid Transit (BRT), involvement of the masses in decision making via media, introduction of traffic radio 96.1 FM station to divulge information pertaining to traffic situation in all areas of the metropolis (see Fig. 2.2).

Presently, road transportation accounts for over 90% of domestic passengers and freight movement on about 30,500 km State owned roads and 129,577 km of Local government roads in Lagos State [16]. There is a daily haulage of 6 million passengers between the Mainland and Lagos Island in about 75,000 unregulated mini-buses. In a bid to reform this situation, the Lagos State Government in collaboration with the World Bank initiated the Lagos Urban transport Project (LUTP), to create an efficient and effective integrated inter-modal mass transit system in the State. The Lagos State Government in its effort to improve the public transport system has through LAMATA deliberately eased out the old para transit buses popularly called *Molues* to give way to brand new and more comfortable buses called Bus rapid Transit (BRT) and LAGBUS plying over dedicated lanes within the metropolis. The BRT Lite scheme commenced operation with a total number of 220 Ashok Leyland high floor capacity buses along Mile 12—TBS corridor from 6:00 am to 10:00 pm daily. Over 900 bus pilots/bus officers and 57 inspectors/monitoring officers were employed for BRT operations with about 106

Fig. 2.2 A typical traffic situation in Lagos during peak period. *Source* [16]

bus-stops [17] (see Fig. 2.3). In mid-2016, about 450 new buses were added to the declining fleet.

The pilot BRT Lite scheme has performed creditably with over 400 million passengers served in the last five years. Available statistics indicate that the Average daily ridership is about 180,000 passengers. This scheme has brought about a reduction of 30% in average fares and fare stability, 40% in journey time, 35% in average waiting time and reduction in exposure to robbery on public transport [17]. Since 2013, the Lagos Metropolitan Area Transport Authority (LAMATA) commenced utilisation of e-ticketing to board BRT buses [15].

All this was achieved by engaging the populace via social media and other useful medium of information dissemination. The introduction of E-ticketing is one of the useful contributions of E-democracy to transport development. Another important area of E-democracy involvements in transportation development in Lagos is the inauguration of the state's traffic radio station (96.1 FM) in a bid to rid the roads of gridlocks. The radio provides traffic news to assist motorists make the right decisions that will in turn help to decongest the roads.

2.3.1.2 Health

The Lagos State Ministry of Health in collaboration with the State's Ministry of Science and Technology has initiated the Lagos State eHealth project which is an

Fig. 2.3 BRT terminals at Tafawa Balewa Square, Lagos. *Source* [23]

interactive Hospital Information System Software in some of the State owned hospital facilities. The Health Management Information System (HMIS) is a broad based modular Hospital Administration System Database solution that allows the operational procedures and patients' flow in health facility to be harmonized for prompt and effective healthcare delivery. The pilot phase of this project commenced fully in January, 2009 at the Lagos State University Teaching Hospitals, (LASUTH), General Hospital Isolo, Lagos and currently on its second phase which involves ten (10) General Hospitals.

With this project, any patient in a Lagos State hospital will be registered on a central computer server and since it is an integrated system, the doctors treating a patient will have simultaneous access to the up-to-date medical records and treatment information. The links to the diagnostic modules ensures that tests can be ordered by doctors online while results can be returned via the same channel [18]. The e-Health strategy covers a wide range of services namely: electronic health records, telemedicine, consumer health informatics, health knowledge management, virtual healthcare teams, mobile health (m-Health), and healthcare information systems [19]. Also, a study by [20] indicated that mobile telephone technology has served as a platform to offer better healthcare management because healthcare services are still being offered to out-patients despite the fact that they are far away from the hospital.

2.3.1.3 Emergency Medical Services

Increasing urbanization and industrialization of Lagos State has led to a considerable increase in the number of road traffic, industrial and marine accidents on one hand and other domestic medical emergencies on the other hand, hence the strengthening of the existing hospital care facilities i.e. Lagos State Emergency Medical Services (LASEMS) and the establishment of the pre-hospital care service i.e. the Lagos State Ambulance Services (LASAMBUS) in 2001 (see Fig. 2.4). Both services run on a 24/7 basis. This has significantly improved the response time and quality of care with attendant improvement in morbidity and mortality rates occasioned by medical emergencies. A dedicated (easy-to-remember) phone hotline '123' was created to link the public, LASEMS, LASAMBUS and the Ministry of Health. This service provides a free treatment within the first 24 h. Fifteen ambulance points throughout the state are provided presently with plans to add five more points in the nearest future. Ten (10) units of mobile intensive care unit (MICU) ambulances (with capacity to administer advanced life support) are in the fleet of ambulances. In addition, the Marine rescue Unit was also established consequent upon increased incidence of vehicles plunging across the bridges into the lagoon [18].

In addition, as shown in Fig. 2.5, the State has launched Nigeria's first medical helipad at the Lagos State University Teaching Hospital (LASUTH). The helipad is to boost life-saving chances of the hospital, and to provide a traffic-free transport for accident victims and others. The helicopter ambulances are equipped with modern ICT gadgets to aid communication for optimum healthcare delivery.

Fig. 2.4 Lagos state ambulance point in Ikeja, Lagos. *Source* [18]

Fig. 2.5 Medical Helipad at the Lagos State University Teaching Hospital (LASUTH), Ikeja, Lagos. *Source* [24]

2.3.1.4 Housing

The way and manner land ownership, land title, documentation and transaction are structured, have a direct impact on the state's economy and general development. Therefore, government must make the processes seamless because land provides shelters, food, minerals and other resources needed for urban development. In view of this, the government of Lagos State commissioned the computerization of the land registry and the launch of the Cadastral Mapping and Geographic Information System (Geo-Spatial Information Survey) scheme in the office of the Surveyor-General of the State [21]. All these are geared towards providing better and effective service to the citizens and the business community at large.

As a result of the introduction of ICT in land registration and other processes that could aid rapid housing development, the Lagos Ministry of Housing has embarked on increasing the numbers of smart houses on a yearly basis. For instance, in 2010, a total of 2892 units of buildings ranging from one to four bedroom apartments were built in various part of the state. In 2011, 643 units of buildings were also built in various parts of the state as well. However, in 2012, 434 units of flats were built compared with the previous years [22]. The beautification of the city through the construction of many relaxation parks, roundabouts and planting of trees and grasses along major highways are parts of numerous efforts of the state government to smartly build the city. The State has completed about 116 beautification projects in the State making the metropolitan Lagos more beautiful and livable.

2.3.1.5 E-Governance-Tax Payments

The Lagos State Government Electronic Banking System of Revenue Cycle Management (LASG EBS-RCM) is aimed to Go-online with the Direct Bank Lodgment System (DBLS) of the revenue collection process and provides information for tax administration and planning while monitoring and coordinating all revenue generating activities of the state. It is an information network system linking tax stations and other revenue agencies to lodgment banks. The system uses an E-Pay messaging system for inter-bank communications by creating and maintaining a database of all revenue collection activities thus allowing for on-line tracking. This system has largely assisted the State in constantly increasing the internally generated revenue base, provides easy administration, monitoring and co-ordination of all revenue activities in the state and has been assisting to identify fraudulent debit, diversion of funds and excess charge on the state accounts [20].

2.3.1.6 Security and Surveillance

Under the flagship 'Rapid Response Squad' (RRS) project, the Lagos State Government has spent several billions of naira on the acquisition of ICT gadgets to assist the smooth operation of RRS. The RRS is a security outfit made up of the police and military created to respond swiftly to any security breach in any part of the metropolis. Recently, over one billion naira was realised as cash donations from various corporate organisations and individuals to support the Lagos State Security Trust Fund (LSSTF) towards advancing a safer Lagos. The fund used to acquire 10 brand new Hilux vehicles and 15 motorbikes were handed over to the Lagos State Police Command as part of the state government's commitment to ensure a secured and investors' friendly state. Likewise, 100 new power bikes, 10 armoured tanks and a helicopter have been acquired by the state government to reinforce security in the state. This is in addition to the purchase of 100 new squad cars for a new initiative tagged Special Operation Service (SOS), which will harmonize community policing in partnership with the RRS.

Given that we now live in an ICT driven world, the state government is already working on a scheme to incorporate technology into the state's security system. Through this scheme, the whole of Lagos would be electronically monitored in order to prevent violent crimes, protect lives and property and to be able to prevent crisis and to manage crime situations at any location in the state. The technological device has the ability to recall events in real time at an accurate date and provide information that may lead to the detection and prosecution of criminal activities. The device is to be used in sensitive parts in the state.

2.3.1.7 Emergency and Disaster Management

The Lagos State Government has provided ambulances with a direct emergency number at strategic positions in the metropolis for immediate response to emergencies. Also, the use of ICT is being deployed in disaster managements such as fire incidences, flooding and other challenges. For instance, the Lagos Fire Service provided several mobile telephone emergency numbers to be reached in case of any fire or related disasters. Also, the Lagos State Emergency Services (LASEMS) deployed the use of ICT to disseminate information on impending flooding in 2014, 2015 and 2016. This measure went a long way to reduce the negative impact of flooding in those years. In addition, estimated properties worth N992.1 billion has been saved in the cause of fire/rescue operations Lagos metropolis, while an estimated properties worth N436.55 billion were lost [9].

2.3.2 Failures of E-Democracy in Lagos

Irrespective of the successes of e-Democracy outlined earlier, there exist some failures as well. These failures are not because E-Democracy is not a viable concept. The shortcomings are as a result of the prevailing conditions in this part of the world. The failure of E-democracy in Lagos has been largely due to poor electricity infrastructure, abuse of the concept where the idea is used to spread hate campaign. Also, the attitude of politicians towards the transparency and sanity which E-voting will bring to the political space is one major failure of E-democracy.

2.4 Conclusion

From the foregoing, e-democracy is thriving in other aspect of the electioneering processes such as E-campaign and the screening of eligible voters using CRM. Another critical area where E-democracy is contributing positively is in the area of e-monitoring of projects. This has largely being helpful in the provision of critical social infrastructure in Lagos metropolis. Therefore, the potential of E-democracy to transform the internal efficiency of government and the relationship of government with citizens is one of the ultimate panaceas that would place Nigeria's democracy on the global democratic map. However, E-democracy in Lagos is currently marked by some level of apathy arising from the constitutional framework and the history of elections in Nigeria. The history of elections in Nigeria has shown that Nigerians cannot rely on the technology as veritable means of installing the kind of leaders they want and by implication in changing the material conditions of their existence. The overarching implication is the need for government to sensitise Nigerians on the imperative of genuine transparency in the electoral process.

Appendix 1: Lagos State Government Ministry Website Addresses

S/N	Ministry	Website address
1	Ministry of Agriculture	www.agriculture.lagosstate.gov.ng
2	Ministry of Commerce, Industry & Cooperatives	www.mcic.lagosstate.gov.ng
3	Ministry of Economic Planning & Budget	www.mepb.lagosstate.gov.ng
4	Ministry of Education	www.education.lagosstate.gov.ng
5	Ministry of Energy & Mineral Resources	www.memr.lagosstate.gov.ng
6	Ministry of Establishments, Training and Pensions	www.metp.lagosstate.gov.ng
7	Ministry of Finance	www.finance.lagosstate.gov.ng
8	Ministry of Health	www.health.lagosstate.gov.ng
9	Ministry of Home Affairs	www.homeaffair.lagosstate.gov.ng
10	Ministry of Housing	www.housing.lagosstate.gov.ng
11	Ministry of Information & Strategy	www.information.lagosstate.gov.ng
12	Ministry of Justice	www.justice.lagosstate.gov.ng
13	Ministry of Local Government & Community Affairs	www.mlgca.lagosstate.gov.ng
14	Ministry of Physical Planning & Urban Development	www.mppud.lagosstate.gov.ng
15	Ministry of Science & Technology	www.most.lagosstate.gov.ng
16	Ministry of Special Duties & Intergovernmental Relations	www.specialduties.lagosstate.gov.ng
17	Ministry of the Environment	www.environment.lagosstate.gov.ng
18	Ministry of Transportation	www.transportation.lagosstate.gov.ng
19	Ministry of Waterfront Infrastructure Development	www.waterfront.lagosstate.gov.ng
20	Ministry of Works & Infrastructure	www.worksandinfrastructure.lagosstate.gov.ng
21	Ministry of Youth & Social Development	www.youthandsocialdevelopment.lagosstate.gov.ng
22	Ministry of Tourism, Arts & Culture	www.tourismartandculture.lagosstate.gov.ng
23	Ministry of Women Affairs and Poverty Alleviation	www.wapa.lagosstate.gov.ng
24	Ministry of Wealth Creation And Employment	www.wealthcreation.lagosstate.gov.ng
25	Lagos State Sports Commission	www.sportcommission.lagosstate.gov.ng
26	Central Business Districts	www.cbd.lagosstate.gov.ng
27	Office of Overseas Affairs and Investment	www.overseasaffairs.lagosstate.gov.ng
28	Office of Civic Engagement	www.civicengagement.lagosstate.gov.ng

Source [8]

Appendix 2: Lagos State Government: Department/Agency/Office Website Addresses

S/N	Department/Agency/Office	Website address
1	Drain Ducks	www.drainducks.lagosstate.gov.ng
2	Lagos State Film and Video Censors Board	www.lsfvcb.lagosstate.gov.ng
3	Hotel Licensing Authority	www.hla.lagosstate.gov.ng
4	Judicial Service Commission	www.jsc.lagosstate.gov.ng
5	Kick Against Indiscipline(KAI) Brigade	www.kai.lagosstate.gov.ng
6	Lagos State Agriculture Inputs Supply Authority	www.laisa.lagosstate.gov.ng
7	Lagos State Agricultural Land Holdings Authority (ALHA)	www.alha.lagosstate.gov.ng
8	Lagos State Agricultural Development Authority	www.ada.lagosstate.gov.ng
9	Lagos State Blood Transfusion Services	www.bloodtransfusion.lagosstate.gov.ng
10	Lagos State Building Control Agency	www.buildingcontrol.lagosstate.gov.ng
11	Lagos State Security Trust Fund	www.lsstf.lagosstate.gov.ng
12	Lagos State Infrastructure Asset Management Agency, LASIAMA	www.lasiama.lagosstate.gov.ng
13	Lagos State Council for Arts and Culture	www.artandculture.lagosstate.gov.ng
14	Radio Lagos	www.radiolagos.lagosstate.gov.ng
15	Eko FM	www.ekofm.lagosstate.gov.ng
16	Lagos State Electricity Board	www.electricityboard.lagosstate.gov.ng
17	Lagos State Emergency Management Agency	www.lasema.lagosstate.gov.ng
18	Lagos State Environmental and Special Offences Enforcement Unit	www.taskforce.lagosstate.gov.ng
19	Lagos State Environmental protection Agency	www.lasepa.lagosstate.gov.ng
20	Lagos State Examination Board	www.examboard.lagosstate.gov.ng
21	Lagos State Ferry Services Corporation	www.ferryservices.lagosstate.gov.ng
22	Lagos State Independent Electoral Commission	www.lasiec.lagosstate.gov.ng
23	Lagos Traffic Radio	www.trafficradio.lagosstate.gov.ng
24	Lagos State Liaison Office	www.liaisonoffice.lagosstate.gov.ng
25	Lagos State Library Board	www.libraryboard.lagosstate.gov.ng
26	Lagos State Metropolitan Area Transport Authority	www.lamata.lagosstate.gov.ng
27	Lagos State Muslim Pilgrims' Welfare Board	www.muslimpilgrims.lagosstate.gov.ng
28	Lagos State Pension Board	www.pensionboard.lagosstate.gov.ng
29	Lagos State Planning & Environmental Monitoring Authority	www.environmentalmonitoring.lagosstate.gov.ng
30	Lagos State Printing Corporation	www.printing.lagosstate.gov.ng
31	Lagos State Public Service Staff Development Centre	www.staffdevelopment.lagosstate.gov.ng

(continued)

(continued)

S/N	Department/Agency/Office	Website address
32	Lagos State Public Works Corporation	www.www.publicworks.lagosstate.gov.ng
33	Lagos State Records and Archives Bureau (LASRAB)	www.lasrab.lagosstate.gov.ng
34	Lagos State Scholarship Board	www.scholarshipboard.lagosstate.gov.ng
35	Lagos State Signage & Advertisement Agency (LASAA)	www.lasaa.lagosstate.gov.ng
36	Lagos State Sports Commission	www.sportcommission.lagosstate.gov.ng
37	Lagos State Teachers Establishment and Pensions Office	www.teachersestablishment.lagosstate.gov.ng
38	Lagos State Traditional Medicine Board	www.traditionalmedicine.lagosstate.gov.ng
39	Lagos State Traffic Management Authority (LASTMA)	www.lastma.lagosstate.gov.ng
40	Lagos State Universal basic Education Board	www.subeb.lagosstate.gov.ng
41	Lagos State University	www.lasu.lagosstate.gov.ng
42	Lagos State University Teaching Hospital	www.lasuth.lagosstate.gov.ng
43	Lagos State Urban Renewal Authority (LASURA)	www.lasura.lagosstate.gov.ng
44	Lagos State Valuation Office	www.valuationoffice.lagosstate.gov.ng
45	Lagos State Water Corporation	www.watercorporation.lagosstate.gov.ng
46	Lagos Television (LTV/LWT)	www.ltv.lagosstate.gov.ng
47	Lagos Waste Management Authority (LAWMA)	www.lawma.lagosstate.gov.ng
48	Local Government Service Commission	www.lgsc.lagosstate.gov.ng
49	New Towns Development Authority	www.ntda.lagosstate.gov.ng
50	Office of Surveyor General	www.surveyorgeneral.lagosstate.gov.ng
51	Office of the Special Adviser on Education	www.osae.lagosstate.gov.ng
52	Central Business District	www.cbd.lagosstate.gov.ng
53	Lagos State Safety Commission	www.safetycommission.lagosstate.gov.ng
54	Lagos State Technical & Vocational Education Board	www.lstveb.lagosstate.gov.ng
55	Lagos State Residents Registration Agency (LASRRA)	www.lasrra.lagosstate.gov.ng
56	Local Government Establishment, Training & Pension Office	www.lgetpo.lagosstate.gov.ng
57	Health Service Commission	www.hsc.lagosstate.gov.ng
58	Lagos State Waterways Authority (LASWA)	www.laswa.lagosstate.gov.ng
59	Office of Public Defender	www.opd.lagosstate.gov.ng
60	Lagos State Physical Planning Permit Authority (LASPPPA)	www.laspppa.lagosstate.gov.ng

(continued)

(continued)

S/N	Department/Agency/Office	Website address
61	Lagos State Market Development Board	www.marketdevelopment.lagosstate.gov.ng
62	Lagos State Public Procurement Agency	www.publicprocurement.lagosstate.gov.ng
63	Lagos State Driver's Institute	www.driversinstitute.lagosstate.gov.ng
64	Lagos State Primary Health Care Board	www.primaryhealthcare.lagosstate.gov.ng
65	Lagos State Parks and Gardens Agency (LASPARK)	www.laspark.lagosstate.gov.ng
66	Agency For Mass Literacy, Adult & Non-Formal Education	www.adulteducation.lagosstate.gov.ng
68	Office of State Auditor-General	www.auditorgeneral.lagosstate.gov.ng
69	Lagos State Audit Service Commission	www.asc.lagosstate.gov.ng
70	Office Of Transformation, Creativity & Innovation	www.otci.lagosstate.gov.ng
73	Motor Vehicle Administration Agency	www.mvaa.lagosstate.gov.ng
74	Civil Service Commission	www.civilservice.lagosstate.gov.ng
75	Lagos State Christian Pilgrims Welfare Board	www.christianpilgrims.lagosstate.gov.ng
76	Law Enforcement Training Institute	www.leti.lagosstate.gov.ng
77	Lands Bureau	www.landsbureau.lagosstate.gov.ng
78	Education District 1	www.educationdistrict1.lagosstate.gov.ng
79	Education District II	www.educationdistrict2.lagosstate.gov.ng
80	Education District V	www.Educationdistrict5.lagosstate.gov.ng
81	Education District VI	www.educationdistrict6.lagosstate.gov.ng
82	Lagos State Water Regulatory Commission (LSWRC)	www.lswrc.lagosstate.gov.ng
83	Education District III	www.educationdistrict3.lagosstate.gov.ng
84	Education District IV	www.educationdistrict4.lagosstate.gov.ng
85	Lagos State Fire Service	www.fireservice.lagosstate.gov.ng
86	Lagos State Mortgage Board	www.lagoshoms.lagosstate.gov.ng
87	Lagos State Material Testing Laboratory, LSMTL	www.lsmtl.lagosstate.gov.ng

Source [9]

References

1. Caldow J (2004) E-democracy: putting down global roots. Institute for Electronic Government. http://www01.ibm.com/industries/government/ieg/pdf/edemocracy%20putting%20dow%20roots.pdf
2. lamata.lagosstate.gov.ng
3. Briony O (2003) The potential contribution of ICTs to the political process. Electr J e-Government 1(1):31–39
4. Azeta A, Azeta V, Olaniyan O, Azeta A, Ayeni G (2015) Implementing an e-democracy system in Nigeria. J Res Dev Manage 4:2015
5. Policy and Legal Advocacy Centre (2012) Social Media and the 2011 Elections in Nigeria. PLAC
6. www.ericsson.com
7. http://www.vanguardngr.com/2016/05/lagos-leads-latest-active-voice-internet-users-survey/
8. http://www.lagosstate.gov.ng/lagos-state-governmentministries-addresses/
9. http://www.lagosstate.gov.ng
10. Jackson N (2007) Political parties, the internet and the 2005 general election: third time lucky. Internet Res 17(3):249–271
11. Williams P, Tramell K, Postelnicu M, Landreville K, Martin J (2005) Blogging and hyper linking use of the web to enhance visibility during the 2004 US campaign. Journal Stud 6 (2):177–186
12. Iwu MM (2008) Electronic voting and the future of the electoral system in Nigeria. The Niger Electoral J 2(1):1–29
13. Alabi MO (2009) Electoral reforms and democratic consolidation in Nigeria: the Electoral Act 2006. CEU Polit Sci J 4(2):278–304. Retrieved from http://www.ceeol.com/aspx/getdocument.aspx?logid=5&id=73488cf8239a482ca21618b0bf0a4e00
14. Ogbaudu F (2011) General election review: experience sharing, lessons learnt and the way forward—the Nigeria Police perspective. Paper presented at the review of elections security during the 2011 General Elections in Nigeria Justice Sector Reform Monograph Series
15. www.lamata.lagosstate.gov.ng
16. www.lagosstate.gov.ng/mepbbc/transportation
17. www.lamata-ng.com/brt/php
18. www.lagosstateministryofhealth.com/programme
19. Aiyegbajeje FO (2016) Mobile telephone usage and physical accessibility to healthcare services in Lagos Metropolis Nigeria. An unpublished PhD thesis submitted to the Department of Geography, Faculty of the Social Sciences, University of Ibadan, p 203
20. Aiyegbajeje FO, Ajayi DD (2016) Mobile telephone technology for better healthcare service provision in Lagos Metropolis, Nigeria. Digital Transformations and Global Society. Volume 674 of the series Communications in Computer and Information Science. Springer International Publishing AG, pp 532–543
21. The Guardian Newspapers 29/01/2009, p 19
22. Digest of Statistics (2013) Lagos Bureau of Statistics, Ministry of Economic Planning and Budget, the Secretariat. Alausa, Ikeja, Lagos
23. www.snitchngr.com/brt-to-commence-e-ticketing-on-tuesday
24. www.naij.com/924854
25. Flavio C, Eleonora P, Alberto P (2005) E-democracy: a solution for disadvantaged territories. In: eceg2005, Conference Proceedings, pp 101–110

Chapter 3
Outline Development Plan for Feroke Municipality in Execution Framework of Internet of People, Internet of Government and Internet of Things

T.M. Vinod Kumar, P. Bimal and Aruna Sri Reddi

Abstract Historically, the urban planning in India has been implemented with a top-down approach where there was little opportunity for citizen participation. The intervention of information technology or 'smart technology' is changing this scenario. More people can voice their opinions on different forums online and expedite the process of participation. Currently, this process is widely used within the realm of private/ personal opinion sharing, and improvement of business performance. Though slowly, different government agencies are also catching up the use of information technology. They are using different online services and telecommunications for police response, birth-death certificate registration, payment for different public services, etc. To maintain the quality of these services, it is important for the governing bodies to have an inflow of citizen participation and opinion sharing as a constant process. The 74th amendment of Indian constitution gives authority to Urban Local Bodies (ULB) to self-govern and plan for future of their respective jurisdictions. Outline Development Plan (ODP) is proposed for a period of 20 years that includes future direction of spatial development and economic aspirations of that jurisdiction as per the amendment requirement of economic development and social welfare. Citizen involvement is critical in every stage of development ODP. Traditionally, public participation meetings were held for general public and stakeholder groups' involvement in the plan development. E-governance and e-democracy, use of information technology for information dissipation, and citizen participation in the plan making is expected to improve the frequency and number of citizens participating. The current study looks at the City

T.M. Vinod Kumar (✉)
School of Planning and Architecture, New Delhi (Spa-D), India
e-mail: tmvinod@gmail.com

P. Bimal · A.S. Reddi
Department of Architecture and Planning, National Institute of Technology Calicut, Kozhikode, Kerala, India
e-mail: bimalp@nitc.ac.in

A.S. Reddi
e-mail: r.arunasri.2008@gmail.com

T.M. Vinod Kumar (ed.), *E-Democracy for Smart Cities*, Advances in 21st Century Human Settlements, DOI 10.1007/978-981-10-4035-1_3

of Feroke for implementing the ODP in the e-governance and e-democracy format. With more than 80% people being high-school graduates and very high literacy and e-literacy based on state sponsored project, the city of Feroke, Kerala, is a good example for implementing the use of information technology in the ODP development. Different approaches such as web-based information, interaction, and web-profile in the social networking sites, emails, and telephonic interviews were used as part of the e-democracy while developing the ODP. Along with it, a traditional household survey using a questionnaire was also used. Delphi was used for analyzing the specific stakeholders and expert opinions. Six areas—smart people, smart economy, smart living, smart mobility, smart environment, and smart government—were identified for implementing various smart technologies. The executive framework for Feroke Municipality ODP adopted Internet of People or E-Democracy (IOP), Internet of Government E-Governance (IOG), and Internet of Things (IOT). The project concludes with spatial proposals, specification for IOP, IOG, and IOT emerging out of six systems of Smart Cities.

Keywords E-Democracy · Outline Development Plan · Internet of People (IOP) · Internet of Governance (IOG) · Internet of Things (IOT)

3.1 Feroke City

Feroke is a medium sized municipality about 10 km south of the city of Calicut (Kozhikode). Kozhikode is the third largest city in Kerala. Feroke has a land area of 1424 ha (14.24 km^2). Feroke was classified as a ‘Gram Panchayat’ as per the ‘Peoples Planning structure’ [1] and got the municipality status in 2014. As per 2011 census, the population of Feroke is 54074, and the population density is 37.95 people per hectare. Feroke demands the attention of urban planners due to its phenomenal growth of built-up space in a short period of 2006–2011 as shown in Fig. 3.1. In the past, Feroke has always been an important centre of historical significance. It was ruled by Zamorin until 1788 when Malabar was conquered by Tipu Sultan from Karnataka. Tipu selected Feroke, which he called Farookhabad to be his capital for Malabar [2]. He began construction of a new fort in Feroke, which he was unable to complete as British defeated him before completion. British took over the control, and established tile industry at Feroke, taking advantage of the perineal river and large deposits of clay in the region. Feroke was well known as the capital of tile industry in the southern India which were never allowed to further diversify and develop under British rule and in the age of industrial revolution, thereby losing its significance in South India. The timber industry was established even before tile industry due to the ease of transporting timber through Chaliyar river from Nilambur forests in the eastern side which helped British develop Rail system around Ferole. Feroke had established itself as an industrial town in the modern period with a chain of tile factories along the Chaliyar river, forming its industrial base.

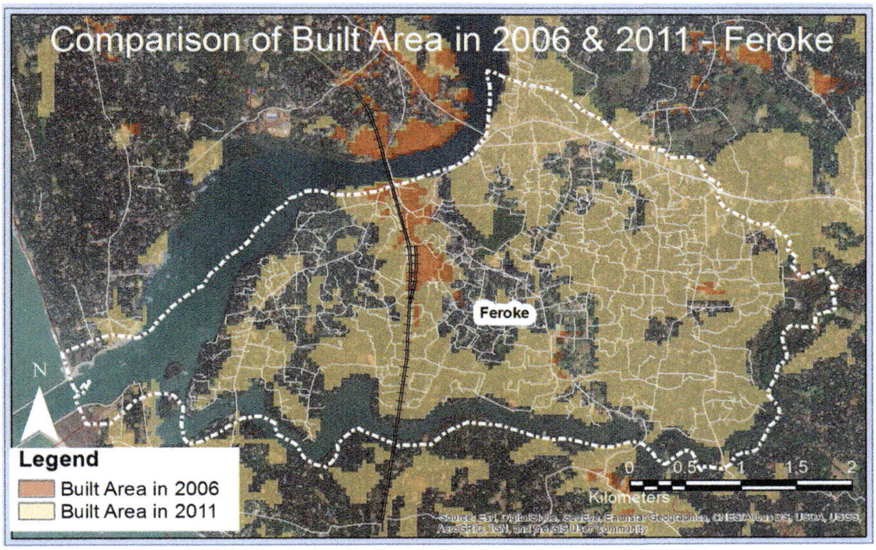

Fig. 3.1 Comparison of built area in 2006 and 2011—Feroke

The peculiarity of the increase of built-up area mentioned earlier is that additional built spaces were in the residential sector, which changed its image as an old industrial town. This sudden growth of residential space and its underlying population triggered the promotion of Feroke from its past status as 'Gram Panchayat' to the status as 'Municipality.' This additional population has overloaded the available infrastructure. Feroke like any other place in Kerala has a homestead style of development, where each house has a sizable piece of land around for at least few coconut tree to grow along with some vegetable. People depend on the shallow wells made in these plots for drinking water as well as watering plants. Biodegradable waste produced at the household was composted in these plots and converted to manure for the cultivation of vegetables and coconut. The sudden increase in demand for housing cut down the availability of land and houses are built with just a narrow strip of land; often this is covered by interlocked tiles which added to the foldability of the area. High density living with inadequate infrastructure has created a variety of problems. The newly formed municipality are faced with a multitude of challenges. Apart from inadequate infrastructure related challenges, they face a challenge in the exercise of democracy as well which is the basis of people's planning the movement of Kerala which will be discussed in detail later in this chapter. The attendance in the 'Gram Sabhas'—the lowest level of citizen's forum where the need for developments in that area are discussed and formulated-was very low when it was Gram Panchayat. This was the major reason NIT Calicut undertook this project in Feroke. This scenario is found to be continuing after it got converted to the municipality. Attendance in ward councils– which are the lowest citizen's forum as per Kerala Municipality Act [3] is found to be very low in reality. This low level of participation poses a fundamental concern about the validity of the glorified 'people's planning' movement in Feroke Municipality. Hence there is a dire need of a system to increase participation of citizen's in the a strengthened democratic planning process which is the subject matter of this chapter.

This chapter is, therefore, an attempt made by National Institute of Technology, to explore possibilities of transforming the present democratic planning process in the framework of smart city technology which we believe can only survive in E-Democracy where all population is involved. The project was formulated as a planning exercise for Master's students of Urban Planning. The project was to prepare an Outline Development Plan (ODP) to convert Feroke to a self-financing Smart City, although it is not a rich city with high per capita income. The project also explored possibilities of enhancing the democratic process by extensive use of ICT-enabled technologies.

3.2 Participatory Planning

The fundamental definition of democracy requires the massive citizen participation in the decision-making process of the government. The definition of 'citizen' and amount of participation differed at different times. Even the existing oldest modern democracy in the USA did not have the voting rights to all its citizens immediately after the independence from the Great Britain. Poor whites, women, and African Americans were restricted to vote in political elections at the beginning of its democratic governance in the USA. However, today the USA is considered as one of the best of nations for its democratic values and citizen participation in the decision-making process of the government. The countries that are considered as advanced are not only economically developed but also have strong citizen participation in the government decision-making process which is considered as the secret of the prosperity of cities and nations.

Until very recently, the citizen participation in governance is limited to voting to select a political representative. It is understood that the political representative is an extraordinary citizen who 'knows all' and takes the right decision on behalf of all he represents who elected him. The reality is far from this ideal. Their political representative is however guided by his political party which is again under the influence of various interest group who may not be interested in developing the area he represents according to the wish of people there. This makes the local population very sceptical of representative democracy practised in India. Some nations have government functionaries that make every decision related to daily life of its citizens. It is believed that the communist governments in the Eastern Europe have had this system. In other countries, such as in some countries in the Middle East, quite often, the governments are formed through monarchy system where there is no citizen participation to elect their representative or in the government. However, where the democracy is the core objective for the government, the governments try to include as many citizens as possible in the decision-making process so that the decisions made do not face negative repercussions in the implementation stage. It is considered as the main characteristic of good governance.

The citizen involvement has been progressing with the development of communication systems. The first generation of communication in the governance has been mostly non-participatory and one-sided information dissemination. Newspapers, advertisements, billboards, campaign by a vehicle with recorded messages, ads in movie theatres are used majorly for information dissemination to large-scale populations. Radio and television are used for instant dissemination of information. The drawback of this communication is that the same communication system cannot be used for mass scale response to the government as part of the citizen participation. In general, the citizen reaction was known only when there is a negative response in the form of protest. Other than this, there was no actual participation in the decision-making process by citizens. In modern days, group emails, and SMS are used for information dissemination at mass scale.

The citizen participation is less and ineffective wherever ICT is not used, such as citizen participation in terms of community meetings, surveys or interviews either personal or by telephone. Gathering information through these methods are time-consuming as well as expensive. Electronic communication, especially the internet has given a solution to this issue. Internet communication is not only instant, but also inexpensive that can reach a large number of people, and contesting parties in election process can communicate with each other at the same time. The first generation of e-participation is email, and e-chat that were used for personal communication, whereas websites such as Priceline, and e-bay were used for commercial interactions where the consumers have the opportunity to choose an alternate product by a source based on the price of the good. Today, the communication channels such as Facebook, WhatsApp or Twitter have become popular in both social as well as political communication. The commercial apps such as Ola or Uber have made it possible for consumers to deal with service providers to travel directly.

Today, no business or political decision is done without collecting stakeholder opinion. As part of the transparent governance, all government agencies are trying to provide information through websites. Even projects are funded through crowd funding where individuals or groups can directly fund the projects. The use of technology not only improved the communication systems for public participation but also enhanced the quality of governance in various aspects of service provision. Government services such as public health, transportation, utilities, and other services such as tax payments are improved due to technical interventions.

3.3 Existing Situation of Participatory Planning in Kerala

Documentary evidence of participatory planning is available as a spatial outcome in the cadastral maps and land schedule of Feroke if you know how to read it with respect to participation. Further evidence emerges in the existing land use map of three Master Plans of Calicut area which also includes Feroke. The degree of participation that was evident in three phases of development of Feroke prior to 1920, 1921–2016 and afterwards differs a great deal.

Prior to 1920, Municipality of Calicut existed which does not include Feroke even as its planning area. Feroke was a village with agriculture economy with paddy cultivation, river and sea-based fisheries and animal husbandry as the main occupation where females also participated more than today. Thereafter, the joint family system the mainstay of agriculture started gave way to nuclear families. This implied that large holdings gave way in Feroke to fragmentation of holdings when it became nuclear families and this continued in all these periods. The system of adoption of survey number of plots divided gives insight to how this division took place. Soon these subdivided plots had a house constructed and a compound wall around in most cases.

The biggest problem faced during the subdivision was the access to footpaths, cycle paths, and roads. The house owners themselves identified a footpath or cycle tracks by donating part of their land holdings without any Government interventions or seeking compensation which can be considered as an outcome of participatory planning and development by beneficiaries. That community of land owners who were far-sighted with an idea of a hierarchy of footpaths created wider paths and others narrower paths. No attention was spared about drainage structure on one side of the paths. There was no concept of a hierarchy of these paths in general which requires wider and wide footpaths in higher levels from the footpath to the cycle path and then to motor-able roads. There was none to educate them on linkage hierarchies.

The resulting footpaths were surveyed using Remote sensing maps and GPS in this project and are given below (Fig. 3.2) for Feroke Municipality.

Historically residential areas were islands within the large extents of paddy fields. These paddy fields progressively reduced to islands in the residential area in the existing land use plan (Fig. 3.3) and even in the proposed ODP.

This was mainly due to the subdivision of land primarily to transfer it to the next generation by parents as per the existing legislation. The current land legislation gave indications of how the land should be subdivided and who should get family house based on whether you are Muslim, Hindu of different castes, or Christians. There was nothing in these legislations to preserve the existing drainage structure in the area of high rainfall or for environmental conservation. This is a testimony of legislative ignorance. These created vast areas of water logging during rains and soil erosion due to the disturbances of drainage structure due to improper property partitions.

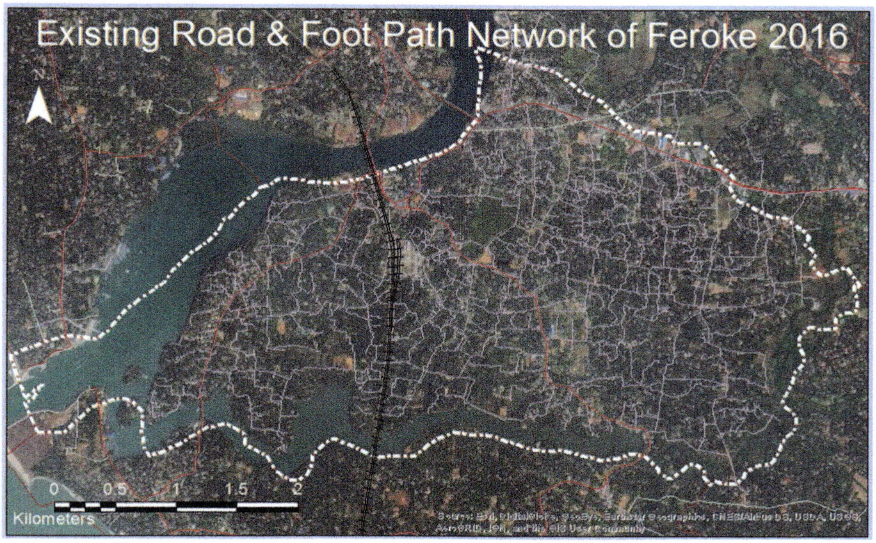

Fig. 3.2 Existing roads and foot path network of Feroke in 2016

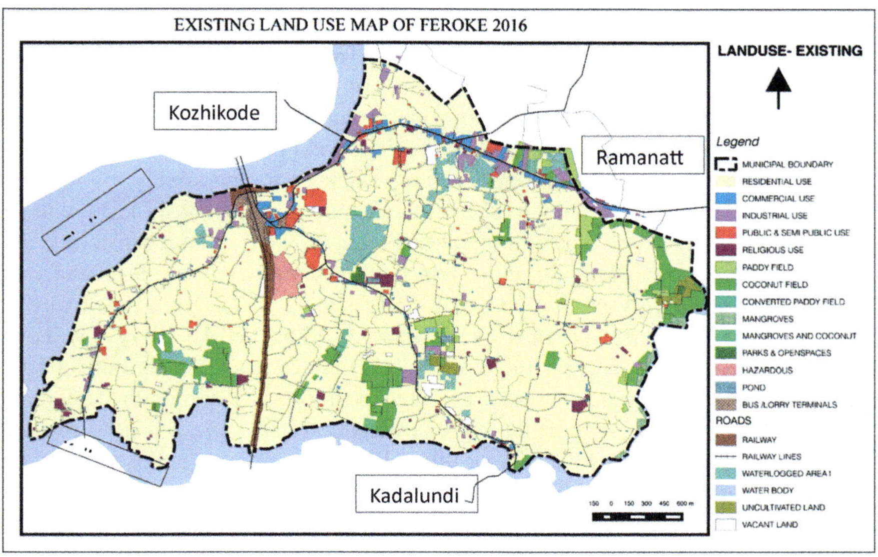

Fig. 3.3 Existing land use map 2016

Fragmentation of plots was less around railway stations, bus stations and other public spaces. Many of them were acquired using the British Land Acquisition Act, and there is no possibility of subdivisions there.

Although local self-government existed during British, Colonial days in Feroke, there was no attempt to study the habitat as a whole and find ways and means to develop it scientifically prior to 1920. There were concerns about public health and environmental engineering, but no attempt was spared to use participatory ways of developing urban areas. There was no effort spared to building upon the experience of participatory footpath making graduated to participatory integrated area planning through participatory public awareness. It can be considered as a great opportunity lost by authors of this chapter.

Prior to the independence of India in 1920, Madras Town Planning Act which was adopted in Malabar as Malabar Town and Country Planning Act 1920 came up which gave a legislation to reshape towns in Malabar spatially. There were Town Planning Scheme and Detailed Town Planning Scheme embodied in 1920 act. There was not much of democracy in planning and participation was not sought since British Colonial rulers felt their bureaucrats knows better what the native Indians want. The government mostly was using the British Land Acquisition Act for the native Indian than detailed town planning schemes requiring a great amount of work for bureaucrats for eliciting wider participation in urban development was absent and hence did not show much of success in urban development?. This approach is very much evident in all Kerala even today which we consider as a bad colonial legacy. Land acquisition act does not treat landowners and Government on equal footing and eminent domain which actually help British than Indians was implemented with the police force. Detailed Town Planning schemes were largely not successful and made largely not participatory in reshaping all plots. This mostly gave approaches to widen the existing paths at Government expenses since they were carrying motor vehicle traffic from the yester years bullock carts and horse carts. This killed the participatory initiative of the local population who gave births to footpaths and cycle paths in Feroke sharing their own lands freely.

In 1993, there was an attempt to deepen the democracy at the local level with the enacting of 73rd and 74th Constitutional Amendment Act (CAA). Prior to that and even after that, Urban Planning and urban land development were a bureaucratic exercise with no or very limited participation of people. There were practically no major urban development projects in Malabar including Feroke. Very few Town Planning Schemes were successfully completed in all towns of Malabar under 1920 acts. 1920 Town Planning Act was not in a position to incorporate democratic planning and development with the high participation of People and need revision immediately. The slow and out of date urban land management practised in Kerala did not give an impetus to change the 1920 Act in tune with 74th CAA immediately. Only in 2016 Government of Kerala could legislate a Town and Country Planning Act in tune with the democratic planning and development at an instance of a court ruling on this subject. Implementing 1920 Act in Feroke and Malabar created a negative impact in Feroke as well as all towns in Malabar region of Kerala. It resulted in the narrow roads which cannot carry present traffic volume, more road accidents cities with any adequate public and semi-public and recreational areas and inadequate parking spaces which affect the shop keepers and other services were the result of not converting the Town Planning Act in tune with CAA. 1920 Act strengthen bureaucrats against community initiatives and if this process was reversed with community gave all powers then Feroke and Malabar should have a better urban structure.

The participatory planning at spatial level was not there to show success in Kerala. However, people's planning movement which essentially assessed felt needs at ward and village panchayat level integrated district and state level that came out in the 1990s was a positive development in democratic planning and development in Kerala.

3.3.1 People's Planning Movement in Kerala

Grass root level participatory planning has been the focus of academic and professional discussion for several decades in India. The practice of people's planning and execution has been there for many decades in Kerala. However, attempt to implement this concept has not been taken up so seriously because of severe obstacles faced to achieve this goal. They are non-availability of planning expertise at the local level, weak and highly centralised administrative setup and inadequacy in administrative procedure, lack of experience, and inadequate database. Because of these limitations, State Government felt it tough to pass on the financial and management functions to the grass root level. However, Government of Kerala took a bold initiative to embark on people's planning movement despite the above limitations. The strategy adopted was to venture boldly based on well thought out procedures. Action learning from people's planning experiences was utilised to straighten the movement in subsequent years. All the time the goal of decentralisation and 73rd and 74th constitutional amendments were kept in mind. The lessons learned are utilised in the following annual plans.

Kerala's people's planning movement demonstrates the art of decentralisation of power to plan manage and make financial decisions in area development by local people within a multi-level planning framework. This experience is fascinating and unique for in any part of the world both developed and developing nation. It is a noteworthy example of a model of translating the 74th and 73rd Constitutional amendment without any reservation or holding on to power at the State headquarters. The belief in local people that is so much lacking in India is the sound basis of such venture. This is in contrast to the then existing centralised administration, which does not trust local people with public finance, management, and planning, and over-administer them with checks, counter check as if they are criminals. This has been by and large non-productive for urban and regional development. At present in many States in India, decentralised planning means planning at district level without massive mobilisation and active participatory decision-making at Municipal Ward and Gram Panchayat level. Kerala's attempt is to empower people and local self-government such as ward, municipality, village, block and district to undertake area development planning and implementation.

The people's movement of planning started in Kerala in the financial year 1997–98. A constitutional amendment has spelt out and confirmed the role and responsibilities of local government such as Panchayat and Municipalities. These institutions are being converted into strong constitutional institutions of planning and governance in Kerala. Mandatory periodic elections and assured share of State Government grant through mandatory Finance Commission recommendations are two pillars of strength for participatory planned area development. They now have the power to decide the developmental activities and implement it. There is considerable scope for improvement in people's planning movement, and this research and development project for E-Democracy for Smart Cities focus on the scope for improvement using technology and tools of Geographic Information Science.

The most significant contribution of the people's movement of planning was the way in which Kerala mobilised the participation of people. Further, State Planning Board in Kerala steered the complex activities of people's planning with their direction, training programs, documentation and evaluation of past experiences. Beneficiary Committees Task Forces, the working group worked side by side with the elected members of local bodies. They showed that they have an important role to perform in people's planning movement. The initial mistrust between elected and local people melted away as the planning movement started showing positive results. All obstacles prevalent for these committees and Task Forces have been removed for their better functioning. In the past, such committees and officials are subjected to the rigour of administrative controls. These controls have not resulted in better governance. The experience shows that more the controls, more the corruption and administrative inefficiencies. The result is exhibited in the websites of Central Vigilance Commission that show never ending a long list of corrupt senior administrators. Even Chief Secretaries the administrative head of state in some states are kept behind jail for corruption so that they do not pollute the rest of population with corruption. It looks like corruption is a part of Indian administration. The government of Kerala took a bold stand that these Task Forces will not misuse power given to them since they are closely watched by local people in their transparent functioning. It is hard to be corrupt when several local persons are involved in decision making for a local level project, which benefits many. Further, there is scope for reviewing decisions at different levels. In a situation where all are equal as far as decision-making power is concerned, then corruption gets reduced.

During 1998–99 Rs. 950 Crores was set aside by the Government of Kerala to implement plans prepared by the local Government. This was 25% higher than that was allocated during the financial year 1997–98. The next budget for 1999–2000 has earmarked for Panchayat Raj/Nagar Palika institutions a sum of Rs. 1020 Crores as plan grant (Rs. 40 Crores for Tribal Sub Plan, Rs. 200 Crores for Special Component Plan, and Rs. 780 Crores for the General Sector Plan). Further, local planning body should mobilise as many local resources as possible other than State Government grant. Earlier the usual practice of State Government Grant does not spend during the financial year was largely known only to Government officers, but now all population of the ward or municipality knows about it since they are part of participatory planning. The first instalment of the second year plan will be available to the local Government only if 90% of the fund allotted is spent. Once it is not available, the whole population gets critical of all in power. Subsequent years the amount of participatory planning got increased year by year, but the administrative direction from State Planning Board changed from time to time. Difficult and tighter administrative directions resulted in less allocated money.

Year	No of projects by LSGIs	Amount in Rs in Crores
2012–13	1,84,658	8594
2013–14	2,06,375	10,680
2014–15	2,22,849	11,916
2015–16	2,19,409	12,185

Source Government of Kerala LSG videos PowerPoint Presentation KILA. LSGI-Local Self Government Institutions

The training programs and the related documentation prepared by the State Planning Board for annual planning was voluminous and of high quality with a practical orientation. This project is in tune with these documents so that GIS can be easily integrated with the ongoing program of annual planning. Kerala Institute of Local Administration (KILA), Thrissur was entrusted to conduct a series of training programs to different target groups for preparing annual plans. The success of the annual planning was mainly due to the effort of KILA supported by the State Planning Board.

These programs, however, were largely rural oriented. No doubt, in the distant past in Kerala more percent of people live in villages. This is changing since 2011 census where more metropolitan agglomerations are growing. The rural-urban dichotomy seems to be non-existent in Kerala with the scattered settlement structure where rural settlement merges with urban. There can be “food shed” similar to watershed within the metropolitan areas supplying the daily needs of fresh vegetable and fish. In conclusion, it may be stated that there is neglect on urban aspects in these training programs of people’s planning movement.

Urban planning approach is different from rural planning. The regional impact of urban development could be agriculture, mining or other primary activities in their hinterland. This often results in the development of the market towns, but marketing is not a primary activity. Urban economic development, on the other hand, is triggered by secondary and tertiary sector than primary sector from within. Secondary sector showed slow development all over Kerala. Kerala has missed Industrial revolution. There is a multiplier effect in an urban economy with the interaction between the secondary and tertiary sector. Therefore, the definition of productive sector in annual planning in Kerala should change from agriculture, animal husbandry and fisheries to urban-related productive activities which was never considered in Kerala. About 30% of plan allocation in annual planning are set aside for primary activity. The amount set aside is often not spent. This is not justifiable in Kerala's annual planning of urban centres where there are not many agricultural activities in cities.

Another aspect of the rural and urban approach is more pronounced in the manner in which financial resource mobilisation is possible for land and urban infrastructure development. Community cooperation is the key for financial resource mobilisation in urban management. Annual Planning is paving the way for high-level community cooperation. Urban development can be self-generating. Finance can be mobilised using urban land as resources in a participatory planning process which were not attempted. This 30% misallocation to urban centres and not using land and community to mobilise resources are largely due to the ignorance of the State Planning Board on matter urban and so-called rural lobby of Kerala politics.

The essential characterises of Kerala's "Janakiya Asuthranam" or people's planning is enumerated below.

1. Towns and Village Panchayats were in a position to know the total annual plan budget they are likely to get in the financial year using a formula that translates Finance Commission's recommendations as interpreted by State Planning Board of Kerala.
2. State Planning Board suggested the percentage split of this budget under several sectors based on State Priorities and policies which were more rural oriented than urban.
3. A series of training program conducted by Kerala Institute of Local Administration (KILA) taught all how to conduct Janakiaya Asuthranam.
4. A voluminous background information for Janakiyya Asuthranam was prepared by State Planning Board and distributed.
5. A timetable of multi-level planning (Ward, Municipality, and District) under people's planning was implemented for the financial grant, monitoring of implementation and other administrative details by the State Planning Board.
6. Task force for different sectors was convened to study and formulate projects.
7. Ward conventions gave opportunities to deliberated problems and solutions in every ward and identify projects and estimate costs and prepare an annual budget.

3.3.2 Participatory Planning in Operation

Important stages of this participatory planning methodology are outlined below:

1. **Needs identification**: The felt needs of the community are identified in Gram Sabha/Ward Sabha meetings. All living in these places were invited, but 10–20% attended these meetings and sometimes it was difficult to get a number of people to attend. These meetings are held in a semi-structured manner with plenary sessions and subgroup sessions dealing with identified specific

developmental issues. The decisions are recorded and forwarded to the Local Governments. Each meeting is chaired by the elected member and has an official as its Coordinator.

2. **Situation analysis**: Based on the demands emanating from the first special Grama Sabha/Ward Sabha and based on developmental data, both primary and secondary, exhaustive Development Reports have been prepared and printed in the case of every Local Government in the State. These reports describe the status in each sector of development with reference to available data, analyse the problems and point out the directions for further development.
3. **Strategy setting**: Based on the Grama Sabha/Ward Sabha feedback and the Development Report, a one-day seminar is held at the Local Government level in which participation of experts, elected members, representatives nominated by the Sabhas, practitioners from among the public is ensured. The development seminars suggest the broad priorities and general strategies for developmental projects to be taken up for a particular year.
4. **Projectisation**: The ideas thrown up by the above three stages are translated in the form of projects by Working Groups at the Local Government level. For each Local Government, about eight Working Groups are dealing with different sectors of development. Each Working Group is headed by an elected member and is convened by the concerned government official. The Vice-Chairman of the Working Group is usually a nongovernment expert in the sector. The projects are prepared in the suggested format outlining the objectives, describing the benefits, explaining the funding and detailing the mode of execution and phasing of the project.
5. **Plan finalisation**: From among the projects, based on the allocation communicated, the concerned Local Government finalises its plan for the year, and this plan is submitted to the District Planning Committees (DPCs) through the Technical Advisory Committees. The Panchayat is free to take up any project, irrespective of its cost, subject of course to the resources actually available and within the sectoral limits.
6. **Plan vetting**: The Technical Advisory Committees at the Block or the District level consisting of official and nonofficial experts vet the projects for their technical viability and conformity with the mandatory government guidelines on planning and costing and forward them to the District PlanningCommitte (DPC). They cannot change priorities or projects; they can only ask for rectification.
7. **Plan approval**: The DPC gives the formal approval to the plans after which the Local Government can start implementation. It is to be noted that the DPC also cannot change the priority of a Local Government. It can only ensure that government guidelines are followed. The Local Government gives administrative approval for implementation projects. Every Local Government has unlimited powers of Administrative sanctions subject only to the limits of its financial resources.

This broad approach has the following execution steps

3.3.3 Participatory Planning Optional Steps

1. Identifying and Entrusting Plan Coordinator
2. Formation of Working Groups for subjects suggested by planning board
3. Preparation of Status Report
4. Discussion with Banks
5. Stakeholder Discussion
6. Plan Formulation on Gram/Ward Sabha
7. Finalisation of Status Report and Suggestions on Projects
8. Preparation of Development Report and Plan Documents
9. Exploring possibilities of Integration
10. Development Seminar
11. Approval of Development Report
12. Preparation of Projects in 'Sulekha' software
13. Finalisation of Plan
14. Approval of Plan
15. Scrutiny and Approval of Projects In 'Sulekha' software
16. Approval of Plan by DPC 'Sulekha' software.

Sulekha software is used to implement annual planning and development at Local Self-Government which is explained below. Before that, the first planning process in 1999–2000 is given in a flow chart (Fig. 3.4).

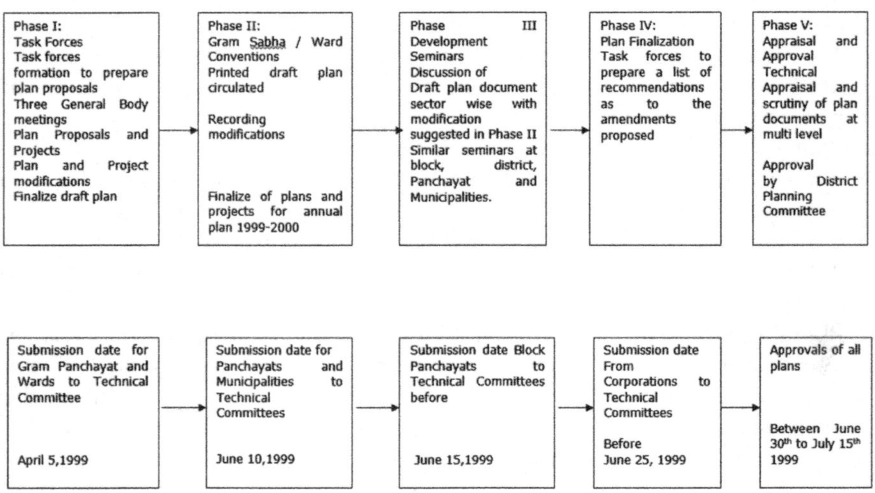

Fig. 3.4 Plan formulation procedure for the annual plan 1999–2000

3.3.4 *Sulekha Software*

Sulekha software is a software used for decentralised planning and monitoring in Kerala. It is in fact computerised centralised planning and monitoring server based system. Kerala participatory planning has now progressed from using hard copy preparation ten years back, shown in the flow chart above to soft copy oriented preparation which can be accessed from anywhere by a hierarchy of sanctioning authorities and those who wants to monitor progress from the server.

Sulekha is the Plan monitoring software developed by Information Kerala Mission for Local Self Government Department (LSGD) of Government of Kerala for the formulation and monitoring of the nearly 2 lakhs annual decentralised plan projects of local governments. Sulekha is being used in Kerala for more than one decade; tailor made to track several steps enumerated above in the plan formulation, appraisal, approval, monitoring, revision processes and expenditure tracking of the plan projects of local governments of Kerala. This web application suite meets the requirements of not only the Local Self Governments but also of the State Planning Board and several ministries and departments of State Government of Kerala.

3.3.5 *Steps in Decentralized Planning in Kerala as Per Use of Sulekha*

3.3.5.1 Planning in Sulekha

Planning as per Sulekha system is a process in public governance that analyses the situation, identifies the needs, sets the priorities, allocates the resources and fixes the targets for achieving the pre-determined objectives in a participatory way in a predetermined structured process. A plan must be a realistic account of the expectations of the local population. Preparation of a comprehensive plan may not guarantee success, but the lack of a sound plan will almost certainly ensure failure. It was as well believed that Gram Sabha, the people's assembly at the ward level, could play a major role as well as ward Sabha in towns. The scope of Sulekha is diagrammatically presented in Fig. 3.5, which is self-explanatory.

The objective of the plan can be increasing production, establishing public facilities, creating job opportunities, reducing disparity among different sections of people, enhancing the capacity of citizens, redressing of physical inabilities, improving knowledge level of the population or so. The process of prioritisation is used in different components of planning such as need identification, choosing focus areas, allocation of resources, and fixation of targets. Success in planning depends on how the planner can use prioritisation skill and strategically use the developing and critical sectors in enhancing the economic development. Proper sequencing of components and strategic timing are other elements that make

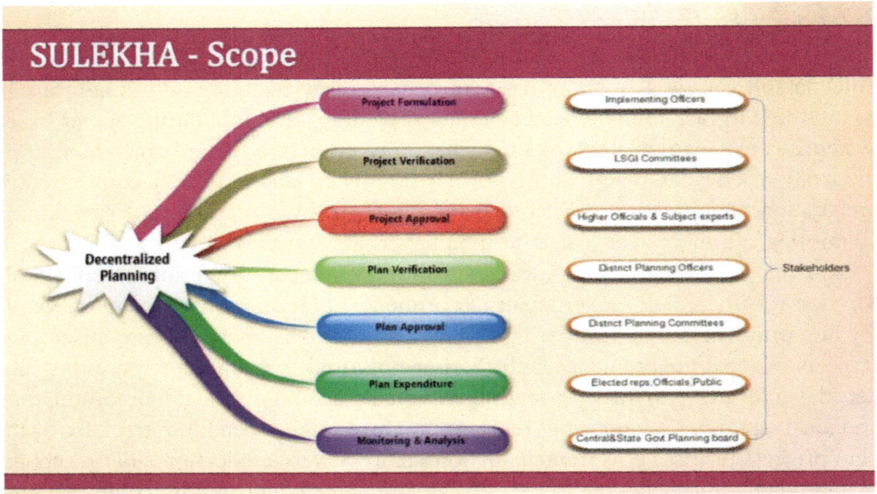

Fig. 3.5 Scope of Sulekha in decentralised planning

planning a success. A plan should reflect economic aspects of sustainability, spatial perspective for fair distribution of benefits and public policy concerns for reducing inequality.

The decentralised plan here refers to the plan of the local governments at the sub-state level at the district or below. Decentralised planning is an integral component of national planning that complement each other. It is nothing but participatory process plan formulation by the local governments as part of the national and state level planning as per Indian Constitution.

People's Planning Campaign

The much acclaimed decentralised planning in Kerala started a campaign called Peoples Plan Campaign (PPC) launched in 1996. It was a pioneering movement towards decentralised planning and governance at the sub-state level. People's planning is probably the most radical development that has taken place in Kerala in the recent past which two coalition Governments that ruled Kerala alternatively and successively accepted.

To break the initial inertia, the campaign started with high-voltage publicity and by involving all stakeholders in it. The campaign later moved on to the staggering phase of institutionalisation and succeeded in evolving a concrete methodology for participatory planning and development at the sub-state level for the first time in any Indian state. However, neither the Campaign nor the institutionalisation process was without problems or threats against its long-term sustenance. The process of decentralised planning is ongoing in the State in some form or other. Administrative process underlying people's planning determined the success and failure of the process and outcome, and expenditure pattern and attendance in the ward and Gram

Sabha meetings determined its performance standards. The Fig. 3.6 shows the login page of Sulekha and Fig. 3.7 describes various search options to easily search and locate a prepared paln.

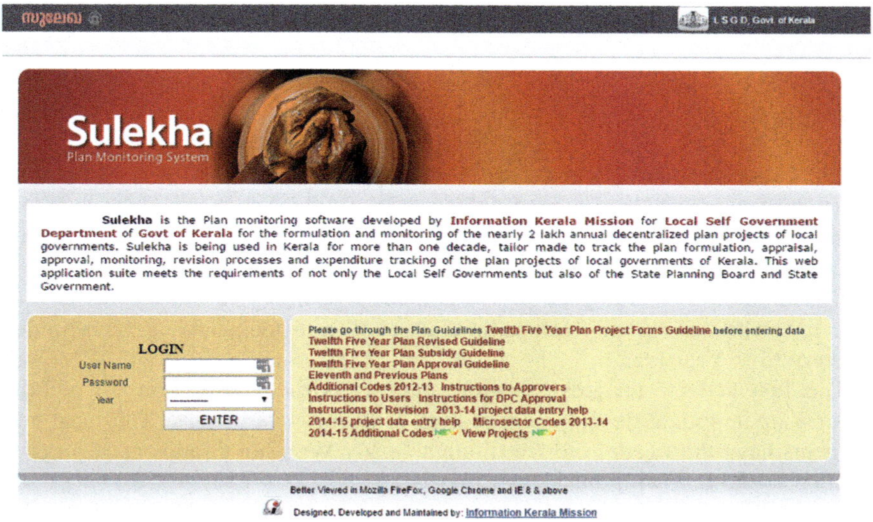

Fig. 3.6 Login page of Sulekha website

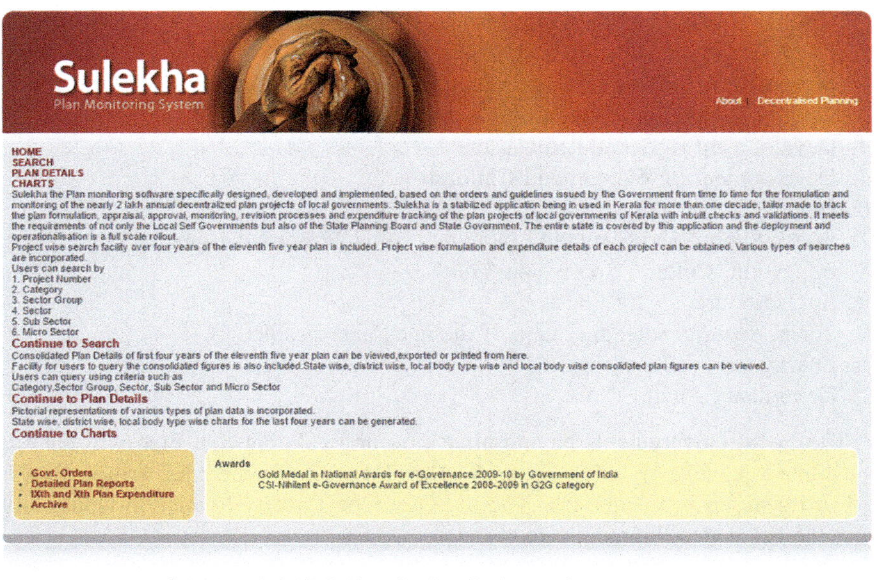

Fig. 3.7 Sulkha website

Steps in Decentralized Planning in Sulekha

Kerala follows a multi-level decentralised plan formulation process that has tremendous de-bureaucratisation potential. The different stages adopted and found successful in this decentralised planning are environment setting, situation analysis, need identification, vision setting, plan formulation, prioritisation, plan vetting, plan approval and plan implementation. Each stage has a few sub-components too.

3.3.5.2 Environment Setting

Creation of Working Groups and making them functional for setting the planning process to move on, are the twin items coming under this.

Constitution of Working Groups was made mandatory for each local government to constitute Working Groups for the sectors, as listed below, at the beginning of every Five-Year Plan.

The first activity towards situation analysis is the constitution of Working Groups for important development sectors in Local governments. The local governments have the freedom to constitute as many Working Groups as required in addition to the following mandatory ones.

Mandatory Working Groups

The mandatory working groups are

1. Watershed Management including Environment, Agriculture, Irrigation, Animal Husbandry, Dairying, Fisheries and related sectors
2. Local Economic Development other than agriculture including local industries, promotion of private and community investment and mobilisation of credit
3. Poverty Reduction including housing
4. Development of Scheduled Castes
5. Development of Women and Children
6. Health
7. Water Supply and Sanitation including Solid Waste Management
8. Education, Culture, Sports and Youth
9. Infrastructure
10. Social Security including care of the aged and disabled
11. Energy
12. Governance Plan.

The Local Governments having allocation under Tribal Sub Plan will have to constitute a separate Working Group for Development of Scheduled Tribes, also. In Urban Local Governments, the Working Group on Poverty Reduction would look after the Slum Development sector too.

Working Group: Its Structure
Each Working Group will have an elected member of the Local Government as its Chairperson. The Working Group on Development of Scheduled Castes should be chaired by a Scheduled Caste Member and the Working Group for Women and Children by a women member. The Working Groups on Watershed Management and Anti-Poverty Sub Plan should be headed by the Chairperson of the concerned local government. A leading expert in the sector, who is nominated by the local government has to function as the Vice-Chairman of each Working Group. The Convener of each Working Group in the local government should be the senior most official devolved to the local government in that sector. Other professional officers and Experts capable of contributing to the functioning of Working Groups can be included in the Working Groups as members. The experts from outside the local government jurisdiction who are willing to work voluntarily in the Working Group can also be included. In short, any practitioner or professional showing interest and activism in the sector can be included in the Working Group.

At least one member of the Community Development Society of the Kudumbashree should be included in each Working Group, and two should be there in the Working Groups on Poverty Reduction, Development of Women and Children and Development of Scheduled Castes. Scheduled Caste Promoters should be nominated for every Working Group. The Working Group has the power to co-opt more member to its fold and to set up Task Forces to perform any task devolved on to it.

Working Group Meeting: Procedure
The Working Group has to meet as frequently as possible and keep a brief record of its deliberations. The quorum for the meetings shall be four including the mandatory presence of the Convener. It should be ensured that representatives of commercial banks participate, in the Working Group on Watershed Management and Local Economic Development, to the extent possible to ensure the bank loans for plan projects. In Local Governments having the Forest areas, officers of the Forest Department and the Presidents and the Secretaries of Vanasamrakshana Smithies shall be inducted into the Working Group on Watershed Management.

Situation Analysis or Status Assessment
Preparation of status reports for every Sector starts with a survey of existing resources, analysis of the situation, and exploration of development potential come under this. Each Working Group will have to prepare a development status report for each sector with the following items of content for the purpose:

1. List of schemes taken up in the sector in earlier Five Year Plans by the Local Governments.
2. Key indicators of physical and financial achievements of the above schemes.
3. A comprehensive list of beneficiaries of the plan projects for earlier plans in the local government.

4. A list of assets created during earlier plans.
5. List out major schemes implemented in the sector by Government or other agencies within the Local Government area.
6. The database relevant to the sector from all available secondary sources.
7. A note on issues in the planning, implementation, and monitoring aspects in the last ten years.
8. Preparation of development problem matrix for different wards or areas in the local government.
9. A note on key issues facing the sector, existing gaps, the local potential for development in the sector, strategies for addressing the issues and achieving the potential for development.
10. Potential projects—for Plan and Maintenance Plan.

Methodology for Working Group Report Preparation
The Working Group may verify records, conduct field visits, discuss with selected beneficiaries of the previous scheme, interact with key stakeholders, conduct surveys or make studies, for the preparation of the working group report. The Working Groups should maintain frequent interaction or sharing among the key members so as to bring about cross-sectoral linkages but should not remain as watertight compartments.

It is necessary to use services of all officials devolved to the local governments in the functioning of respective Working Groups. A framework of the Working Group Report was also developed in the form of a template to serve as a guide for the preparation of the report. Working groups shall continue to work even after the preparation of annual plan till the final stage of plan implementation.

The Working Group reports containing a microcosm of the plan set the basic framework for local development. The working group reports will be acceptable to people in the next stages if they are in tune with the people's perception.

3.3.5.3 Need Identification

Identification of the needs of the people is the next step in the process. The need identification encompasses a wide range of consultations and consolidation of ideas. It starts with the holding of stakeholders meeting for gathering their needs.

Consultation with Stakeholders
The next component is holding consultations with the principal stakeholders in each local government as enlisted below:

- Farmers and agricultural workers.
- People engaged in industrial activities and services (both traditional and modern) including workers.
- All the Area Development Societies.
- Headmasters and key PTA office bearers.
- Anganwadi workers and Mothers' Committee Chairpersons.

- All Hospital Management Committee members of the Government Hospitals within the Local Government (of all three streams) and key medical professionals within the Local Government from the NGO and the private sector.
- Youth Clubs, youth organisations and activists and functionaries of the literacy and library movements, eminent persons in the field of arts and culture and representatives of disabled groups.
- Vanasamrakshana Smithies and environmental activists.
- Political parties and trade unions.
- Citizens consultation: In the Gram Sabha/Ward Sabha.

The basic purpose of this meeting is to gather development needs. Before holding the meeting of Grama Sabha/Ward Sabhas, there would be sufficient environment creation and information dissemination by using the media. Six Facilitators consisting of three women nominated by the Area Development Society of Kudumbashree and one woman and two men identified unanimously by the Local Government concerned should be trained for and involved in each Grama Sabha/Ward Sabha. The agenda for discussion in the Gram Sabha/Ward Sabha should be a semi-structured questionnaire type one covering key development issues within the Local Government as a whole. After the plenary session, the participants should divide into breakout groups based on development sectors for deeper discussion and assemble in the summing up session to consolidate the discussion in all the break-out groups.

In the discussion, the Grama Sabha/Ward Sabhas would be asked to list out development priorities ranked in the descending order of preference. In the case of infrastructure like roads, buildings, irrigation schemes, water supply schemes, electrification the Grama Sabha/Ward Sabha should put forth norms for prioritisation of beneficiaries for the whole Village Panchayat/Municipality/Corporation. The Gram Sabha/Ward Sabha should prepare suggestions on maintenance of assets separately.

The following records shall be collected and maintained by the Secretary of the local government to ensure strict compliance with the instruction.

1. Photographs
2. Attendance register showing details like House No., address, age, whether male or female, whether belonging to SC, whether belonging to ST, occupation, etc.
3. Record of discussions of breakout groups on development needs
4. Recommendations of the Grama Sabha/Ward Sabha in a consolidated form.

In the case of Block Panchayats, Grama Sabha like sessions would be held with all elected members of Village, Block and District Panchayats within their jurisdiction; in the case of District Panchayats this exercise would be limited to Standing Committee Chairpersons of Grama Panchayats and elected members of Block Panchayats and the District Panchayat.

3.3.5.4 Goal or Vision Setting

Every plan should be based on a development vision derived after wide discussion and consultation.

Preparation of Vision Cum Development Report
Next step is the preparation of a two-part document Vision and Development Report prepared by a team consisting of officials and resource persons based on the outputs from the Working Groups and Gram Sabha meetings held as above. The report may appear like the City Development Plan under JNNURM. The District Panchayats would not have a separate Development Report and Vision Document. The District Planning Committee would prepare the Development Report and Vision Document for the whole district.

The vision document would set the objectives, output or outcome of planning attempt. The Vision Document part would go beyond five years and give the vision of the overall development of the Local Government as well as development in different sectors.

The Development Report would focus on the development situation in the Local Government in respect of various sectors along with an overall assessment of development based on all data available locally and collecting additional data required especially for the purpose. Development Report would summarise the strategies being followed by the local Plan, the key project interventions and their outcomes as assessed by the Working Groups and project beneficiaries.

3.3.5.5 Preparation of Draft Plan

Each Local Government would estimate the broad allocations for different sectors and call a meeting of Working Groups and arrive at a consensus on key strategies and priority of schemes. The local government should set the priority sector or sectors for the entire plan periods or annually synchronise the local plan with the state and national plan and focus on building up necessary pre-requisites. This meeting should finalise the draft project proposals emerged from the Development Report. The draft Five Year Plan document should consist of the following chapters.

1. Development scenario of the local government.
2. Efforts during the past ten years.
3. Success and failures.
4. Physical and Financial achievements and outcomes in the Ninth and Tenth Plan.
5. Strategic vision of the local government.
6. Summary of possible projects sector wise within each sector giving the existing scenario, the intended scenario, size of the gap and the intended phases of filling up of the gaps with monitorable targets—separately for Plan and Maintenance plan.

7. Allocation of resources sub-sector-wise.
8. Write up on Anti-Poverty Sub Plan and Destitute Plan, Women Component Plan, Plan for Special Groups and Special Component Plan and Governance Plan -only very brief summaries.
9. Maintenance Plan (summary only).
10. Write up on credit linkages.
11. Write up on integration.
12. Write up on monitoring arrangements intended.
13. The Anti-Poverty Sub Plan, Governance Plan, Maintenance Plan and wherever applicable, the Tribal Sub Plan should be separate documents with full details presented.
14. Consensus Building in Development Seminar.

A Development Seminars of the local government as a whole would be held by involving the key representatives from different walks of life and professionals including two representatives—one male and one female—nominated by each Grama Sabha/Ward Sabha to evolve a consensus on development strategy. The Draft Plan and Maintenance Plan would be discussed in these seminars through group discussions. The gist of the Development Report and Vision Document would be circulated. The Development Seminar would thus fine-tune the specific strategies to be followed and will fix the priority. Each Local Government should have a key development theme for the Plan as a whole or for each of the five years in relation to the broad themes for the district developed by the District Planning Committee.

The following records of the development seminar need to be maintained.

1. Photographs
2. Attendance Register showing details like House No., address, age, whether male or female, whether belonging to SC, whether belonging to ST, occupation, etc.
3. Record of discussions of breakout groups
4. Recommendations of the Seminar.

The elected Councils of local governments would meet along with key members of Working Groups and consider the suggestions and recommendations of the development seminar and make suitable refinement in the priorities, strategies, and allocations.

3.3.5.6 Stakeholder Discussion for Plan Refinement

A second stakeholder discussion of modifying the allocation for each sector, setting sectoral priority and other aspects on the basis of Gram Sabha consultation and Development Seminar Discussions need to be conducted.

3.3.5.7 Preparation of Detailed Projects

The Working Groups would then prepare detailed projects in the prescribed formats by limiting the estimated expenditure within the pre-determined allocation. The project proposals should be submitted to a web-based support system to ensure standardisation, easy approval, and consolidation at the State level.

The quality of projects is directly related to the quality of functioning of the Working groups. While preparing projects they should have a full understanding of the experience of last ten years. Working Groups would also be responsible for proper technical assessment in matters of the suitable type of irrigation projects, a mix of measures in watershed management, source sustainability in water supply programs, technological and managerial soundness in sanitation projects, etc. The Working Groups should give an accurate assessment of cost as well as environmental implications of projects having such implications. The number of projects should be reduced significantly by completely avoiding small and low-impact projects and a thin spread of resources.

Plan Finalisation and Submission for Approval
Plan finalisation would be done by the Local Government. The following documents should be submitted along with the plan for approval by the District Planning Committee.

- The documents relating to the Grama Sabha/Ward Sabha, Working Groups and Development Seminar
- Two printed copies of the Development reports and two CDs
- Prescribed expenditure statements for the Five Year Plan 2002–07
- Eleventh Five-Year (2007–12) Plan Document
- Master Plan Document on Watershed development
- Anti-Poverty Sub Plan
- Tribal Sub Plan (wherever applicable)
- Maintenance Plan
- Governance Plan
- Statistical Annexes
- Details of own revenue included in the Plan
- Resolutions of the Local Government approving the Plan Document Nos. 3 to 9 shall also be given in the electronic form to be developed by IKM

3.3.5.8 Plan Vetting

The Technical Advisory Groups (TAGs) created at the district and block level will vet the plans. TAGs are constituted at the district level for Block Panchayats and District Panchayat, at the Block level for Village Panchayats and at the Municipality/Corporation level for Municipalities/Corporation. The members are chosen from among the experts from departments, professional colleges, academic institutions, public/private sector organisations, NGOs, Bankers and from among retired personnel and practitioners as members. The District Collector would be the Chairman and District Planning Officer the Convenor and Secretary of district level TAG. The Secretary concerned, would be the Convener and Secretary of other TAGs. The TAGs should have subgroups for different sectors. The Chairpersons of TAGs other than the District and the Chairpersons and Conveners of sub-groups of all TAGs would also be decided by the DPC. DPC would also fix the quorum for the subgroups, and it shall not be less than three including one non-official.

Functions of TAGs

The functions of the TAGs are:

1. Ensuring that local government plans are in accordance with the mandatory guidelines issued by Government particularly in relation to investment ceilings for the broad sectors, subsidy limits, sectoral guidelines, priorities to various groups, ineligible categories for assistance, etc.
2. Ensuring that the plans are in accordance with prevailing technical guidelines.
3. Verifying whether the costing is appropriate and the phasing is reasonable.
4. Giving suggestions for innovative plans and integrated projects, which Local Governments may accept if they so desire.

The TAG does not have any power to change the priority determined by a Local Government or to force a local government to take up a particular scheme or work. Any dispute regarding acceptance of a scheme at the Block/Municipality/Corporation TAG may be referred to the District TAG for decision.

The TAG subgroups should go through every project in detail, visit sites if required and make suitable recommendations to the DPC. A checklist for vetting different kinds of projects would be incorporated into an electronic process of appraisal. In case the TAG identifies any problem with Local Government projects it should hold discussions with the elected head and the implementing officer concerned of the Local Government and sort out matters across the table. No plan, which does not have the required allocation for the mandatory schemes, should be forwarded to DPC.

The appraisal of projects by Technical Advisory Group should be done scrupulously to ensure the quality of projects. The District Collectors can initiate penal action against those members of Technical Advisory Group who recommend projects without proper scrutiny.

The Local Governments would submit their plans in one lot to the Secretary of the TAG concerned and obtain a receipt. The Secretary of TAG should conduct a quick preliminary scrutiny and then divide the projects among different subgroups of the TAG and pass them on to them for detailed scrutiny. The TAG shall not take more than 10 days for vetting the Plan of a local government.

The draft Plan of District Panchayat and Corporations are to be submitted to a State Level Technical Group after vetting by TAG. Technical Advisory Groups should submit a special report to the DPC on their general assessment of the quality of projects along with suggestions on improving the quality of implementation.

3.3.5.9 Plan Approval

The projects vetted by the Technical Advisory Groups should be considered in detail by the DPCs. Technical Advisory Groups may be asked to present their assessments before the DPC and a considered decision taken. DPCs should go beyond ensuring adhering guidelines to verifying whether the plans of local governments match with the priorities outlined in their vision documents as well as the district vision. DPCs cannot give ad hoc clearances or conditional clearances. If DPCs view that modification of a Local Government plan is necessary, the discussion should be held with key representatives of the Local Government including the elected head. If there is a dispute that cannot be sorted out locally, the matter needs to be referred to the Decentralization Co-ordination Committee at the State level.

A summary of the approved plan in an electronic form containing details like allocation, implementing officer, physical targets, implementation charts, etc., would be given to the office of the DPC for approval. The DPC would issue an official order containing those details. The order would be used for fund release by treasuries and monitoring purposes by the DPC.

3.4 The Importance of Developing Outline Development Plan for Feroke Municipality for Activating E-Democracy

Spatial Planning of Feroke Municipality being a complex, hierarchical and multi-level spatial decision-making exercise, the political and democratic process of planning is not sacrificed on the altar of technicality and complexity. A democratic, participatory planning of a complex town planning task does not mean oversimplification. Technical temper in which spatial information are systematically analysed to reach rational decision making by community shall be encouraged through E-Democracy and public consultation. This is because of spatial planning, unlike participatory planning practice in Kerala discussed in Sect. 3.3 does not merely assessing felt needs under limited information not supported by most up-to-date location specific spatial information and attribute data for making a decision by the local community.

State of Kerala practices Multi-level Governance for designated constitutional areas having power for local self-government and a prefixed share of consolidated funds of the Indian Union and the State of Kerala as per the Indian Constitution and based on the recommendation of Finance Commissions from time to time. Since they use public funds, they are expected to ensure its rational utilisation based on most up to date information. They also have freedom to mobilise funds from various sources to implement plans prepared. These constitutional areas adopt representative democracy with elected members for a five years' period of majority rule in participation with the citizen of these areas. The success of representative democracy is measured by how well the representative can represent the needs of the community he represents.

These are areal units of spatial Governance. Indian constitutions mention area levels such as the state, district, Metropolitan area, Municipality, wards and Gram Panchayats. In multi-level planning framework, Feroke appears in the Kozhikode District Plan and Metropolitan Area plan as well as the area plan of Kozhikode Municipal Corporation Area Plan. These higher-level plans give vertical functional linkages of these multi-level areas and are involved with strategies of spatial economic development in higher levels such as integrated infrastructure, and municipal services can be planned, managed and executed holistically.

The twelfth schedule given in the 74th constitutional amendment gives power for Feroke Municipality to plan and execute development in an area under their jurisdiction respecting the democratic rights of municipal wards. Department of Planning, National Institute of Technology Calicut adopted Outline Development Plan approach for area planning of Feroke Municipality. Outline Development Plans (ODPs) are a form of local structure planning undertaken to provide guidance and flexibility for the future development of Feroke Municipality. ODP is not comprehensive planning but strategic planning based on SWOT analysis. Thereby a new structure of Feroke is designed to serve the people of Feroke better based on a detailed spatial analysis of community participation. ODP of Feroke is prepared in conjunction with the community, local experts, homeowners, residential welfare associations, and elected members of Municipality of Feroke and the relevant Officials of Government Agencies.

An ODP ordinarily consists of a plan supported by a report and technical appendices. These documents are available in public domain through a website, prepared in consultation with the community and can be viewed on the City's website prepared by National Institute of Technology, Calicut (www.odpferoke.co.in). This website itself is interactive, and anyone who logs in can put forwards his or her point of view regarding the development of and management of Feroke (Fig. 3.8).

An ODP does not change the zoning under present use of land under a Zoning Scheme. However, zones depicted on an ODP provide for the same permitted uses as those depicted in the Scheme. NIT Calicut has used mixed land use in a various proportion of mixed-use strategy to achieve this goal. In essence, an ODP provides an interim flexible measure to allow development to proceed and reduces the need for formal Scheme Amendments. NIT Calicut Prepared existing land use based on recent Remote sensing data available and updated by field surveys. It may be mentioned here that aerial photography survey is the preferred database supported by Total Station Surveys, but NIT Calicut did not use it since these are mostly required in the Urban Land Management stage or Detailed Town Planning

Fig. 3.8 Home page of ODP website for Feroke

Scheme stage which is not executed by NIT Calicut. The Plan uses a variety of surveys in addition to land use survey such as Household survey as well as transport and traffic surveys and ODP is prepared based on secondary official documents. Household surveys were conducted with the local community involved and covered all municipal wards with an adequate number of household sample per ward.

ODP provide future direction for the development of Feroke Municipality by designing an appropriate urban structure and related land use and infrastructure by augmenting the existing infrastructure. ODP makes an excellent reference tool for landowners both existing and proposed. Although ODPs are flexible, any subdivision and/or development are required to be in accordance with an adopted and sanctioned ODP as per the existing legislative and administrative procedures in Kerala. This means they have to follow zonal codes of ODP and building codes of Kerala Building by-laws. At the time of development and/or subdivision, it is a good practice to assess and collect a once-off site cost contribution may be payable toward Common Infrastructure Works for the area by Feroke Municipality. This is not in practice today in Feroke, but people totally depends on Government budget which often fails to materialise resulting in the inadequate provision of infrastructure resulting in the poor urban area. The payment of contributions, where required, ensures that area-level infrastructure in town is able to be provided in a coordinated manner to support the overall development of the area. Unfortunately, such practices are not here in Kerala which practices a client consumer democracy as if all city development shall be done by meagre tax revenue and this tax revenue is from an insignificant tax base of the total population of India. The mindset of people of Feroke shall change to incorporate that users pay for what they get if at all they desire a comfortable life. This can be achieved by the democratisation of ODP and better awareness and education on urban development.

All cost contributions discussed above are different, due to different land areas, environmental conditions and scope of physical works required which shall be jointly assessed in a democratic way. Generally, they are based on either a per lot or per hectare basis. Feroke Municipal Council staff can advise people as to the current cost contributions applicable. Whilst Feroke Council staff may be able to provide cost estimates; it is difficult to accurately determine future contributions until the ODP has determined the full scope of infrastructure works.

Various areas will be identified on ODPs for public purposes as shown in the Proposed Land use of Feroke Municipality (Fig. 3.9). There will be a change in the percent of land uses in existing with the proposed. Urbanisation Policy of Kerala Government proposes Urban Compaction and densification resulting in less percentage of the residential area from the present. The land allocated for the existing land use for open space and recreation has been found to be inadequate, and this land use has been increased with a higher allocation of land. This is based on multi-level consideration. Kozhikode area plan proposes less than standard open and recreational area in the recently prepared Master Plan of neighbouring Kozhikode area. The needs of the people of Kozhikode for recreation and open spaces can be met in nearby Feroke if it is provided with more open space and recreation for the use of Tourist from inside and outside of Kozhikode. NIT Calicut has used a democratic process of Consultation with Association of Feroke Residential Welfare Association as well as the elected Municipal counsellors, and Various Task Force Members of Feroke Municipality. There are areas in Feroke which are required to serve particular functions, such as drainage, conservation or recreation. At the time of development or subdivision, the land is shown for such a purpose is required to be ceded to the Municipality. This will be implemented using approved land management techniques in the Town and Country Planning Act 2016 of Kerala like Land pooling and land reconstitution, accommodation

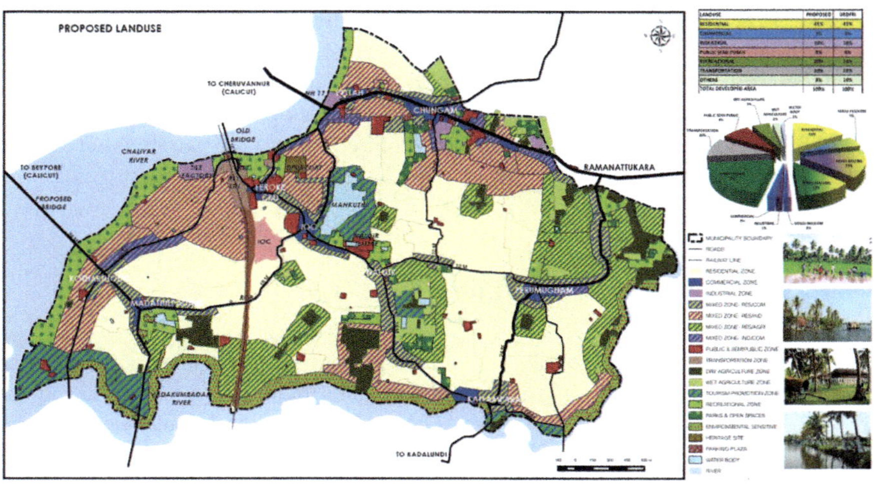

Fig. 3.9 Proposed land use for Feroke 2036

reservation and transfer of development rights. Where a landowner is required to provide land at a rate greater than the standard rate, compensation is often available at the time of subdivision or development.

Where land is required to be provided for Regional Open Space, compensation and acquisition will generally be negotiated directly with the Municipal Government. In Feroke, land owners and developers are typically required to give up some of their lands, or cash equivalent free-of-cost upon the subdivision of land into two or more lots, to provide land for the future recreational needs of residents. In certain areas, open space contributions may be greater. This is generally due to the need to acquire areas other than public open space through the ODP, most commonly, core conservation and drainage areas.

The amount of public open space required to be given up by each landowner is determined through the ODP.

- Where the amount of Public Open Space required being given up free of cost equals the amount shown on the ODP for a particular parcel of land, no public open space contribution is payable.
- Where no public open space is shown upon land about to be subdivided/developed, the landowner must provide an equal contribution in cash. These funds are used to acquire and upgrade public open space close to the originating land.
- Where the amount of Public Open Space required to be given up under the ODP is greater than the standard contribution rate, the landowner at the time of subdivision may be able to claim compensation for the land required in excess of the standard contribution rate.

Contributions towards Common Infrastructure Works are separate to Public Open Space requirements. However, they may be paid at the same time, or one may be used as a partial credit towards the other, depending on the circumstances.

There is no obligation upon landowners to subdivide and/or develop their land. Property values and development trends may make it attractive for land to be developed, but there is no obligation to do so. There may be an instance where Council has to undertake works/acquire land in the public interest. In these cases, negotiations and arrangements may be required ahead of subdivision and/or development of land.

3.4.1 Democratic Approach to ODP Preparation

National Institute of Technology Calicut adopted a democratic approach to the preparation of ODP, which the study group felt is the first step in reaching the objective of E-Democracy for ODP. The following are the elements of the Democratic Approach.

Fig. 3.10 Images of fast-forward organised by Mathrubhumi and Feroke municipality

1. The study group should study the problems and possible solution from those local residents who had intimate knowledge of Feroke who were living and working there for a long time. This was achieved by the Fast Forward event (Fig. 3.10). The local newspaper Mathrubhumi "Fast Forward" gathered all of them for a luncheon meet in a hotel to discuss the problems and prospects of Feroke City where the NIT study group was also invited. Those who attended do not represent any particular political party, income group, cast, religion, sex or any other categorisation you can think about. They all lived in Feroke, hold a position such as an engineer, doctor, teacher, college professors, a local journalist and so on constantly interacting with the people of Feroke. They all openly discussed in this congregation the problems of Feroke and attempted to find a solution in terms of alternate suggestions. NIT ODP team were given time to discuss their project and were determined to find out the goals and objectives of development of Feroke using Delphi technique which is discussed in the Feroke ODP website. The team could not complete the required sample for Delphi analysis in this meeting. This was one of many meetings to fill in Delphi forms. Those who could not complete and submit were directed to the website NIT team had prepared to fill in the form.
2. In front of the elected members of Municipality, NIT ODP team had two opportunities to discuss the ODP Project and elicit participation of all for successful completion of ODP. While the first was centred around the ODP preparation, goals and objectives identification, and various surveys that will be conducted in collaboration with the officials and elected members (Fig. 3.11), the second presentation was to discuss the final report of ODP to get the view of all relevant elected members like chairperson, deputy chairperson and various functionaries of task force team leader among elected Municipality members. This was also an opportunity to work out goals and objectives of ODP by Delphi method by filling in the forms.

Fig. 3.11 Delphi for arriving at goals and objectives with municipal councillors

Fig. 3.12 Delphi for arriving at goals and objectives with residential welfare association representatives

3. In addition to the elected members of the Feroke Municipality, Association of Residential Welfare Association convened a meeting to discuss Draft ODP prepared by NIT Calicut and also fill in for Goals and Objective using Delphi (Fig. 3.12). All critical observations of members' present were incorporated in the plan. After formally submitting the plan to Municipal Chairperson, Association of Residential Welfare Association convened a large gathering in the Town Hall and asked Professor T.M. Vinod Kumar to formally hand over the Report to recently elected Member of State Legislative Assembly (Parliament of Kerala State) who was also earlier the Mayor of Kozhikode Municipal Corporation. This is an indication of ownership of the ODP by all residents of the Feroke Municipality since they organised the event fully on their own (Fig. 3.13). This large gathering was attended by Chairperson and Deputy Chairperson of the Feroke Municipality and all other team leaders of sectors in the municipality all prominent citizen of Feroke and common people.

Fig. 3.13 Presenting ODP to the MLA and public

4. Democratic Planning was emphasised in surveys with about 500 households covered by a household survey conducted by NIT planning team with the active association of local residents. The latest satellite imagery was used to digitise Feroke base map and an existing land use survey was conducted by NIT Planning team accompanied by local people. Lanes are the most important part of the walkway for people of Feroke which was surveyed using GPS. Traffic and transport surveys were also conducted such traffic volume, parking, Origin-destination study and so on. All of such surveys were conducted with the active participation of local residents.

3.4.2 E-Democracy for ODP Preparation

People of Feroke Municipality is fully prepared to undertake E-Democracy for ODP preparation and execution. Primary school attendance is 100% in Feroke. Feroke is both fully literate and e-literate Municipality such that at least one member of the household can use a laptop and the internet as part of Government of Kerala Program. There is not much of distinction between the male and female's achievements in this city as regards to preparation for E-Democracy. They are equal as far as e-literacy and literacy are concerned. Although laptop ownership and penetration of broadband network have not reached every household, a minimum of one mobile phone with internet capability is there in every household. Higher internet users in e-commerce and industries shall require higher bandwidth which is not there. Although Akshaya Centres which actively participated in making people of Feroke fully e-literate, they do not have necessary software and training program in e-democratic implementation of ODP execution. Further mobile phone oriented apps are to be developed and made available and Internet of Things, Internet of People and Internet of Governance of Feroke based on needs of Feroke to be in place which can only be possible if centres of higher learning in and around Kozhikode prepare and provide them. This means nearby Feroke College, University of Calicut, and National Institute of Technology Calicut have a greater role in the days to come for implementation of E-Democracy for ODP. This project will execute ODP of Feroke, leaving the mobile app developments and IOT development for the concerned coders and developers in NIT and Calicut University.

Our approach to ODP preparation has been presented elsewhere in the chapter. These approaches will considerably change if we implement E-Democracy in the preparation of ODP.

ODP conventionally has been based on once in 20 years' data collection and resort to all sorts of projection of variables for benchmark years of five years. The experience shows that none of this projection and even population projection has been correct in all the ODP and Mater Plans prepared all over Kerala over sixty years' period as observed by authors of this chapter. E-Democratic ODP preparation will be based on constantly updating the database of all variables of Feroke by electronic means. Remote sensing data for the mapping of land use will be used continuously depending on the next available remote sensing data within few days. People of Residential Welfare Association will be there who will be trained how to update land use data and incorporate in the constantly changing GIS database. Traffic data will be collected by sensors and will be available in Feroke data cloud for accessing by everybody for all purposes with or without a fee for many apps to be developed for Internet of Things, Internet of Governance or Internet of People. The household sample used in this project gets transformed to household census annually. This will be filled in by households and uploaded to Feroke Web site for further processing and presentation in the web from time to time. This can trigger in Facebook media deliberation by local residents attached to the website of Feroke designed. Goals and Objectives of the development of Feroke shall be assessed by all households periodically and be the motivators to develop apps that attain these goals. Strong linkage of Feroke with centres of higher learning in Kozhikode will give them the capability to analyse these data and provides them with apps that can be used in conjunction with Internet of Things, Internet of People and Internet of Governance to deal with emerging issues in Feroke based on the constantly updating database of Feroke.

3.5 Development of ODP for Feroke City in IOT, IOG, and IOD for Achieving E-Democracy

The ODP for Feroke was intended to convert Feroke to a self-financing smart city. It is found that people of Feroke have all the potentials to transform themselves to Smart-People, which is the most important component in the building of the smart city [4]. The economy of Feroke is in a transition mode and has all the indications that it can be easily upgraded to a smart economy [5, 6] if there is a conducive environment available. The ODP is intended to evolve this conducive environment and form the six elements of the smart city namely Smart People, Smart Economy, Smart Mobility, Smart Environment, Smart living and Smart Governance. ODP envisages to develop this conducive environment through a set of tools. These tools can be classified to Internet of Things (IOT), Internet of Governance (IOG), Internet of Democracy (IOD). The following section briefly explains the solutions proposed for each element of the smart city based on various surveys conducted in Feroke and its participatory analysis and conclusions drawn.

3.5.1 Smart People

The ODP intends to upgrade the citizens to smart people by empowering each individual, by understanding his or her capabilities and limitations, and opening up opportunities to develop the skills required by them to fit in the community. It also intends to make the public spaces and facilities accessible to all individuals especially for marginalised communities in the present system. The strategies include inclusive planning and design of streets, parks, public places, civic facilities, sanitation infrastructure and transport modes, starting women-centered industrial projects, promoting women entrepreneurship and provide skills development training, the creation of geriatric/especially skilled/physically challenged care infrastructure in govt. The sector and preferential housing of the marginalised.

3.5.1.1 IOD Tools

A mobile application is proposed, as par of the survey and analysis to make the differently abled (blind) more independent and self-reliant. The application is expected to be controlled by speech and communicated back to the user through speech synthesis. Helpline numbers with facilities of the speak-and-dial facility and the app may avail facilities of hiring volunteers to guide them in malls (Calicut), bus terminals, commercial centres of Feroke etc. Booking prepaid services for transportation-taxi, auto-rickshaw, bus by E-banking facilities which can be done by scanning thumb impression for the username and a pattern for a password. Booking an appointment with doctor or booking tickets can be done by opening the app, touch anywhere on the screen which can make the app start talking about the categories, selection of Category can be done by repeating the name of the category. Similarly, the headings under each category can be selected in a similar way.

3.5.1.2 IOG for Safety

The IOG tools bring smartness to the governance. The application will have an SOS button to send pre-set messages to the defined contacts. In the same app, the user can report incidents like crime, traffics, harassments, domestic violence, corruption, smoking in public areas, etc. There will be a provision to find nearby places like a police station, bus station, ATM, hospital, etc. Apart from the obvious advantage to the user, the application collects data regarding the incidents using the location and other sensors in the mobile. Such data, combined with other information available with the smart city ICT, can be effectively used in diagnosing the problems, predicting safety hazards, and taking effective actions against such incidents.

3.5.2 Industry

The ODP proposes to encourage the existing industries through various strategies and at the same time it recognises the need to revitalise the traditional industries through innovations to enable to tap the potential of the global market. There is promising potential in the active household based industries in Feroke. Tourism potential is also remaining untapped. The ODP recognises that there is ample opportunities for one side and skilled and technical workforce available on the other side, but there is still an issue of a proper interface. The smart city expects people to be more dynamic, and capable of learning new skillsets as the industries demand changes.

3.5.2.1 IOD for Industries—Workforce Interaction and Skill Development

The proposed IOD application for this purpose is expected to have capabilities to collect the requirements for jobs from various industries, and subsequently, alert the individuals with matching qualifications. It may also rely on available historical and real-time data to propose a skill improvement programme for the individual job seekers to improve their prospects. This will also help the proposed skill development centres to form new courses matching with the demand and disseminate the training sessions in custom made fashion to match with the requirements of the public. The schematic process is presented in Table 3.1 and Fig. 3.14.

Table 3.1 IOD for industries

Input	Process	Output
Industries • Registration number • No. of employees • Production • Specialisation, skills	• Individuals, dealers and entrepreneurs from anywhere can register themselves • Discussion forum—start-ups, Issues and solutions • Volunteering for CSR works • Start-ups—sharing of idea, presentations, seeking funding and marketing • Skill development: people of one trade can undergo skill development in another trade through this application	More productive Efficient interaction between industries and people Promotes innovation and research
Dealers • Registration number • Location • Products		
Entrepreneur • Aadhar card number • Skills • Market areas and products		
Individuals • Aadhar card number • Qualifications • Skills • Area of interest		

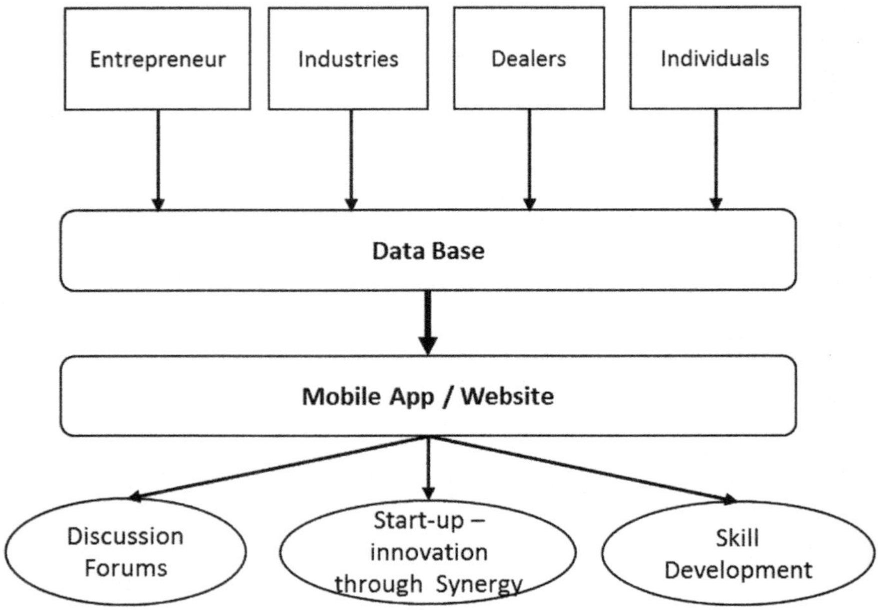

Fig. 3.14 Schematics of IOD for industries

3.5.2.2 IOD for Self-employment

The casual labourers for building maintenance and other minor jobs are always an issue in Feroke as it is for any other city in Kerala. Though there is sufficient demand for such jobs, and many skilled people are available, they tend to charge full wage for even works for few hours as they may not be able to get another work on the same day. Similar issues can be solved easily by the deployment of IOD system for self-employment. Many of the utility systems in the buildings will become smart and systems can diagnose themselves and can raise a warning when they reach their expected service period, or when faults are identified. Such maintenance requests can be directed automatically to the available workforce with the required skill sets. They may, in turn, take up the work and may issue a quote for the work to the owner, who may award the work to the selected party. This will enable the labourer to plan his daily activities and cover the maximum possible in a day, and hence he will be able to charge only for the number of hours worked. A feedback mechanism will ensure the improvement of quality through competition.

3.5.2.3 IOD for Tourism

Tourism is an untapped potential for Feroke. There many locations and activities of interest to the domestic as well as international tourists. The ODP proposes to encourage the public to extend homestays and community events like local temple festivals, cultural programmes of youth clubs, Soccer tournaments, etc. for tourists. An IOD application is proposed to integrate such events and facilities. The application may assist tourists to plan their available time, by suggesting places they can visit, suggest events they can attend and likely places to stay.

3.5.2.4 IOT for Tourism

This is an appropriate system of sensors employed at various tourist locations, and boating centres, which keep track of a number of users at any time. This count helps in the management of disasters like boat capsize, drowning, etc. This count will also assist in judging advertisement potentials in such places.

3.5.2.5 IOT for Agriculture

The increased demand for residential land, high labour wages, segmentation of agricultural land, and negative sentiment against mechanisation in the past by certain political parties contributed to the vanishing of agricultural activity from Kerala. Kerala relies on nearby states for the agricultural supplies. However, the recent increase in incidents of cancer, diabetes, etc. were identified as effects of indiscriminate use of pesticides and chemical fertilisers in these states. This has created a sentiment among people to start their cultivation of vegetables, which resulted in rooftop cultivation etc. OPD identifies this as a positive move and promotes urban agriculture, by proposing IOT and IOD tools for promoting agriculture within the urban space. The solutions range from the automated irrigation system, which senses the need for irrigation through soil moisture sensors and communicates to the owner through the mobile. There are solutions proposed to monitor nutrient requirements in the soil and suggest suitable bio fertilisers and solutions market their surplus products easily. The marketing can be done for even small quantity. This helps residents to cultivate different vegetables, and exchange them within the community so that everyone will get virtually any vegetables he or she require. The transactions can be even barter, which will encourage every member to produce some vegetables. The IOD app serves as a marketplace for such transactions and can keep a record of the barter system.

3.5.3 Smart Mobility

Many of the transportation related issues faced by Feroke can be traced to the demand for vehicular traffic through the narrow roads. Most of the roads within Feroke was formed by nominally widening erstwhile footpaths. These footpaths are very efficient if a suitable mode of transport is used. Hence the ODP proposes to encourage the use of bicycles and promote walking. The strategies proposed to include identification of cycle tracks and promoting them to the major traffic route by developing commercial land use along with this route.

3.5.3.1 IOT and IOD for Parking

ODP also propose to widen few selected roads and to develop dedicated parking area near to the central Feroke. Still, the parking will be a deficit, but ODP proposes to manage the parking through an IOD application which allows the users to reserve parking slots automatically and helps to find the nearest available parking places. To achieve this an IOT devices are to be deployed to monitor the parking places automatically.

3.5.3.2 IOD and IOG Tools for Efficient Bus Transportation

An IOD application is proposed for public to identify the availability of public buses to travel to a given destination. This application works by assimilating location information from the running buses, and their designated routes. The application will be capable of proposing itinerary of travel, to optimise either time of travel, the cost of travel or distance of travel. This can also be extended to other modes of transport like auto-rickshaws, taxi and trains. IOG tools are proposed to manage bus route identification and assignment. Currently, the routes are decided by the bureaucracy as required by the political powers. The data collected through the above mentioned IOD tools for bus travel can be used to decide the effectiveness of the present route assignments, and GIS models can be developed to propose alternate routes which are good for the larger public.

3.5.4 Smart Environment

Smart environment deals with the application of technology for responsible utilisation of environmental resources and management of wastes produced by the city. This majorly involves utilisation of water, power, Green cover, and disposal of wastewater, and solid waste. The city meets its larger part of the water demand by ground water, from individual wells. This is augmented by the new JICA funded large-scale water supply scheme that covers the whole district. OPD also promotes identification of the responsible use of smaller water supply schemes. As the water is consumed from multiple sources, it is important to monitor the quality of water.

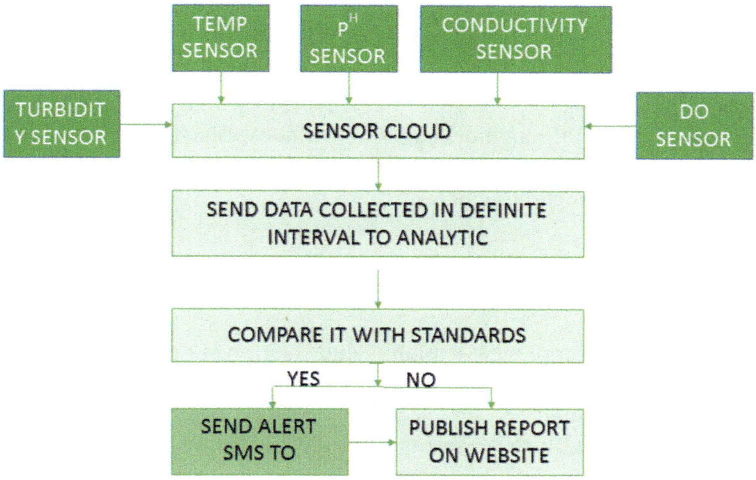

Fig. 3.15 IOT for water quality monitoring

3.5.4.1 IOT for Water Quality Monitoring

An array of water quality monitoring sensors is proposed to be distributed throughout the area, which is IOT enabled and constantly communicates the quality of water available. Smart water sensors analyses pH, dissolved oxygen (DO), oxidation-reduction potential (ORP), conductivity (salinity), turbidity, temperature and dissolved ions (Na^+, Ca^+, F^-, Cl^-, Br^-, I^-, Cu^{2+}, K^+, Mg^{2+}, NO^{3-}), compare it with standards and will send details to the owner or authorities. This regular supply of information helps to analyse the change in pollution components in the water and alerts the public and the authorities to take suitable action to identify the source of pollution and take necessary steps to manage that. The schematic diagram is given in Fig. 3.15.

3.5.4.2 IOG and IOT for Water Usage Monitoring

JAICA funded water supply scheme is using smart meters and are managed by SCADA based control system. Other schemes for water supply also may be monitored through a similar system and may be connected together at the city level. This will enable to forecast the demands during weekends, seasons, etc., and may be used to arrange for the additional supply when it is required. The smart meter readings can be monitored in real time by the user, and the IOG application can give suggestions based on the usage pattern.

3.5.4.3 IOT for Waste Management

Smart bins with small battery/solar powered wireless sensors that are attached to the underside of the container lid can sense the volume of garbage in the bin. It can also communicate this level through cellular/wireless networks. The alerts from various places will be analysed by the fleet management system to determine the best route to be taken by the collection vehicles.

3.5.4.4 Energy

Alternate sources of energy were explored, and solar power is found to be the most suitable for Feroke. However, the throughput is reduced considerably during the monsoon season in Kerala. Smart electricity meters are proposed to monitor the usage of electricity and also to account the electricity given back to the system during peak solar hours. This will reduce the need for storage facilities, and will also reduce the usage of hydropower during day time.

3.5.5 Smart Living

Smart living details about the schemes and programmes proposed for improving the living standard of citizens using smart technology. Community health management schemes of Kerala were well known as they achieved very high standards with a nominal budget outlay [7, 8]. Recently Govt. of Kerala has decided to revive its 'Family Doctor' scheme under the brand name 'Ardram' [9, 10]. The scheme assigns the doctor in the primary health centre as the first point of contact for any health related matter of the family within the limit of the primary health centres. There are many kinds of health trackers available in the market, which collects vital health measurements like blood pressure, sugar level and activity of the user and transfer the same to the mobile phone on a periodical base. If such devices can be made available to the needy, the 'Family Doctor' will be able to get the full health statistics of the patient, and will be in a position to do the needful. An IOD application can be designed to alert the doctor automatically if the patient has unusual variation in the health statistics, and may automatically arrange for the necessary first aid on the spot.

3.6 E-Democracy Tools for Public Participation in the ODP Preparation Process

The above examples show that the IOD tools are applicable to any decision-making process for the public to the public. The process of ODP preparation will become more efficient if conducted in a scenario where IOD tools are already deployed. The following section discusses the conversion of ODP preparation to an IOD process. The planning process is essentially a cyclic and continuous process which involves the following major stages.

1. Identification of issues faced by people and identification of priorities
2. Identification of goals and Objectives
3. Extensive data collection
4. Draft plan preparation
5. Publication of the ODP for larger public opinion
6. Legalisation of ODP
7. Implementation of projects
8. Review.

This process may be accepted for the implementation through e-democracy platform. Potentials of existing social interaction tools itself may be used for the e-democracy, to begin with. Specific requirements for interaction tools will gradually evolve over time. The advances expected in the process may be classified into three generations.

3.6.1 *First Generation*

The first generation is characterised by the application of general digital communication techniques for the interaction and keeping all the back end processes the same. The citizens and stakeholders participate through elected representatives. Data collections are done with the help of publicly available repositories and through manual data collection process.

3.6.2 *Second Generation*

Exploitation of social media tools, and usages of dedicated tools like website, or mobile applications to for assimilation of public opinions and to encourage public participation. Dedicated tools may be employed to automate assimilation of feedbacks. Availability of data collected for prior purposes and reuse of such data for the planning process. Stakeholders participate in the process through elected representatives, but options will be available for the participation of the larger body.

Extensive application of technology to generate scenarios for illustrating the future of current decisions taken.

3.6.3 *Third Generation*

The third generation will be characterised by extensive exploitation of social media along with digital communication for the planning process. Most of the data requirements of the ODP preparation will be satisfied by mining the huge repository of data generated by the galaxy of IOT sensors and other IOD and IOG applications. Stakeholders are participated directly, without any elected representative. There will be automated backend processing tools for processing the feedbacks from the larger public through multiple sources to precipitate the general sentiment, similar to the methods used by large scale companies like Google, Microsoft and Facebook to judge the user satisfaction. Extensive use of simulation techniques to generate alternate scenarios. Individual stakeholders will be able to try out alternate decisions and figure out the end results themselves through simplified visual representations.

The expected evolution of ODP process in an IOD framework is given in Table 3.2.

3.7 Conclusions

There are two E-Democracy Models; Consumer E-Democracy and Prosumer E-Democracy. In Consumer democracy people consumes the tax revenue for development expenditure of Feroke after all administrative and debt servicing expenditure are paid for. The resulting funds are not sufficient to meet the growing demand for public funds for urban development of Feroke. An analysis of figures available in Sulekha website shows that Kerala by and large depends on Government funds for Urban Development as a full-fledged consumer democracy. Feroke shall strive for a Prosumer democracy. In a Prosumer democracy, the democratic capability is shared to bring about a better income generating economic activities that are shared between a citizen of Feroke in addition to consuming tax revenues for Government for urban development. Part of this income generated shall form a complementary fund for urban development of Feroke by paying what is consumed for to generate a self-reliant Feroke. This fund undoubtedly shall be higher than Government funds.

In addition to Government, there are at least two actors in the urban development of Feroke. They are Malabar Chamber of Commerce and Industries and Academia like National Institute of Technology, Calicut, other institutions of higher learning in Calicut and Calicut University. They are not recognised by Government of Kerala as partners of urban development. So much so Metropolitan Planning

Table 3.2 ODP preparation process in IOD framework

	Major processes in ODP preparation	Traditional method	e-democracy platform		
			Generation 1	Generation 2	Generation 3
			Application of digital communication for the planning process	Exploitation of social media along with digital communication for the planning process	Exploitation of social media along with digital communication for the planning process
			Manual back-end processing	Manual or semi-automated back-end processing	Automated backend processing
			Participation of stakeholders through representatives	Reuse of data collected for another purpose to avoid unnecessary duplication of effort	Mining of the required information from the huge repository of data generated through IOD, IOT tools
				Participation of stakeholders directly of through representatives	Direct Participation of stakeholders
1	Identification of issues faced by people and identification of priorities.	The survey, Meeting of representatives of stakeholders	Meeting of stakeholder representatives	Facebook like media and Manual consolidation of feedbacks	Identification of issues through Big Data analytics of the enormous data collected over the previous plan period through sensors and plethora of social media
			Discussion forums on the website	Dedicated web pages in Municipality website for two-way public interaction	Face to face meeting only if necessary
2	Identification of Goals and Objectives	Delphi, Meeting of representatives of stakeholders	Delphi conducted over email communication with representatives of stakeholders	Delphi conducted through the dedicated web or mobile applications and virtual	Draft goals and objectives are formed automatically through dedicated big data analytic tools

(continued)

Table 3.2 (continued)

Major processes in ODP preparation		Traditional method	e-democracy platform		
			Generation 1	Generation 2	Generation 3
			Use of cloud solutions like Google forms to facilitate updating of feedbacks	meetings organised through social media	Modifications and suggestions are sought from each citizen, and voting is taken as the approval
3	Extensive data collection	Household survey, Manual survey for aspects like transportation, land use, buildings, etc.	Through extensive use of remotely sensed data and digitised vector data from sources like Landsat, Open Street Maps, etc.	The data collection from databases created by various agencies time to time	Automatic sourcing of information for each entity from IOD, IOT and IOG tools deployed
			Manual surveys when necessary		Crowdsourcing for missing information
					Prediction of missing data from the available unstructured data
4	Draft plan preparation	By planners, in consultation with focus groups	Prepared by planners in consultation with focus groups, with the help of scenarios already prepared	Prepared by planners through assimilation of feedbacks from the larger public on draft proposals, illustrated with alternate scenarios	Prepared by planners, with the aid of simulated city models, which allow citizens to adjust control parameters, and visualise the results
					Citizens work as explorers of alternate scenarios and reports fruitful alternate solutions
					The suggested solutions may be further optimised through techniques like Genetic Algorithm or Ant Colony Models

(continued)

Table 3.2 (continued)

Major processes in ODP preparation		Traditional method	e-democracy platform		
			Generation 1	Generation 2	Generation 3
5	Publication of the ODP for larger public opinion	Publication through the internet, printed media, and key features through newspapers	Published through the website, publicised through social media	Not required as the previous stage itself involves the larger public	Not required as the previous stage itself involves the larger public
6	Legalisation of ODP	Approval by the legislative assembly after the approval by the local self-government	Existing procedure will continue	Existing procedure will continue	Approval by the elected representatives may become less relevant as each individual is able to vote their approval
					Vetting by an expert committee may be given importance
7	Execution of projects	Municipality allocates budget and executes the projects by inviting contracts for various subtasks	Publication of budget in the municipality website, E-Tendering for projects	E- Budgeting - Budgeting by elected members based on detailed statistical information, E-Tendering	Participatory budgeting by the public through dedicated IOD tools
					E-Tendering with participatory assessment of the execution
8	Review	Not in practice. Ad hoc solutions are made from time to time	Manual assimilation of feedbacks received through direct or social media	Automatic assimilation of feedbacks on a continuous manner through certain specified channels like feedback boxes in websites, emails, etc.	Review as a continuous and automatic process
					The review is done by analysing the maintenance requests, Crime and accident data, demographic change, land use change, etc.

Committees, District Planning Committees, State Planning Board, Municipalities do not have seats reserved for memberships of these executing committees. They served in this project Feroke Municipality free of cost, unlike Government where administrative expenditure and debt servicing is deducted from the developmental expenditure. For example, the ODP prepared and presented to people of Feroke was free of cost since the Marginal cost of working on the ODP was zero to National Institute of Technology, Calicut. This in no way belittles the role of Government of Kerala who has their unique constitutional responsibility for Tax collection and Tax sharing for Development which academia and the Chamber of Commerce and Industries cannot perform but what Government of Kerala can allocate to Feroke the aspiring Smart City is not at all sufficient.

The ODP preparation was executed by a nine-member study team and three member faculty but largely assisted by people of Feroke in a very massive scale. People of Feroke was there with the team to collect information from 500 households, conducting various other surveys such as traffic surveys. Without the active participation of people of Feroke, the planning team cannot achieve what we have done. All of them conducted with near zero marginal cost.

ODP generally prepare a spatial plan to implement through regulated development. Detailed spatial and non-spatial survey research opened up problems and potential of development of Feroke Municipality. This resulted in a spatial plan and a set of priority action plans. Feroke ODP here makes a departure from conventional ODP. This departure is in the implementing framework.

Implementing Framework adopted by NIT Calicut for ODP is conceived as IOP, IOG and IOT. National Institute of Technology, Calicut or Calicut University is fully capable of deploying IOT, IOP and IOG. IOP and IOG essentially involve the development of mobile based apps to be used by people of Feroke. IOT involves developing one or more IOTs to perform the work stipulated for the activity as discussed in this chapter.

The Department of Planning and Architecture, NIT Calicut, had exhibited capabilities in this direction. The author of Chap. 16 Swapnil Shrivastav of this book is working in a start-up which produces drinking water simulating rainfall in a confined space using a series of IOTs for drinking water supply.

The development of IOT, IOP and IOG for this project means for National Institute of technology breaking open departmental walls and collecting a series of teachers and students to produce the apps or IOT configurations. NIT Calicut is in a position to perform this and present it to Feroke Municipality at the marginal cost of near zero at zero price. This approach is far superior to Government of Calling Tenders and awarding contract for the lowest tender or calling foreign collaborators and foreign loans for Smart Cities.

Bimal. Aruna and I, the three authors of this chapter carefully selected several teachers of other departments in the National Institute of Technology, Calicut such as Electronic Department, Transport Engineering and Electrical Engineering Department and made a presentation of this project. This is because of a request from Professors Chen Feng and Zheng Liu of University of British Columbia, Canada and Authors of Chap. 17 of this book. The task was to compete for an

Indo-Canadian Project administered jointly by Government of India and Canada on the development of IOTs for Smart Cities. During the presentation of study teams, those assembled came out with the design of many more IOTs and IOPs and IOGs discussed in our presentations. This documentation does not present all these because our study teams did not discuss with the people of Feroke for their agreement.

Acknowledgements The work explained in this chapter is based on one semester work of the second semester post graduate students of the Department of Architecture and Planning, NIT Calicut during 2015–16. The authors would like to gratefully acknowledge the efforts of our students, Ms. Anu Paul, Ms. Asha Devadas, Ms. Ashikha Raoof, Mrs. Fathim Rashna Kallingal, Ms. Maya K.P., Ms. Nunna Tagore Sai Priya, Mr. Prabhakar Uniyal, Mr. Sarath Chandran P., and Ms. Shabana Yoonus. We are grateful to our faculty colleagues, Dr. Naseer M.A, Dr. Anilkumar P. P., Dr. Mohammed Firoz C., Ar. Chithra K., and Ar. Lakshmi Manohar for helping the students with special lectures on their respective expertise during the programme. We would also acknowledge the whole-hearted help extended by the Municipality of Feroke, represented by Mrs. T. Suhrabi (Chair Person), Mr. Asif (Standing Committee Chairman), Ms. Sabeena (Councillor) and Mrs. Sarada C.P. (Ward Member). We would also like to extend a special acknowledgement to Mr. Mohanan master (Coordination Committee chairman of residential associations in Feroke) who was always ready to provide us any help we required throughout the project.

References

1. Oommen MA (2004) Deepening decentralised governance in rural India : lessons from the people's plan initiative of Kerala. Centre for Socio-economic and Environmental Studies, India
2. Menon DM (1999) Houses by the sea—State-formation experiments in Malabar, 1760–1800. Econ Polit Wkly 34(29):1995–2003
3. Government of Kerala (1994) The Kerala municipality act. India
4. Kumar TMV (2015) E-governance for smart cities
5. Vinod Kumar TM, Bimal P, Mohammed Firoz C (2016) e-design of umami by smart people for smart economy. In: Vinod Kumar TM (ed) EdSmart economy in smart cities: international collaborative research: Ottawa, St.Louis, Stuttgart, Bologna, Cape Town, Nairobi, Dakar, Lagos, New Delhi, Varanasi, Vijayawada, Kozhikode, Hong Kong, Springer Singapore, Singapore, pp 391–442
6. Vinod Kumar TM (ed) (2016) Smart economy in smart cities. Springer Singapore, Singapore
7. Nabae K (2003) The health care system in Kerala—Its past accomplishments and new challenges. J Natl Inst Public Health 2(52):140–145
8. Padmaja K (2005) An evaluation of primary health care system in Kerala. Cochin University of Science and Technology, Kochi
9. Government of Kerala (2016) Health, family welfare—government of Kerala, India. [Online]. Available https://kerala.gov.in/health-family-welfare. Accessed 25 Dec 2016
10. Kerala state to revive family doctor concept—medical dialogues. [Online]. Available: http://medicaldialogues.in/kerala-state-to-revive-family-doctor-concept/. Accessed 25 Dec 2016

Chapter 4
E-Democracy in New Delhi Municipal Council: A Case of Smart City Planning Processes

Ashok Kumar

Abstract New Delhi Municipal Council is one of the rare urban local bodies of the National Capital Territory of Delhi, which is governed by Government of India through an organizational arrangement led by senior civil servants who are assisted by professionals like engineers, architects, city planners and others. Normal democratic regimes and routines such as elections and re-elections are non-existent in the working of the New Delhi Municipal Council. As is well known, in the rest of the country, democratic processes are well and truly established since the early 1950s hailing India as the world's largest democracy. With the declaration of the area under the jurisdiction of NDMC as the first smart city in Delhi, the highly planned and designed city by Edwin Lutyens aspires to become a benchmark for the global smart capitals. Covering the geographical area of 43.7 km^2 (2.95% of the total area of Delhi), and population of only 0.3 million (1.79% of the entire city state), NDMC expects to achieve the smart global capital status when one also considers that huge investments per person are being made year after year. The key issue that I would like to answer in this chapter is: Can New Delhi Municipal Council unleash e-governance including e-democracy, and pave the way for e-democratic governance in the future for its citizens. I would like to answer this key question by analysing three empirical cases apart from an examination of the NDMC Smart City Proposal with a view to assessing how far the smart city processes have promoted democracy in general and e-democracy in particular (citizen engagement) in the NDMC area leading to transparency, accountability, and ease of doing business. All these are core pillars of Government of India's urban and economic policies. In the end some concluding are made.

Keywords E-Democracy · E-governance · Citizen engagement · Building permissions

---

A. Kumar (✉)
Department of Physical Planning, School of Planning and Architecture,
4 Block B, I.p. Estate, New Delhi, India
e-mail: dak13april1962@gmail.com

T.M. Vinod Kumar (ed.), *E-Democracy for Smart Cities*, Advances in 21st Century Human Settlements, DOI 10.1007/978-981-10-4035-1_4

4.1 Introduction: An Overview of the Smart City Proposal of NDMC

Flagships projects capture the imagination of the policy framers, private sector multinational companies, anyone concerned with planetary urbanism, and policy scholars who critically examine such policies. India's own smart city policy known as the Smart Cities Mission is no different. It has become the talk of the town throughout India. Indian capital city is abuzz with smart city talk. A large number of seminars and conferences bear witness to the centrality of this flagship project in the urban debate. The Smart Cities Mission has become the mission of seminar and conference attendees. Even the BRICS conference held at Vishakhapatnam, Andhra Pradesh hosted a special session on the smart cities. It appears the smart city idea has caught the imagination of all concerned.

A State Level High Powered Steering Committee (HPSC) set up under the chairmanship of the Chief Secretary, Government of the National Capital Territory of Delhi (GNCTD) recommended that New Delhi Municipal Council should be developed as a smart city. On 27 August 2015, Ministry of Urban Development, Government of India, accepted the recommendation for participation of New Delhi Municipal Council in the second stage of the Smart Cities Mission popularly known as the City Challenge round.

Prior to this New Delhi Municipal Council hired KPMG Advisory Services Private Limited for the preparation of the Smart City Proposal. After deliberations, the Council of the NDMC approved the Smart City Proposal on 7 December 2015, which was supposed to be submitted to the Ministry of Urban Development on 15 December 2015 for inclusion and consideration in the City Challenge competition. Once the NDMC cleared the City Challenge stage, the Ministry of Urban Development would allocate funds to the NDMC. But before this, the NDMC Smart City Proposal has to be approved by the State Level High Powered Steering Committee. So there is some role that the state government could play in shaping the Smart City Proposal. However, as far as GNCTD is concerned, Lieutenant Governor is more powerful than the elected government as per law currently provides, the fact that has been fully reinforced after the recent court judgment.

In May 2016 the second list of 13 smart cities announced by the Ministry of Urban Development. On 29 January 2016, the same Ministry declared the list of 20 cities, which would be funded under the Smart Cities Mission. New Delhi Municipal Council, the only smart city of the GNCTD also found place in that list of 20 cities. These 20 cities were selected on the basis of the feasibility of the proposal, cost-effectiveness, result orientation, citizen participation, strategic plan, vision and goals, etc. Total funding amounting to Rs. 50,802 crore for 20 cities and towns over five years is made available by central government. Out of this amount Rs. 38,693 crore will be spent on area development and Rs. 12,109 crore on 56 pan-city solutions [1]. Area level projects are preferred over city-wide projects. Now that NDMC has been selected as one of the smart cities in the GNCTD, the

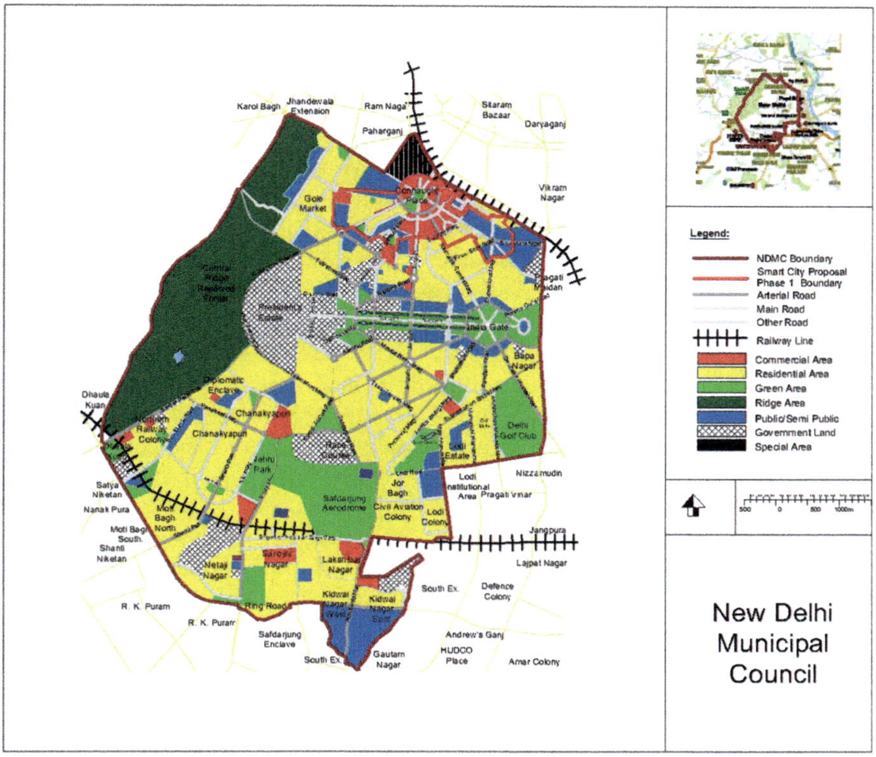

Fig. 4.1 New Delhi municipal council. *Source* Adapted from [2]

crucial question is—why this part of the city—why not any other part, may be less influential and more deprived.

This issue becomes important because the NDMC contains only 3% of the total population of Delhi and a very small area of the metropolis (see Fig. 4.1). Moreover, the NDMC area is highly developed physical and social infrastructure with no slums. At the same time the NDMC has been continuously receiving large annual budgets per head when compared with the other three municipalities of Delhi, previously known as Delhi Municipal Corporation.

New Delhi Municipal Council is a unique case of government by unelected bureaucrats whereby the municipality reports to central government through LG and elected state government has very little say in municipal affairs. Additionally, the Smart Cities Mission would compel implementation of the smart city projects through SPV, an appointed body largely made up of non-elected members. It will be run like a corporation under the Companies Act 2013. This does not promote the democratic cause. If the Smart Cities Mission has to be implemented democratically through elected local governments, it should at the minimum be done in consultation with the elected state government.

Before I move forward, it is important that the term e-democracy is briefly explained. In popular imagination e-democracy implies provision of various kinds of services through internet based electronic media. Municipalities claim to graduate from democracy to e-democracy when they are able to provide several of their services through electronic media. For example, New Delhi Municipal Council (NDMC) website reports 11 services that are provided online including smart city related services, building approvals, and online payment of property taxes. Similarly other municipalities in Delhi and throughout the length and breadth of country have resorted to online provision of services to their citizens. There is no doubt that online provision of services makes life easier for the concerned citizens but the key concern is does it also promote e-democracy. To answer this question one needs to clearly define e-democracy. E-democracy like democracy should be viewed as a relationship between governments and citizens whereby desires and aspirations of citizens get reflected in policies and programmes. Electronic media should be treated as instruments for facilitation of deliberations and debates leading to formulation of concrete decisions and polices reflective of the desires and aspirations of its citizens (see [3]). Whyte and Macintosh [4] reinforce this idea: "E-democracy relates electronic delivery of the processes of democratic representation to practices of communication, consultation, and participation in public decision-making". Citizens are situated at the heart of e-democracy while governments and electronic media remain the vehicles for designing decision making processes and policies soaked in democratic ethos leading to visible shift of power from governments and machines to the citizens. Empowering the citizens could be regarded as the chief objective of e-democracy, only more effective than traditional democratic processes and policies emerging out of these processes. Now I will analyse whether 'citizen engagement',—a critical aspect of inclusive smart city building process—promotes e-democracy.

4.2 Citizen Engagement

Public participation is not new to city planning and development. Sherry Arnstein is the most cited and celebrated author of the classic paper titled 'A Ladder of Citizen Participation'. As one moves from bottom to top and traverses the ladder upwards, the degree of participation increases leading to the top of ladder called 'citizen control' where "have-not citizens obtain the majority of decision-making seats, or full managerial power" [5]. More recently more complex understanding of public participation has been developed by Fung [6] who argues that for effective public participation we need to focus on the three dimension of participation. First dimension is a clear understanding of the participants who participate in a decision making situation implying whether participation is voluntary to all citizens or by invitation for selected elite. Second dimension of participation is about the nature of participation i.e. whether participatory processes are discursive, deliberative or only informing. Third dimension of public participation is how what gets decided through participatory processes (information receiving, discussions, and deliberations) links up with policy making and implementation.

At about the same time a well-known planning scholar Sandercock [7: 437] noted that public participation in planning has become critically important due to the following reasons:

- The crisis of the expert knowledge
- Failure of the city-building professionals
- Crisis of the representative democracy
- Issues of the neglected and marginalized citizens
- Recognition of difference
- Politics of difference.

Inness and Booher have argued that involvement of public in planning could achieve a number of purposes. First, decision makers want to find out public's preferences that can play a part in their decisions. Second, public participation can improve decisions by incorporating citizens' local knowledge into the calculus. Third, public input can also advance fairness and justice. Fourth, participation is used to secure legitimacy for public decisions. Fifth, participation is done because the law requires it [8: 422–423].

Complexities of effective citizen participation have been highlighted ever since challenges to rational planning began to emerge in 1960s. "The idea of citizen participation is a little like eating spinach: no one is against it in principle because it is good for you" [5]. Governments and power-holders drop this virtuosity when poor city dwellers begin to demand redistribution of decision making powers and the myth of participation explodes. In spite of the fact that it does not take much for public participation to turn into pure tokenism, "public participation is usually considered an unalloyed good. This emphasis on the inherent desirability of public involvement is part of a tradition which seeks to 'open up' planning processes to democratic scrutiny and to expand the scope of public involvement as an integral part of improvements in policy delivery" [9]. They also highlight two rationalizations of public participation in environmental planning. First rationalization is about "the democratic right to be involved in the public policy process and the importance of all barriers to such involvement being reduced or withdrawn. The emphasis here is on enabling access to the policy process, encouraging the take-up of that access and ensuring that such participation makes a difference to policy outcomes. The policy process is seen as a locus for the articulation of values and preferences on policy options, and public participation is a means of bringing the pattern of values and preferences represented within the policy process closer to that existing within society as a whole" [9]. "The second rationalisation focuses on the effectiveness of policy delivery and considers how public participation can assist in producing a 'better' policy outcome" [9].

Whatever its rationalizations, public participation implies 'efforts to increase public input, oriented primarily to the content of programs and policies'. Inclusion means 'continuously creating a community involved in coproducing processes, policies, and programs for defining and addressing public issues'. Both are dimensions of 'public engagement' [10: 272]. Public engagement therefore embraces both participation as well as inclusion widening the scope of peoples' involvement in planning and development of cities. Here public engagement is viewed as transformative and empowering in terms of its impact on policy making and redistributive in nature, specifically for those located at the margins of decision making regimes. But even before the term citizen engagement became commonplace "participation without redistribution of power is an empty and frustrating process for the powerless. It allows the power holders to claim that all sides were considered, but makes it possible for only some of those sides to benefit. It maintains the status quo" [5].

Today both in city planning and administration, we have moved beyond public participation and engagement because policy making and implementation is seen as a collaborative work where "the coproduction approach assumes that service users and their communities can—and often should—be part of service planning and delivery" [11] (see [12] for similar arguments in city planning). Terming coproduction as 'we government' Dennis Linders proposes three dimensions of coproduction—citizen sourcing, government as a platform, and do-it-yourself government. I take ideas of [11], Quick and Feldman [10] and Linders [13] together and view public engagement as a situation where policy planning and implementation processes, institutions, and outcomes are coproduced by public agencies and local communities.

Public engagement is centered on ten core principles. These include begin by listening; attend to people's leading concerns; reach beyond the "usual suspects"; frame issues for deliberation; provide the right type and amount of information at the right time; help people move beyond wishful thinking; expect obstacles and resistances; create multiple, varied opportunities for deliberation and dialogue; respond thoughtfully and conscientiously to the public's involvement; and build long-term capacity as you go [14] (see Table 4.1).

4.2.1 Smart City Led Public Engagement in India

Public participation in planning in the Indian cities is provided through state mandated planning laws. Generally statutory public participation takes place only after a draft of a city plan is prepared by a planning agency without much involvement of the citizens. Citizens come into view only when a draft planning agenda is already set by a planning agency largely led by politicians, bureaucrats and planners. But there are few exceptions to this trend. For example, the state of Goa involves public in planning even before drafts of various kinds of plans are prepared. Kerala similarly engages public in the plan making processes prior to the preparation of drafts of development plans. Whatever may be the degree of influence the citizens are able to exert; public participation in plan making processes has been designed as a city wide phenomenon covering all citizens. It is a different

Table 4.1 Ten core principles of public engagement

S. no.	Core principles of public engagement	Explanation of the principle
1.	Begin by listening	Effective listening involves being attentive to peoples' concerns, aspirations, knowledge base, misperceptions, and initial solutions
2.	Attend to people's leading concerns	Leaders and policy makers should acknowledgment and address peoples' concerns even when they do not match with their own concerns
3.	Reach beyond the "usual suspects"	Find ways to include or represent those who are traditionally excluded from decision making processes
4.	Frame issues for deliberation	Use language that people can understand to address public concerns, keep the technical details to a minimum, and present people with information for weighing different options
5.	Provide the right type and amount of information at the right time	Present to people carefully selected non-partisan information for effective deliberations and do not overload people with information. Overload of information overwhelms and disempowers people
6.	Help people move beyond wishful thinking	Trade-offs should be highlighted so that people can look for multiple ways of achieving certain outcomes
7.	Expect obstacles and resistances	Multiple opportunities should be provided for absorbing information about the trade-offs and working through problems
8.	Create multiple, varied opportunities for deliberation and dialogue	Inclusivity is central to dialogue and deliberations. Multiple opportunities for people to learn, talk, think and act about problems contributes to inclusivity
9.	Respond thoughtfully and conscientiously to the public's involvement	Public engagement is incomplete with policy makers responding adequately to peoples' involvement including letting public know how their inputs will be used in policy making to foster respect and trust
10.	Build long-term capacity	Public engagement is more than public problem solving. It is about better reach out, working collaboratively in order to build common ground

Source Center for Advancement of Public Engagement [14: 3–5]

matter that given the present design of institutionalized participation, a minuscule part of the citizenry participates. One reason for non-participation and lack of interest in plan making activities is the lack of trust citizens have in the policy makers where people feel participation may not be able to bring about transformative and empowering changes. Citizens' indifference to planning get exacerbated when they see plans finalised and approved without citizens even getting a chance to know how their inputs were considered before excluding them from approved plans. So the existing arrangement of public participation in the draft of a city plan is unlike public engagement as defined above.

Citizen engagement in the preparation of Smart Cities Proposals under the Smart Cities Mission further restricts public participation and this has various dimensions which could be discussed as 'citizen exclusion points' from the public debate and deliberations for influencing and delivery of public planning policies. Under the Smart Cities Mission citywide (pan-city) plans are given lesser importance as indicated through the award of scores than the Strategic Action Plans which include three type of projects—retrofitting, renewal and greenfield development [15]. As C. Rammanohar Reddy has shown: "The focus is almost entirely on the area-based plan. In fact as much as 81% of the total outlay in the proposals for the 33 cities is on area-based plans" [16]. This kind of citizen engagement excludes a vast majority of the citizens from the participatory processes since projects are meant for specific geographical areas, and it is increasingly becoming clear that all people living in a smart city project area may also not get a chance to participate in an area plan. Even when citizen participation is solicited for a pan city proposal citizen engagement is "achieved through citizen consultations, including active participation of groups of people, such as Resident Welfare Associations, Tax Payers Associations, Senior Citizens and Slum Dwellers Associations" [15]. This indicates that citizens would be participating indirectly through these organizations in identifying their needs and aspirations leading to probable misrepresentation of citizens' desires.

Misrepresentation has resulted in much avoidable controversy that is generated through a number of smart city proposals in various cities. For example, in Dehradun people have been agitating because the state government plans to build a smart city over the existing tea estate measuring 1900 acres. Echoing the sentiments of the public at large after Dehradun could not make it to the second list of smart cities, Mahesh Bhandari, President, Doon Residents' Welfare Front noted: "The authorities did not do their homework and failed to realise that the citizens want retrofitting for the whole city and not just of few parts. The proposal could have been made more innovative and eco-friendly like other cities but that was not done" [17]. In new Bhopal 333 acre area in Shivaji Nagar is slated to be redeveloped for ultra-modern apartments and business centres. People have been asking why develop Shivaji Nagar which is already better developed than several other areas in the city [18]. Citizens appear to view the smart city proposals as misplaced; not representative of the needs and aspirations of the general public.

Another dimension of 'citizen exclusion points' is the setting up of the Special Purpose Vehicle. The Smart Cities Mission will be implemented by an organizational arrangement dubbed as the Special Purpose Vehicle. The SPV is a limited company incorporated under the Companies Act 2013, and will be promoted by equity share holding of 50% each by the state government and an urban local body. Private sector and financial institutions could hold equity shares in the SPV if equal share of equity shares between the state government and urban local body are maintained. The SPV will perform several functions. "The SPV will plan, appraise, approve, release funds, implement, manage, operate, monitor and evaluate the Smart City development projects. Each Smart City will have a SPV which will be headed by a full time CEO and have nominees of Central Government, State Government and ULB on its Board" [15: 12 and 35–36]. The SPV must ensure that it is a credit worthy organization in the market, which is able to raise resources from the market. The SPV is allowed to implement projects through joint ventures, subsidiaries, public-private partnerships, turnkey contracts, etc.

This kind of organizational arrangement is highly focussed on efficiency, productivity and profitability, which on its own terms is not to be derided. But after 74th amendment to the Constitution of India enforced in 1993, elected municipalities have a major role to play in city building and maintenance including land use planning as listed in the Twelfth Schedule to the Constitution. However, in the Smart Cities Mission, the elected urban local government appears to have little or no role in the implementation of the Smart Cities Mission except some participation of selected mayors and municipal commissioners or chief executive of ULBs in the State Level High Powered Steering Committee and very limited participation of some ULBs in the Board of Directors of the SPV. Central government expects that "the rights and obligations of the municipal councils with respect to the Smart City project [are delegated] to the SPV" [15]. If these organizational arrangements were to be implemented, it will create a large '*democratic deficit*' and even larger democratic deficit in cities like Delhi where municipalities directly report to central government and not to the state government.

On the other hand, national level Apex Committee headed by the Secretary, MoUD with representatives from related ministries and organisations will approve proposals for the Smart Cities Mission, monitor their progress and release funds. This Committee will consist of Secretary, Housing and Poverty Alleviation, Secretary (Expenditure), Joint Secretary, Finance, MoUD, Director National Institute of Urban Affairs, Chief Planner, Town and Country Planning, select Principal Secretaries of States and select CEOs of SPVs as members with Mission Director as the Member Secretary. Further, the Representatives of organizations such as UN Habitat, World Bank, TERI, Centre for Development of Advanced Computing, Centre for Smart Cities Bangalore and other bilateral and multilateral agencies and urban planning experts could be invited [15]. On similar lines state level committees headed by respective state Chief Secretaries would be established with the exception that such committees would also include selected mayors and municipal commissioners or chief executive of ULBs, and heads of concerned line departments. This bureaucratic arrangement at central and state level may appear necessary and interventionist, but it is appropriate for disbursal of funds and monitoring of projects under the Smart Cities Mission. In this backdrop, now I turn to the case of New Delhi Municipal Council.

4.3 Public Engagement in the New Delhi Municipal Council

New Delhi Municipal Council (NDMC) is governed by appointed civil servants who are accountable to central government. At the very basic level public engagement would warrant that eligible voters are able to elect their own representatives and are empowered to elect new representatives periodically if they do not fulfil the aspirations of the citizens. Elaborate provisions for elected urban local government are made through the seventy fourth amendment to the Constitution of India. But this constitutional amendment has not been made operational in the NDMC area. Next level of public engagement emerges from the provisions to prepare local area plans at the scale of a ward whereby people are entitled to frame planning policies for their own neighbourhoods. Public engagement at the local area plan level has been anything but effective because organizations responsible for the preparation of these plans treated these planning exercises as consultancy projects rather than socially transformative projects.

New Delhi Municipal council in the Smart City Proposal (Smart City Challenge—Stage 2) has adopted a variety of citizen engagement methods to involve citizens and residents in the area (see Table 4.2). These could be classified into to two categories—high-tech engagement and low-tech engagement. High-tech engagements ranged from online consultations, multimedia presentations (Mygov, Twitter,

Table 4.2 Different methods of citizens' engagement by NDMC

S. no.	Methods of citizen engagement	Involvement of groups
1.	Face to face unstructured consultations	Interactions with 1056 citizens that included RWAs, students, slum dwellers, traders associations, government employees, diplomats, and vendors
2.	Face to face structured consultations	RWAs, students, slum dwellers, traders associations, government employees, diplomats, and vendors also consulted using a questionnaire administered through Tablets and fill-in forms
3.	Online consultation	Discussion forums, polling, online questionnaires, essay writing competitions and logo design administered through MyGov.in, www.ndmc.gov.in, www.smartnewdelhi.in, twitter and the Facebook
4.	Nukkad Natak	21 street plays in all major markets and slums, reaching out to large population, for effective participation in the NDMC smart city consultation process and also to spread information on key aspects of smart city initiative
5.	Multimedia presentations	Used for generating awareness with RWAs, students, traders associations, diplomats, NDMC council members and government employees
6.	Mobile polling	10,800,000 touch points with citizens through SMS campaign with 6700 responses

Source New Delhi Municipal Council [2, 19: 20]

Facebook, and its other websites) and mobile polling. Low-tech engagements included unstructured face to face interactions of officials and various groups, and Nukkad Natak or street plays. Impact of both high-tech and low-tech citizen engagements are unclear from the text of the Smart City Proposal (Smart City Challenge—Stage 2) of the NDMC.

What is clear about citizen engagement in the Smart City Proposal (Smart City Challenge—Stage 2) is the fact that the NDMC commits itself to the maximum public participation in setting up 'the vision and goals'. NMDC claims to "achieve informed high quality consultations, ensuring inclusion of diverse groups and communities in the city" [2: 19]. A number of key stakeholders have been involved in the process of preparation of the Smart City Proposal for the NDMC area. These include hotel and traders association; diplomats—embassies and high commissions; economically weaker sections and slum dwellers; women; elected representatives; resident welfare associations; students from the schools and colleges located in the NDMC area; and the public at large. It is asserted that 95% of the residents have been contacted through resident welfare associations resulting in a total of 150 h of face to face consultations. Further 10,800,000 mobile SMS (also called touchpoints) were sent to 600,000 people but only 6700 plus responses received [2: 19].

A number of key points emerge about the citizen engagement exercises carried out by the NDMC. First, even on an easy to access electronic platform like sending SMS through mobiles, only 0.062% responses were received by the NDMC, which is very low rate of citizens' responses by any standards. Second, face to face consultations in NDMC were carried out through resident welfare associations, which are regarded as representatives of the middle classes and elite groups, rather than low income slum dwellers who are treated as a nuisance (see [20, 21]). What is strange in the Stage 2 plan is the fact that NDMC has proposed a system of volunteers to ensure citizen engagement on a regular basis ignoring completely that people in the present circumstances are unwilling to engage with the municipality in its smart city endeavors.

Further the NDMC claims that it has constructed its 24 smart city features on the basis of the feedback provided by the residents of the areas through the processes of citizen engagement. In other words, citizen engagement has helped build vision and mission for the city, which in turn has helped NDMC in identifying 24 smart city features. These 24 smart city features of the NDMC Smart City Proposal reiterate normal problems of any city with additional label of 'smart'.

Another aspect of citizen engagement in the Smart Cities Mission is the division of marks in the second stage. Marks are awarded out of a total of 100 which are further sub-divided into three parts—city level (30), area based development (55), and pan-city solutions (15). Out of a total of 100 marks, a maximum of 16 marks could be awarded for citizen engagement. Marks decrease as one moves from city level (10), area based development (5), and pan-city solutions (1) [2: 6]. From this sort of marks distribution, it is apparent that more marks are awarded for citizen participation for city level planning and fewer marks are awarded for project level planning and engagement.

Another aspect pertains to the Special Purpose Vehicle being set up for implementation of the Smart City Proposal in NDMC area. Although 74th amendment to the Constitution of India has facilitated decentralized urban local government throughout the country, political devolution with executive powers in the hands of elected representatives have not yet truly happened. Urban local bodies still implement policies framed at central and state levels, and do not really make policies, a critical element of any autonomous urban local government. Financial dependence of three municipalities on Government of Delhi and resulting conflict between Delhi government and the municipalities highlights this point well. A good beginning has been made by decentralization of urban local government throughout the country, but attempts are required to be made to make these bodies autonomous with clear leadership roles assigned to locally elected politicians. A system of directly elected mayors entitled to appointing groups of professionals is required for efficiently running cities and towns. NDMC is already run by a committee of appointed bureaucrats under the oversight of Government of India. After the implementation of the Smart City Mission policies, further corporatization will manifest itself in the form of the Special Purpose Vehicle in the NDMC.

4.4 Three Case Studies of Citizen Engagement

This section will explore the cases of citizen engagement—the heart of democratic practice. The first case involves empirical work in three distinct areas of NDMC—a middle income group area, a small squatter settlement, and a large slum—carried out by the author in 2016.

4.4.1 Safdarjang Squatter Settlement

Safdarjang with a pin code 110003 would be normally regarded as a prestigious address. But not quite so for some people who live underneath the Safdarjang Flyover—named after the local airport located close by. Railway line passes below the flyover forming a boundary while roads on all other sides envelop the area. Approximately 80 huts made of polythene and other precarious materials provide shelter for estimated 450 persons (see Fig. 4.2).

It is well into afternoon when we arrive in area. Children are playing around and younger folk are sitting in groups, some playing cards, others simply looking into the sky rather hopelessly and aimlessly. It is clearly on display that at least in this area youth energy is being wasted (Fig. 4.3). Unaware of the risks, youth spend their time sitting on the railway lines. This spare time represent no leisure, it is a tragedy of unimaginable proportions.

Fig. 4.2 Safdarjang squatter settlement. *Source* Author (November 2016)

Fig. 4.3 Gazing aimlessly into nowhere. *Source* Author (November 2016)

As we sit down to talk with the residents of this squatter, we see few menfolk are carrying some wares for sale. Most of them sell what is locally known as *nada* and *rassi* (rope). By selling these and other items, one person is able to earn about Rs. 300 daily, not enough to run a family of five to six members. As we continue our discussion with voluntarily gathered menfolk, problems and issues begin to tumble one by one. Unemployment and housing remain the predominant concern of squatter dwellers. When I ask about their place of residence, they claim to have come from Sitapur in Uttar Pradesh in 1985. They report that they came to Delhi for the purposes of seeking employment and sadly have failed to get a decent employment here in Delhi. But when compared with their previous residence in Sitapur, they are largely satisfied.

We look for reasons for their satisfaction and we find that kids do indeed go to school in a nearby New Delhi Municipal Council school located two to three kilometres away near Race Course Road. Apart from education, there is no problem of water although quality of water is an issue before 8 a.m. after which clean water is supplied. However, this satisfaction turns into dismay as we talk about sanitation. No toilet is located inside the area. One toilet with 8 seats has been recently constructed by the NDMC and made available to the residents since the last one year. Open defecation was the norm before this. This free of cost community toilet is located outside the area along the road in front of the squatter settlement (also see Fig. 4.4). Residents claim that 20 toilet seats are needed but only 8 are provided causing major problem every morning. Even out of 8 only 4 seats are made available at a time by the private contractor to whom the NDMC has outsourced

Fig. 4.4 Sanitation facility outside the squatter settlement. *Source* Author (November 2016)

this facility. The contractor keeps doors shut on other 4 toilet seats. Location of community toilet makes it unsafe for use by kids and women specifically during the night. Although it must also be said that available toilet seats were clean and water was available for flushing and cleaning. In addition power also remains one of the biggest problems.

Hesitantly I ask whether any NDMC officer visited this place in regard to smart city mission. Without wait all of them say "No one from the NDMC ever visits us. Occasionally they come to displace us from here. But we come back after they have left after destroying our huts and ruining our meagre belongings".

4.4.2 Veveka Nand Camp

A local road separates Veveka Nand Camp and the Austrian Embassy on Chandra Gupta Marg. On the other side NDMC Service Centre Building Maintenance Bapu Dham office is located. Some 40 years ago, people from West Bengal came to settle in what is now known as Veveka Nand Camp. At present approximately 1000 households or over 50,000 persons live in Vivekanand Camp. This is indeed a large slum located in the heart of NDMC area (see Fig. 4.5).

Sulabh International—a social service organization is tasked with the provision of basic sanitation and bathing facilities in one large complex. About 20 toilet seats are constructed in this community toilet block. Condition of community toilets

Fig. 4.5 Veveka Nand Camp. *Source* Author (November 2016)

Fig. 4.6 Open defecation in front of Sulabh Sauchalaya. *Source* Author (November 2016)

during our visit was very bad, toilets being unclean full of human excreta. We also found few kids defecating in the open into the drain constructed along a main road (Fig. 4.6). Tankers are used to provide water to the slum residents. This slum appears to be more stable in comparison to Safdarjang Squatter Settlement. Power is made available to slum dwellers.

This area is dominated by Muslims, who are predominantly engaged in *kabadi* work. Entire slum is filled with the materials bought from rag pickers. We also saw several families cooking food by burning wood. But some families also have access to cooking gas. Kids go the NDMC School located at Malcha Marg. Large solid waste bins are placed before the slum for collection of garbage from the area (see Fig. 4.7).

4.4.3 B.K. Dutt Colony

Inadvertently as I enter into the Anjuman-e-Haideri Dargah, I am welcomed by two old men sitting underneath a large tree located besides a large car parking area. Soon we are introduced to a young man who is the care taker of the Dargah. A part of Dargah is being used as a residential area for about 20 families. I am fully engrossed with my quarries with Mr. Zahrul Hasan, Manager of the Dargah, when I

Fig. 4.7 Kabadi spread all over the slum. *Source* Author (November 2016)

am disrupted by a group of people who appear to be the residents of the B.K. Dutt Colony (rechristened after the name of a freedom fighter), which was previously known as Karbala, a Muslim religious gathering, and also a city in Iraq. Earlier the place was also known as Ali Ganj.

After my discussions with the Manager of the Anjuman-e-Haideri Dargah, I move towards the group that disrupted my conversation with the Manager. The group informed that they rehabilitated in 1950 in Karbala. All of them being Hindu came from Pakistan (specifically from Sargodha and Multan) after the partition of the country in 1947. Today 1038 households live in this middle income area measuring about 4 acre (see Figs. 4.8 and 4.9).

A bitter fight is being fought between the management of the Dargah and the residents of B.K. Dutt Colony. Tazia (a procession) from Jama Masjid in Old Delhi was to the Anjuman-e-Haideri Dargah was undertaken on 21 November 2016. Both sides complained that other side violently interferes during such events. Muslims claim that Hindus of B.K. Dutt Colony throw stones at the procession while Hindus complain that theft increases during such occasions. Basically it is a fight for land and not a communal tension between two communities. The matter is pending in a court in Delhi.

President of the Citizen Welfare Forum B.K. Dutt Colony RWA (Registered) Mr. Harbans Dunkall gave me a letter of 18 May 2015 that he wrote to the Chief engineer (Civil) of the NDMC. In this letter 8 concerns of the area have been raised and action is sought. I reproduce the contents of the letter below:

Fig. 4.8 Anjuman-e-Haideri Dargah, Karbala. *Source* Author (November 2016)

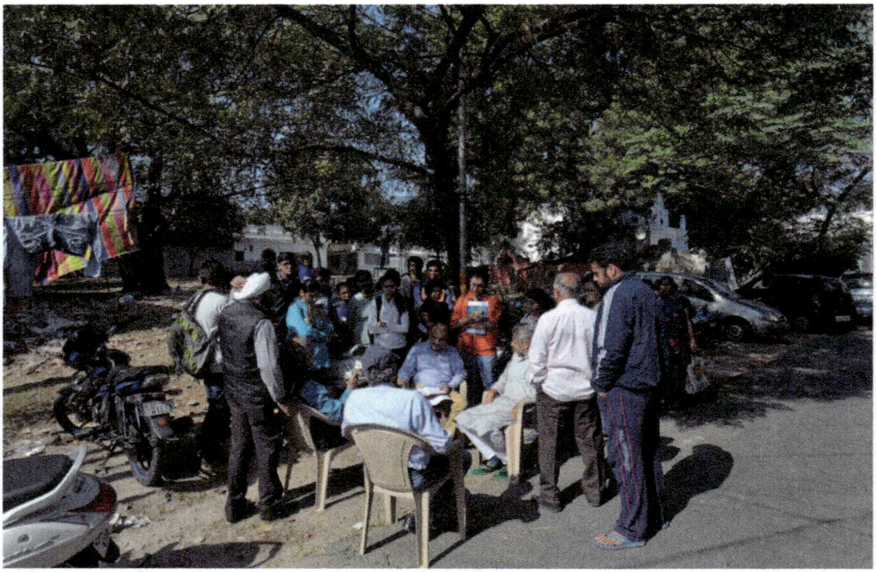

Fig. 4.9 A discussion with RWA members of B.K. Dutt colony, NDMC. *Source* Author (November 2016)

- Water harvesting system is urgently needed to collect the rain water as total rainy water is waste in the drainage. It will not only keep the colony neat and clean but also raise the underground water level.
- The drainage system of the colony is not proper. Water logging can be seen during rainy season, which is approaching i.e. June end. Hence, it is requested that drains, etc. be cleaned before rainy season starts.
- The water pipelines in the colony are very old and out served their life and require immediate replacement.
- Public conveniences both for men and women are required at different locations.
- Porta cabin is needed for RWA members for smooth functioning.
- Boards for writing names of Executive members and rules and regulations of the society are to be made.
- Colony layout maps and direction boards are required to show exact location of houses in the colony.
- Roads of the colony are in bad shape and require immediate repair.

But it should be mentioned that internal roads in the colony has been freshly repaved by the NDMC (see Fig. 4.10).

When asked whether anyone from NDMC came to discuss with you about the smart city proposal, which deals with many of the above mentioned issues, the group informed that three meetings were held with the NDMC officers in which RWA was also represented. During these meetings we have highlighted our problems pertaining to deteriorating condition of internal roads, security, street lighting, water (supplied only for two hours), rainwater harvesting. But no action

Fig. 4.10 Roads in B.K. Dutt colony. *Source* Author (November 2016)

Fig. 4.11 Entangled wires in B.K. Dutt colony. *Source* Author (November 2016)

has been taken so far. Security appears to the biggest challenge the area is facing. For example, in one instance 128 car batteries were stolen from the area. So for improving security within the area, the RWA has recently installed 168 CCTV cameras out of their own funds that are collected from all the households at the rate of Rs. 500 per household per quarter. Water meters are yet to be installed in the area. Residents are charged a fixed rate of Rs. 49 per month per household, which is too little compared with meter based water rates. The smart city proposal of the NDMC has not yet arrived in B.K. Dutt Colony as can also be seen from interlaced wires (see Fig. 4.11).

4.5 Conclusions

The three case studies discussed above show that democracy and e-democracy are on the decline due to planning processes adopted under the Smart City Proposal of NDMC. Citizen engagement has been restricted to machine led engagement. As apprehended in our previous work, these cases show that smart cities project incapacitates e-democracy particularly for the vulnerable groups because of its neoliberal leanings. The Smart City Proposal of NDMC does not really affect 10, 000 households living in slums in the NDMC area except that it reinforces their exclusion from the planning processes. However, the "problem of squatters" is recognized in the Smart City Proposal [19]. But as the three cases show, it appears that "smart solutions may not touch the lives of the urban poor as the poor do not have access to smart technologies and even if they have the ability to use smart technologies, they simply will not be able to afford such solutions" [1]. This is largely because smart cities are meant to obtain other purposes: "Economic growth is the key objective, technology is the key driver, knowledge production is the key resource, and high quality of life is the key promise" [1] but only for the few.

Citizen engagement in the smart city proposals is constrained by the very design of the Mission resulting in a number of 'citizen exclusion points'. The first 'citizen exclusion point' that militates against citizen engagement is the Mission's main focus on small area projects rather than pan city planning. Project cantered planning excludes much of the area and population of a city from engaging with the local state. The second 'citizen exclusion point' is that citizen engagement takes place through groups like resident welfare associations, which clearly work against the interests of the urban poor further disengaging them from the state. The third 'citizen exclusion point' is that the Mission's implementation would take place through a corporate body like the Special Purpose Vehicle whose design is inimical to participation and citizen engagement. The fourth 'citizen exclusion point' relates to the Mission's technology focus that prevents the urban poor from public engagement. All the four citizen exclusion points indicate efficiency has won over equity when both could go together.

References

1. Kumar A, Sarkar PK (2016) Making Delhi a smart city: economic buoyancy with spatial justice. In: TM Vinod Kumar (ed) Smart economy in smart cities, Springer, Singapore
2. New Delhi Municipal Council (2016a) Smart city proposal (Smart city challenge—Stage 2), New Delhi Municipal Council, New Delhi
3. Marche S, McNiven JD (2003) E-government and e-governance: the future isn't what it used to be. Can J Adm Sci 20(1):74–86
4. Whyte A, Macintosh A (2002) Analysis and evaluation of e-consultations. e-Service J 2(1):9–34
5. Arnstein S (1969) A ladder of citizen participation. J Am Inst Planners 35(4):216–224
6. Fung A (2006) Varieties of participation in complex governance. Public Adm Rev 66:66–75
7. Sandercock L (2005) The democratization of planning: elusive or illusory? Editorial. Plann Theor Pract 6(4):437–441
8. Innes JE, Booher DE (2004) Reframing public participation: strategies for the 21st century. Plann Theor Pract 5(4):419–436
9. Rydin Y, Pennington M (2000) Public participation and local environmental planning: the collective action problem and the potential of social capital. Local Environ 5(2):153–169
10. Quick KS, Feldman MS (2011) Distinguishing participation and inclusion. J Plann Educ Res 31(3):272–290
11. Bovaird T (2007) Beyond engagement and participation: user and community coproduction of public services. Public Adm Rev 67:846–860
12. Healey P (1997) Collaborative planning: shaping places in fragmented societies. Macmillan Press, London
13. Linders D (2012) From e-government to we-government: defining a typology for citizen coproduction in the age of social media. Gov Inf Q 29:446–454

14. Center for Advancement of Public Engagement (2008) Public engagement: a primer from public agenda. Public Agenda (1):1–10
15. Ministry of Urban Development (2015) Smart cities mission—transform-nation, mission statement and guidelines. Ministry of Urban Development, New Delhi
16. Reddy CR (2016) To locate the scam, use GPS. Outlook LVI(33):80–81
17. Budhwar Y (2016) Dehradun has yet again failed to make it to the smart city list. Time of India (Dehradun City), 24 May 2016
18. Ghose D (2016) Why Shivaji Nagar, why not a less developed place? Ask Bhopal residents. Firstpost. 30 Jan 2016
19. New Delhi Municipal Council (2016b) https://www.ndmc.gov.in/smart_city_proposal.aspx. NDMC, New Delhi
20. Ghertner AD (2011) Rule by aesthetics: world-class city making in Delhi. In: Roy Ananya, Ong Aihwa (eds) Worlding cities: asian experiments and the art of being global. Blackwell, Oxford
21. Ghertner AD (2011b) Gentrifying the state, gentrifying participation: elite governance programs in Delhi. Int J Urban Reg Res 35(3): 504–532

Chapter 5
E-Engagement in Hong Kong

Sujata S. Govada, Frank Wong, Timothy Rodgers, Ho Yan Claudia Chow and Krista Busa

Abstract Citizen engagement has taken many centuries to evolve from its roots during the Greek period, developing along with the concept of democracy to its current form as it is practiced today. Participation in the planning process emerged in Great Britain in the 1940s, and began to become institutionalized in 1960s. This participation increased in popularity during the 1980s and 1990s as it was seen to result in better decision making, and eventually became known as participatory planning. Citizen involvement continues to evolve and varies with specific context, but is generally aimed to influence the policies and decision making process that shapes the communities and cities of those citizens. More recently, the tools of public engagement are transforming considerably with the advancement and use of ICTs during the digital age, which has led to the emergence of internet-enabled online public engagement, also known as 'e-engagement'. Planning in Hong Kong has evolved rapidly over the last two decades, transforming from a top down decision making process to a more participatory process involving active citizen engagement that has become more pronounced since the handover from British Rule in 1997. Since then, with a renewed sense of ownership, the citizens of Hong Kong have become more interested and involved with the planning and the political processes. However, through more public engagement and increased participatory planning, there can be seen positive changes in various urban processes in Hong Kong such as urban renewal, heritage conservation, harbourfront development. In Hong Kong, information is readily available and accessible both in English and Chinese and in printed and in online formats.

S.S. Govada (✉) · F. Wong · T. Rodgers · H.Y.C. Chow · K. Busa
UDP International, Institute for Sustainable Urbanisation, Chinese University of Hong Kong, Suite 9B, Queens Centre, 58 Queens Road, Wan Chai, Hong Kong
e-mail: sujata@udpcltd.com

F. Wong
e-mail: frank@udpcltd.com

H.Y.C. Chow
e-mail: claudia@udpcltd.com

K. Busa
e-mail: krista@udpcltd.com

T.M. Vinod Kumar (ed.), *E-Democracy for Smart Cities*, Advances in 21st Century Human Settlements, DOI 10.1007/978-981-10-4035-1_5

In addition to participatory processes, the use of e-engagement is becoming more prevalent for numerous development studies and projects in Hong Kong. The Kai Tak Development Case Study demonstrates that although the planning process has been underway for more than two decades, community involvement has been instrumental in its planning and development to date. There is now a considerable amount of e-engagement in the Hong Kong planning process and various tools are being used and developed to engage people in more creative ways during both formal and informal engagement processes. With the emergence of the digital age and through the use of ICT, e-engagement has become an integral part of citizen engagement. E-engagement has the potential to transform the public participatory process through specific events such as workshops and forums, and the continuous engagement of the public over the internet. ICT also creates the opportunity to be more open, transparent and inclusive by creating a platform for on-going engagement to develop greater trust between the community and NGOs with the public and private sector. While much progress has been made, e-engagement in Hong Kong can be improved further through the use of various tools and social media available to more effectively engage the wider community rather than only hearing the voices of the vocal minority. This will further remove barriers to the formation of public, private and community partnerships and will help develop Hong Kong into a more affordable, walkable, liveable, sustainable and resilient city for its citizens, as well as become a model for other cities in Asia and the world.

Keywords Citizen engagement · E-Engagement · Participatory · Planning process · Digital age · E-Engagement tools Hong Kong

5.1 Introduction

5.1.1 Evolution of Citizen Engagement

The roots of citizen participation can be traced back to ancient Greece where citizens gathered in Athens's central square to share their opinions and vote for public matters. Therefore, the concept of citizen participation began to develop together with the concept of democracy. Over time, the concept of participation has evolved with a continuously expanding definition. However, it has taken many centuries to develop into the bottom-up, early engagement approach to public participation we see it today.

The 1940s in Great Britain saw planners engage the public in reconstruction efforts after World War II, creating new techniques to communicate with people including mobilizing publicity, measuring public opinion, and organizing exhibitions and forums while experimenting with new visual strategies [1].

Citizen participation started to emerge early on with a more enlightened dialogue between citizens and public governments [2, 3]. Arnstein's critical view on existing public engagement worked as an enabler to change the traditional ways of how to engage the public in the decision-making process and increase citizen participation. However, striving for complete citizen control should not be the ultimate goal, rather it is the engagement of various stakeholders and creation of public, private, and community partnerships which ensures the development of better communities and cities that are more walkable, liveable and sustainable.

In the western world, citizen engagement began to become institutionalized in the 1960s. During this time, political participation and civic participation were distinguished as two separated things. Political participation was defined as a voluntary and lawful action of an individual which aimed to influence functioning government officials and their political choices, or the decisions of the political system on different levels. Alternatively, political participation could also be defined as citizen and non-governmental organizations (NGOs) active involvement in the policy-making processes to achieve a more favourable decision for a particular social group. It is characterized by citizen participation in referendums and elections, as well as involvement in policy-making processes.

The greatest expansion of political participation can be linked to the 1980s and 1990s. Civic engagement developed as the ability to cooperate and come together to achieve a common objective. The increasing number and role of voluntary associations displayed the ability of citizens to build a state of independent association to defend its own interests; an important factor in the development of civil society. Therefore, civic engagement was (and still is) considered as the level of civil individual self-organization where citizens gather in associations or interest groups to express their views on the political aspect of life.

It should be noted that up to this period participation in voluntary organizations and social activities were dealt with outside the structural framework of political participation. However, during the 1980s and 1990s boundaries between political participation and public participation started to weaken largely because of the increasing role of those interest groups.

The differentiation between civic engagement and political engagement began to blur with the formulation of participatory understanding, and there was no attempt to set strict boundaries between political participation and civic participation, as it was thought that the most important thing was participation itself [4]. Various forms of participation include: political participation (voting and membership of an organization on the local committee); participation in voluntary associations (trade unions, sports clubs, religious organizations, cooperatives, etc.); participation in informal activities (card games, reading newspapers, social networks or singing in the choir); and interpersonal trust.

Despite some existing opposition, civic participation is currently mostly studied together with political participation. This approach does not seek to structure the population activities by their very nature, indicating that there are no single, coherent individual engagement features, and highlights the direct participation meaning.

It should be considered that in the current context it is hard to distinguish political participation from civic participation. Often supposedly civic topics such as education, culture and other spheres of life become political issues.

Also, traditional political participation manifestations often result directly from civic participation.

The 1980s and 1990s proved a turning point for the urban planning field as well. Citizen involvement in the planning process started to gain more popularity, becoming known as Participatory Planning [5]. Based on the assumption that open participatory processes lead to better decisions, typically planners engaged directly with various stakeholders to gain support, build consensus, and identify acceptable solutions for the plan to be implemented.

5.1.1.1 Advancing Towards E-Engagement

In the last decade of the 20th century, we have witnessed the development of a new term 'E-democracy'. E-democracy is the use of information and communication technologies (ICT) to support the democratic decision-making processes. ICTs play a major role in organizing and informing citizens in various forms of civic engagement. ICTs are used to enhance active participation of citizens and to support the collaboration between actors for policy-making purposes within the political processes of all stages of governance [6, 7].

E-democracy can exist in various forms by which to engage and interact with citizens. Technologies such as electronic mailing lists, peer-to-peer networks, collaborative software, web pages, Internet forums and blogs have allowed the public to access information and engage in discussions regardless of their location. After the 2000s, social media sites, such as Facebook, Twitter, LinkedIn, Instagram, WordPress and others, started to play an increasingly important role in communicating with the public and giving them a platform to be heard. Citizens began to influence policy making within government, by communicating and making their voices heard and also working together to develop solutions for improving their own local communities.

Participatory Planning has been an integral part of community development in the Western world, with the idea of citizen engagement in the planning process remaining the same. However, the methods and tools of how to engage the public are evolving rapidly as a result of the coming of the digital age, this concept could be termed as 'E-engagement'. In 2007, forms of informal communication between local governments and citizens began to emerge in conjunction with the traditional formal and official forms of communication. The launch of a Facebook page by the Reykjavik Metropolitan Police (Iceland) on January 1, 2007 demonstrates a turning point in the communication between government bodies and its citizens. With this project, police attempted to enhance the perception of safety within the city, and improve interactions between citizens and the public body. Many cities, institutions and governments have followed their example, as it is now a common practice for public bodies to have social media accounts to disseminate information and engage citizens on current topics and issues. Equally important, these social media accounts have enabled and empowered citizens to ask questions and interact with the public government in an informal manner on issues they feel important or relevant.

With the continued proliferation of ICT and social media into the 2010s, E-engagement has become self-evident. So, over time, awareness of public participation and its forms have evolved, ever-expanding the concept. The manner of communication has changed from a highly bureaucratic, time consuming form of engagement to a more democratic, informal two-way interactive communication.

It is noted that E-engagement should not be distinguished from Civic Engagement, as public participation includes E-engagement, especially in this digital age. However, E-engagement only gained popularity at the end of 1990s and beginning of the 21st century. Furthermore, after the first decade of the digital age, public participation has transformed itself with other forms, models and techniques.

ICT creates the opportunity for a government to be more open, transparent and inclusive by creating open online communication and collaboration between the public sector, private sector, professionals, academia and the general public. The responsibility of gathering information and making decisions is shared between those with technological expertise and those who are the decision-makers. Greater public participation in the collaboration and development of ideas and policies makes decision-making more inclusive and democratic. Although ICT helps promote the idea of pluralism within a democracy, bringing new issues and perspectives, it also encourages the younger generation to participate in city and political issues, but may be limited in engaging the elderly.

Technological tools have been increasingly available over the years, emerging with the internet since mid-90s, and by the early 2000s almost everything has its own website to disseminate information and interact with users. Some key tools that have had a large impact on public and private decision-making processes include:

- Data analytics
- Data visualization
- Simulations such as agent-based social simulation
- Open government—document digitisation
- Internet
- Social media

Civic engagement plays a more significant role after the digital age emerged. With the discovery of the internet and its increased speed, accessibility and availability, it has further helped develop and improve the idea of how important it is to engage citizens in decision making process. Although the aim of engagement remains the same, the channels and tools of how to engage citizens have changed completely as ICT changed the way people, communities and cities work and behave.

ICT plays a major role in organizing data and providing access to information and real time displays informing citizens during various forms of civic engagement. ICTs are used to enhance interaction and active participation of citizens and to support the collaboration between actors for policy-making purposes within the political processes of all stages of governance [6, 7].

ICT changed the understanding of an open public government principle, which falls under the freedom of information principle. Due to ICT, it is finally possible to make the governance process more transparent and easily accessible by the public during planning processes. Digitizing and publishing public documents starting from land use plans to the minutes of last committee meetings on city web pages has ensured that the information is easily accessible by the public.

It also changed the whole communication pattern as it facilitates more connections and allows interaction with many people in a very short amount of time. It started with something simple such as an email and developed into various ways of easy communication such as mobile apps. ICT made the public government one click away from it's citizens and allowed the world to become more efficient by saving time and administrative resources. ICT also has the ability to cut down traditional bureaucracy, as in the case of some public governments that have decided to alter their bureaucratic structure by cutting down several layers while digitizing the public documents, creating various communication channels, and offering different public services with forms that are accessible online.

5.2 Public Participatory Processes in Hong Kong

5.2.1 Evolution of Public Participation in Hong Kong

Hong Kong is a compact and dense vertical city with a population of over 7 million residing in a land area of just over 1104 km^2. Three decades ago there was little community engagement in Hong Kong, with a more top-down planning approach taking precedent. In more recent years, however, a recognition of the importance that community engagement plays in the planning process is prominent, yet still evolving.

Following the handover of Hong Kong from British rule in 1997, there has been a renewed sense of ownership among Hong Kong people of their city, as up until then it was a borrowed place on borrowed time. This became more evident in 2006 during the demolition of the Star Ferry Pier in Central to accommodate land reclamation along the Harbourfront. The demolition was met with public outcry and demonstrations as it was evident that the Star Ferry Pier held significant cultural and historical value to the citizens of Hong Kong. However, having previously moved the Star Ferry Pier in Central once before back in 1957 to accommodate land reclamation, the Hong Kong Government did not expect the controversy it received from the public as they believed they had adequately indicated in the plans that the Star Ferry Pier in Central would need to be relocated. The backlash emerged more from the lack of public consultation on plans for the demolition, relocation or preservation of the pier, and blocking of motions from the Legislative Council to suspend demolition pending further consultation.

There is also a growing dissatisfaction among local people with the urban renewal process in Hong Kong. This began with the founding of the Land Development Corporation (LDC) in 1989, a governmental body that acted similarly to a private developer, involved in the redevelopment of older buildings with greater development potential to be replaced with high end luxury developments. The Urban Renewal Authority was formed to succeed the LDC with more resumption powers and a four pronged approach of Redevelopment, Revitalization, Rehabilitation and Preservation. However, the dissatisfaction stemmed from the LDCs predominant focus on redevelopment while ignoring heritage conservation, leading to the transformation of the older historical urban districts and the displacement of many local people in Hong Kong. This was when public engagement in Hong Kong really began to emerge. Civil society groups such as the Central and Western Concern Group and Designing Hong Kong have emerged to become active voices in shaping public policy in Hong Kong.

The reclamation of the Victoria Harbour has also been a contentious issue among the government, public, and conservation groups. Although reclamation of Victoria Harbour has occured numerous times in the past century, the last 20 years saw reclamation efforts become much more extensive. This led to the creation of the Society for the Protection of the Harbour in 1995 and the enactment of the Protection of the Harbour Ordinance in 1997 to formalise Victoria Harbour as a special public asset and natural heritage of Hong Kong. In 2003, the Citizen Envisioning @ Harbour (CE@H) group was formed by sixteen academic, professional and civic organizations who came together to approach the then Development Secretary to form a Harbour Roundtable to help improve the harbourfront planning process in Hong Kong. As a result of the discussions, a Harbourfront Enhancement Committee was established in 2004, and was later replaced by the Harbour Commission in 2010, which aimed to work together with the community and private sector to create a harbourfront that ensures the design, development and management are effectively integrated and consulted on by all parties and community stakeholders and is likely to become the Harbourfront Authority in the future. In 2012, discussions began about the proposed establishment of a Harbourfront Authority, "an independent statutory body with its own financial resources and the mission to press ahead with enhancement of the harbourfront in a holistic manner" [8] by "possessing an innovative mindset and a more flexible management approach to realise the long-term objective of a world-class harbourfront that is attractive, vibrant, accessible and sustainable" [8]. The development of a Harbourfront Authority process has incorporated a consultative approach to by holding two Public Engagement Exercises, in addition to multiple public forums and a mini-movie competition.

On the one hand, information is quite readily available in Hong Kong in English and Chinese both in printed and on-line formats, providing easy access for the local people. Public participation is possible through various channels with multi-stage community engagement, including E-engagement for various projects and policy initiatives, but can be a lengthy process usually lasting from one to two years to complete. However on the other hand, although there are ample opportunities for public participation, it is not too clear whether the public opinion and input have significant influence on the final decision. So while, E-engagement is prevalent and complements the direct engagement with the public, it is not always clear if there are results on the ground as it seems to take a long time for plans to be implemented.

In Hong Kong, all planning projects now require a public engagement process to be involved in various stages of the project from the initial planning stages, through to the construction and commencement stages. Public engagement is undertaken most typically during the envisioning stage, the planning stage and the final stage of the project as highlighted in the Kai Tak Development case study below.

5.3 Case Study of Public Engagement in Hong Kong

5.3.1 Kai Tak Development

Kai Tak is most commonly known for the old Kai Tak International Airport, in commission from the 1920s to 1998. It is located to the West of Kowloon Bay and Northwest of Kwun Tong, together forming the Southeast part of Kowloon District. The former Kai Tak Airport had a unique geographic location, reclaimed from the harbor, it was surrounded by tall buildings and rugged mountains, resulting in it being selected as the 6th most dangerous airport in the world in 2010 (Fig. 5.1).

Fig. 5.1 Kai Tak airplane. *Source* Kai Tai Planning Review: Public Consultation Digest (1)

The Kai Tak Development was previously known as the South East Kowloon Development back in 1991, and over the past two decades various development schemes have been proposed, with one of the most controversial discussion points regarding future reclamation of the harbour. The Hong Kong Government started to collect public views and considered public comments as important inputs towards the development. The planning process viewed the transition from no public participation at all to some public consultation as a major part of the development process. The public participation engagement successfully reduced proposed reclamation of the Kai Tak Development from about 299 hectares in 1998 to today's "zero reclamation".

5.3.1.1 History of Kai Tak Development

The planning of Kai Tak began in the early 1990s, before the closure of the Kai Tak Airport. The South East Kowloon Development (SEKD) planned to focus on the Kai Tak Airport site and proposed land reclamation at Kowloon Bay, the Kai Tak Approach and Channel, and the Kwun Tong Typhoon Shelter [9]. A wide range of proposed land use and redevelopment frameworks were carried out focusing on the development of housing, commercial, open space and high-tech industrial components. In 1993, the South East Kowloon Development Statement Study (SEKDS) proposed reclamation at the adjacent water bodies and to develop more specific planning principles for the Kai Tak Airport site, as well as to prepare an Outline Master Development Plan (OMDP) for the Feasibility Report for SEKD to further discuss the housing demand issue (Fig. 5.2).

The Feasibility Report was completed in 1998, advising to approach development in phases in order to integrate developments for the early development packages. The scheme was published in the statutory Outline Zoning Plans (OZP), however, more than 800 objections from the public were received due to the proposed reclamation at the nearby water bodies [10]. It was at this time that citizens initially raised the concern of the top-down planning development process, encouraging the government to hold public consultation exercises throughout the entire planning process instead of only publishing one-way information on the webpage to inform the public but not hear their views. Due to the significant amount of negative comments regarding the proposed reclamation, a revised development scheme was created and in 2001, and a Comprehensive Feasibility Study of the Revised Scheme of South East Kowloon Development was prepared to address the opposition from the public by revising Outline Concept Plan and reducing the extent of reclamation (Fig. 5.3).

From the South East Kowloon Development in 1993 to the Kai Tak Development in 2007 and the ongoing Energizing Kowloon East Office, Hong Kong citizens are successfully getting more involved in the planning process to comment on and influence future implementation through increased access to information, and availability of platforms to communicate and engage with all relevant stakeholders.

Fig. 5.2 Cover of South East Kowloon development statement. *Source* http://www.pland.gov.hk/pland_en/p_study/prog_s/sek_09/website_chib5_eng/pdf/SEKD_DC_1993_EXESUM.pdf

5.3.2 1993 South East Kowloon Development Statement Study

In 1991, the Executive Council of the Hong Kong Government had endorsed the Metroplan Selected Strategy, which claims, "to provide a framework within which both public and private sector agencies can formulate detailed plans and development programmes with the common aim of making our city a better place in which to live and work" [11]. One of their major recommendations was the reclamation projects in close proximity to the harbourfront areas. Then in 1993, the South East Kowloon Development Statement (SEKDS) Study commenced laying out more concrete and specific planning principles of the conceptual ideas from the Metroplan Selected Strategy. The SEKDS Study was then carried out in preparation for the reclamation and construction of about 300 hectares of adjoining areas of Kowloon Bay to create 580 hectares of potential land available for the new Kai Tak Development [9]. The Study provided a master landscape plan, frameworks of land-use transport, redevelopment needs of the existing urban areas, and the potential environmental impact resulting from development.

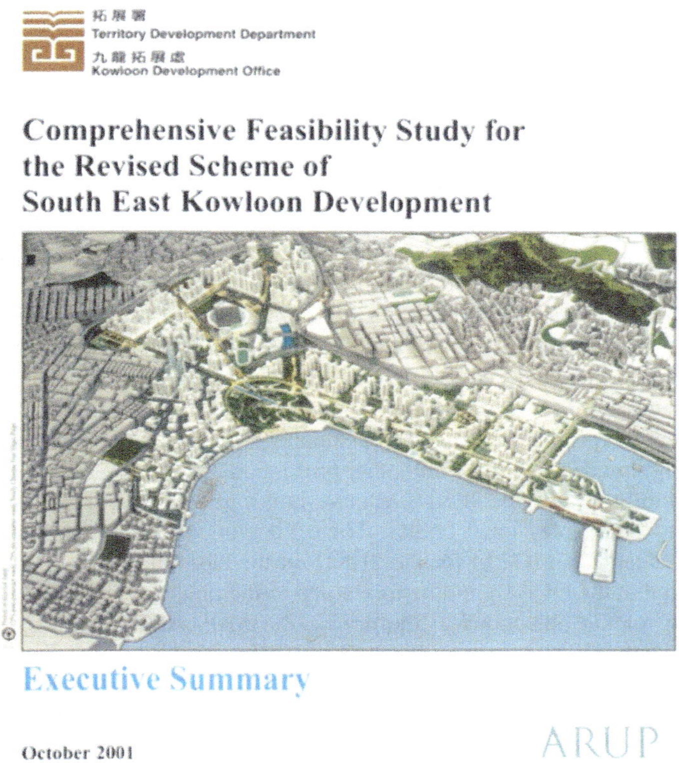

Fig. 5.3 Cover of comprehensive feasibility study for revised scheme of SKD. *Source* http://www.pland.gov.hk/pland_en/p_study/prog_s/sek_09/website_chib5_eng/pdf/SEKD_CFS_2001_EXESUM_ENG.pdf

The first stage of the Study covered baseline studies to analyse the existing conditions of the old Kai Tak Airport site, the potential reclamation area of Kowloon Bay, and examine the current situation of the existing built-up areas of the adjacent districts. The second stage covered various development options and the evaluation of the study area in order to prepare a Preferred Development Concept that represents the optimum land-use/transportation arrangement for the Study Area [12].

In Stage 3, a draft Master Landscape Plan was prepared with land-use and transportation proposals, redevelopment proposals and revenue estimates while in Stage 4, a Final Report was completed. The SEKDS Study has prepared detailed chapters covering the Outline Master Development Plan, Land Use, Environmental and Engineering aspects based on traffic congestion, pollution, water and air quality and reclamation work etc., the approach of Urban and Landscape Design. The SEKDS Study prepared a comprehensive framework for the reclamation and redevelopment of Kowloon Bay and surrounding urban area. However, public participation was not involved throughout the whole study process.

5.3.3 1998 Feasibility Study for South East Kowloon Development

In 1998, a Feasibility Study for South East Kowloon Development (SEKDFS) then began as a preparation for the draft Kai Tak (North) Outline Zoning Plan No. S/K19/1 and the draft Kai Tak (South) Outline Zoning Plan No. S/K21/1, under Sect. 5 of the Town Planning Ordinance (Fig. 5.4). It reviewed the framework of the SEKDS Study including the two planned reclamation phases for Kowloon Bay, updated the OMDP, established the engineering feasibility, developed an implementation programme and carried out preliminary design for development works (Territory Development Department; Kowloon Development Office 2001). However, the proposed reclamation of about 299 hectares in the draft OZPs received several objections from citizens during a two-month exhibition period. The Government therefore held a public consultation with the Legislative Council Panel on Planning, Lands and Works and other professional bodies on the planning issues of the Feasibility Study for SEKD. After several rounds of discussions between the government and professional bodies, Town Planning Board (TPB) considered an Outline Concept Plan (OCP) for the SEKD on the basis of public views collected, and agreed that the OCP should be forwarded to the objectors to provide a basis for the hearing of the objections to the draft OZPs [10]. After reviewing the public comments and objections of the draft OZPs, TPB agreed to amend the OZP on the alternative land use proposals for SEKD along with a public consultation forum and consultation with Legislative Council Panel on Planning, Lands and Works on the Outline Concept Plan.

5.3.4 2001 Comprehensive Feasibility Study the Revised Scheme of South East Kowloon Development

The Government carried out a Comprehensive Feasibility Study for the Revised Scheme of South East Kowloon Development to conduct a public forum and consultation with Legislative Council Panel on Planning, Lands and Works, TPB and other professional bodies on the Preliminary Layout Plan (PLP) prepared under the SEKDS Study, which collected a wide range of support from the public. After the first public consultation forum, the Government had continuously held public consultation forums in different districts e.g. Kowloon Bay, Wong Tai Sin District and Kwun Tong to collect public opinions on the planning of SEKD in order to prepare for the revised OZPs [10]. In 2001, the draft Kai Tak (North) Outline Zoning Plan No. S/K19/2 and Kai Tak (South) Outline Zoning Plan No. S/K21/2 were published under section 7 of the Town Planning Ordinance and the proposed reclamation was reduced to about 133 ha [13]. The OZPs were published for a public inspection period and in 2004, the judgment of the Court of Final Appeal on the draft Wan Chai North Outline Zoning Plan handed down in January 2004,

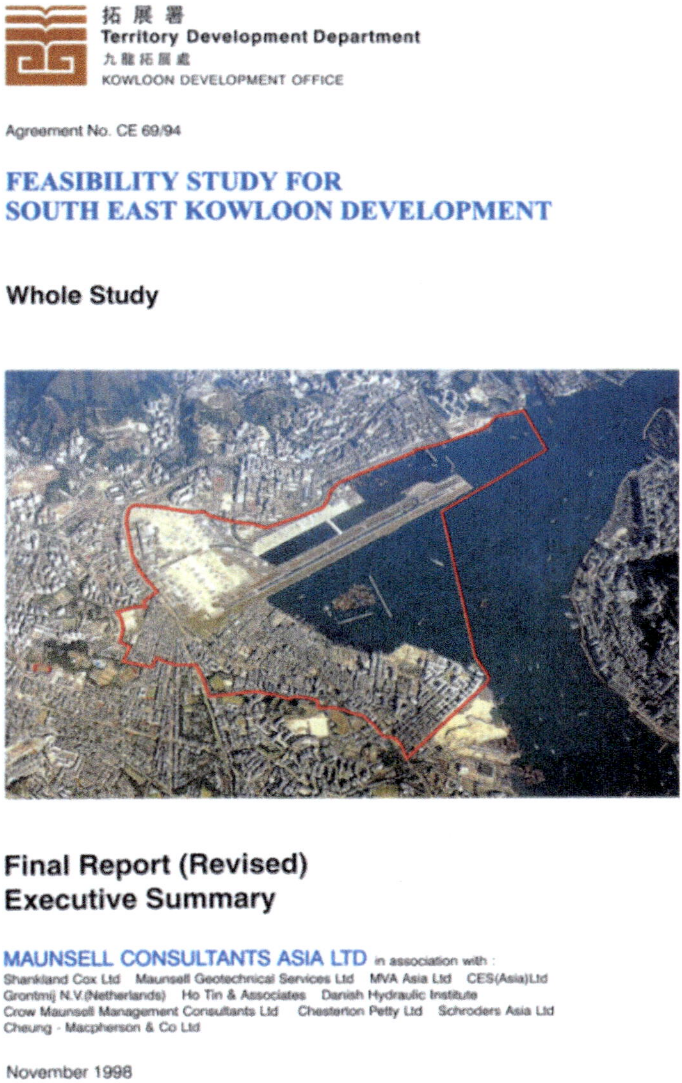

拓展署
Territory Development Department
九龍拓展處
KOWLOON DEVELOPMENT OFFICE

Agreement No. CE 69/94

FEASIBILITY STUDY FOR
SOUTH EAST KOWLOON DEVELOPMENT

Whole Study

Final Report (Revised)
Executive Summary

MAUNSELL CONSULTANTS ASIA LTD in association with :
Shankland Cox Ltd Maunsell Geotechnical Services Ltd MVA Asia Ltd CES(Asia)Ltd
Grontmij N.V.(Netherlands) Ho Tin & Associates Danish Hydraulic Institute
Crow Maunsell Management Consultants Ltd Chesterton Petty Ltd Schroders Asia Ltd
Cheung - Macpherson & Co Ltd

November 1998

Fig. 5.4 Cover of feasibility study for South East Kowloon development. *Source* http://www.pland.gov.hk/pland_en/p_study/prog_s/sek_09/website_chib5_eng/pdf/FS_SEKD_ExeSum1998_Eng.pdf

the presumption against harbour reclamation which can only be rebutted by meeting the overriding public need test.

At this point of the SEKD, public consultation sessions were able to create a platform for effective and productive discussions and communication between the Government and local citizens for considering and realizing the public needs.

Fig. 5.5 Public forum in stage 1 public engagement of Kai Tak development. *Source* http://www.pland.gov.hk/pland_en/p_study/prog_s/sek_09/website_chib5_eng/english/index_e.html

Since then, public consultation forums and engagement activities have become a major part of the planning and development process in Hong Kong (Fig. 5.5).

5.3.5 2004 Kai Tak Planning Review

The Kai Tak Planning Review was carried out in order to review the proposed planning of the area in a comprehensive manner with "no reclamation" as the starting point, in addition to examining the relevant interface and connectivity issues with nearby districts [11]. The Study area of the Kai Tak Planning Review comprises the old Kai Tak Airport site, existing waterfront areas at To Kwa Wan, Ma Tau Kok, Kowloon Bay, Kwun Tong and Cha Kwo Ling. The objective of the Kai Tak Planning Review was to conduct a comprehensive planning review of the "zero reclamation" concept and develop a new scheme for the old Airport site as guidance for its redevelopment. The Kai Tak Planning Review was divided into three stages: Stage 1 Kai Tak Planning Review, Stage 2 Engineering Review, and Stage 3 Statutory Planning Process where public engagement activities launched at the first stage to collect the community's aspirations regarding the conceptual vision of the future implementation for Kai Tak, and as a preparation for stages 2 and 3.

5.3.5.1 Public Engagement

Public engagement is important in town planning especially in the planning process as it allows citizens to voice their concerns and be part of the decision-making process for the future development of their local community. The continuous public consultation activities throughout the Kai Tak Planning Review aimed to gather community support and general thoughts on the key issues, as well as promote ownership on the study proposals. Therefore, three stages of Public Participation Strategy have been carried out to enable a more detailed public engagement structure for collecting ideas from the public. Stage 1 is in response to the Inception Phase to collect community visions of Kai Tak including the study scope and approach, planning objectives and development components; Stage 2 is in response to the Review Phase, with the discussion and comments on land use proposals for the Outline Concept Plans; and Stage 3 is in response to the Plan Preparation Phase, containing the presentation of the study findings and consolidation of the recommended development concept in form of a Preliminary Outline Development Plan (Fig. 5.6).

In addition, a study website as part of the Planning Department was launched to enable a convenient channel for the promulgation of supporting background information, consultation materials and study reports. Members of the public can easily access the materials on the website and review all documents from the different stages of public consultation. It also allows people to make comments in response to the study, which has cultivated an "E-engagement" relationship between government bodies and local citizens in Hong Kong.

Face-to-face meetings and discussions and written materials are no longer the only forms of public engagement activities to rely on. As technology continues to advance further, the internet allows a more convenient way to encourage people to share their feedback on the Government's Studies and enhances the public-government interaction and engagement throughout the Study process.

5.3.5.2 Stage 1: Study Scope and Approach

The first stage of public consultation commenced in 2004, lasting over a period of two-months with the main purpose "to engage the community early in the beginning of the study process with a view to building common visions for Kai Tak as early as possible" [10]. It focused on the discussion of the vision of Kai Tak, views of the background, study methodology, key development components, overall participation framework and the opportunities and constraints. Over 1200 territorial and local organizations were informed of the inauguration of the exercise and were invited to the relevant public activities. The intention was to arouse public engagement interest in this planning review, encourage public involvement and submission of comments [11]. During the period of Stage 1, over 500 participants were recorded with active responses and support in the activities which included three public forums and a community workshop. In order to facilitate the discussion, more than 20 sessions of consultation and discussion meetings were carried out between the advisory bodies, professional and stakeholder groups. Many of the participants submitted written comments, development concepts and specific proposals on Kai Tak, and more than 200 submissions were received. Due to the large

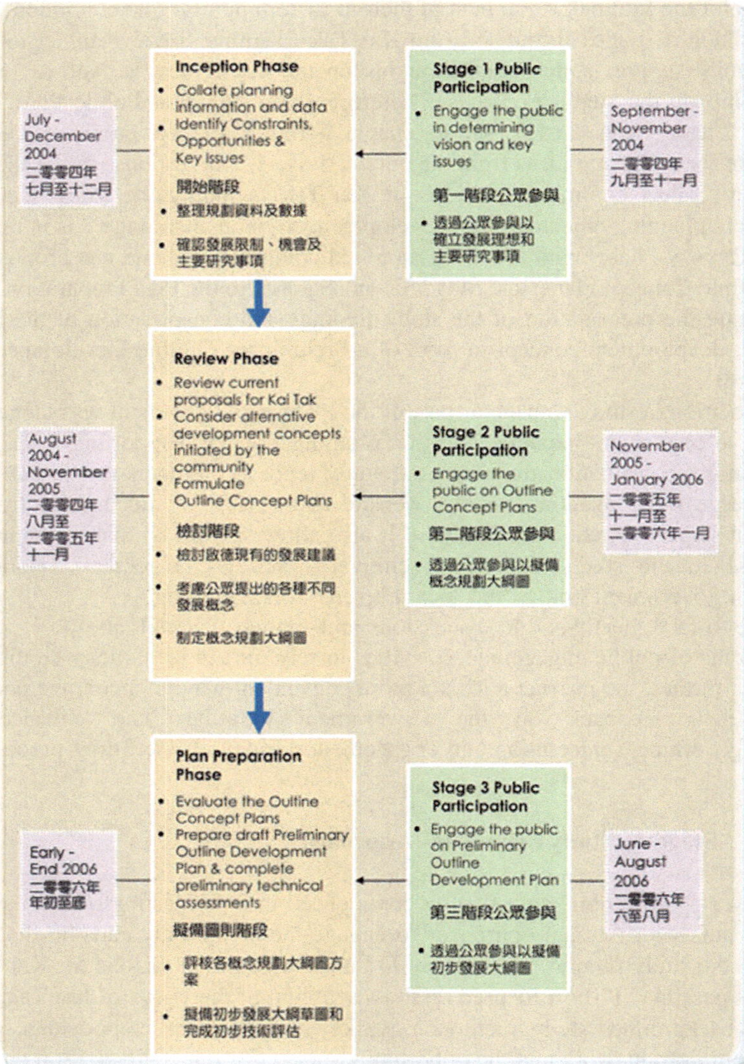

Fig. 5.6 Public participation strategy of Kai Tak development. *Source* http://www.pland.gov.hk/pland_en/p_study/prog_s/sek_09/website_chib5_eng/english/index_e.html

quantity of concerns on the harbourfront development, a sub-committee of Harbourfront Enhancement Committee was created to address the community's aspiration for greater involvement in planning and design of the harbourfront area. It is responsible for providing further comments and advice on the SEKD Review before gathering the public feedback and the preparation of the Outline Concept

Plans for the next stage of public participation. These processes of public participation and engagement have allowed local citizens to express their thoughts on key principles and issues of the Kai Tak Development as the plans themselves are being developed.

The fundamental idea was to enhance the waterfront developments while retaining the heritage of the former airport operation at Kai Tak. People wanted to bring back the memory of landing at Kai Tak and the harbour to add to the social value of the area and enhancing the quality of life. Other participants have suggested that Kai Tak could become a hub for sports, recreation, tourism and entertainment uses. Some of the planning principles brought up during the engagement forum related to the importance of a people-oriented planning approach in the Kai Tak Development, integration of new developments with the existing communities in surrounding areas, and an approach to achieve sustainable and well-balanced developments and high-quality living [11]. Another key contention issue was the idea of reclamation of the waterfront area. Residents nearby expressed not to pursue further reclamation in the Harbour Area or the Kowloon Bay area, which was taken into great consideration by the planning and development team.

The Kai Tak Forum has provided the opportunity for an engaging and mutually beneficial discussion between the community and government on the comments and proposals received in the Stage 1 Public Participation, as well as the responses from the Consultants and Government Bureaus/Departments [14]. The forum concluded that the redevelopment of Kai Tak should focus on the strategic planning context in order to formulate a broad strategic framework for the planning of Hong Kong over the next 25 to 30 years [14].

The Kai Tak Development Stage 1 Public Participation has encouraged public engagement activities to raise the general public's awareness of the planning and development issues of Kai Tak. It has also enabled the community to visualize their thoughts and opinions more formally, and more importantly to develop a sound basis for a dialogue between the local citizens, grassroots organization and stakeholder groups on future implementation and growth of the Kai Tak Development.

5.3.5.3 Stage 2: Outline Concept Plan

In 2005, the Stage 2 Public Participation was carried out as the second public consultation to present the OCPs of land use themes developed in the Stage 1 Public Participation to the community. During Stage 2, three public forums were launched to invite public comments on the development concepts for Kai Tak, which will serve as inputs for the Preliminary Outline Development Plan (PODP) for further discussion in the community [15]. Several public engagement activities were held to conduct suggestions from the community including a public forum, three district forums, and three topical forums on Multi-purpose Stadium, Cruise Terminal and Kai Tak Approach Channel. Prior to the Forum, about 160 participants joined a site visit to Kai Tak to familiarize themselves with the opportunities and constraints of Kai Tak development.

The key issue discussed during Stage 2 Public Participation was obviously the reclamation since many of the participants had strong aspirations to adopt "no reclamation" as the basis for the future Kai Tak Development, while some have no objection to potential small-scale reclamation. Numerous reclamation issues were discussed with the vision, planning principles, and development themes and components, resulting in 3 concept plan options produced based on the comments and suggestions of the public.

5.3.5.4 Stage 3: Preliminary Outline Development Plan

In 2006, the Stage 3 Public Participation included 3 District forums with an exhibition of the Kai Tak Preliminary Outline Development Plan proposals. As the last stage of the public consultation period, it aimed to collect public views on the draft PODP before it was finalized. The comments and suggestions covered plenty of topics including the Kai Tak approach channel, metro park and open space network, sports hub, cruise terminal, tourism node, and transport and connectivity. The public engagement activities in Stage 3 provided platforms for the general public and the stakeholder groups to have conversations with the Government on the planning and development of the future uses of Kai Tak. A 3D animation model and panoramic views of the Kai Tak Development were uploaded to the Kai Tak Planning Review website for general use. Also, at the end of each stage of public participation, a summary report of comments and proposals received were published on the website in response to the public comments (Fig. 5.7).

These three stages of public engagement have revealed that public engagement and consultation sessions in the planning process today are done online in addition

Fig. 5.7 Public forum in stage 3. *Source* http://www.pland.gov.hk/pland_en/p_study/prog_s/sek_09/website_chib5_eng/english/index_e.html

to physical events such as exhibitions, workshops and forums since all the materials and information can be uploaded to a website easily accessed by the general public, also providing an instant and interactive way for feedback and responses to comments relating to issues surrounding the planning and development of a major infrastructure project.

5.3.5.5 Kai Tak Office (KTO)

The KTO was set up to lead the coordination and implementation of KTD. The Head of KTO, H(KTO) is formed by professional staff to monitor the progress of KTD to fulfill public aspirations for the design. Moreover, the professional team attempted to facilitate a smooth implementation in design development and provide a coherent public interaction by enhancing green features in KTD, conserving heritage and surrounding old areas, and directing the design of KTD infrastructure works to public consultation, while achieving the planning vision of KTD [16].

5.3.6 Energizing Kowloon East Office

Back in 20th Century, Kowloon East comprised the former Kai Tak Airport, Kwun Tong and Kowloon Bay Business Areas, and has since experienced rapid growth from the manufacturing industries of the industrial age in Hong Kong with the creation of many jobs to drive Hong Kong's prosperity during the 20th Century. However, after the relocation of the Airport from Kai Tak to Chek Lap Kok, and the migration of South China's manufacturing base from Hong Kong to Mainland China, the area has lost its past vibrancy, leaving a large stock of industrial buildings not fully utilized. Also in the late 20th Century, Hong Kong started to develop into an international financial city with emerging professional service sectors. This evolution resulting in many multi-national companies choosing Hong Kong as the location for their Asia Pacific regional headquarters, boosting the demand for prime office spaces, which is struggling to be supported by the traditional Central Business District (CBD) of Central, Hong Kong.

Therefore, the Energizing Kowloon East Office (EKEO) was set up to target community engagement and public participation efforts in the visioning and planning of the regeneration of Kowloon East into another premier CBD for Hong Kong to support the continued economic growth of the city's financial and professional services in order strengthen and maintain the city's global economic competitiveness, This was achieved through numerous phases of public consultation, various public engagement activities, and recreational events carried out to engage the community and collect local citizens' opinions. In addition, an informative website containing all relevant materials such as the planning and feasibility studies, conceptual master plans, and outcomes of public engagement forums regarding

Kowloon East was created for people to access information about the progress of the project. Up until today, the public consultation for Kowloon East events are still ongoing as part of various studies to collect feedback from local people with different expertise and experiences. Some have reflected that the government has spent over two decades of planning process from the SEKD to the Kowloon East Development in response to public concerns, preparing concept plans for future implementation; however, progress on actual large-scale transformation has not been physically achieved as evidence that the public engagement efforts will be realized.

In conclusion, the government has worked effectively throughout the years in terms of addressing the public opinion and suggestions regarding the Kai Tak Development, most successfully achieving the reduction of proposed reclamation from over 300 hectares to "zero reclamation". However, some may consider it an ineffective process due to the fact it has taken over two decades solely for the planning and preparation for development. During the planning process of SEKD to EKEO, there is no doubt that public engagement has entered a digital world where people can much more easily access the interest to find all the material and information related to the proposed project and location in question. "E-engagement" has altered the traditional concept of public consultation in city planning issues, and has established its own role in the future of all a short and long-term planning processes.

KTD Timeline:

1991	Various reclamation projects in the Harbour areas recommended by Metroplan Selected Strategy
1992	The **South East Kowloon Development Statement Study** to translate the Metroplan concepts into more specific district planning objectives and plans commenced
1995	**Feasibility Study for South East Kowloon Development** commenced
September 1998	The draft Kai Tak (North) Outline Zoning Plan No. S/K19/1 and the draft Kai Tak (South) Outline Zoning Plan No. S/K21/1 were gazette. The proposed reclamation was about 299 hectares. During the two-month exhibition period, objections against the draft Outline Zoning Plans were received. Most of them objected to the extent of the proposed reclamation
November 1998	Public consultation forum and consultation with Legislative Council Panel of the Feasibility Study for South East Kowloon Development were held
May 1999	Town Planning Board considered an Outline Concept Plan for the South East Kowloon Development on the basis of public views collected, and agreed that the Outline Concept Plan should be forwarded to the objectors to provide a basis for the hearing of the objections to the draft Outline Zoning Plans
June 1999	Town Planning Board heard the objections and collected views on the Outline Concept Plan from the objectors. The Board agreed to amend the Outline Zoning Plans pending the completion of a further study on the alternative land use proposals for South East Kowloon Development

(continued)

(continued)

June–July 1999	Government conducted a public consultation forum and consultation with Legislative Council Panel on Planning, Lands and Works on the Outline Concept Plan
November 1999	The **Comprehensive Feasibility Study for the Revised Scheme of South East Kowloon Development** commenced
2000	Government conducted a public consultation forum and consultation with Legislative Council Panel on Planning, Lands and Works, Town Planning Board, Kwun Tong District Council, Wong Tai Sin District Council and Kowloon City District Council on the Preliminary Layout Plan prepared under the South East Kowloon Development Comprehensive Feasibility Study. The Preliminary Layout Plan was widely supported
June 2001	Government conducted consultation with Environmental Impact Assessment Subcommittee of the Advisory Committee on Environment, Kwun Tong District Council, Wong Tai Sin District Council, and Kowloon City District Council on the findings of South East Kowloon Development Comprehensive Feasibility Study
July–August 2001	Government conducted consultation with Wong Tai Sin District Council, Kwun Tong District Council and Kowloon City District Council on the revised draft Outline Zoning Plans
August 2001	The draft Kai Tak (North) Outline Zoning Plan No. S/K19/2 and Kai Tak (South) Outline Zoning Plan No. S/K21/2 were gazetted under section 7 of the Town Planning Ordinance. The proposed reclamation was reduced to about 133 a
January 2004	The judgment of the Court of Final Appeal on the draft Wan Chai North Outline Zoning Plan handed down in January 2004, the presumption against harbour reclamation can only be rebutted by meeting the overriding public need test
July 2004	**Kai Tak Planning Review** commenced. The focus of this study is to review the planning of the area in a comprehensive manner with "no reclamation" as the starting point

5.4 E-Engagement in the Hong Kong Planning Process

In Hong Kong, the town planning process and related matters are governed by The Town Planning Board (TPB). The TPB is a statutory body that aims to promote the healthy, safe, convenient and general welfare of the community through the systematic preparation of plans for the layout of urban and rural areas in Hong Kong as well as the suitable types of buildings to be developed.

Part of the planning process requires the TPB and Planning Department to publicise draft plans for public inspection and consultation, in order to receive representations and comments from the general public. This allows for public input and subsequent amendment of plans to align with overall community needs and goals. In the 1960s before the use of technology, paper plans were presented to the public. Paper plans are much less accessible to the general public due to the need to be physically present, and in addition are significantly limited to those who have the technical expertise to understand such plans. With the evolution and adoption of technology in the late 20th Century, the TPB transitioned to digitized plans in the 1990s, utilizing technologies such as Geographic Information Systems (GIS). These technological advancements and transition to digitalization have enabled the planning process in Hong Kong to become much more accessible and readily available to the general public, in forms that are simpler to understand from a layman's perspective. Furthermore, digitizing the information as allowed the utilization of the Internet to disseminate the relevant information to the general public to be retrieved at any time.

In the past 20 years, advancements in technology have enabled the evolution from physical to digital models and now to 3D models and animations for public engagement and consultation purposes. The use of 3D models and animations over physical models allows for real-time imaginable land use scenarios for the public to clearly identify and visualize the potential future impact that a new development will have on the area. Hong Kong is in fact, the first place in the world to introduce 3D environmental impact assessment (EIA) models as a tool for public engagement and consultation. By developing a 3D visualization, it is easier for the general public to understand complex projects and issues, as well as the effect of mitigation measures prior to construction and implementation. Previously, the use of 2D hardcopy plans or text to describe and illustrate impact and mitigation options were sufficient for informing and engaging industry professionals and a technical audience. However, it significantly limited the accessibility and knowledge of this information for the general public, as it was difficult for the layman to adequately understand such technical plans and to give suitable comments or suggestions. These interactive 3D visualisations also allow for online demonstrations of fictitious projects as well as the ability for the public to try out the 3D process by themselves to experience this higher level of engagement and the ability to make use of the tools to offer suggestions and alternative mitigation options [17].

5.4.1 Development of E-Engagement Tools in Hong Kong

Although technological advances in the digital age has made it easier for citizens to have increased access, availability and understanding of different steps of the planning process, the next step is to enable citizens to participate, contribute and share their views and opinions during the planning process itself. Before the ability to actively engage and participate in the planning and decision making processes, the Hong Kong Government was weak in providing suitable channels for citizens to provide comments and voice their opinions on specific planning projects. The Hong Kong Government successfully uses social media and other channels to "inform, involve, and collect views" from the public, but is not seen to necessarily take these views into formal consideration, limiting the empowerment of citizens in public policy decision making [18]. The opportunities to increase these channels and the types of engagement have grown in recent years, especially in the context of the Town Planning process.

Both small and large-scale development or infrastructure projects in Hong Kong have begun to incorporate various E-engagement platforms and channels to inform, consult and collect comments from the general public. Firstly, there is a significant amount of information to be found online. The TPB has created the Statutory Planning Portal, an interactive website that disseminates statutory planning matters to the general public [19]. It allows users to interact with a map of Hong Kong that contains information such as zoning classifications, the ability to alter the data presented on the map, as well as documents relating to all planning proposals, permissions and rezoning applications and representations. Through the Statutory Planning Portal, the general public has readily available access to information for all planning applications, as well as data about Hong Kong's statutory plans. Prior to the portal, this information could only be obtained in person by visiting the Planning Department offices (Figs. 5.8 and 5.9).

In addition, the Planning Department has its own website where it publishes further information relating to planning matters in Hong Kong, including dates of events and schedules and reports on public consultations. Documents such as press releases, presentations, and educational materials are also provided, as well as links to other affiliated websites such as the TBP and Statutory Planning Portal. The Planning Department website provides information to educate the public on parts of the planning process in Hong Kong so as to facilitate the engagement of the public on town planning matters that were previously not easily available or accessible to those outside of the government or the town planning profession (Fig. 5.10).

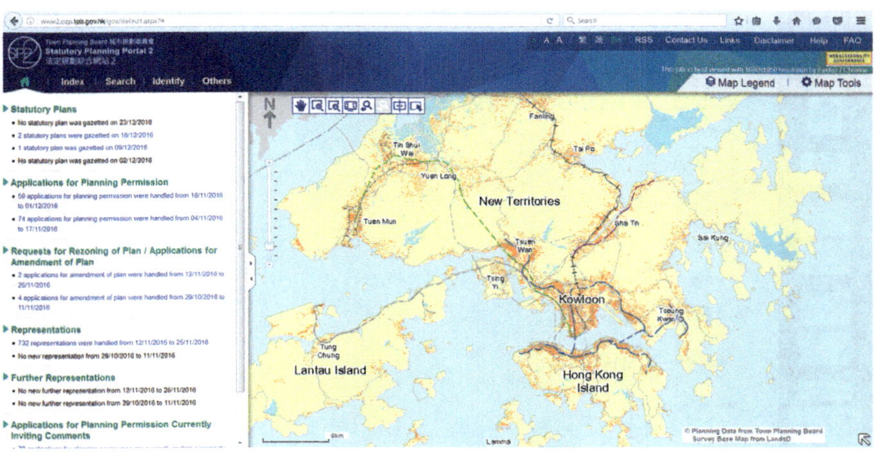

Fig. 5.8 Hong Kong statutory planning portal. *Source* http://www2.ozp.tpb.gov.hk/gos/default.aspx?#

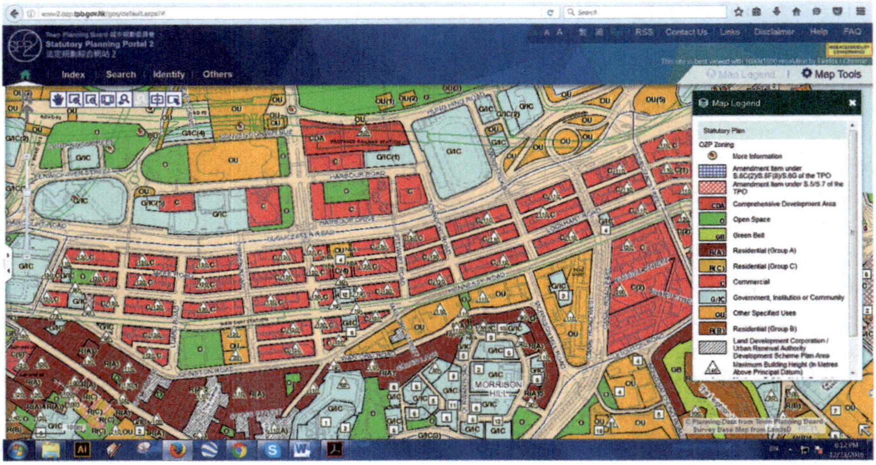

Fig. 5.9 Detailed view of information provided by the statutory planning portal. *Source* http://www2.ozp.tpb.gov.hk/gos/default.aspx?#

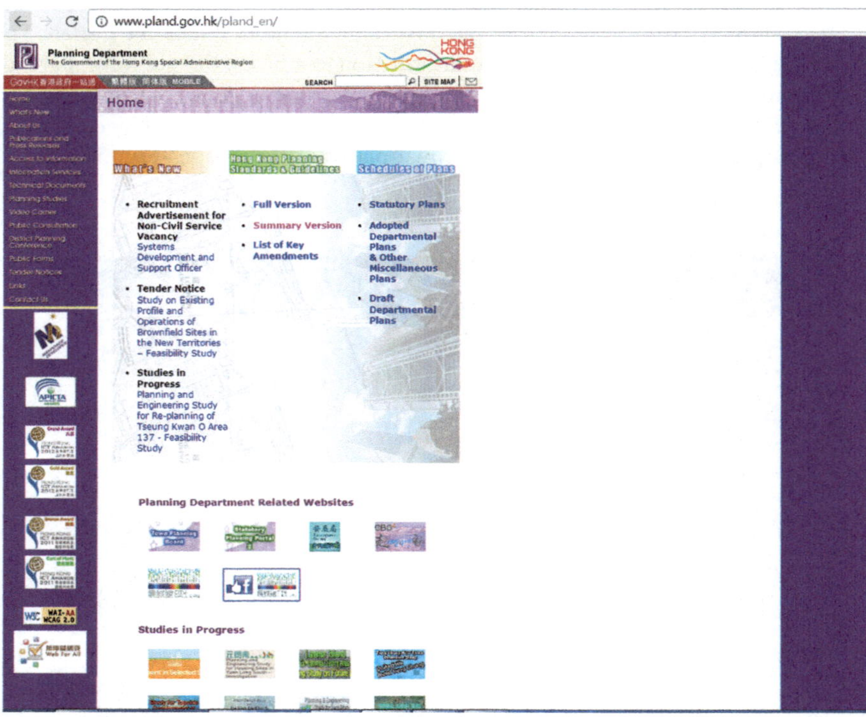

Fig. 5.10 Hong Kong planning department homepage. *Source* http://www.pland.gov.hk/pland_en/index.html

Websites and the internet have allowed for new forms of public engagement that can be accessed by anyone anywhere given a computer and internet connection. This has created new channels and platforms to engage and interact with the public about planning proposals, studies and projects. The three-main e-engagement tools that are now readily available are:

1. Websites dedicated to specific projects from inception to completion, where all relevant information, graphics and documentation is uploaded and available to the general public, and key stakeholder contacts are clearly stated. The use of 3D visualisations and video animations have also enhanced the quality of information disseminated to the public, who can make use of these technological advancements to better understand all aspects of the project.
2. The ability for the general public to submit their views and comments on specific projects or planning proposals online through a website or interactive application, such as Public Participation Geographic Information Systems (PPGIS). This allows a much more accessible and straightforward way for citizens to voice their opinions or concerns on a certain process or aspect of a planning project, which in turn encourages greater public participation in the planning process. Submitted views and opinions are all taken into consideration and reviewed for the next stage of the planning process.
3. From the perspective of urban planning and design firms, the use of the internet has enabled planning proposals and applications to be submitted entirely online, in addition to all related information materials about the planning process such as the Town Planning Ordinance and Planning Standards and Guidelines. This reduces the need to physically print out and hand in submissions in person, saving both time and the environment.

5.4.2 Websites for Planning Development Areas and Studies

Many of the large-scale planning and development projects in Hong Kong now have their own website developed dedicated to that one project, in order to disseminate publicity and planning materials, informative graphics and videos, the schedule of dates of project progress and community engagement events, as well as contact information and ways the general public can voice their own views on the projects. Examples of these websites are the "Tung Chung New Town Extension Study" and the "Planning and Engineering Study on Future Land Use at Ex-Lamma Quarry Area at Sok Kwu Wan, Lamma Island—Feasibility Study" and the Hong Kong 2030+. These project specific websites provide an easily accessible platform for members of the public to understand more about the project in a graphic and user-friendly way, and facilitates the increased engagement of the general public on future planning related matters.

These websites provide a one-stop-shop portal for all information directly related to the proposed planning study or project, as well as links to other related sites and

Fig. 5.11 Website homepage of feasibility study for project on Lamma Island. *Source* http://www.ex-lammaquarry.hk/

Fig. 5.12 Webpage showing media from a stage 1 public engagement on a study for the extension of Tung Chung New Town. *Source* http://www.tung-chung.hk/eng_highlights.html

downloadable documents. It effectively enables the general public to easily and conveniently locate any information about the study or project they may require, and is used to disseminate important materials such as 3D visual animations, videos and photos to publicise the benefits of the project and allow the public to better understand it. By creating and marketing these websites, it provides an additional channel for citizens to access the project information, and a platform to learn and give their own suggestions [20, 21] (Figs. 5.11 and 5.12).

5.4.3 Online Submission of Comments and Suggestions

Another key technology development and adoption by the Planning Department in Hong Kong is the use of Public Participation Geographic Information Systems (PPGIS) for district-wide planning studies and large-scale developments such as new town developments and large residential developments. PPGIS works by producing digital maps, satellite imagery and other spatial and visual tools in both 2D and 3D to facilitate public engagement through enhancing the understanding, awareness and involvement of the general public to these geographic spatial and planning discussions. It also provides an interactive platform for citizens to leave comments and suggestions on existing planning and development proposals in Hong Kong [22]. PPGIS uses the Internet and Google Maps satellite images to overlay planning information, concepts and proposals that help the general public effectively and conveniently visualise and understand the planning proposals of the study area. It essentially presents all the information in a clearly identified and recognisable manner, which can be easily understood and interpreted by those with no professional design or planning training, in order for them to comfortably engage and encourage them to submit more valuable, in-depth and insightful comments and concerns regarding the specific planning proposal.

PPGIS was first introduced in Hong Kong in 2011 as a Pilot Scheme for the Lau Fau Shan Area Improvement Study, whereby the planning concepts/proposals were overlaid onto satellite images to "facilitate visualization and understanding of the proposals on the internet" [23] (Figs. 5.13 and 5.14).

Subsequently, for its innovative and successful use of technology in engaging the public, the Planning Department won an award at the 2011 Hong Kong ICT Awards for Best Public Service Application, signifying the importance and future potential of online interactive platforms such as PPGIS to become commonplace in encouraging the public to participate in planning discussions and to submit their own comments and suggestions [24].

Although not commonplace at the moment, PPGIS is being optimised to successfully function in all major future planning development studies and applications as part of the consultation stage. Furthermore, in the future the PPGIS system will be enhanced to integrate Social Media Platforms such as Facebook, for it to become even more effective and versatile in engaging with the public [25].

5.4.4 Online Submission of Planning and Review Applications

In addition to improving the engagement and participation of the public on existing planning proposals and applications through tailored website and online comment submission platforms, advancements in ICT have also enhanced the engagement capabilities of individuals on the project proposal proponent side by allowing

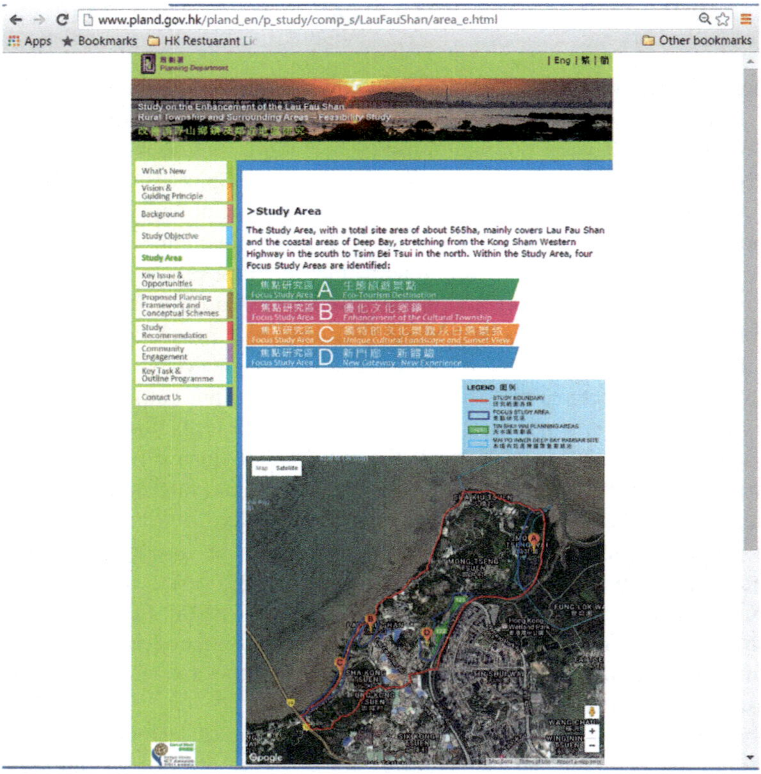

Fig. 5.13 PPGIS integrated onto project website. *Source* http://www.pland.gov.hk/pland_en/p_study/comp_s/LauFauShan/area_e.html

planning applications to be submitted online via the Electronic Planning Application Submission System (EPASS).

EPASS was created as an additional channel for the public to make planning/review applications under certain sections of the Town Planning Ordinance that relate to planning application matters, as well as for the subsequent submission of further information or additional documents for these applications, online to the Town Planning Board in electronic format [26]. Prior to the use of EPASS, all planning/review applications had to be submitted in person or by post with the sufficient number of copies required. EPASS can effectively enhance the capabilities and accessibility of the public and organisations to submit planning applications, as its service is available 24 h 7 days a week, compared to submitting in person during office hours or the time lag of postal services [27]. Physical submission of applications in-person or by post is still acceptable, so EPASS is providing an alternate choice for the public rather than replacing the conventional method with a newer method. By having the choice of a physical or electronic submission, it does not limit the capabilities of those with limited access to the internet or lacking the technological knowledge (Fig. 5.15).

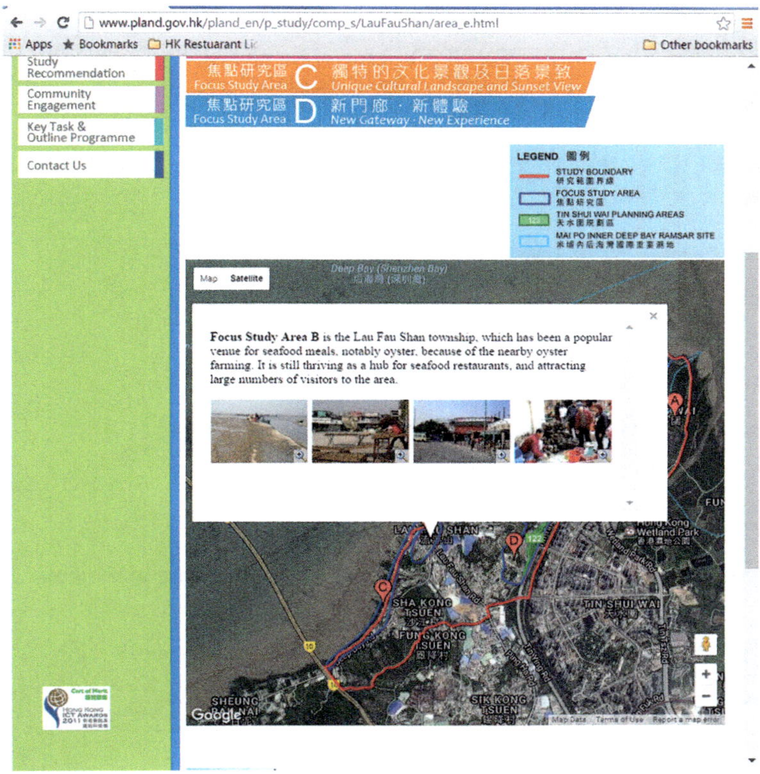

Fig. 5.14 Detailed view of interactive capabilities of PPGIS. *Source* http://www.pland.gov.hk/pland_en/p_study/comp_s/LauFauShan/area_e.html

The limitations of using EPASS to submit planning applications online is that there are certain requirements to adhere to in order for applications to be processed successfully by the system. These specifications include filling in the required e-forms (interactive PDFs), submitting the application documents through EPASS, ensuring that the total size including attachments does not exceed 10 MB, and that it is not more than 20 pages excluding those about particulars of the application and authorised agent, as well as the checklist of documents. Additionally, the TPB states that any documents, plans or images submitted online in colour will only be printed in black and white on A4 sized paper for application processing purposes, severely limiting the scope of possibilities with e-submission. If the applicant wishes their application to be processed in colour, using paper sizes other than A4, in addition to being submitted through EPASS, separate hard copies of the documents, plans or images must be submitted in person or by post to the TPB with the sufficient number of copies required, which can be up to as many as 90 copies for an "Application for Review under section 17" [26] (Fig. 5.16).

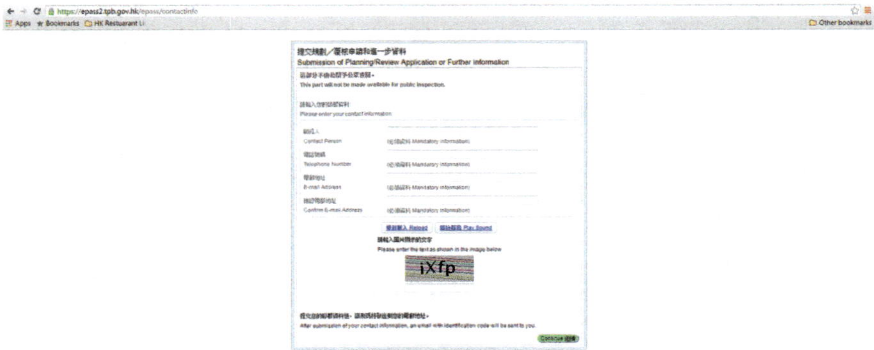

Fig. 5.15 Start page for e-submission of planning review application. *Source* https://epass2.tpb.gov.hk/epass/contactinfo

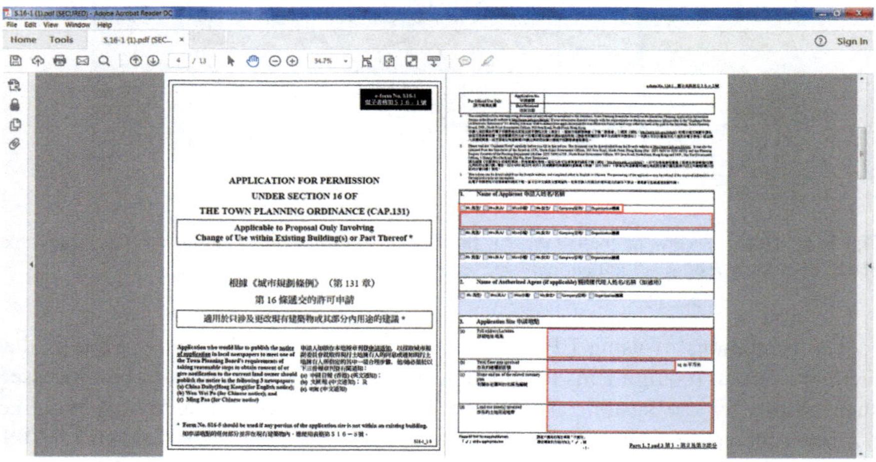

Fig. 5.16 Example of the interactive PDF e-form that is to be submitted through EPASS. *Source* http://www.info.gov.hk/tpb/en/eforms/esub.html

While submitting planning applications online through EPASS may seem complicated, the TPB offers video tutorials on its website whereby they walk users through the individual steps of submitting an application through EPASS, which can sufficiently aid those who lack confidence or understanding of the steps to submit electronically on the Internet (Fig. 5.17).

The development and use of EPASS has been highly effective and successful in enabling a wider range of public individuals and organisations to become involved

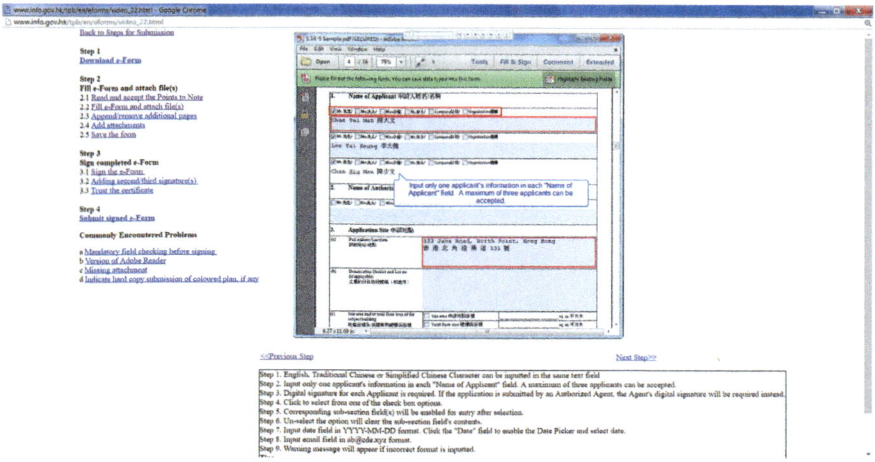

Fig. 5.17 Online video tutorial for submission of e-forms for planning applications. *Source* http://www.info.gov.hk/tpb/en/eforms/video_22.html

in planning matters in Hong Kong by increasing accessibility to, and convenience of, the planning submission process.

More recently a comprehensive mobile app named 'My Kowloon East' (MyKE) was launched by EKEO to connect with the information of Kowloon East development by using ICT. MyKE promotes the development and implementation of smart city initiatives and the concept of 'Walkable Kowloon East'. To implement smart cities initiatives, MyKE highlights the available Wi-Fi Hotspots, EV Charging Facilities, iPostal Stations, smart infrastructure, and smart lockers, enhancing the experience of tourists in East Kowloon area [28].

MyKE has a Facebook comment plugin function, allowing interaction between government and public. Not only can the government announce information of EKD via MyKE, the public can also give comments to KTO through Facebook. With a community sharing platform, 'Happy Share', being a social gathering point, it engages different people and creates critical mass and allows a spillover effect to surrounding areas. In addition, the feature 'Easy Parking' shows the available parking lots nearby to make it easier for people while searching for parking spaces. MyKE utilizes GeoInfo Map providing basic information such as temperature, UV index, humidity and carbon dioxide emission as some supplementary information to achieve the concept of sustainability [29]. MyKE also developed a Green Map highlighting Green Buildings in Kowloon East, to provide diversified and environmental friendly information for the public. Another feature 'Self-guided Tour' shows the existing landmarks nearby and suggested a 'Health and Leisure route', encouraging a healthy lifestyle of walking while allowing people to explore East Kowloon themselves. Moreover, to promote the vitality of the city, the guided tour shows the jogging route to allow different health activity in East Kowloon area (Figs. 5.18 and 5.19).

Fig. 5.18 Screenshot of MyKE mobile app. *Source* https://itunes.apple.com/us/app/my-kowloon-east-myke/id1170321129?mt=8&ign-mpt=uo%3D2

5.5 More Informal Citizen Engagement

In addition to formal channels of public engagement on town planning matters explored above, there are also other informal channels that have emerged from a bottom-up approach by community activists and planning advocates. These informal channels of engagement can be in the form of social media sites and forums, but also community events such as Magic Carpet in Hong Kong, and mobile applications like Neighborland in North America and idiscover city walks in Hong Kong and the region.

In Hong Kong, Magic Carpet is an outreach program launched in 2013 by the School of Architecture and School of Journalism and Communication of the Chinese University of Hong Kong to encourage the combination of daily life, community engagement and urban design. The idea behind the Magic Carpet is to transform a street or open space into an outdoor public space for an evening, for the people to interact and to showcase movies about the local neighbourhood and to encourage discussions about the creation and use of pop up public space in Hong Kong. The films are produced by local secondary students who prior to making the

Fig. 5.19 "Easy Parking" feature of MyKE mobile App. *Source* https://itunes.apple.com/us/app/my-kowloon-east-myke/id1170321129?mt=8&ign-mpt=uo%3D2

films, participate in workshops held for the students to guide and prepare them in encouraging public engagement through conducting video-interviews with people in the community. Moreover, at Magic Carpet events, students of the new MSc Urban Design program participate and interact with community members and the general public who are able to interact with each other to share ideas and opinions on how to re-envision the possibilities of public space together, successfully empowering them in planning matters and building stronger bonds between each other [30] (Fig. 5.20).

The first Magic Carpet project was successfully held in Sai Ying Pun in 2013 and also in 2014, and subsequently held in Tin Shui Wai in 2015, and more recently To Kwa Wan in 2016. By engaging the local community of various districts around Hong Kong in participating in discussions about their local urban environment, it effectively encourages and empowers citizen engagement in planning related matters in Hong Kong, and the community views and comments gathered from informal events such as these are just as valuable to formal comments, and can themselves become inputs to higher authorities such as the Planning Department and the Development Bureau.

Fig. 5.20 Magic carpet event in Hong Kong. *Source* https://www.cpr.cuhk.edu.hk/en/press_detail.php?id=1622

In the near future, Hong Kong's informal public engagement platform could be as convenient and accessible as a mobile applications such as idiscover city walks by Urban Discovery in Hong Kong and the region that helps people experience the city and raises more awareness of critical urban issues such as urban renewal and heritage conservation. While Neighborland popular in the US and helps engage people in more creative ways by empowering them and raising awareness of critical urban issues. Neighborland allows people to submit their comments and suggestions about their built environment or an urban project, which are then sent back to civic leaders or those in charge of a certain project. Neighborland is an online "software that empowers civic leaders to collaborate with residents in an accessible, participatory and enjoyable way", by providing an online platform to engage the general public on desktops, tablets or mobile devices [31]. It allows organisations to reach a much wider audience for public consultation matters than traditional in-person surveys or interviews, and facilitates community engagement by allowing people to submit insights and comments through channels such as workshops and public installations, text message, Twitter or on the Neighborland application or website. The increased number of possible engagement channels significantly improves the engagement capability of projects using this approach. Furthermore Neighborland provides map-based insights, allowing everyone to visualise exact locations of comments and suggestions on an interactive map, and even facilitates users to up-vote and subscribe to other's ideas to get updates on that particular issue. Lastly, for the project leaders, Neighborland conveniently aggregates all collected and submitted data, and can provide reports and data trends to make it

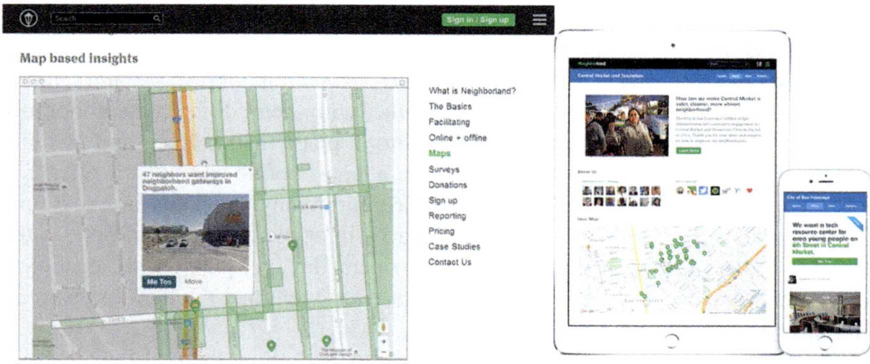

Fig. 5.21 Neighborland's map based insights and example of tablet and mobile applications. *Source* https://neighborland.com/

easier for the organisers to make decisions, allocate resources and know how to keep participants involved in discussions about their local community (Fig. 5.21).

5.6 Conclusion—The Future of E-Engagement

Civic engagement in the planning process is critical in shaping better cities. E-engagement has the potential to make the process more transparent and accessible for more people to participate in the planning process. Rather than project-specific public engagement, participation can occur continuously throughout the year through online channels such as social media. An ongoing discussion in which the public can share their views and discuss topical city planning issues will be more meaningful for the people and the city in future years. One limitation is that E-engagement may be limited to those who are educated enough and have access to technology. As such, both traditional physical engagement and E-engagement should function hand-in-hand to complement one another to work towards building greater social capital.

Since 2004, social media sites such as Facebook, Twitter, LinkedIn, Instagram, WordPress, Yelp and others, have started to play an increasingly important role in communicating and engaging with the public. Around the year 2007, traditionally official and formal ways of communication between local governments and citizens began to transition towards informal communication such as online forums and social media sites. It is now relatively common practice for public bodies and government departments to have their own social media accounts. It is also common for citizens to ask questions and interact with public government in an informal manner using social media.

Mobile Apps is another tool that has developed alongside smartphones and are increasing the level of public participation in everyday life. There are many different mobile apps that enable the dissemination of public information directly to citizen's fingertips. For example, there are apps for public transportation schedules, leisure and recreational activities, healthcare, education, police, weather, home affairs and more, to make living easier. Together with them, different neighborhood apps were discovered to help to engage public in place maintenance and planning process. Many cities have apps that work on a smart phone and allows citizens to make a suggestion or a notice for maintenance/improvement needed in a real time. For example, if a citizen is walking on a sidewalk at the evening and notices that one of the lantern lights are not working, he opens the app, puts on his location as the app is connected with the GPS, and writes a comment that the light in this place should be replaced.

Social media sites became so popular because they allow the real time communication. People can get their answers to questions fast and easy. A message in the Social Media is delivered with a visible date and time tag. This allows citizens to keep track on how up to date the information is. After the 2010s, E-engagement became self-evident to the point that public engagement encompasses e-engagement.

People tend to be more willing to use big platforms as a one stop shop for their daily services simply because they find it easier. This makes the e-engagement process more efficient, more accessible and less time consuming for citizens. In this way, citizens would not only contribute to the government, but they would start to communicate and work together to develop solutions for improving their own local communities.

Going forward, it is important to improve current ICT models as to how their web pages function, interact with the public, and enable the engagement of more people faster and easier. ICTs have to a certain extent increased the transparency in the decision making process, yet challenges still remain. How do we make sure that the people who are engaged are actually participating in the decision making process for the interest of the greater public and not solely their own private interests? How do we make sure that those who are most actively engaged in the public participation process are actually the ones who will use the space the most? It is important to take into consideration that the population is diverse and well represented, to ensure decisions are not influenced by a vocal minority, and to figure out how to engage other groups such as the elderly and younger generations, such as children.

Given that people in Hong Kong are very vocal in making their views heard through means such as protests and demonstrations even though Hong Kong is not as democratic as people would like it to be. The ever increasing cost of living, lower salaries and lack of affordable housing being some of the critical issues as a result of which there is a lot of dissent especially among the younger generations as there is lack of hope for their future in Hong Kong. This led to the recent demonstrations with regard to the "umbrella movement" which were also supported by ICT social media was used as well.

It is also important to note that E-engagement in Hong Kong is still in its early stages, where the information is available and accessible on the websites but the use of social media for public engagement purposes by the public sector has not fully caught on as yet. Although Hong Kong has a very active civil society, the whole engagement process needs to become less project based and more community based, and should be on-going with engagement occuring throughout all stages. One stop shops for all engagement needs of citizens will make the engagement process even more effective. The aim should be to try and engage people of all walks of life and ages, but more importantly to see results on the ground that address the views, opinions and aspirations of the community. It is important to build greater trust among the public sector, the community and its people, with a more positive attitude and approach towards Hong Kong which will require all sectors of society to come together to make Hong Kong an even greater global city.

References

1. Cowan SE (2010) Democracy, technocracy and publicity: public consultation and British planning, 1939-1951. http://escholarship.org/uc/item/2jb4j9cz. Accessed 10 Dec 2016
2. Mesarovic MD (1960) Control of multivariable system. MIT Press, Cambridge and Wiley, New York
3. Arnstein SR (1969) A ladder of citizen participation. J Am Plann Assoc 35(4):216–224. doi:10.1080/01944366908977225
4. Putnam R (1995) Bowling alone. J Democracy 6(1):65–78. doi:10.1353/jod.1995.0002
5. Lang R (1986) Achieving integration in resource planning. In: Lang R (ed) Integrated approaches to resource planning and management. The University of Calgary Press, Calgary, pp 35–39
6. Kubicek H (2007) Electronic democracy: achievements and challenges. Final Report on the ESF Research Conference in Vadstena, Sweden, 21–25 Nov 2007
7. Macintosh A (2006) eParticipation in policy-making: the research and the challenges. Exploiting the knowledge economy: issues, applications and case studies. IOS Press, Amsterdam, pp 364–369
8. Hong Kong Harbourfront Commission (2016) Final report on the proposed establishment of harbourfront authority in Hong Kong. In: http://www.hfc.org.hk/hape/doc/en/news/Report_to_CE_w_appendices.pdf. Accessed 30 Dec 2016
9. Planning Department (2004) Agreement No. CE 4/2004 (TP) South East Kowloon Development Comprehensive Planning and Engineering Review Stage 1: Planning Review (Feasibility Study). 3
10. Planning Department (2007) Planning history of Kai Tak. In: Kai Tak planning review. http://www.pland.gov.hk/pland_en/p_study/prog_s/sek_09/website_chib5_eng/english/index_e.html. Accessed 23 Dec 2016
11. Planning Department (2003) 3 review of metroplan form and function. In: Pland.gov.hk. http://www.pland.gov.hk/pland_en/p_study/comp_s/metroplan/metro_finalreport/ch3.htm. Accessed 23 Dec 2016
12. Planning Department (1993) South East Kowloon Development Statement Executive Summary. 1–15
13. Planning Department (2007) Planning history of Kai Tak. In: Kai Tak planning review. http://www.pland.gov.hk/pland_en/p_study/prog_s/sek_09/website_chib5_eng/english/index_e.html. Accessed 23 Dec 2016
14. Planning Department (2005) KAI TAK Report of Stage 1 Public Participation: Community's vision for Kai Tak. 1–43
15. Planning Department (2006) KAI TAK Planning Review Report of stage 2 Public Participation: Outline Concept Plans. 1–46
16. (2008) Our Mission. In: Kai Tak Development. http://www.ktd.gov.hk/eng/index.html. Accessed 29 Dec 2016

17. (2016) HK pioneers 3D tools in environmental impact assessment. In: Info.gov.hk. http://www.info.gov.hk/gia/general/brandhk/171004e1.htm. Accessed 29 Dec 2016
18. Fung M (2011) Evaluation of the implementation of social media of Hong Kong Government. Thesis. doi:10.5353/th_b4678036
19. (2016) Statutory Planning Portal 2 (TKO). In: Www2.ozp.tpb.gov.hk. http://www2.ozp.tpb.gov.hk/gos/default.aspx?#. Accessed 29 Dec 2016
20. (2016)In: Ex-lammaquarry.hk. http://www.ex-lammaquarry.hk/. Accessed 29 Dec 2016
21. (2016) United we build Tung Chung: welcome. In: Tung-chung.hk. http://www.tung-chung.hk/eng_index.html. Accessed 29 Dec 2016
22. Sieber R (2006) Public participation geographic information systems: a literature review and framework. Ann Assoc Am Geogr 96(3):491–507
23. (2011) Planning department annual report, 1st ed. 1–86
24. (2011) The Hong Kong Institution of Engineers. In: Hong Kong ICT Awards 2011—best public service application award. http://it.hkie.org.hk/en_it_news_details.aspx?ID=40&&TypeName=What%27s+new. Accessed 29 Dec 2016
25. (2012) Planning Department Annual Report, 1st ed. 1–87
26. (2015) Electronic planning application submission system (EPASS). In: Town planning board. http://www.info.gov.hk/tpb/en/eforms/esub.html. Accessed 29 Dec 2016
27. (2015) EPASS- FAQ. In: Town planning board. http://www.info.gov.hk/tpb/en/eforms/faq.html#q1. Accessed 29 Dec 2016
28. (2016) Smart KE iPEMap. In: Mapping.arup.com. http://mapping.arup.com/smartke/en. Accessed 30 Dec 2016
29. (MyKE) MRegion D (2016) My Kowloon East (MyKE) on the App Store. In: App Store. https://itunes.apple.com/ph/app/my-kowloon-east-myke/id1170321129?mt=8. Accessed 30 Dec 2016
30. (2013) CUHK Organized 'Magic Carpet: Re-envisioning Community Space in Sai Ying Pun'. In: Communications and Public Relations Office. https://www.cpr.cuhk.edu.hk/en/press_detail.php?id=1622. Accessed 29 Dec 2016
31. (2016) Neighborland. In: Neighborland. https://neighborland.com. Accessed 29 Dec 2016

Part II
E-Democracy Domain Studies

Chapter 6
The Indian Theory of E-Democracy and E-Governance and Its Linkages with the Smart Cities Mission

Ashok Kumar

Abstract A theory is a simplified cognitive construct of a certain aspect of the reality built for the purposes of explaining that reality in a reasonably simplified form. Internal coherence is a central feature of a structure of any theory. One significant aspect of today's city and urban planning in India is the reality of the smart city. Government of India is steadfast in developing 100 smart cities. A large number of seminars, conferences, and panel discussions are held since the last two years in the major cities of India involving global and local policy makers, business people, and academics examining and presenting policy elaborations, assessments, analyses and critiques. In view of this author, so far a theory of E-Governance specifically in the context of the Indian Smart Cities Mission has not been built and this paper is an attempt to fill this gap. One caveat however is necessary—a unique E-Governance theory for a smart city is necessary and possible for the Indian city given the special circumstances of Indian urbanization and its unique historical and cultural settings. The E-Governance theory for a smart city can be constructed in the form of a number of prominent nodes including use of information and communication technologies for enabling economic, social and political transactions, selective commodification and intense development of places within cities, intensification of inequalities within the city as conceived and defined by census, and development to be carried out through projects disconnected from city plans. In the understanding of this author, E-Governance theory should prominently focus on issues such as fairness of decision making processes in governments, businesses, NGOs, and most importantly decisions taken by the private sector and decisions taken by governments under the influence of the private sector. Second, E-Governance in Indian cities should focus on the nature of the technological agenda being pursued by governments and the private sector supported and promoted by big national and international think tanks. Third, the Indian E-Governance

A. Kumar (✉)
School of Planning and Architecture, 4 Block B, IP Estate, New Delhi, India
e-mail: dak13april1962@gmail.com

T.M. Vinod Kumar (ed.), *E-Democracy for Smart Cities*, Advances in 21st Century Human Settlements, DOI 10.1007/978-981-10-4035-1_6

theory should also focus on reducing the material inequalities in the cities specifically related with access basic services such as clean drinking water, sanitation, education, health, and decent housing for all citizens. Fourth aspect of E-Governance theory is that it should encourage and promote economic activities that involve the city's urban poor. No smart city can have an adequate theory of E-Governance without simultaneously looking after its citizens because E-Governance theory in a smart city is essentially for citizens' welfare.

Keywords E-Governance · E-democracy · E-government · E-business · E-arenas

6.1 What Is Theory?

Apart from innovating and developing new ideas, theories unearth unexplored relationships by distilling patterns based on the analyses of social processes embedded in lived human experience. In his magnum opus 'The Constitution of Society: Outline of the Theory of Structuration', Anthony Giddens elaborates: "Social theory has the task of providing conceptions of the nature of human social activity and of the human agent which can be placed in the service of empirical work that of the social sciences in general: the illumination of concrete processes of social life" [1: xvii].

John Friedmann argues about the omnipresence of some or the other theory. He points out that "there is no planning practice without a theory about how it ought to be practiced. That theory may or may not be named or present in consciousness, but it is there all the same. Thus, when we argue that planning *ought to be* in or *reflect* a general or public interest, we have in mind a theory of planning [2]. Eric Reade insists that "… a theory is not a theory at all, until it has been used in practice over a considerable period of time" [3]. Inseparability of planning theory and planning practice has been widely debated and accepted in planning literature. But more recently Neil Brenner made an important intervention by outlining critical urban theorist should believe in the following four elements [4]:

- they insist on the need for abstract, theoretical arguments regarding the nature of urban processes under capitalism, while rejecting the conception of theory as a 'handmaiden' to immediate, practical or instrumental concerns;
- they view knowledge of urban questions, including critical perspectives, as being historically specific and mediated through power relations;
- they reject instrumentalist, technocratic and market-driven forms of urban analysis that promote the maintenance and reproduction of extant urban formations; and
- they are concerned to excavate possibilities for alternative, radically emancipatory forms of urbanism that are latent, yet systemically suppressed, within contemporary cities.

Agreeing with Neil Brenner, I accept that planning theory is historically specific emancipatory project presently dominated by free markets where power relations are made invisible and considered normal. My own attempt in this chapter will focus around some of the issues raised by Neil Brenner.

Nonetheless, a diversity of views emerges about a definition and scope of theory. After reflecting upon these contested views, we can move forward to argue that a theory is a simplified cognitive construct of a certain aspect of reality built for the purposes of explaining that reality in a reasonably simplified form. Internal coherence is a central feature of a structure of a good theory.

Efficient working of cities has been a major concern of central, state and local governments in India. Introduction of E-Governance for effectively managing the Indian cities propelled by the internationalization of the economy and society in the early 1990s has increasingly become popular in the public and policy discourses. Smart city policy was first included in a national political party's election manifesto during the national elections of 2014. In order to fulfill this political promise, central government initiated the Smart Cities Mission along with several other such missions for rejuvenating the Indian city.

So far 33 smart cities have been selected after a due process for funding by central government. Thus one significant aspect of today's city planning in India is the reality of the smart city mission. A large number of seminars, conferences, and panel discussions are held since the last two years in the major cities of India involving global and local policy makers, business people, and academics examining and presenting policy elaborations, assessments, analyses and critiques. In spite of intellectual discourse, the idea of smart city remains under-theorized. Particularly in view of this author, so far a theory of E-Governance specifically in the context of the Indian Smart Cities Mission has not been built and this paper is an attempt to fill this gap. One caveat however is necessary—a unique E-Governance theory for a smart city is necessary and possible for the Indian city given the special circumstances of Indian urbanization and its unique historical and cultural setting.

Two kinds of theories could be developed for E-Governance in the context of smart cities. One possibility is that an E-Governance Theory for a smart city can be constructed in the form of a number of prominent nodes including use of information and communication technologies for enabling economic, social and political transactions, selective commodification of the city and intense development of places within the city without prominently explaining entrenched social and spatial inequalities. Alternative planning theory, which this author develops, is by giving a place of prominence to the challenges of Indian urbanization and building a planning theory for meeting directly these challenges.

In the understanding of this author, E-Governance theory should majorly focus on issues such as fairness of decision making processes in governments, businesses, NGOs, and most importantly decisions taken by the private sector and decisions taken by governments under the influence of the private sector. Second, E-Governance in Indian cities should focus on the nature of the technological agenda being pursued by governments and the private sector supported and promoted by big national and international think tanks. Third, the Indian E-Governance theory should directly focus on reducing the material inequalities in the cities specifically related with access basic services such as clean drinking water, sanitation, education, health, and decent housing for all citizens. Fourth aspect of E-Governance theory is that it should encourage and promote economic activities that involve the city's urban poor. No smart city can have an adequate theory of E-Governance without simultaneously looking after its citizens because E-Governance theory in a smart city is essentially for citizens' welfare. As several authors in an edited volume 'Cities for People, Not for Profit: Critical Urban Theory and the Right to the City' underscore that cities are meant for the general public rather than few capitalists [5]. Before I provide the brief outline of the E-Governance theory for the Indian smart city, let me make a brief sketch of the structure of the Indian planning practice.

6.2 Multiplicity in Indian Urban Planning

Urban planning is a critically significant function of the local government. However, many forms of organizations are involved in urban planning practice including development authorities, municipalities, slum boards, state town and country planning departments, etc. with varying roles and jurisdictions. For example, development authorities like Delhi Development Authority is entrusted with the task of bringing about orderly development of the entire area of the National Capital Territory of Delhi, and is established under a central government planning statute. State level development authorities also perform planning and development functions, and could be found, among others, in the states of Punjab and Haryana. City level development authorities exist as well in several parts of the country like Uttar Pradesh, Madhya Pradesh, Rajasthan, Karnataka, Odisha, Tamilnadu, Karnataka, Andhra Pradesh, etc. Then there are planning agencies that function at the state level but have restricted functions domain, for example, City and Industrial Development Corporation tasked with urban and industrial development of the state of Maharashtra. State level town and country planning departments are those organizations, which expectedly work for the entire state but particularly, prepare master plans for towns and cities where few or no planners exist. Slum boards have been largely stripped of their bigger role that they played earlier. Most slum boards now focus on housing for the urban poor including the low income groups and the economically weaker sections.

More recently planning function is also being delivered by the private sector as private entities get involved in the process of building cities for the rich and middle classes as well as elite neighborhoods. Colonization through the private sector is widely practiced in the state of Haryana and other states are not very far behind. For example, role of the private sector in building entire cities has come to prominence. For instance, Lavasa, a planned hill city set in the Sahayadri Mountains, is being built by Hindustan Construction Company near Pune. Several private companies such as Ansals, DLF and others have built large townships in Gurugram (previously known as Gurgaon). Foreign players are also beginning to make an impact in urban development as Indian economy is further deregulated and opened to international markets.

It is abundantly clear that a multiplicity of public, private, and public-private partnership agencies have crucial roles to play in Indian urban planning and development. Public agencies of all hues and colors have not greatly succeeded in democratizing planning processes. These agencies have not been able to encourage city dwellers to participate in their decision making regimes. For example, during the preparation of the Master Plan for Delhi, 2021 when suggestions and objections were invited by the Delhi Development Authority from the public, only over 7000 people participated out of the total population of nearly 17 million people, and literacy rate being 86.21% as per 2011 census [6].

So far we have not even mentioned the highly influential role that central government and state governments play in urban planning by making policies and through funding various kinds of projects. So many mission programmes and other planning policies have been recently framed by central government including the Smart Cities Mission, Atal Mission for Rejuvenation and Urban Transformation (AMRUT), Swachh Bharat Mission, Heritage City Development and Augmentation Yojana (HRIDAY), National Mission on Sustainable Development, and several others. While missions and policies made by higher levels of government promote centralization, e-government could reduce the impact of centralization on local governments by making continual dialogue between local, state and central governments a common place.

6.3 Explaining the Five Es

Before I start building the prime features of the Indian Theory of E-Governance and start establishing its relationships with city planning in general and the smart cities in particular, it is crucially important that I define the central terms being used in this paper and make my position clear about them.

(a) **E-Government**

Carrizales [7: 13–14] shows that e-government must have four functions. As outlined below, these four functions include e-organization, e-services, e-partnering, and e-democracy.

- E-organization: "E-organization is defined here as the use of technology such as e-mail and intranets for internal government efficiency and effectiveness. It also refers to the integration of technologies for horizontal and vertical communication between agencies and departments via the Internet".
- E-services: E-services is defined here as the use of technology for external government efficiency and effectiveness in providing services.

- E-partnering: E-partnering is defined here as the use of technology for external government efficiency and effectiveness in working with public and private organizations, including businesses, other municipalities (horizontal relationships), schools, hospitals, and public and nonprofit entities.
- E-democracy: E-democracy seeks to increase participation in government. E-democracy is defined here as the use of technology for increasing citizen participation in government decision making.

(b) **E-Democracy**

According to Coleman and Norris [8: 7] "E-Democracy is not about paying speeding fines over the Internet (that is e-government); it is about consulting on whether the speed limit on a particular stretch of road should be raised, lowered or left as it is. It may, one day, be used as a way of empowering citizens in the process of making major national decisions". When it comes to e-democracy, people remain at the center of decision making processes. In this line of thinking Whyte and Macintosh [9] "E-democracy relates electronic delivery of the processes of democratic representation to practices of communication, consultation, and participation in public decision-making". Marche and McNiven [10] argue that "E-democracy is a technology—mediated relationship between citizens and their governments from the perspective of potential electronic deliberation over civic communication, over policy evolution and in democratic expressions of citizen will".

E-democracy is a sub-set of E-Governance, which includes virtual interactions among the political classes, democratically elected governments (including urban local bodies) on one hand, and the citizens on the other through electronic media including the ICT. These interactions could take a number of forms such as e-election, e-referenda, e-participation, e-deliberations, e-consultations, e-administration, e-petitions, and e-campaigning, etc.

(c) **E-Business**

This is a situation where most of the economic transactions are undertaken with the help of electronic and telecommunication media. Global cities are the frontrunners in this direction where electronic and telecommunications play foundational role in control as well as coordination roles [11].

(d) **E-Society**

E-society refers to a society where citizens increasingly rely on digital technologies for day to day living, and society's needs and aspirations are met by governments and businesses through the use of internet and web-based technologies. Buying and selling takes place through the in.

(e) **E-Governance**

E-Governance involves e-government, e-democracy, e-business, and e-society. E-Governance has the potential to improve decision making processes by linking governments, the private enterprises, and the civil society in an efficient manner through participation enabled diverse electronic media. E-Governance could potentially liberate citizens from unaccountable and less transparent governments, large corporations and specifically from urban local bodies, where citizens have to interact on a daily basis for the effective delivery of services. The solution is found in the increased participation of the citizens in the affairs of the government, the private sector and the third sector through electronic media. E-Governance is useful from the point of public, businesses, employees and government [12: 290–291].

In relation to public:

- Tax Payments
- Paying Bills
- Voter Information
- Birth and Death Certificates
- Reservations and Status Information
- Complaint Tracking
- Video on Demand
- Public Information.

In relation to Business:

- Supplies
- Procurement
- Proposals and Bids
- Data Sales
- Permitting and Licenses
- Inventory Management.

In relation to Employees:

- Intranet
- Document Management
- Executive Info System
- Enterprise Resource Planning
- Geographic Information Systems
- Customer Relationship Management.

In relation to Government:

- Information Exchange
- Reporting
- Regional Planning
- Transfer of funds
- Joint Power Agreements.

6.4 E-Democracy and the Smart Cities Mission

Bringing about e-democracy in Indian cities generally is fraught with dangers because vested interests could easily manipulate digital technologies for their own private ends. To develop e-democracy in a smart city is even more difficult because tenants of the idea of smart city agitate against the very idea of democracy. Some of these are elaborated below.

6.4.1 Institutional Structures

The Smart Cities Mission is presented as an inclusive flagship programme. It is argued in the policy document that the Smart Cities Mission requires smart people where smart people would actively participate in governance and reforms. However, it appears that it is no more than a ceremonial participation in governance because the Smart Cities Mission of India requires that a smart city would be constructed under the overall control and supervision of a newly created organization called the Special Purpose Vehicle. Participation is expected in defining the smart city, decisions on deploying smart solutions, implementing reforms, and oversight during implementing and designing post-project structures. States and urban local bodies will play *key supportive role* in the development of smart cities. Simultaneously the policy also provides that Special Purpose Vehicle (SPV) will enable participation through ICT and mobile based tools [13].

The SPV will be a limited company incorporated under the Companies Act, 2013. The Special Purpose Vehicle will be headed by a full time CEO (subject to the superintendence, control and direction of the Board of Directors under section 53 of the Act). The SPV board will have nominees of central, state, and urban local government. The SPV will have state and urban local governments as promoters. Both promoters will have 50:50 equity. Private sector can have shares in this company provided the state and ULB together to have a majority shareholding and control [13]. The equity between governments can range from state government and urban local body with 25 plus 30% at the minimum level with private sector with equity of 45%.

Board of Directors of SPV will have representatives of central and state governments, ULB and independent directors, functional directors, and a CEO. Divisional Commissioner or Collector or Municipal Commissioner or Chief Executive of an Urban Development Authority will be a Chairperson. Central government representative will be a Director on the SPV Board. CEO of SPV will be appointed with the approval of MOUD [13].

The SPV will plan, appraise, approve, release funds, implement, manage, operate, monitor, and evaluate the smart city development projects. Projects could be executed through joint ventures, PPP, turnkey contracts, etc. The SPV may appoint Project Management Consultants for designing, developing, managing and implementing area based projects [13].

The Smart Cities Mission intends to make governance citizen friendly through online provision of services. It is also provided that urban local governments should form online e-groups to listen to people. Policy idea is that "… all the city residents feel there is something in it for them also" (Ministry of Urban Development [13: 7 and 9]. The rhetoric of the Smart Cities Mission does not point towards citizen involvement in government's decision making processes. To that extent, E-Governance is expected to be thwarted by this policy.

6.4.2 Projects Versus Plans

The Smart Cities Mission proposes pan-city programmes and area specific programmes. Strategic Action Plans (retrofitting, renewal and greenfields) are given more scores than pan city programmes. My own research reveals that nearly 81% of total outlay is earmarked for area specific programmes in the proposals of 33 cities that are declared smart cities by Government of India so far.

This kind of asymmetrical arrangement of plans and projects certainly hastens lop-sided development in a city whereby certain areas could get more material resources in the form of projects than others, and these areas could very well be those which are already resource rich. Uneven development, which is a natural outcome of urbanization under capitalist conditions, is further perpetuated through project centered mission programmes.

6.5 Chief Characteristics of the Indian Theory of E-Democracy for a Smart City

6.5.1 Fairness in the Decision Making Processes

Fairness of decision making in governments, businesses, NGOs, and most importantly decisions taken by the private sector and decisions taken by governments under the influence of the private sector and global players like the World Bank, IMF etc. should become the chief concern of the Indian theory of E-Democracy for a smart city. New Delhi Municipal Council is largely a settlement inhabited by the elite public servants and foreign missions. But there are several small and large slums. All income groups live in other smart cities too. Income diversity makes the idea of fairness even more relevant and urgent. In this context we can think of at least three dimensions of fairness:

(a) **Procedural Fairness**

It relates to the process by which decisions are made—the steps to be followed before, during and after decisions are made i.e. how decisions are made.

(b) **Substantive Fairness**

It relates to the fairness of the decision itself as spelt out by law or other institutions, e.g. promotes rights and avoids oppressions.

(c) **Relational Fairness**

It relates to how people are treated during decision making process, e.g. courteous, timely, clear communication, etc. [14].

We could also think of distributive justice as fairness as shown by Walzer. First, it is believed that the shared economic, social and cultural infrastructure shapes our private lives and privatization takes us to least democratic government. Second, a system of communal provision is promoted and it is argued that the welfare state must meet needs of the people. Third, equality of opportunity means that the slope of ambition and advantage has to be less steep so that even weaker persons are able to reach their dreams. Fifth, strong democracy should be promoted where political power is widely available to all citizens [15] also see [16]. So fairness in all its forms and manifestations is the nerve center of the theory of E-Governance in smart cities.

6.5.2 Nature of the Technological Agenda

Second, E-Governance in Indian cities should focus on the nature of the technological agenda being pursued by governments and the private sector supported and promoted by big national and international think tanks. Smart Cities programme of the IBM larger initiative of the Smarter Planet and involvement of other private giants such as Siemens, Cisco, etc. have been promoted certain technologies like sensors' driven dashboards in order to instrumentalize cities. Smart cities themselves are viewed as complex systems, which need to be controlled through information technologies. Batty et al. [17] succinctly explain: "Cities however can only be smart if there are intelligence functions that are able to integrate and synthesise this data to some purpose, ways of improving the efficiency, equity, sustainability and quality of life in cities. In FuturICT, we will research smart cities not simply in terms of their instrumentation which is the domain of both large and small ICT companies who are providing the detailed hardware and software to provide what some have called the operating system for the smart city, but in terms of the way this instrumentation is opening up dramatically different forms of social organisation" [17: 482].

6.5.3 Reducing the Material Inequalities in Cities and Towns

Third, the Indian E-Governance theory should also focus on reducing the material inequalities in the cities specifically related with access basic services such as clean drinking water, sanitation, education, health, and decent housing for all citizens. Material inequalities are manifested in the form of five challenges of urbanization that we comprehensively discussed in our latest contribution [18]. These challenges are:

- The challenge of urban infrastructure with a sanitation focus
- The environment Remonstrance
- The mobility challenge
- The governance challenge
- The challenge of urban planning.

Radical substance and content of this theory should be assessed in terms of how far the Indian Theory of E-Governance is able to explicate the above mentioned urban challenges.

6.5.4 Promoting Informal Economic Activities of City's Poor

Fourth aspect of E-Governance theory is that it should encourage and promote economic activities that involve the city's urban poor. No smart city can have an adequate theory of E-Governance without simultaneously looking after its citizens because E-Governance theory in a smart city is for citizens' welfare. The core issue is how strength of IT and ICT could be deployed to support informal sector businesses. This may include use of information technologies for identification of spaces for hosting informal sector businesses, grant of permissions or licenses for carrying out such businesses, distribution of commodities and services for sale to designated informal spaces, maintenance of hygienic environments in the informal premises, payment of rents for hiring business spaces, facilitation of payment to workers involved in the informal sector, and reduction in police harassment, which is to be separated carefully from enforcement of genuine rule of law which is a critical part of e-democracy.

6.5.5 Inter-organizational Coordination

Given that the function of urban planning will be delivered by a multiplicity of organizations as mentioned above, any Indian theory of E-Governance must also take into account how coordination among the public, private, partnership agencies, and the third sector would be undertaken. This theoretical site becomes even more important as we are aware about the lack of coordination within government planning agencies. One development agency remains blissfully unaware of the related activities being undertaken by the other agencies. This continuously happens between Delhi Development Authority on one hand and the Delhi Municipal Corporation on the other (see [19]).

Countrywide this is a common problem as new organizations are created. For example, after 73 and 74 amendments to the Constitution of India several organizational entities came into existence that clearly affect delivery of the planning function. For example, in urban area three forms of municipalities were created. In addition, Metropolitan Planning Committees and District Planning Committees under the Seventy Fourth amendment to the Constitution of India were also set up notwithstanding the fact that Metropolitan Planning Committees have not been set up in the country so far with exception of Kolkata Metropolitan Planning Committee. Add city level development authorities (for example in Madhya Pradesh and Uttar Pradesh) and state level development authorities (for example in Punjab and Haryana) further complicate the issue of inter-organizational coordination.

In this maze of organizations—elected and unelected—e-democracy would suffer because there is no transparency of decision making processes, and issues like policy framing and implementation that is reflective of citizens' aspirations and desires takes a back seat.

6.5.6 Deliberation and Debate

So far we have only seen electronic and communication interface between governments, citizens, businesses, partnerships of various forms, and the third sector. Mostly this is of informing kind whereby stakeholders are able to secure information from the governments and municipalities. We have also seen that governments provide a large array of services to the citizens through electronic portals where citizens avail the services through the internet. Services like payment of property tax, payment of water, cooking gas, and power bills, issuance of birth and death certificates, etc. to name just few of them can be done through the internet. Complaint mechanisms have also been set up through dedicated application based arrangements whereby citizens can take a photograph, say, of a garbage pile, and send that to the concerned authorities on a GIS enable application. For example, New Delhi Municipal Council has launched NDMC 311 app, which can help citizens to secure 11 sets of services. Indian e-democracy requires *e-arenas* for discussion, deliberations and debate, which are rare because to effective such electronic spaces must involve feedback mechanisms. As an OECD research report insists that "feedback on citizens' input to e-consultation can, hopefully, contribute to the overall transparency, accountability and openness of government" [20].

For this author even more crucial question is: how these electronically embedded technologies can provide some sort of social, economic and spatial centrality to the poor and marginalized citizens, who live in the peripheral places of our cities and towns. If IT and ICT could achieve political ends of e-democracy of the urban poor, achieving other ends could be relatively easier. Achieving this goal will promote e-democracy because a majority of the people in Indian cities are either poor or lower income groups by any reasonable global income standards. Income poverty and e-democracy cannot be made bedfellows as both repel each other.

6.6 Concluding Remarks

Theories are useful tools and indeed play important instrumental role by making city planning endeavours effective and egalitarian. Planning theories also play transcendental role by promoting certain kind of ethics and moralities. E-Democracy and E-Governance Theory for the Indian Smart City sketched in this chapter is also expected to play this double role. A majority of the citizenry living in poor and desperate territorial circumstances deserve to be made part of any theoretical movement because cities are built for people in general rather than commodities manufactured for specific markets. Therefore a critical role played by democratically elected governments is required to be made central to any theory of e-democracy and e-governance for the smart cities.

More significantly information technologies should be used to reduce spatial inequalities. E-democracy and e-governance is a good idea but should it be used only for collection of taxes and for issuing of certificates. This appears to be a thin idea of e-democracy and e-governance. Even it is a less appropriate use of technology. Information technologies instead could be also used to meet ends of spatial justice. For example, during the building o neighborhoods, roads and bus stops, etc. we could plan and design these places in multiple ways that these sensors and sophisticated cameras seamlessly collect data to reduce crimes against vulnerable social groups. Among other things, technology is a political tool and its use cannot be left uncontested in the cities and towns of the country.

References

1. Giddens A (1984) The constitution of society: outline of the theory of structuration. Polity Press, Cambridge
2. Friedmann J (2003) Why do planning theory? Plan Theor 2(1):7–10
3. Reade EJ (1987) British town and country planning. The Open University, Milton Keynes
4. Brenner N (2009) What is critical urban theory? City 13(2–3):198–207
5. Brenner N, Marcuse P, Mayer M (eds) (2012) Cities for people, not for profit: critical urban theory and the right to the city. Routledge, London

6. Census of India (2011) Provisional population tables, Census of India, Government of India, New Delhi
7. Carrizales T (2008) Functions of e-government: a study of municipal practices. State Local Gov Rev 40(1):12–26
8. Coleman S, Norris DF (2005) A new agenda for e-democracy, Forum discussion paper no. 4, Oxford Internet Institute, The University of Oxford, Oxford
9. Whyte A, Macintosh A (2002) Analysis and evaluation of e-consultations. E-Ser J 2(1):9–34
10. Marche S, McNiven JD (2003) E-government and e-governance: the future isn't what it used to be. Can J Adm Sci 20(1):74–86
11. Sassen S (2005) The global city: introducing a concept. Brown J World Aff XI(2):27–43
12. Barthwal CP (2003) E-governance for good governance. Indian J Polit Sci 64(3/4):285–308
13. Ministry of Urban Development (2015) The smart cities mission, Ministry of Urban Development, Government of India, New Delhi
14. Ombudsman M (2009) Achieving fairness: your guide to dealing with government. Manitoba Ombudsman, Manitoba
15. Walzer M (2007) Thinking Politically: essays in political theory. Yale University Press, New Haven, CT
16. Walzer M (1983) Spheres of justice. Basic Books, New York
17. Batty M, Axhausen KW, Giannotti F, Pozdnoukhov A, Bazzani A, Wachowicz M, Ouzounis G, Portugali Y (2012) Smart cities of the future. Eur Phys J Spec Topics 214: 481–518
18. Kumar A, Sarkar PK (2016) Making Delhi a smart city: economic buoyancy with spatial justice. In: Vinod Kumar TM (ed) Smart economy in smart cities. Springer, Singapore
19. Kumar A (2000) Some problems in the coordination of planning: managing interdependencies in the planning of Delhi. India, Space Polity 4(2):167–185
20. OECD (2003) Promise and problems of e-democracy: challenges of online citizen engagement. OECD Publications Service, Paris

Chapter 7
SMART DUBAI: Accelerating Innovation and Leapfrogging E-Democracy

Ashmita Karmakar and Ummer Sahib

Abstract Democracy means a million-different transaction of information to a million-different people, so does E-democracy with the help of information technology system and social media. At present the world is experiencing a dramatic "E-democracy evolution." In this evolution, the role and interests of the current and future activities of all sectors—government, public and citizen is undergoing an immense transformation. The transformation is rapid and continuous for many cities of the world and Dubai is one such city which is progressing steadily in the sector of online activities to achieve the greater goals of E-democracy. Innovation has played a major role in the region. With immense support from the ruler and his extremely progressive vision, Dubai residents has shown readiness and activeness that has supported government smart strategies and in return, the government took responsibility by continuously creating jobs for the citizens. This chapter is a presentation of a framework which shows how UAE and Dubai has developed a strategy in many sectors including the field of non- oil sectors where smart technologies and innovations of government and resident's collaborative approach has resulted in the strong bonding between the public and private sector and eventually became responsible in the growth of the country. The chapter illustrates the potential role of government in introducing smart technologies and letting the residents enable the transformation of traditional identities into state of the art-digital identities. Further, the chapter highlights the impact of oil prices and the government's renewable energy initiatives to keep the pace of development unaffected and transform the country into a green country through its effective policies and world class project development by the year 2030 using underline strong bonding of government and its people- setting a unique feature of democracy.

A. Karmakar (✉)
Villa 15, Street 2, Meadows 8, Emirates Hills, Dubai, UAE
e-mail: ashmita.karmakar@gmail.com

U. Sahib (✉)
Informap Production LLC, Office 502, The Fairmont,
Sh. Zayed Rd., Dubai, UAE
e-mail: us@informap.ae

T.M. Vinod Kumar (ed.), *E-Democracy for Smart Cities*, Advances in 21st Century Human Settlements, DOI 10.1007/978-981-10-4035-1_7

Keywords Smart dubai · E-democracy · Strategies · Vision · Innovation · Smart technology · Public private participation · Renewable energy · Expo 2020 · Hyperloop

7.1 Introduction

E-democracy is the process where communication technologies and internet has provided opportunities to transform the 'Government-Citizen' relationship in a new way, contributing to the achievement of good governance goals. Information technology has increased the involvement of citizens in the process of governance at all levels through on-line discussion groups that enhanced the development and effectiveness of the system. The government is able to provide better service in a fast, efficient and effective manner at a significantly low cost and become more accessible to the public at large.

E-democracy refers to activities that increase citizen involvement including virtual town meeting, open meeting, cyber campaigns, feedback polls, public surveys and community forums such as e-voting. It is also the political dialogue in which citizens and the community in general, engage in the political process.

Typically, the kinds of enhancements sought by E-democracy proponents are framed in terms of [1]:

- Making processes more accessible
- Making citizen participation in public policy decision-making more expansive and direct to enable broader influence in policy outcomes
- Increasing transparency and accountability
- Keeping government closer to the consent of the governed.

The UAE has advanced from Rank 42 of 193 in 2005 to rank 29 according to the 2016 UN E-Government Readiness Survey. This chapter will highlight those fields by which the country has achieved its status and focus on those government sectors that has revamped its websites by integrating information and services into a single gateway where its services can be easily located. Dubai government has re-branded itself and came up with the digital system for sharing its information with the residents, visitors, citizens or the end users and enable them to make use of the website for availing services or sending feedbacks. Furthermore, the government took steps to provide participatory features on one of its ministry websites making it one of the few government sites in the Middle East that offer an open discussion forum.

The creation of smart government is further accentuated by inventing various digitization, introduction of applications and innovative processes for providing services to the people. The chapter will discuss those innovations and applications as provided by the UAE government for residents to carry out business transactions on a daily basis.

This chapter will also focus on the use of renewable energy and for the path to become a smart sustainable city. In this way the government of this country although following monarchy is involving its people for participating with innovative and smart ideas and letting the country to set a unique example of E-democracy.

The people of Dubai are blessed with a visionary leader armed with a great team with diverse leadership qualities and backed by highly motivated youth who are dared to scale the limits of human excellence. UAE is the one of the first Countries that has a Minister of Tolerance, Minister of Happiness and a Youth Minister. Being City Planners and long-time residents of Dubai, the authors are fortunate enough to witness Dubai becoming a smartest city and report the transformation as it unfolds. This chapter will delight the readers with Dubai's smart innovations and E-democracy initiatives to serve as a model for any smart city.

This chapter is split into 5 sections;

Section 7.1, describes E-democracy and its implementation in Dubai
Section 7.2, covers factors leading to the uniqueness of the City
Section 7.3, illustrates Dubai government's smart innovations
Section 7.4, showcases Green City initiatives, and finally
Section 7.5, reflects on how these innovations foster E-democracy.

7.2 Why Dubai Selected for Study?

All global rankings place Dubai at the top in the Arab world, known for its innovative smart city projects. Following sections give an overview of the city that illustrates its unique position.

7.2.1 Political Set up

The UAE's political system combines traditional and modern with a federal system of government including the Supreme Council, the Council of Ministers (Cabinet), a parliamentary body-the Federal National Council (FNC) and the Federal Supreme Court as an independent judiciary. The Supreme Council elects a president and vice-president from amongst them to serve for a renewable five-year term in office [2].

Unique federal government: Dubai is subject to the federal law of the UAE but retains the right to administer its own internal affairs and enjoys certain other exclusive rights. There are federal codes of law applicable to civil, commercial, civil procedure, companies, intellectual property, immigration, maritime, industrial, banking and employment law. The laws enacted by the Ruler of Dubai relate to matters which are more administrative in nature, such as the establishment and operation of government affiliated entities. Dubai's courts comprise a Court of First Instance, a Court of Appeal and a Court of Cassation.

7.2.2 *Demographic Profile*

Demography plays a dominant role for E-democracy since the target population is the end users of innovation and ultimately the prime drivers for the country's progress. As Dubai stands out as the most progressive and smart city of the region, its mix of population can be seen as an opportunity and also a challenge for its progress and success. The factors given below summarises the character of Dubai's demographic pattern.

Highest concentration of population: Dubai being the hub of trade and commercial activities, hospitality, education, healthcare and unlimited events and leisure activities it has the highest concentration of population [3]. The high population figure and mix of various nationalities make Dubai a unique city for exploration. As per the Dubai Statistical Center, the current resident population is 2,655,819 (updated November 2016). The active day time population is about 3,408,000. This population comprises of people working in Dubai but staying outside, tourists, sailors and other temporary residents [4].

Working population: The potential of the city is very well reflected by the young population between the age group of 25–35 years. Most of this population is under working sector and with high affordability as well. The young population is making Dubai techno savvy and increasing the adaptability of smart techniques and innovations [5]. The immigrant population belongs to Asia, Middle East and North African (MENA) region and the white collar sector comprises of people from the United Kingdom, United States, Australia, Canada, various Western European countries in the high-skilled sectors such as the oil and gas industry as well as banking and finance.

High affordability: Dubai also ranked as the first leading city destination for international visitor spending—tourists spent over 40 billion U.S. dollars there in 2016 with the average of AED 3000 spent by each tourist as per Dubai Tourism Statistics and Trends [6].

Population mix: UAE's population can be classified under three major sectors. They are the Emiratis/ the citizens, the expatriates- comprising of the working population and the visitors who come from all over the world for business or tourism purpose.

- **The citizens**: At present the Emirati population in the UAE is about 1.4 million to 2.0 million which is 20% of total population of UAE. The citizens are blessed with huge support from the government and their high affordability make them gadget savvy and acquire new and innovative products.
- **Expatriates**: white and blue collar: Owing to the increase of investment and job opportunities in Dubai, there is a considerable number of expatriate population

contributing to the total population of UAE with most the expatriates residing in Dubai. In 2013, the total population of UAE was recorded to be 9.2 million with 7.8 million of expatriates and 1.4 million Emiratis or Nationals [7].
- **The visitors**: The number of tourists in Dubai increased year-on-year between 2010 and 2016, reaching over 15 million in 2016. This made Dubai the fourth most visited city destination worldwide after Bangkok, London, and Paris according to the MasterCard Global Destination Cities Index. The estimated visitor's population is expected to reach about 20million by 2020.

Political support: The existing political structures appear to suit the locals as well as the expatriates. The distribution of huge revenues in the form of social and economic infrastructure, high salaries, a high standard of social services, such as health and education, has raised the standard of living for Dubai.

Dubai's active participation in many regional and international associations such as the Arab League, the United Nations, the Non-Aligned Movement, the Arab Gulf Cooperation Council, and the Organization of the Islamic Conference and its good relations with Western democratic countries, have been traditionally warm. Dubai has enjoyed a political stability, resilient nature made this place more hospitable, organized and safe to live in.

7.2.3 Social and Human Resources

Investment in people: This is considered as the 'wealth of the nation' and has been a primary focus of the UAE Government. Efforts are being made to develop human resources, empowerment of women and provide social welfare to the more vulnerable in society. Despite the demographic imbalances brought about by urbanization and immigration, UAE continues to be a tolerant, open, caring society that cherishes its traditional roots.

UAE rose to ranked forty-first in the Human Development Index (HDI) compiled by the UN Human Development Report for 2005, compared to its forty-ninth position in 2004. This progress was reflected by the rising income per capita, an increase in government expenditure on education and health services, a higher percentage of women participating in the work force, a low rate of post-partum and infant mortality, and the elimination of dangerous diseases.

Secured employment for citizens: The UAE nationals account for a very small percentage of the total workforce especially in the private sectors. Therefore, the creation of job opportunities for national graduates in the private sector is a situation that the Federal Government is anxious to address.

The establishment in 2001 of the National Human Resource Development and Employment Authority (Tanmia), an independent body was created whose main objectives are to create job opportunities for UAE nationals, especially in the private sector, reduce the unemployment ratio, enhance the skills and productivity of the national workforce through relevant training, undertake policy oriented research on a multiplicity of labour-market issues and make recommendations to the Government.

Involvement of citizens in trade unions: Labour Law authorizing the Labour Ministry to issue the necessary regulations facilitating the establishment of trade unions was submitted to a ministerial advisory council. The memorandum suggested that membership of trade unions should be open to UAE nationals and expatriates, but there should be full membership for citizens and associate membership for expatriates.

Uplifting labour sector: The Labour Ministry has taken significant steps in immediate processing of visas and cancelling of bank guarantees, to remedy difficulties over the prompt payment of worker's salaries, working and living conditions, and the regulation of labour agencies that recruit construction and other temporary workers, especially those that illegally charged for visa applications, workers thereby incurring crippling loans.

The housing conditions, health and safety of labour accommodation camps are monitored by an innovative technology that uses high-resolution satellite imagery. This is an effort to fill a big gap in the needs of an under-staffed inspection department.

The government is planning to construct specialized 'workers cities' for temporary workers to introduce the concept of affordable housing and low density housing and offer property ownership to the workers as well.

The Government of UAE has launched Retirement Pensions and Benefits Fund website to provide a wide range of useful information and services to its members comprising insured individuals, pensioners and beneficiaries in the government, semi-government and the private sectors, throughout UAE.

Ministry of Labour and Social Affairs with the network of Ministry-supported social centers of General Women's Union along with the government supported social welfare and rehabilitation centers assist the disabled citizens for free and provide subsidized housing.

Rights for physically and economically weaker sector: The UAE Red Crescent (RCS) is the largest charitable organization in the country and a major international relief agency created and their volunteers to contribute to social welfare services and development programs covering such areas as medical aid, student sponsorship, disability, special care, health care, the welfare of prisoners and lump-sum aid programs.

Free food and drinks are offered to the needy people at the kiosks in the public places.

Believing that handicapped persons possess skills and capabilities that could be utilized in economic and social development, the Government has established care centers for persons with special needs. It has also expanded and enhanced vocational training programs for the handicapped so that they can acquire professional skills, build their self-confidence and facilitate their integration into society.

The draft of a ten-page law enshrining the rights of the disabled was prepared that sets out general principles on the rights of all persons with special needs.

Job guarantee through quota in the public and private sectors, increasing accessibility to public buildings and residences and integrating people with special needs into public and private schools and even pension schemes after their retirement.

Creation of a body called DTCM that works round the year to understand the sensitivities related to special needs tourism and constantly works towards achieving those goals.

Care for the youth: Recognizing that the youth of today are the future of the country, the UAE has implemented child-centred programs, enacted laws and acceded to regional and international conventions protecting children's rights, including the UN Convention on the Rights of the Child.

The UAE government has set up many health care centers, medical centers, child care centers and centers for mother and child. Introduction of compulsory education for both sexes at primary level and free education for all stages from kindergarten to university.

New laws were enacted to ensure the safety and protection of children within the family or in national institutions that are concerned with providing care to orphans and abandoned children and to protect children from all forms of exploitation, the latest of which is the law that specifies a minimum age of 18 years for jockeys in camel races.

The UAE has made significant contributions in cash and in kind to the UN agencies concerned with children, and extended direct assistance to developing countries and countries affected by natural disasters and armed conflicts, where children suffer from poverty, illiteracy, violence, displacement, sexual exploitation, trafficking and recruitment in wars, to assist these countries in improving their social and economic conditions and creating a world fit for children.

Care for tourists/ visitors: Dubai's Tourism Vision for 2020 is a strategic roadmap with the key objective of attracting 20 million visitors per year by 2020, doubling the number welcomed in 2012. The Vision was approved in May 2013 by the ruler of Dubai. This strategy outlines what needs to be delivered for the city to effectively drive and serve this visitor growth with multiple initiatives covering regulatory policy, infrastructure development, product offering enhancement, and destination marketing investments. The overall goal of these initiatives and the strategy is to position Dubai as the 'first choice' for the international leisure and business traveller [6].

Dubai has created a body called DTCM that empowered by the Government to issue legislations and instructions from time to time that govern Dubai's tourism industry. The focus of DTCM is to maintain the market share in existing source, increasing market share with high growth potential and increasing the number of repeat visits. The aim of DTCM is to ensure a high level of visitor and customer experience across the full tourism value chain. It is responsible for the classification hotels, hotel apartments and guest houses; issue, renew, amend and cancel licenses of all tourism related activities such as tours; issue permits, for entertainment events, conferences and exhibitions, among others.

DTCM is responsible for administrative tasks related to the issuance of permanent/temporary tourism activity permits, follow-ups and inspections of such authorized activities to ensure their compliance with the laws and regulations implemented in Dubai [8].

7.2.4 Expo 2020 Host

For Dubai and the UAE, hosting Expo 2020 is a high calibre exhibition that could fuel the future pillars of the economy of the region with tourism, construction and hospitality sectors set to see monumental growth [9].

Expo 2020 will be a transformational event for Dubai and its neighbouring region in terms of:

- An Economically-Driven, Technologically-Focused Vision of UAE/Dubai's Development and Diversification
- Building up multiple Opportunities and a greater role for the private sector
- Signifying strategy of international integration and openness in the region
- Becoming the hub of region's young, growing population + Road into Africa and SE Asia

Bringing "Mobility, Sustainability and Opportunities" with efficient resource usage, greater connectivity between people and places, and broader opportunity for all [10].

7.3 E-Democracy Evolution in UAE

Unlike many other parts of the world, the Middle East Region has limited experience with public participation in policymaking. Political stability, with wider cultural and socio-economic conditions, varies greatly throughout the region—presenting each country with its own set of unique challenges in terms of fostering enhanced citizen participation.

The UAE enjoys relative political stability, and is ranked among the top in the World Happiness Index. Rulers take strong care of the country's more than a million citizens, and there is an apparent equitable distribution of wealth. Therefore, political apathy is high. As the population grows, popular movements increase throughout the region and around the world, it is important that all Emiratis, particularly more socially excluded groups like the poor, illiterate and less abled, be included and have a greater voice in the running of the government. The challenge for the Government is to pave the way for democratic participation whilst avoiding the type of massive social unrest that other Gulf States have recently witnessed.

In other words, the challenge is to create a successful model of democratic nation building in a region that is virtually lacking in democratic traditions.

It is within this context that the leadership of the UAE took the bold and conscious decision to embark upon a far-sighted initiative to establish democratic governance structures and foster enhanced public participation in decision-making. So far the following processes are implemented:

- Introduction of E Voting in FNC where citizens are registered with National ID Cards
- E- Readiness approach by the Government
- Open data (in digitized version) government information
- E Participation or Citizen engagement
- Public consultation
- Public participation in Innovation.

Introduction of ID Cards and E Voting system in FNC Election:

Embracing this challenge as an opportunity, the leadership took the decision in 2006 to introduce elections for citizens to sit on the FNC as the first step in a wider effort to embed democratic participation into a way of life for all Emirates. The government knew that the ultimate success of this long-term effort would depend upon the extent to which the Emirati population had a strong sense of confidence and pride in the new democratic system. Thus, rather than replicating an often-discredited paper-based system, it decided to use the country's ground breaking National ID Card as the cornerstone of an innovative new e Voting system that was tailor made to install transparency and trust.

At the time, the number of participants in the E-voting process though was limited to 6596 citizens in a government selected 'electoral college'. A number which increased by 20-folds to 135,308 participants in the 2011 FNC election process in the UAE.

E-readiness:

The United Arab Emirates (UAE) is one of the countries that have made tremendous strides in advancing its E-readiness global ranking from 2004 to 2005. The gain is largely attributed to a revamped national site that integrates information and services into a single gateway where its services can be easily located.

Introduction of open ended discussion forum:

Furthermore, the government took steps to provide participatory features on few of its ministry websites, and offered an open-ended discussion forum. An interesting feature on the UAE gateway is the organization of the site by topics that are geared towards addressing the needs of the end-users, integrating information, services, and transactions under separate sections for residents, business, visitors, and government, thereby enhancing its interactive presence.

Introduction of open data:

Acknowledging the importance of government data availability in the age of open and big data, the UAE did manage to achieve several milestones in providing citizens with public information and access to informational services, reaching for example, a high score of 93% according to the UN's 'E-information' indicator.

E-participation:

Until recently, there has never been a formal or even informal process in the UAE where the public was able to communicate directly and systematically with the government—the government could take part alone in decision making, services design or delivery.

- Public rating of government services:

Today, citizen engagement and E-participation are strongly acknowledged as vital processes enhancing government efficiency and sustainable development in the country. The UAE government has taken expedited steps towards engaging larger segments of the public by utilizing the ubiquitous ICT in the country. The impact of these public participation initiatives was clear on the levels of public satisfaction. A regional study conducted by MBRSG's Governance and Innovation Program found a clear co-relation between levels of citizen engagement on government services and levels of satisfaction with the public services provided. Across all government sectors, the research showed that service ratings submitted by customers who said that their governments regularly asked them for suggestions and input into the service design process were considerably higher than those submitted by their counterparts who said that they were never asked for input.

Recently, the UAE ranked 13th globally in the UN "E-participation index" which measures the process of engaging citizens through ICTs in policy and decision-making in order to make public administration and service provision participatory, inclusive, collaborative and deliberative.

- Public interaction through social media:

The UAE society is widely recognized as 'tech-savvy' and the government in the UAE has traditionally been active in pursuing reform and modernization methods through technological means, especially in terms of public service delivery. The positive public response here towards using social media for citizen engagement in public service in the UAE is not surprising. Earlier findings of multiple public surveys conducted by MBRSG's Governance and Innovation Program confirm the positive perception held by the public generally towards interacting with government entities through social media. Most respondents in that national research project agreed that the benefits presented by engagement through social media for service delivery included: improved quality of services, increased service accessibility, inclusiveness and reduced costs.

It is clear that people in the UAE have high expectations and positive views of the advantages of using social media for service delivery. The receptiveness of UAE customers to social media driven citizen engagement efforts appeared to be high.

Similarly, government employees in the UAE hold positive views towards utilizing technology and social networking technologies to enhance government operation. The UAE's public sector also indicates that government employees in the UAE have long realized the potential of social Web 2.0 tools. A majority of respondents in that study considered utilizing social networking tools a key tool for fostering collaboration and trust within and between government agencies. This perception continued to be evident that citizen engagement through technology, and specifically social media, was viewed positively by government employees individually and institutionally.

The Ruler of Dubai, an active social media user himself, when he thanked his six million followers on Twitter, Facebook, Instagram and Google + and called social media channels an "open parliament that cannot be ignored or misled". The organizing team realized this vision stating "the trend communities through Social Media. The UAE is well positioned to be the number one government when it comes to social media interaction with its community.

E consultation:

Process evaluation aims to identify the main problems encountered, whether the consultation reached the target group and the level of participant satisfaction. Evaluating the impact of consultation provides an estimation of whether participants' input had an identifiable impact on the content of the final policy decision. Through public feedback, the health and education sector was proved to be the weakest sectors of UAE, thus improvement of those sectors became the primary aim of the government.

The targeted public for obtaining feedback was not limited to online participations but also obtained from representatives of students, teachers, parents, school administrators, doctors, nurses and other specialists in the health and education sectors were also invited to share their feedback in person during the Cabinet retreat. Evaluation results were well communicated to the Cabinet whose outcome initiated a number of projects.

Public participation in innovation:

UAE is widely recognized as a global leader in enhancing the quality, efficiency and effectiveness of public sector services through creative approaches. Since its inception in 1971, the United Arab Emirates has constantly been distinguished as an icon for innovation and creativity, enhancing its social and economic status and transforming into a primary destination for talents and businesses in record time.

The government has managed to develop multiple innovations in public service delivery through taking advantage of the country's high internet penetration rates and the high level of technology acceptance in society. The population of the region which comprises of mostly young people with high affordability has welcomed the change and readily accepted the technological transformation.

The progress that has also been achieved by the UAE in this field resulted from the adoption of creativity and innovation methodology by government institutions across the country. The 'Emirates Center for Strategic Studies and Research' (ECSSR) has played an important role in promoting the culture of creativity and innovation through its various activities and events.

The later section will highlight the types and details of innovation that has taken place in the UAE.

Views of UAE people on selection of leaders:

Many Emiratis were upbeat at what they called "a great leap" the UAE's leadership has taken to extend political rights to nearly 25% of the citizens eligible to vote. They said they feel they are very close to free direct elections, hoping the next parliamentary term in 2015 will see every citizen practising his or her full political right.

UAE to achieve better shape:

With the national elections including a large pool of citizens will contribute to the building of a successful parliamentary experiment, paving the way to free, direct elections and enhancing the democratic experience in the country. The electoral rolls, including more than 129,000, Emiratis eligible to vote in and run for the Federal National Council (FNC) elections makes almost 25% of citizens eligible by age going to vote in the elections, the UAE will certainly be in better shape.

Hope for a different future:

At present about 129,274 citizens can cast their vote to elect 20 members of the 40-seat FNC, while the other half will be appointed by the nation's Rulers. In future there is a hope of the upcoming parliamentary experiment involving the development of the election system, which will in turn lead to direct general elections.

Participation of all age:

The electoral rolls were mostly made up of the young generation, with 74% younger than 39 years of age. As many as 31,452 citizens or the biggest segment on the rolls were of the age group 25–29 years.

Achievement of greater powers:

The citizens are hoping constitutional changes for greater powers for the FNC and election law to "accommodate the evolution of the UAE as a democratic state." The significant increase in the number of Electoral College members in 2011 underlines the commitment of the UAE and its leadership to further promoting political participation in the UAE. Registration of nominations will be undertaken at 13–14 polling stations across the country, compared to seven in 2006. About 46% of citizens on the rolls are women, almost three times the numbers in 2006 elections.

Safe voting:

Voters with Emirates identity card can take part in the upcoming elections. E-voting is used to ensure a safe, effective and convenient election process.

7.3.1 Benefits of People-Government Participation

- Citizen Engagement: Involve the community in government decision-making processes and contributing to government strategies
- Development: Encourage citizens to participate in developing health and education sectors in the country
- Ideation: Contribute to establishing a "Government Innovation Laboratory" in the cabinet retreat, where members of the cabinet can review and discuss the ideas submitted by the public
- Citizen Centricity: Encourage the citizens to participate in making positive changes to their lives and contribute to improving the critical sectors such as health and education
- Collaboration: Utilize the possibilities offered by information and communication technologies (ICT) to collaborate with the public for better services
- Partnership: Create a creative partnership between the public-sector agencies and the citizens through modern communications modes especially, social media
- Trust: Build a stronger level of trust between government and people better information exchanges using social media, in a serious shift from the one-way communication flow [11].

7.3.2 Outcomes of the Participation

- New solutions for existing and identified public service problems: The public produced new solutions to solve identified problems in health and education public services
- Identifying new problems in existing public service: Many new problems in existing services were identified through public participations
- Identifying indirect negative impact of existing services: The brainstorming session provided the government with invaluable insights and awareness about unintended negative impact of existing government services, which can be mitigated
- New public services: The process was an excellent platform for ideation of new government services. Ideas for new services were provided by the public to be developed by the government

- Doing things better: The public generated innovative ideas that can enhance government functionality in certain services
- New communication channels between the government and the public: The government feels that it now enjoys a direct communication link with the UAE community, unseen anywhere in the history.

7.4 Smart Dubai Framework for Innovation

Dubai started its path towards smart city transformation way back in 2000 by starting the Dubai E-government initiatives. This matured gradually by 2007 to bringing E-governance to many of the government departments with many services going online led by the Smart Dubai Government (DSG).

In 2013, Dubai government set up a Smart Dubai vision and instituted a higher committee consisting of various department stakeholders to layout a comprehensive strategy for smart Dubai. In 2014 the Smart Dubai Executive committee was formed that included all government department stakeholders and the private sector to carry out the execution of smart Dubai city strategy. This culminated in 2015 to give a final legal and institutional framework for Smart Dubai.

The visionary leadership of Dubai recognized the need for a well-defined organizational and legal framework to realize the vision of making and sustaining Dubai the smartest city in the world. Furthermore, institutionalising and establishing a funding source to manage, sustain and provide fast innovative services in a smart city has been the part of the vision. This section examines various organs of the framework, establishing strategic partners and their roles in achieving that vision.

7.4.1 Legal and Organization Framework

Dubai established a legal framework by issuing four new laws and two decrees [12] in December 2015 to establish Dubai Smart City Office, smart city establishment, formation of the Board of the Dubai Smart City Office with a renewable term of three years and appointing a Director General, Dy. Director General and CEO for the Smart city office.

The Dubai Smart City Office is established to facilitate Dubai's efforts to transform itself into a smart city with the responsibilities of:

(a) developing the overall policies and strategic plans with regard to ICT and smart government
(b) supervising and providing guidance for the smart transformation process
(c) Approving joint initiatives, projects and services that facilitate the smart transformation in coordination with the concerned government entities.

The Law also authorises the Dubai Smart City Office to enter into partnerships with any organisation within and outside the emirate to implement best practices in the field, approve relevant plans and budgets in coordination with the Department of Finance and other government departments and propose legislative amendments to encourage and empower the public and private sectors in the field of smart transformations.

The new Law also authorises the Dubai Smart City Office to establish, develop and manage a central database of all the information pertaining to smart transformation initiatives as well as to establish a portal for smart applications and outlining key performance indicators (KPIs) to assess the implementation of relevant policies and procedures. It will also create innovative e-services and develop the necessary infrastructure that meets international quality standards and best practices.

7.4.2 Dubai Future Agenda

Dubai Future Agenda is a strategic framework for Dubai future Foundation and an integrated plan to strengthen Dubai's position of becoming the smartest city. It is in alignment with the FutureAgenda.org [13] that explores the key issues facing society over the next decade. Dubai Future Agenda will focus on supporting and developing individuals, organizations and sectors that includes more than 20 initiatives to enhance the leadership of Dubai and the UAE.

Dubai Future Agenda includes three main pillars: Individuals, Organizations and Sectors:

(a) Individuals pillar aims to build the capacities in culture, knowledge and communities of individuals in making the future. A range of initiatives and training programmes will be launched targeting individuals in the government and the private sector within the UAE and abroad to prepare them for future leadership and explore future in strategic sectors. Further, this pillar will develop initiatives such as 'Mostaqbal Portal', which is a scientific platform to publish reports and visual material about developments in future science and the latest innovations in strategic sectors. There will also be regular 'Schools of the Future' reports that will highlights the future of the educational system in the short and medium term.
(b) The Organizations pillar will enable organizations to drive innovation within their sectors. The future Agenda also aims to support and strengthen the role of future sectors of the national economy. The Future Agenda will focus on engaging various organizations from the public and private sectors by establishing 'Future Teams' to conduct research and apply the latest future technologies. The agenda also includes the establishment of 'Future Laboratories' specialized in the analysis of strategic challenges of the government sector and the development of innovative solutions to solve these challenges in partnership with international companies and research institutions.

(c) Sectors Pillar: The third pillar of the future agenda will enhance the role of the existing sectors of the future based on innovation, science and technology in the national economy through four main programmes. The *first programme* is future of sectors that aims to organize interactive exhibitions inside and outside the UAE to explore of the future of sectors in the long term and develop the concepts that will re-invent these sectors such as the annual Museum of the Future during the World Government Summit. The *second programme* is Future Partners programme which will be a network of global partners in the areas of innovation, research and development, and establishing specialised companies in consulting, modelling and future design. The *third programme* called Future Cities will work in cooperation with government entities to launch specialised strategies and pilot projects to achieve global leadership for the UAE within the main innovation sectors. The *fourth programme* is Future Making programme that will establish value-added new sectors economically and socially, through initiatives such as Future Incubators and Future Endowment Fund.

7.4.3 Dubai Future Foundation (DFF)

The idea of Dubai ruler to institute DFF [14] is to revive the "House of Wisdom" established by the Arabs in the 9th century AD as a global model for sciences and a platform for innovators from all over the world. The concept of recreating wisdom through Dubai Future Foundation has become relevant in the 21st century to keep up with the changes and shape the future to serve humanity. An endowment fund of US$272Million was allocated to invest in innovation and perceiving the future of strategic sectors in the UAE. DFF is constituted as a trust with a board of trustees headed by the crown prince followed by an MD and CEO.

DFF has launched many initiatives since its inception in August 2015. Museum of the Future is one of the most important projects that include laboratories of innovation in health, education, energy, smart cities, transportation and a Museum of Future Inventions. Other initiatives include the Observatory of the Future, the Office of the Future and the Future Accelerators Programme. DFF has also signed several agreements with government entities, companies as well as academic and international organizations such as Mohammed bin Rashid Space Center, IEEE, Shenzhen Foundation for International Cooperation, UNESCO, IDA, General Electric, and others.

The Foundation has also set up many awards at the international level, including the UAE Drones for Good, the UAE AI & Robotics Award for Good, as well as the world's first 3D printed office in Dubai. The Foundation has also established the World Federation of Future Sports as an organization to oversee the future sports sector and advance research and development in the field of future technologies.

DFF also founded the Global Blockchain Council and has also been instrumental in developing future-focused strategies such as the Dubai Autonomous Transportation Strategy and Dubai 3D Printing Strategy.

7.4.4 Dubai Future Accelerators

This is an initiative of Dubai Future Foundation to attract the brightest and most innovative global minds to work on future challenges in seven key innovation sectors which are health, education, renewable energy, transportation, infrastructure, safety and security and advanced technology.

This initiative aims to provide an integrated programme to accelerate the development of the work of local, regional and global emerging companies working in the field of future technology within key innovation sectors. It also looks to contribute in expanding and facilitating global cooperation in the field of innovation, as well as offering an original platform for emerging companies to work with the relevant government entities and provide solutions to the toughest challenges in their respective sectors.

The Dubai Future Accelerators [15] provides an intensive 12-week program pairing top companies and cutting-edge entrepreneurs backed by powerful Dubai partners to jointly design and create breakthrough solutions and prototypes for rapid deployment across Dubai. This is Dubai's answer to the challenge innovators are facing whereby Innovation is happening at a faster pace than ever before, but the courage to test these ideas is not keeping pace. Real world applications are lagging and truly transformative and disruptive ideas often lack the political and financial support essential for testing and deploying at a larger scale. Dubai Accelerators have identified following 7 Key partners who have the area of knowledge to appreciate an innovative idea, the courage to realize them and the budget to roll out.

(a) Dubai Holding: Deploy digital solutions across hospitality, real estate and telecommunications industries that reduce paperwork, increase transaction speeds, and increase customer satisfaction by an order of magnitude (Blockchain, digital personalization, etc.)
(b) Dubai Electricity and Water Authority: Reduce water and energy consumption in the home, in production and distribution, and/or in commercial application.
(c) Dubai Roads and Transport Authority: Prototype automated transport systems that could cut congestion in half and reduce transport-related CO_2 emissions across all modes, but in particular focusing on freight and fleet mobility systems such as taxis and public transport.
(d) Dubai Municipality: Test automated, recycled, and nature-inspired building materials and solutions that use energy and CO_2 including robotics, automation, bacteriological, mycology, algae, biomimicry, regenerative systems, sand-based 3D printing materials, and others.

(e) Dubai Health Authority: Harness Dubai's remarkable genetic diversity to enhance diagnostic speed and effectiveness using genomics, analytics, telepresence and personalized medicine.
(f) Dubai Knowledge and Human Development Authority: Test 21st century assessment systems and personalized learning solutions that work across curriculums with an emphasis on Arabic language education, STEM(Science Technology Engineering Maths), coding and other disciplines.
(g) Dubai Police: Test integrated behavioural, genetic and biological systems for identifying, tracking and sharing information on criminals, in particular across borders and between jurisdictions.

This supportive platform provided by user agencies for R&D centres and innovators from around the world to work on cutting-edge technologies like AI and robotics, genomics, 3D printing, distributed ledgers, bio mimicry and biotechnology to tap "21st century opportunities" is commendable and worth emulating for cities who are keen on promoting innovation.

7.4.5 Smart Dubai Strategy

Dubai Government has a well-articulated smart city strategy in line with the UAE strategy for the future to be implemented in 3 phases; short, medium and long-term, each with clear benchmarks and performance indicators. As part of the strategy, future prototypes for sectors including healthcare, education, social development and the environment are under development. Every government department has a 'Director of the UAE Strategy for the Future,' who is mandated with driving future planning. This makes the UAE the first government in the world to have such a position within all its departments with a view to predict, analyse and implement highly effective action plans for accelerated development.

The Dubai Smart Government Strategy 2014–2017 focuses on smart connected services, lean administration, an open government and an enabling environment with the vision to be a world class smart government for the 21st century. Together with all Dubai Government entities and the private sector, the strategy aims to connect systems, devices, services and customers in one connected, collaborative government.

Building on a strong e-government foundation, the Dubai Smart Government department is focusing its efforts to digitalise its services, creating greater efficiency and better customer-centric service delivery. Over 95% of Dubai's most important government services have been made available on mobile devices. Through a single identification number, citizens will be able to access all government services and a personal account page with updates on transactions, pending procedures and outstanding payments. Transactions via the mPay platform have doubled in 2015 to reach Dh230.68m ($62.8m), according to Dubai Smart Government.

7.4.6 Smart Dubai: Strategic Partners

Smart Dubai has chosen 12 strategic partners [16] to achieve the smartest city vision. These strategic partners are the true drivers of innovation. This partnership ensures inter-departmental cooperation, non-duplication of efforts and synergy to realize the goal with a collaborative approach.

(a) Dubai Electricity And Water Authority to create sustainable innovative world-class utilities
(b) The Executive Council to formulate strategic plans and enforcement of local and federal laws
(c) Dubai Health Authority to provide an accessible, effective and integrated healthcare system
(d) Roads and Transport Authority to plan and provide an effective and integrated transport system
(e) Dubai Tourism to plan, supervise, develop, and market Dubai's tourism industry
(f) Dubai Police to enforce the law and maintain security and safety of the community
(g) Dubai Municipality to plan, design, build and manage city infrastructure, facilities and services
(h) Dubai Smart Government to provide government services to citizens, residents and visitors
(i) Department of Economic Development to set and drive the economic agenda of Dubai
(j) Dubai Electronic Security Center to make Dubai the safest electronic city in the world
(k) Dubai Silicon Oasis Authority to host advanced electronic innovation, design and development
(l) Dubai Design District to provide a regional ecosystem for creative designers.

7.4.7 Smart Dubai: Vision and Mission

The vision of smart Dubai resonates with the ultimate purpose of democracy stated by Plato and Aristotle. ***To make Dubai the happiest city on Earth***. Happiness is the keyword. Everything that smart Dubai does is centered on this ultimate goal of achieving people's happiness. Dubai recognizes people as Citizens, residents and visitors. The visitors are emphasized in view of EXPO 2020 and the target of 30 million visitors in the next 5 years.

To achieve the happiness vision, Dubai has outlined its mission to use technology as the platform for solutions playing the role of enabler, rather than the principal goal. The path to achieve this is to embrace technology innovation for making Dubai the most efficient, seamless, safe and impactful experience for its people. There is an aggressive drive to make Dubai the world's leading Smart city by 2021.

The Smart Dubai vision consists of 4 pillars, 6 dimensions, 100 initiatives and 1000 e-services.

The four pillars are; Leadership, Readiness, Smart phone usage and Availability of Infrastructure.

The six dimensions are; Transport, Communication, Infrastructure, Electricity, Economic Services, and Urban Planning.

The smart city model is built upon six main pillars: Mobility, People, Economy, Environment, Living and Governance [17].

The targeted 100 smart Dubai initiatives are continuing to be rolled out one after another in quick succession with the completion date set Dec 2017.

Similarly, the target of 1000 e-services mainly by way of Apps is lined and being rolled out before set date. Currently, the apps are deployed under 1 dashboard, 3 login boards under 11 categories by 22 government entities.

7.4.8 Dubai—Smartest City Performance

UAE and the Dubai government carry out an on-going evaluation [18] using 52 national key performance indicators to measure accomplishments with the baseline of UAE national agenda 2021. As of 2016, 62% has been achieved. In order to fulfil the indicators, UAE government has set up a National Agenda Execution Team that that comprise 550 members of government across all sectors and departments responsible for fulfilling the indicators. The teams are under the direct supervision of the Prime Minister to intensify the efforts, mobilise assets and redouble the work in the coming period to achieve 100% by the year 2021.

In 2016, UAE maintains its number 1 position in e-services and m-services indices of the UN E-Government survey in the Arab world, 3rd in Asia and 8th globally.

In a study on 10 gulf cities released by Huawei and Navigant in October 2016 [19], Dubai is ranked first in the smart city index. The UAE leads the Arab world in the United Nations' e-services index [20] and the country also figures among top 10 countries globally. The UAE has moved up four places in its position in the e-smart services index of the E-Government Development Index, or EGDI, issued by the United Nations Department of Economics and Social Affairs.

Dubai Smart Services Key Performance Indicator Wheel

7.4.9 *Key Performance Indicators of Innovation*

Economies increasingly compete based on product differentiation, of which innovation is a key driver. Successful innovation yields new products, goods and services that are superior in terms of utility, quality and price. Innovation can also take the form of new production and delivery processes and organizational structures that lead to improved quality and/or increased efficiency and ultimately higher profit margins and perhaps new profit streams. Innovation also has social implications as the process is largely interactive, involves intensive learning, and is embedded in various social and business networks. Innovations thus generate positive spill overs throughout the entire value chains that produce them, creating further opportunities to improve competitiveness across an economy.

The centrality of innovation to competitiveness has become the key driver of productivity and economic growth, thereby departing from the traditional emphasis on the accumulation of physical capital. In this regard, knowledge refers to the cumulative stock of information and skills involved in connecting new ideas with commercial values, developing new products and processes, and therefore doing business in a new way. This may be called knowledge for innovation or innovative knowledge. The ability to ensure that knowledge, generated internally or externally, adds value depend on how firms and other organizations are linked through business, academic and social networks.

Innovation and competitiveness are closely linked in the minds of policy-makers and has become a central policy priority around the world. The five pillars of innovation capabilities as per the Innovation Efficacy Index (IEI) tracks the various economic capacities for innovation and measures their actual performance. The five pillars of innovation capability are based on the framework that makes innovation systems function effectively: accessing, anchoring, diffusing, creating and exploiting knowledge. The capacity to execute these five functions (described in the diagram below) is the key factor that varies across countries.

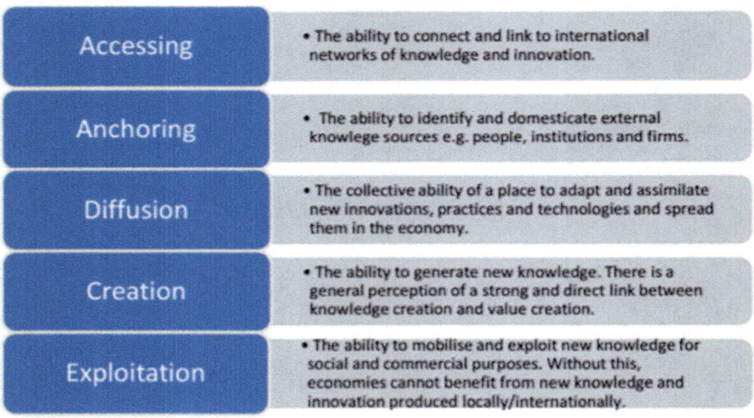

Problems related to innovation can be solved using the functions of any of these five capabilities. An effective and efficient innovation environment provides all the functional capabilities listed above. The IEI develops indicators to measure the capacity and effectiveness of the innovation environment across all five capabilities. A high or low score on a given indicator is not simply good or bad, rather it can indicate consistency with policy objectives and can be used to track the success of policy targets. Accordingly, a high score on an area of policy priority is a positive indicator, while a low score in a non-priority area is a neutral indicator. The table below identifies the five key pillars of innovations with their corresponding capacity indicators and performance indicators.

Pillar	Capacity indicators	Performance indicators
Access	Internet users per 100 people	Value chain presence
	Total broadband per 100 people	Breadth of international markets
	Extent of business internet use	
	Prevalence of trade barriers	
	Infrastructure	
Anchor	Days for starting a business	Presence of clusters
	Number of procedures	FDI technology transfers
	Regulatory quality	Inward FDI
	Protecting investors	Foreign born immigrants
	Foreign ownership restrictions	
Diffusion	Literacy rates	Firm level technology adoption
	Quality of education system	Technology awareness
	Availability of scientists and engineers	Production process sophistication
	Extent of staff training	ICT goods imports
	E participation index	
	Gross capital formation	
Creation	Quality of R&D institutions	Scientific publications per capita
	Intellectual property protection, including anti counterfeit measures	Patent filings per capita
	Gross domestic expenditure on R&D	
Exploitation	Venture capital availability	GDP per capita
	Quality of management schools	Service exports
	Local equity market access	Industry value added
	Government procurement of advanced technology products	Service value added

The UAE has made substantial progress in building their innovation capacities. After many years of relying on natural resources to drive economic growth, this country has now focused on making investments in innovation to catch-up with more developed economies. While the UAE exhibits a strong innovation capacity base, particularly in accessing, anchoring and diffusing knowledge, its creation capabilities have significant potential for growth. In this regard, the UAE continues to focus on input factors that will enhance its knowledge creation capabilities as well as its knowledge commercialization and exploitation efforts.

The UAE's Capacity to Access Knowledge: The UAE boasts a very strong capacity to access the resources it needs for innovation. This is reflected by the presence of advanced information and communication technology (ICT) infrastructure, a culture of ICT use, and a supportive regulatory framework that encourages international business and trade. These factors support the UAE's so-called Accessing Capacity and enable the country to link to international networks of knowledge and innovation, as well as to advanced markets. According to the 2012 GII rankings, the UAE is ranked 37th internationally in total innovation.

The UAE's Capacity to Anchor Value Creation Activities: The UAE has demonstrated a relatively strong anchoring capability despite its fairly young economy. It is ranked 62nd internationally. Anchoring is an important indicator of how well a country is engaged in two-way flows of learning and investment. The UAE boasts favourable attributes such as reduced time and procedures for starting a business; efficient government agencies that permit and promote private sector development; and the strength of laws providing investor protection, as well as a visa system that facilitates the inflow of necessary skills from overseas.

The UAE's Capabilities in Diffusing Knowledge: For an economy to fully benefit from the strengths of its accessing and anchoring capabilities, it needs a population and a workforce that are well educated and highly skilled. These are essential preconditions for the spread of learning from accessing and anchoring activities. Diffusion capacity facilitates the spread of value creation activities using innovation. The UAE has made great strides in this area, especially if one compares current education attainment levels with those of previous generations.

Investment in education has also helped close the gender gap as well as the gap between urban and rural areas. For example, the educational opportunities offered recently to Emirati women in the Emirate of Abu Dhabi, the ratio of female to male students, up to the secondary stage, increased from 95% in the academic year 2000/2001 to 98.7% in 2009/2010. These levels have been accompanied by an improvement in the quality of the education and training on offer.

Another area, in which the UAE has achieved remarkable success, is the level of technology adoption. In the Global Information Technology Report, the UAE was ranked 30th (in 2012) in the Networked Readiness Index, which measures the degree to which economies across the world leverage ICT for enhanced competitiveness. This positive result is largely attributable to a strong commitment on the part of the UAE Government to developing a world-class ICT infrastructure.

The UAE's Capabilities in New Knowledge Creation: The ECC indicates that, in the last five years, the number of registered UAE patents has almost doubled. The UAE now generates the highest number of patents per capita in the Gulf Cooperation Council (GCC) region. This largely attributable to increased spending on R&D and the launch of initiatives such as the UAE Offset Program (subsequently the UAE'S Industrial Development Program). Such efforts are an integral part of the UAE's future development, as expressed in the national strategic policy of Vision 2021. Consequently, the UAE's capacity to produce scientific and technological output has increased dramatically in recent years, as reflected by the rise in the number of local and foreign scientists and researchers working in the UAE.

As per the latest Global Competitiveness Report, the UAE is ranked 15th internationally in terms of the availability of scientists and engineers. There has also been an increase in the number of programs, funds and awards geared towards scientific and technological activities. Examples of research and innovation awards that reward top-quality talent include the Young Emirati Innovators Prize, the Patent Filing Award, the University-Industry Research Collaboration and the Zayed Future Energy Prize. The latter offers a USD 4 million prize the world's largest award for innovation in the development of sustainable energy solutions.

Amongst the GFCC countries considered in this report, the UAE is ranked 4th in terms of its existing capacity to create new knowledge. The UAE has witnessed the mushrooming of institutes of higher education over the last decade, some of which have a strong research component. Growing from only nine universities in 2000, the UAE is now home to more than seventy-five universities and institutions of higher education. As is the case with all countries, however, the UAE cannot afford to specialize across all scientific fields. Most its research and scientific activities have concentrated on key and strategic pockets of science and technology (S&T), such as renewable energy and health sciences.

The UAE's Capacity to Generate New Economic Value: One major challenge facing governments and companies around the world is the valorisation of knowledge. Several constraints on knowledge valorisation are often cited, notably the lack of an entrepreneurial culture, a weak Venture Capital (VC) base and a punitive insolvency law, among others. The UAE, overall, performs above average in its ability to exploit resources for the creation of new value. Two areas in which the UAE excels are the effective use of skilled international talent and government financial support for S&T activities.

Venture Capital activity has been increasing in the UAE, despite the facts that its business environment is a relatively young one, and that some businesses tend to be large family-owned firms, which are less likely to seek investments from VC firms. Other factors that have traditionally limited the supply and demand of VC activity in the region are constraints on foreign ownership in certain sectors, and the absence of preference shares frequently used in western VC finance. Therefore, compared to other asset classes of private equity and buy-outs, VC remains a less popular funding option in the UAE.

Supporting Entrepreneurs in the UAE: Encouraging entrepreneurs is a main plank for the UAE, and is well recorded in Abu Dhabi and Dubai. The work has recently been undertaken by government agencies to improve the climate for small businesses. Moreover, many banks in the UAE now have dedicated centres for Small and Medium-sized Enterprises (SMEs) and offer special loan packages. Government initiatives aimed at supporting entrepreneurs include the Khalifa Fund for Enterprise Development in Abu Dhabi and the Mohammed Bin Rashid Establishment for Young Business Leaders in Dubai. The Mohammed Bin Rashid Establishment for Young Business Leaders offers office space, administration, incubation facilities, funding and training for Emirati businesses. Various funding, training and support services, the Khalifa Fund for Enterprise Development offers a range of programs that combine entrepreneurship and social development such as public programs rehabilitate prison inmates, providing them with an opportunity to enter the business sector and be reintegrated into society. Other programs aim to preserve the local heritage.

7.5 Why Is Innovation Vital for E-Democracy?

Open government and smart city apps have disrupted the centuries old democratic norms of people approaching government for availing its services, voicing their concerns and seeking redressel. Interaction in a smart city is switched whereby the government has the opportunity of being ever present on people's fingertips through a smart phone app. The government's main challenge therefore shifts to ensuring people's inclusive participation and helping them climb the participation ladder from unawareness to impact participation across all age groups, abilities and cross sections.

To accomplish this feat using the state of the art Information and Communication Technologies requires creative innovation. In all smart cities around the globe, innovation, co-creation and co-management has taken the central stage for citizen engagement, particularly the youth. Innovation motivates people to engineer out of the box solutions for the various problems faced by the residents in a city. Innovation in a smart city is therefore not a luxury, but a critical means of keeping up with changing circumstances and opportunities. Making people smart and enabling them to co-create leads to perfection and to realize the full potential of the human resource leading to a happier city.

To sustain innovation, businesses attract and retain the most creative and productive minds. Innovative governments also need to attract talent, perform efficiently and continually upgrade their systems and services. They have to empower their citizens to develop their full potential and become drivers in transforming their city to a smart city.

7.5.1 Smart Dubai: Accelerating Innovation

UAE recognizes innovation as a cornerstone of social and economic development and key to promoting economic growth, increasing competitiveness and providing new job opportunities. Believing that innovation is the future of human investment, the UAE Leadership emphasizes its importance across all sectors through the UAE Vision 2021: "Innovation, research, science and technology will form the pillars of a knowledge-based, highly productive and competitive economy, driven by entrepreneurs in a business-friendly environment where public and private sectors form effective partnerships." The Cabinet directed all federal government bodies to boost cooperation and to revise their policies to develop a nurturing environment for innovation. The following initiatives [21] have been taken to achieve the objectives.

(a) Creation of a new job post—CIO of Innovation—for every government department.
(b) Established Mohammed bin Rashid Centre for Government Innovation which aims to develop government operations and enhance UAE's competitiveness, making UAE one of the most innovative governments around the world.
(c) The UAE Cabinet approved the designation of 2015 as the Year of Innovation, in a plan to make the UAE a world innovative leader.
(d) In order to foster innovation, UAE launched the idea of innovation week and innovation prizes to be held every year, starting November 2015. It is one of the largest innovation initiatives in the world and a consolidated effort by government, the private sector and individuals to help create a widespread culture of innovation in the UAE and strengthen its position as a global hub for innovation. The activities touch upon different themes such as exhibitions that showcase innovative products and services, the launch of new national initiatives, hackathons and competitions, interactive programs such as labs and workshops, engaging talk sessions and fun evening activities in public places.

Dr. Tim Jones, the programme director of Future Agenda [22], the world's largest open foresight initiative, highlights 3 three business innovations for UAE [23].

(a) **Education Revolution**

Broader access to improved education is seen as a major catalyst for empowerment, sustained economic growth, overcoming inequality and reducing conflict. Many see that we need an education system fit for the digital revolution, not one still based on practices from a century or so ago. The idea of 'teacher-less' classrooms where learning comes from the likes of YouTube, Coursera.com and multiple 'MOOC's is rapidly gaining traction and, with it, the shift in the role of the teacher from the source of content and information to that of coach and curator. Implicit within this is a change in standards and accreditation away from traditional exam bodies to a world where knowledge and skills are gained from multiple sources. While many are keen to accelerate this shift and help better prepare the next generation for the world ahead, the challenge in some regions is in finding parents willing to let their own children be the pioneers, and experiment with this new model. As governments seek to improve the human capital of their nation, balancing this is key.

(b) **Energy Storage**

Energy storage, and particularly electricity storage, is now seen as the missing piece in the renewables jigsaw. If solved, it can enable truly distributed solar energy as well as accelerate the electrification of the transport industry. Core to this is the overnight, localised storage most often associated with batteries and an area where lithium-ion and air technology is rapidly improving performance. However, it is not just batteries that are undergoing innovation as there are multiple other options for alternative energy storage including hydrogen, hydro, heat and momentum. Bringing to the market a low-cost, highly efficient solution that can be adopted in many applications is a focus for many researchers and success will drive a major change across the whole energy landscape.

(c) **Data Marketplaces**

Many see that data is now becoming a currency. It has a value and a price, and therefore requires a market place. An ecosystem for trading data is now emerging and several leaders see that anything that is information can be represented in new data marketplaces. Especially as we shift from bi-lateral sharing of data to a more open and transparent approach, how a stock exchange for data would work is being debated around the world. The challenge is that unlike a simple time vs. value market as exists for shares or commodities, a data market place would have an extra two dimensions of variance. Many would only want to access data and not own it and the value that one party would place on a particular piece of information will not necessarily be the same as another party. Solving this conundrum is no easy task, but if achieved it could change the way business and society interact.

All three of these are clearly major global opportunities but also areas with significant potential to help move things forward. Whether by proactively testing new approaches to education, investing in the most promising energy storage technologies or even creating the supportive environment for a data marketplace.

7.5.2 Global Innovation Index (GII)

The UAE ranked first [24] among Arab countries and 41st worldwide in terms of overall performance on the index, up from second in the Arab World and 47th globally in 2015.

The 2016 Global Innovation Index measured the performance of 128 countries and economies in the area of innovation which represents the majority of the global workforce and global GDP. The index is based on 82 sub-indices which measure innovation in detail and provide a clear understanding of how it drives economic growth and prosperity.

The index results attribute the outstanding progress of the UAE to its sustainable achievements in innovation inputs, especially the strengthening of institutions and market sophistication. The UAE has one of the best political and legislative environments in the world along with outstanding development in information technology, communications, infrastructure, and innovation linkages.

7.5.3 National Innovation Strategy

In 2014, UAE launched a National Innovation Strategy [25] with a seven year plan that aims to make the UAE among the most innovative nations in the world. The objective of the National Innovation Strategy is to stimulate innovation in seven sectors where innovation plays an essential and integral role in achieving each sector's key strategic objectives. These seven sectors are: Renewable energy, Transportation, Education, Health, Technology, Water and Space. The strategy will fulfil the demand for a constant flow of new ideas, as well as innovative leadership using different methods and tools to direct the change.

UAE's National Innovation Strategy requires all government agencies to cut spending by 1% and dedicate the savings to research and innovation.

A new Dh2 billion (USD 544 Million) fund was recently announced to provide financing for start-ups and to support them in transforming ideas into projects. The Ministry of Finance will guarantee commercial loans to finance projects, and draw in support from commercial banks and investment funds.

3 distinct innovation types are identified: Business Innovation, Technology Innovation and Science Innovation. The goal of the UAE's Science, Technology and Innovation policy is to achieve robust science and technology-based innovation as described below.

7.5.4 Science, Technology and Innovation (STI) Policy

Framing the STI policy [26] is a turning point in UAE's march towards progress, economic diversification and prosperity and aims to prepare UAE for a post oil world. STI policy forms the roadmap to make UAE among the most innovative countries in the world. It will allow UAE to transition into a knowledge and innovation based economy and to enhance scientific and technological innovations in its public, private and academic sectors. The STI policy identifies 24 focus areas which are chosen considering present and future national needs, aligning with international trends and UAE's capabilities and unique assets.

24 selected focus areas are; Education Innovation and Technology, Health Information Technology and Bio-informatics, Public Health, Non-Communicable Diseases and Wellness, Biotechnology and Genomics, Water Management and Economics, Solar and Alternative Energy Technology Systems, Space Sciences, Cubesats and Nano satellites, Cyber security, Semiconductor Process Development, Robotics and Artificial Intelligence, Smart City Applications and Solutions, Architecture and Urban Design, Arabic Digital Technology, Financial Services Technology, Petroleum Geosciences, Internet of Things and Big Data, Additive Manufacturing (3D Printing) Advanced Building and Construction Materials, Food Security, Transportation Logistics, Analytics and Security, Aerospace Advanced

Materials, Manufacturing, Maintenance and Testing, Commercial Unmanned Aerial Vehicles and Autonomous Vehicles.

7.5.5 Smart Dubai Innovations

Innovation is a key driver in a smart city to enable the youth, in particular, to find creative ways to solve various challenges to improve the life experiences of its residents and visitors.

Since 2015, Dubai has accelerated the pace of Innovative projects and is expected to continue. The following sections give an insight on these innovations.

7.5.5.1 Museum of the Future

To mark 2015 as the year of innovation, Museum of the Future [27] is built to showcases how technology over the coming years will transform government services. The museum's motto "See the future, create the future" reflects a new approach to government innovation. Rather than just displaying exhibits or publishing reports, the new institution will use design, technology prototyping and foresight to create real examples of change. The Museum is the first of its kind, and represents a leading example of entrepreneurial governments embracing change and creating futuristic visions for a better world. The Museum will host innovation labs focusing on health, education, smart cities, energy and transport and a permanent section on the greatest global innovations. It will provide a platform to demonstrate and test the latest inventions from world-leading tech giants and up-coming start-ups, as well as collaborations between companies and leading research institutes and universities. The museum is set to open by the end of 2018.

7.5.5.2 Smart Dubai Platform

Announced in March 2016 [28], the Smart Dubai Platform is a digital backbone powering the city in the near future, in partnership with Du, the telecom operator. The Smart Dubai Platform will be the central operating system for the city providing access to city services and data for all individuals, private sector as well as public sector entities. This will facilitate:

(a) collecting data from thousands of new sensors to be installed across Dubai on roads, parking lots, building entrances, public parks, curbs etc. or in digital devices such as smartphones, analyse the information and then share it live in real-time.

(b) exchange of open and shared data between the public sector, private sector and individuals, creating new opportunities for engagement in accordance with the Dubai Data Law, contributing to the smart city future of Dubai.
(c) enhance synergies between government and private sector services, optimising cost and delivering time saving benefits to both organisations and customers alike.
(d) augment real-time and data-driven decision making capabilities for the city government, enabling city leaders to engage in community-wide dialogues and analyse rich city data across multiple dimensions.
(e) leverage the latest technological innovation to facilitate the exchange and delivery of city data and services, to make Dubai a more seamless, safe, efficient and impactful city experience for all residents and visitors.

Billed as the "digital backbone powering the city in the near future [29]," the new Smart Dubai Platform will begin operation in the third quarter of 2016 and will help achieve smart city targets in six key pillars; smart life, smart transportation, smart society, smart economy, smart governance and smart environment. NeXgen's Yousef Khalili and Andrew Rippon outline the benefits and pitfalls [30] of Smart city platform as follows:

Efficiency: The Smart City Platform brings the ability to make governments more efficient in their growing need for technology systems. It provides a private or hybrid cloud environment where any number of government systems can be hosted and improve timelines to providing services. Cities need no longer invest in new data centres and hardware as they can use the City cloud. The Smart City platform, enhanced with ubiquitous wireless networks, can create a true IoT environment where sensors are automatically detected on the network and their data plugged into the platform. By centralizing control functions into a city control centre, day to day running of the city becomes more efficient and new enhanced capabilities are brought forward, such as proactive management. This function adds agility to city governments, to the benefits of their stakeholders.

Interoperability: Most cities have a myriad of systems today, from road tolls to licence registration to planning to financial management and beyond. The Smart City Platform empowers cross system or entity workflows and data exchange through the use of controlled data stores, APIs and guaranteed messaging.

Data governance: As more of a city's interactions become dependent on digital systems, data governance becomes essential. The Smart City Platform is the physical implementation of data laws and guidelines that enables sharing data between government entities and onwards to the private sector safely.

Entrepreneurship: The Smart City Platform is a technology incubation platform for encouraging start-ups and to help deliver services for the city through crowdsourcing. If the city has problems to solve, they can be offered to the market by exposing data and technology infrastructure or tools that would enable any size of private entity to respond, down to the individual entrepreneur or developer.

Transparency: Through the sharing of data and interoperability of systems, cities gain efficiency and the ability to bring about new services for residents, visitors and companies. Transparency is achieved through data dashboards, as well as the release of catalogues of data and new applications that reduce barriers for people and create new opportunities. This is especially true where real time data is used to give more timely transparency when there is the case of emergency or event management.

Prediction of needs: Using Big Data analytics, cities are in the position to more accurately predict the resident's needs and provide services even before the residents know they need them.

New uses of data: New applications and uses of data can be discovered through the combination of data sets that were not previously intended to combine. Examples of this include the use of company registration data to predict CO_2 creation and determining traffic impacts from event registration data.

The potential security and budgetary issues are:

Security: In effect a smart city platform becomes a honey pot for malfeasants as the multiple systems they had to hack before are now wrapped up into one. This is why security has to be a key concern at every layer of the platform, while empowering the open use of data.

Budgetary issues: Legacy thinking can result in a new platform built on old principles that do very little to improve city resident's lives and consumes large budgets. A root and branch approach needs to be taken to adopt modern methodologies as well as systems.

Ultimately the importance of a Smart City platform lies in the benefits that it can bring to residents, visitors and companies within a city. By making current services more efficient and enabling new outcomes, the Smart City Platform becomes the central lever of execution for city government.

7.5.5.3 Driverless Vehicles

Masdar City in Abu Dhabi has been using autonomous vehicles for ferrying passengers locally. Dubai implemented driverless vehicles in April 2016 on experimental basis in downtown Dubai carrying 10 passengers and has been extended to business bay district over a 650 meters long track. Dubai targets 25% of cars to be driverless on Dubai roads by 2030 and is framing rules and legislations for implementing smart mobility strategy for deploying autonomous vehicle for the first and the last mile transportation in residential, business centers and parks.

The car-booking service Careem and Next Future Transportation, part of Dubai smart government initiative, showcased driverless electric pods for the mass-transit systems of the future. The pods can move individually or attaching themselves to other pods forming a modular bus that allows passengers to move freely from one pod to another. The pods can pick up passengers on demand and drop off to different locations as the pods can attach and detach according to passenger's destination selection. This driverless mass transportation system is efficient, safe and carbon friendly.

7.5.5.4 Hyperloop

Dubai will be the first city in the world to implement fully operational high speed land transportation for goods and passenger transport by 2020. Hyperloop [31] is a broadband of transportation that uses pressurised capsules on air cushions in giant reduced pressure tubes, simulating an aeroplane. It can reach a speed of 1200 km per hour, lesser (USD 40 Million per km^2) than a metro, energy efficient, on-demand and safe, reducing travel time between 2 cities that are within 2–3 hours of today's automobile driving limits. Cities in the UAE are well poised to take advantage of this technology. Roads and Transport Authority, Dubai who is implementing this project expects hyperloop to bring all UAE cities within 10–12 min reach, thus enabling business and resident's higher mobility and freedom to live in any city. This can be further stretched by residents using smartphone app to order driverless self-driving pods that can provide door to door service picking passengers and connecting to hyperloop.

7.5.5.5 3D Printing and Smart Buildability Index

In line with international technology trends, Dubai is the first mover on embracing 3D printing. The world's first 3D printed office and first 3D-printed Smart Palms are done in Dubai. The 3D-printed office provides valuable insights on the real application of 3D printing technology wherein a 3D-printer, equipped with an automated robotic arm measuring 20 ft high, 120 ft long and 40 feet wide was used to print a 2690 ft^2 building with the labour cost reduced by more than 50% compared to conventional buildings of similar size. Benefits of 3D printing include 80% reduced construction costs, 70% lesser time to build, 50% man power, 25–30% of current emissions and 30–60% lesser construction waste.

UAE has standardised 3D printed buildings with a smart buildability index [32]. It aims to standardise building designs that adopt modern building techniques such as 3D printing and use locally manufactured steel structures, pre-fabricated pillars and walls. The Index seeks to promote the use of techniques and materials that accelerate construction processes, increase productivity and rationalise the use of unskilled labour.

7.5.5.6 UAV—Drones

UAE has given a big thrust on UAV and has developed several applications on using Drones for safety, security, environment monitoring and research. Dubai Police and Roads and Transport Authority are using drones to monitor traffic, check traffic accidents, survey condition of roads, structural analysis, event monitoring, monitoring projects and mapping. Sharjah police drone program called sakhab is used to prevent suicide from high rise building by using a camera and sim card to talk to suicide person and negotiate them to safety. Environmental protection

Agency, Abu Dhabi has used drones to monitor the high population of turtles in Kalba's (Eastern UAE) mangroves, which are notoriously difficult to count. UAE government has instituted Drones for Good award with a prize pool of over one million USD to promote civilian drone technology to serve humanity. Security is the main concern however in using drones and the UAE government is working on legislative framework for commercial drone usage.

7.5.5.7 Robotics

In order to attract students and young scientists into innovation in robots, UAE has introduced a number of challenges and awards. The UAE National Robotics Challenge is held by Abu Dhabi Education council every year to select teams to participate in World Robot Olympiad (WRO). In 2016 [33], 375 teams comprised of 1,500 students participated in the challenge that aims to promote creative thinking, improve communication/cooperation skills and expose students to latest trends in Science, Technology, Engineering and Mathematics (STEM).

The UAE AI & Robotics Award for Good with a prize of USD 2Millon aims to encourage research and development of innovative solutions in artificial intelligence and robotics to meet existing challenges in the categories of health, education and social services.

7.5.5.8 The Global Blockchain Council

Dubai Museum of the Future Foundation has established the global blockchain council [34] to promote using next generation of banking technologies to enhance UAE's position as a leading centre for innovation and knowledge economy. The council consists of 46 members drawn from government entities, international companies, leading UAE banks, free zones, and international Blockchain technology firms. To promote blockchain and digital currencies, the council holds an annual conference hosting high-profile individuals, workshops, awareness sessions, hackathons, and start-up conferences.

Blockchain is defined as a digital protocol to carry out transactions and transfer money using a wide network of computers around the world requiring digital currencies, accurate documentation and absolute trust between all stakeholders. Known primarily as the public ledger of Bitcoin transactions, blockchain help maintain tamper-proof lists of all data records and enables secure value exchanges for money, stocks or data access rights. Unlike traditional trading systems, no intermediary or central recording system is needed to track the exchange. Instead, all parties deal directly with each other. Blockchain can be both private and public that enables people to choose whether to share or withhold the personal details of the stakeholders. Transactions on Blockchain can go through only if all the members approve thereby limiting the chances of fraud and money laundering. As the digital currency cannot be forged or damaged, it can be moved across borders with ease and also facilitates the shopping process across social media and websites.

Booz Allen in its report [35] highlights Blockchain as integral to smart cities and digital economy. The main benefits are reduced cost of transfers and near real-time payment, increased data integrity and efficiencies through automation, provide time-stamped verification and attribution, protection from hacktivism and cyber-crime, digitally transparent transaction systems, plays an important role in incubating regional start-up ecosystem and support for industries such as healthcare, transport and finance. Smart city governments can also extract tangible value from vast amounts of data that can be used to meet changing consumer demands at minimal additional costs.

Dubai blockchain council has initiated a number of pilot projects [36] to promote blockchain technology; '*BitOasis*' that focuses on Bitcoin exchange to secure '*flexidesk*' contracts and registration processes using blockchain ledger. 'Trade Flow' will provide digital financial transaction services to reduce costs, increase operations security, unify transaction procedures and provide super-fast services round the clock. '*Dubai Points*' which aims to incentivise tourism by awarding points to tourists when they visit a tourist destination. '*Digital Wills*' to register wills using blockchain links in order to enable clients transfer their properties to beneficiaries using a fast and secure process while complying with local regulations. '*Operation Kimberley*' that will modernise the verification process of diamond sources and their trade mechanisms and to issue global certificates related to the diamond industry by coordinating the activities of its 81 member countries. '*Health Records*' that aims to create a unified digitised health records in hospitals and clinics to provide a modernised record system that is fast and efficient and capable of saving many lives.

7.5.5.9 Private Innovation and Incubation Centers

(a) **Visa launches RIC**: Visa in partnership with TECOM opens Regional Innovation Center in Dubai [37] designed for technologists, financial institutions and developers to co-create the next generation of payment and commerce applications. This center will serve the Central and Eastern Europe, Middle East and Africa (CEMEA) regions and is part of a network of new innovation centers. Other flagship innovation centers are in San Francisco and in Singapore. Visa envisions encouraging and fostering innovation, entrepreneurship and digital advancements. This will support the local start-ups and entrepreneur community through mentorship and guidance programs, Leveraging each other's networks for the promotion of digitally advanced and innovation related solutions, Collaborating on pilot programs, tests and trials within the TECOM Group Co-launching Design Thinking workshops and supporting local hackathons and developer workshops.

TECOM has initiated in 5 Incubation arms in 2013 to support the growth of entrepreneurs and start-ups across different verticals thus enforcing a solid entrepreneurial ecosystem in the region as well as the launch of the Innovation Hub which will see over 15,000 business professionals prosper to stimulate innovation and excellence across technology, new media, education and life sciences.

(b) The **Emirates Group, along with co-founders GE and Etisalat** are propelling a start-up incubator in the aviation travel-technology [38] field for young entrepreneurs and university students in the UAE called “Intelak,” which means “taking off” in Arabic. The initiative will contribute to the country’s innovation ecosystem and help develop local talent in high-tech fields. Structured as a competition, the program is open to Emirati-led teams of 3–5 people who propose innovative ideas that will make air travel simpler, better or more exciting at any point along the travel experience—from the moment one leaves home through the airport to flying in the aircraft.
(c) **GE and the UAE Ministry of Economy** have signed a cooperation agreement [39] to support the UAE government in embedding innovation into its organizational culture. GE, whose cultural ethos and heritage is founded on innovation helps spark innovation through open innovation challenges targeting local issues, localized research facilities, regional innovation centers, region-specific white papers and locally held GE Garages workshops.
(d) **Huawei and Dubai South** has signed an MOU [40] to Collaborate on Smart City Innovation for bringing future-city capabilities and services to Dubai South’s Business Park Free Zone. Huawei’s Smart City and Safe City Solutions will provide enhanced services to residents and businesses in the Park. Using New ICT, governments can enable a properly connected safe city where visualization and collaboration are embedded into the city’s infrastructure to maximize public safety. This enables the safe city to become a critical component in smart city development
(e) **UN centre of excellence in Dubai** [41]. This centre will provide training activities, research analysis and publications that focus on education, technology and transport which can be compared with best practices around the world.

7.5.5.10 E-Democracy Innovations

There are a number of other innovations worth mentioning that directly benefit residents.

(a) **Unified Traffic Control** [42] introduced by Dubai’s Roads and Transport Authority (RTA) considering the doubling of vehicles from 2006 (0.7 M) to 2016 (1.4 M). The unified Traffic control will connect 408 traffic signals wirelessly to optimize junction traffic light sequences based on traffic volume. Road users are now able to see traffic congestion, road works and accidents on their mobile in real-time in order to adjust the travel plans.

(b) **Al Raqeeb System** from RTA is a smart system to detect exhaustion signals in bus drivers while sitting behind the wheels. Meant to provide added safety to public transport riders, the Raqeeb system has reduced fatigue or exhaustion related cases among bus drivers by 88%.
(c) **Real bus departure information** from RTA provides 4 screens distributed in DSO show 4 lines of real bus departure information. This will be rolled out in other areas as well.
(d) **Taxi Hailing** company careem much like uber has tied up with Dubai Taxi agency to book taxi using a mobile app. User can either book the vehicle of their choice in advance or instantly and track the car in real-time. Payment can be made either by cash or deducted from registered credit card.
(e) **Smart Energy Grid** [43] initiated by Dubai Energy and Water Authority (DEWA) plans to install 250,000 smart meters in residential, industrial and commercial properties by 2018. Consumers are able to monitor electricity and water consumption, tailor utility usage to their needs and reduce wastage through mobile apps. Residents are also motivated to install photovoltaic panels on their rooftops, collect solar energy and sell power back into the grid.
(f) **First charging station for electric vehicles** is introduced by Dubai Silicon Oasis Authority (DSOA) [44] with the goal of Install Seven Additional Stations by 2017.
(g) **Smart Lamp Posts** installed at Tech Park by DSOA in cooperation with Dubai civil defense allows People to report emergencies. Using motion sensors, the smart street lights will provide 25% visibility until triggered to full power by approaching vehicles and pedestrians. This allows for less energy consumption, prolong the life of electric bulbs, and reduce operational costs and carbon impact. Additional sensors can be added to the street lights for collecting and sharing information and data on the surrounding environment, such as pollution and weather conditions or even relay CCTV, advertising messages and/or safety warnings.
(h) **Smart Waste Management System** [45] from DSOA is achieved by installing sensors in garbage bins that would send alerts when the bin was full and needed emptying. With this new technology, garbage collection vehicles would take trips to empty the bins only when full, thus reducing their total number of overall trips leading to lower operating costs and a reduction in CO_2 emissions resulting in a cleaner environment. In addition to measuring how full the bin is, the system also gives an overview of the total weight and volume of the waste. Using this data, DSO plans to implement further technologies on how to reduce overall waste in the future. DSO today is monitoring 95 bins through the smart system and aims to increase the number to 120 by the end of 2016.
(i) **CloudOne** [46] announced by Dubai Smart Government offers an enterprise-grade, scalable, agile, multitenant infrastructure as a service (IaaS) cloud computing solution for government services. This service provides a fully automated environment for the approval and provision of on-demand IT resources, saving time and reducing the cost of IT management in government agencies.

(j) **Media Cloud Services** offered by Du is Middle East and North Africa's first media Cloud services aimed to simplify media management for broadcasters, content producers and providers.
(k) **Smart City Diploma** [47] Rochester Institute of Technology (RIT) Dubai, a non-profit global campus of the RIT in New York, started offering a Diploma comprises of a series of modules related to the pillars of a smart city including: Economy, Governance, ICT, Environment, Health, Education, Security, Mobility and Infrastructure, and People.
(l) **Master's degree in Smart City** [48] Offered by RIT Dubai with Polytechnic University Spain. This unique and innovative master's degree is the first of its kind in the region. With 1000 government services expected to go smart in 2018, this course explores what cities are and what they can be. It goes beyond the concept of a smart city to also cover important urban, social and sustainability issues to ensure a comprehensive and holistic approach to the cities of today and of tomorrow.
(m) **Intel Jointly Set up Smart City Center of Excellence** in the Middle East that will support the development of a smart city center of excellence and innovation within Dubai Technology Entrepreneurship center (DTECT). The center implemented the region's first water saving subsurface irrigation system [49] in line with its tri-pillar CSR strategy. The initiative aims to reduce the current irrigation water consumption levels and related operational costs by almost 40%.

7.5.6 Dubai Design District (d3)

The d3 [50] is a greenfield benchmark for districts around the world, showcasing smart initiatives designed to improve the experiences of its visitors, tenants and business partners. An array of smart service initiatives including smart parking metres, smart energy metres and smart waste management combine to truly bring a new degree of sustainability, resource management and energy efficiency to life at d3. With additional security through smart access control and more innovative ways of utilising data through data virtualisation, d3 stays ahead of customer and market trends and therefore enhances the experience of its commercial and retail business partners and better serves d3 as a community.

Since 90% of Middle East design market is driven by fashion, furniture and architecture, d3 hosts the in5 design business incubator. This will act as a catalyst for start-ups and students in design to prototype using 3D printers, electronics and modern machinery working with metals, wood, plastics and textiles.

d3 will usher in a new wave of innovations in IoT and digitisation that are already part of a smart city. D3 will serve as a smart city benchmark for the rest of the world's emerging economies to follow. Through the 21 smart initiatives introduced at d3, Dubai wants to change the way business is conducted across sectors and to deliver a greater quality of life and experience for millions of people.

7.5.7 Building Smart Cities in Other Countries

In recognition of Dubai's leadership in smart city experience, Dubai Holding was awarded the multi-billion dollar project [51] to establish a Smart City in Seoul, South Korea. Covering an area of 51 million square feet, Smart City Korea will be strategically located between the two cities of Seoul and Incheon, close to two international airports.

In February 2016, Dubai Holding inaugurated Phase 1 of Smart City Kochi's operation in Kerala, India. It is a city that is designed to cater to knowledge services and information technology.

The company also inaugurated Smart City Malta in collaboration with the Maltese government in 2014.

7.6 Dubai's Green City Initiatives

The global energy landscape is rapidly changing. Increased energy demand in emerging markets, uncertain energy prices and international pressure to reduce carbon emissions, mean governments and businesses see the economic and security benefits of shifting to clean energy. The unprecedented growth of energy demand during Dubai's infrastructure development has increased the need to find a prudent strategy for the security of supply and the diversification of fuel sources.

The UAE has long been an important supplier of energy and is now becoming an increasingly relevant consumer of energy as well. In its efforts to accelerate the development of additional hydrocarbon reserves and in its efforts to contribute to the development and implementation of alternative energy sources, the UAE hopes to continue its long tradition of responsible energy stewardship.

Increased energy demand in emerging markets, uncertain energy prices and international pressure to reduce carbon emissions, mean governments and businesses see the economic and security benefits of shifting to clean energy. The United Arab Emirates (UAE) is at the heart of this shift—the same as it was at the heart of the last energy boom. And just as its massive oil reserves provided a springboard for the incumbent energy revolution, its huge solar resources and cutting-edge technology development are shaping the region as a hub for the new, clean energy revolution.

The UAE government agencies are seeing the benefits of renewable energy, the most influential companies—looking to secure low cost power supply—are also taking advantage of the abundance of solar energy in the country.

Since the formation of the United Arab Emirates in 1971, energy and natural resources have been closely linked to its growth. Today, sustainable energy has become a key factor in our efforts to promote economic, environmental and social growth, and the key driver for sustainable development and the preservation of the environment

The Dubai Supreme Council of Energy is tirelessly working to achieve the initiative of HH Sheikh Mohammed bin Rashid Al Maktoum, Vice President and Prime Minister of the UAE and Ruler of Dubai, entitled 'Green Economy for Sustainable Development,' to promote the use of renewable sources of energy.

The Supreme Council of Energy has formulated the Dubai Integrated Energy Strategy 2030, with the purpose of diversifying sources of energy, so that by 2030, gas will provide 71% of Dubai's total power output, nuclear energy will provide 12%, Clean Coal 12%, and Solar Energy will provide 5%. We are also working to reduce energy demand by 30%. The Council organized the Dubai Global Energy Forum in 2011 and in 2013, to highlight the most prominent issues of energy, environment and sustainability

An annual State of Energy Report, Dubai 2014, which has been launched by the Supreme Council of Energy. This report is one of the key factors of continuous progress to achieve the objectives of the Dubai Integrated Energy Strategy 2030. This report defines the key issues, required plans and important analysis of the sustainable energy sector in Dubai.

On January 15, 2012, HH Sheikh Mohammad bin Rashid Al Maktoum, the Vice-President and Prime Minister of the UAE and Ruler of Dubai, announced the launch of a long-term national initiative to build a green economy in the UAE under the slogan 'A Green Economy for Sustainable Development'. This initiative aims at making the UAE one of the global pioneers on the green economy. HH Sheikh Mohammad bin Rashid Al Maktoum noted that "our goal from this national initiative is clear, that is, to build an economy that protects the environment as well as an environment that supports the growth of the economy". For growth to be 'green' it must be based on a low-emissions development model and on more efficient use of energy and other natural resources; these are two key pillars of the UAE strategy for achieving a Green Economy [52].

7.6.1 UAE Vision 2021

This vision specifically mentions the need to develop and promote sustainable energy as one of the key challenges facing the UAE and an important opportunity for the future. It notes that "we want the UAE to sustain its drive toward economic diversification, as this is the nation's surest path to sustainable development in a future that is less reliant on oil." In the face of climate change and its elects on current and future generations, and with the vision for an economy based on sustainable energy, UAE Vision 2021 supports initiatives to catalyse more sustainable use of energy and to be instrumental in the battle against climate change.

The UAE is committed to playing its part in developing and implementing solutions to reduce its carbon footprint while leading new innovations to achieve the goals of scaling-up sustainable energy solutions.

7.6.2 Dubai Strategic Plan 2015

To ensure proper focus on sustainable development in the context of Dubai's considerable economic growth, Dubai Strategic Plan 2015 aims to integrate sustainable energy solutions into energy supply and demand, to meet the Emirate's growing needs in a sustainable manner. It also highlights the concept of excellence as a core to Dubai's government policy to ensure that it leads by example on sustainable development initiatives.

7.6.3 Dubai Integrated Energy Strategy 2030

The Dubai Integrated Energy Strategy supports the vision of the Dubai Supreme Council of Energy to emerge as a global and regional best practice and role model in development and use of sustainable energy. As per this strategy, improving energy demand efficiency in Dubai could close the energy supply gap by up to 40% in 2030. The most substantial efficiency improvement opportunity is in the buildings, commercial and residential sector which is Dubai's largest consumer of energy. In addition to improving the overall efficiency, Dubai is also pursuing an energy supply diversification strategy. The Energy Diversification Strategy, part of the Dubai Integrated Energy Strategy 2030, has defined targets for renewable energy to supply 1% of Dubai's energy mix by 2020 and 5% by 2030. This has recently started with initial pilot solar initiatives which will serve as a base for developing large-scale solar starting in 2020, once the technology has attained cost-competitive levels.

(a) **Creation of jobs**

The government also recently launched its Green Growth Strategy to create 160,000 new jobs by 2030, which aims to bring a total boost to the country's GDP of up to 5%. The country's low carbon efforts are also illustrated by the fact renewables projects worth a combined US$1 billion are currently being developed.

(b) **Public private participation**

The ruler and the Crown Prince of Dubai and Chairman of the Dubai Executive Council, has launched the Dubai Green Economy Partnership to position Dubai firmly amongst the global cities that are leading their transition to green economies. A group of committed public and private sector leaders convened to realise the vision of the Crown Prince of Dubai. Their aim was to identify how best to encourage market leadership, innovation and investment in people, in a collaborative manner. A few months later, they formed a strategy framework and a clear roadmap to guide our actions through 2015.

This integrated strategic direction is driven by Dubai's energy key-players, who have shaped a common vision supported by a defined roadmap that will regulate the market and draw investors in clean energy technologies. Dubai is promoting the principles of Public Private Participation (PPP) and introduced laws to boost the market dynamics for partnership on several projects, beginning with the solar and clean coal power generation.

In addition, the regulatory framework for district cooling and Energy Service Companies (ESCO) is underway to ease implementation where the efficiency of buildings can be attained.

To make renewable energy more attractive, the Dubai Supreme Council of Energy and the Regulatory and Supervisory Bureau has developed the Feed-in-Tariff and Technical Codes for Independent Solar Power Producers and seek to have them ratified by the Government of Dubai before the end of 2013.

(c) **Inclination towards the use of solar power**

Solar power offers the strongest economic opportunity. Other technologies such as geothermal and hydrogen power are viable complements in the UAE's renewables journey, but due to falling costs coupled with the region's vast resources, solar is now accepted as the key competitive alternative to oil in the UAE.

The UAE has so far only set modest targets to increase renewable energy. Its biggest cities, Abu Dhabi and Dubai aim to provide just 7% by 2020 and 15% by 2030, respectively, of their energy needs through renewables. But these relatively low renewable targets should not be mistaken for a lack of ambition. Based on the increasing capacity of renewables in the region, compared to other members of the Gulf Cooperation Council (GCC) and even the MENA (Middle East & North Africa) region as a whole, the UAE is in fact already leading the way.

7.6.4 Practice in the Field of Clean Energy

The government's UAE Vision 2021 clearly sets out its goal to reduce the nation's dependence on fossil fuels while increasing the use and development of renewable energy. [53].

(a) **Use of energy efficient lighting**

Dubai has witnessed its first large efficient lighting initiative in 2014. Dubai Carbon distributed compact fluorescent lamps (CFL) along with information on energy savings in households throughout the Emirate. The awareness campaign was to provide approximately 800,000 11 W and 23 W CFLs free of charge.

- Use of LED lamps for outdoor commercial lighting such as advertising boards
- Introduction of Bio-fuel technology in one of their vehicles (pilot study).
- Introduction of Electric cars for in-compound transportation.
- Introduction of Solar powered speed cameras by Dubai Police.
- Emirates Energy Star to help Dubai Police calculate their footprint from electricity use, while monitoring and controlling their buildings more efficiently.

(b) **Application of green building standards**

Applied 'Green Building' standards and codes for all their new buildings that are to be approved by the Dubai municipality to finalize the design/ planning process for construction. The Supreme Council of Energy's strategy to reduce energy consumption in existing buildings by 30% [53].

(c) **Use of solar power**

Dubai Electricity and Water Authority (DEWA) and Saudi Arabian power company ACWA will build a 200-MW solar power plant providing electricity at 5.85 cents per kilowatt-hour—the lowest solar electricity price in history. Once complete, the Mohammed bin Rashid Al Maktoum Solar Park will be the UAE's biggest. Generation will be further boosted by Dubai's recent announcement to invest US$3 billion to increase capacity from 1 gigawatts (GW) to 3 GW. The park will also be home to a renewable technology research and development center, as part of the city government's aims to make Dubai a renewables hub.

The government's future energy company Masdar—a huge investment which in itself signals the UAE's desire to be a leader in the clean technology market—recently opened one of the world's largest concentrated solar power farms, called Shams 1. The facility has a 100 MW capacity, which will supply power to 20,000 homes in the UAE. The project is a joint venture between Masdar, French Total and Spanish Abengoa Solar, and when fully constructed, will account for almost 68% of the region's renewable energy capacity.

Masdar also has a joint venture with BP called Hydrogen Power Abu Dhabi, to build the world's first hydrogen-fuelled power plant to run on natural gas, limiting emissions by using carbon capture and storage technology. The hydrogen power plant will generate around 400 MW electricity, and could provide more than 5% of all Abu Dhabi's current power generation.

Another critical to the region is access to fresh water, so Masdar City is currently testing desalination plants powered by renewables, with a goal of having commercially viable desalination facilities throughout the UAE by 2020. Although this technology is crucial to the region, the UAE also recognizes its global importance as climate change puts increasing pressure on current freshwater resources.

The City is also planning to build the Gulf's first geothermal energy facility, with cooperation from Icelandic company Reykjavik Geothermal—another way to explore the full range of renewable energy options available.

The major telecommunications companies 'Du' and 'Etisalat' are leading in the local commercial sector. Du has just completed its fifth solar site installation which is well on its way to becoming 100% renewable, and joins four more sites in the

UAE that are already running on 100% solar energy. Etisalat has large solar stations at 50 locations in the UAE, as well as solar and wind power in Egypt and Afghanistan.

Leading businesses from many other industries in the UAE are also beginning to invest heavily in renewables. ICT-enabled services company, Pacific Controls Systems is using solar PV systems in its Dubai headquarters that generate around 50 kW of energy to power its own lighting.

Dubai Airports and DEWA have partnered to build 100 rooftop solar panels that will supply 48.8 MWh—about two thirds of the building's power—to the airport [54].

(d) **Dubai smart grid**

In line with the Smart Dubai vision, the Dubai Electricity and Water Authority (DEWA) developed a comprehensive Smart Grid strategy with the aim to connect renewable energy sources and promote environmental initiatives. DEWA has been a Strategic Partner of Smart Dubai since inception in 2014, and through the Smart Grid contributing to making Dubai a smart city.

The Smart Grid consists of three initiatives designed to ensure seamlessly accessible and nonstop connected water and electricity services that meet daily living requirements of supply while enabling monitoring to facilitate smart and environmentally friendly use of resources.

The first Smart Grid initiative aims to integrate renewable energy in form of solar power into houses and buildings. In supporting the leverage of given and foremost free, energy sources, DEWA encourages environmentally friendly while economically driven behaviour within Dubai. While personal consumption can be covered autonomously through solar panels, DEWA enables to feed surpluses into the common city-wide grid.

Within the second initiative Smart Meters will be installed into private and public spaces. These Meters provide automatic and detailed readings of individual consumption enabling end users to better understand and manage their personal DEWA bills. 200,000 such Smart Meters will be installed and become fully operational by January 2016.

In another, third, initiative, DEWA commits to create an infrastructure for electrically driven vehicles to be charged. Following the environmental initiative to incorporate solar energy into the Dubai grid, this initiative is another green goal DEWA realises.

Through the Smart Grid DEWA is supporting to turn His Highness' vision into reality by offering new smart initiatives and services to the community of Dubai [55].

(e) **Solar park**

The Mohammed bin Rashid Al Maktoum Solar Park was announced in January 2012 in line with the vision and directives of His Highness Sheikh Mohammed bin Rashid Al Maktoum, Vice President and Prime Minister of the UAE, and Ruler of Dubai, to enhance the sustainable development of Dubai. It also supports the Dubai Clean Energy Strategy 2050 to make Dubai a global centre of clean energy and green economy. The strategy also aims to provide 7% of Dubai's energy from clean energy sources by 2020, 25% by 2030 and 75% by 2050. DEWA is managing the Solar Park, which is the largest renewable-energy project on a single plot in the world with a planned production capacity of 5000 MW upon completion in 2030 [56].

(f) **Use of LED lights**

The Roads and Transport Authority (RTA) has started replacing traffic lights in Dubai with halogen bulbs by LED power-saving technology. This drive comes as part of the RTA's safety and environmental push to enhance the traffic safety level and keep abreast of the green economy drive. The project will be carried out in phases and phase one is now under way. The bulbs will also save around USD 236,000 a year in expenditure, of RTA's Traffic and Roads Agency.

The new (Light Emitting Diode) lighting system will last longer (about ten years) while using the least power. The low-maintenance bulbs are also more reliable and efficient than conventional lighting [57, 58].

(g) **Use of solar power to light up Dubai parks and streets**

Dubai Municipality aspires to construct public parks, neighbourhood gardens and public squares by relying entirely on the solar energy sources to operate all facilities of these parks as well as operating the lighting elements and irrigation systems for trees and green spaces (Zero Energy parks). The General Projects Department at Dubai Municipality has excelled, using environmentally friendly components and technologies aiming to create a suitable environment to help optimizing the use of power with innovative designs that take into consideration the use of solar energy.

The streets and roads of Dubai dazzle in the night are also considered for the use of solar powered lights and use of LED for the roads and traffic signals as an environment friendly drive.

(h) **Becoming a sustainable city**

For a city to be sustainable it needs to encompass the three main pillars of sustainability: Environmental, Economic and Social sustainability. Ultimately, the key goal of any sustainable city should be to provide residents and workers with the highest quality of life together with the lowest environmental footprint.

A good sustainable city is one where people want to live, work and visit.

Dubai, also recently announced plans for additional sustainable projects such as the Desert Rose, a 14,000-ha smart city expected to accommodate 20,000 plots for Emiratis and will cost 20 billion AED to build. There is a pattern emerging from these developments—the United Arab Emirates is already emerging as a global leader in sustainability. This pattern became increasingly apparent, after Dubai recently set targets for itself to become one of the most sustainable cities by 2020.

The growing development boom is clearly going to set a challenge for Dubai to achieve its sustainability ambitions by 2020. A key challenge is educating private and public stakeholders about the main principles of building sustainably, without increasing the built cost. Cost is currently the main deterrent factor for most developers, as it is a common belief that building sustainably means more cost, thus reducing profit margins, which for developers is understandably an outright turn-off.

From the beginning of each project it is important for the client and consultant team to develop a series of sustainable initiatives and targets. These targets will help guide the design in the pursuit of the sustainability goals. They can be divided into categories such as: Water, Health and Well-being, Energy, Materials, Pollution, Ecology, and Waste.

If we take water as an example, the ultimate design intent for any sustainable project would be to minimise net water use by increasing efficiency and re-use, which can be achieved by [59]:

- Recycling water and by encouraging the collection and re-use of rainwater to reduce the demand for potable fresh water.
- Looking at highly efficient and smart irrigation systems and storm water design.
- Installing smart water efficient fittings, water metres and appliances.
- Maintaining and improving the quality of ground and surface water.
- Low-water use landscaping.

7.7 Smart Dubai: E-Democracy Initiatives

E-democracy is the use of technology for strengthening the mechanisms of democratic decision making. While E-voting achieves automation in the voting process, E-participation involves citizens in the policy process giving rise to two different sets of relationship between citizens and policy makers. The first is a vertical relationship in which policy makers use the speed and immediacy of information technology networks to consult citizens on various policy issues. The second is a more complex, horizontal and multi-directional interactivity in which citizens and groups will be able to make use of information available from multiple online sources to bring pressure to bear on their government. The latter may also take the form of e-activism or the use of ICTs by civil society organisations to advocate their view points and influence political or policy process.

Dubai has carried out a number of initiatives to foster E-democracy with the central theme of making residents and visitors happy.

7.7.1 Minister of Happiness, Youth and Tolerance

Happiness as a measurement first planted its seed in the public imagination in 1972 when it was declared that Bhutan, a Buddhist kingdom on the Himalayas' eastern edge with a population of just 790,890 [60], had created a Gross National Happiness index. Its principles were based on equitable social development, cultural preservation, conservation of the environment and promotion of good governance and the rest of the world were quick to sit up and take note.

Since then, four national governments [61] namely Bhutan, Ecuador, UAE and Venezuela have appointed ministers of happiness responsible for coordinating their national efforts.

UAE in its twelfth cabinet of 2016 [62] appointed Minister of Happiness, Minister of Tolerance, Youth Minister and Minister of The Future. This is a big leap taken by the UAE President and Prime Minister of the UAE for national development based on core values, led by youth and focused on a future in which everyone achieves happiness. These new portfolios intend to empower people and not hold power over them. Within a smart city framework, it will nurture an environment of E-democracy in which people create and enjoy their own happiness. Through this, UAE also sets a precedent to governments in the region for revising their role and creating an environment in which people can achieve their dreams and ambitions instead of the current environment where government is meant to control. Shaikha Lubna, the Minister of State for Tolerance said [63] "Happiness and tolerance are values, not services, an integral part of all of us". "It doesn't matter who you are, what you are, or where you're from. The UAE provides a common path for everyone. It is the land of dreams and opportunities." This is a very strong commitment where 85% expatriates co-exist in harmony.

7.7.2 Tolerance Institute and Award

Dubai with 200+ nationalities and many languages, the spirit of tolerance has gained lot of importance. Dubai launched [64] the International Institute of tolerance and Shaikh Mohammed Bin Rashid Award for Tolerance in October 2016 as a global initiative with the aim of encouraging tolerance, co-existence and honour champions of tolerance. The Tolerance award is categorized into Human thought, literary creativity, arts, youth projects and new media aimed to reinforce tolerance in the Arab region. The tolerance institute will offer expert views on policies for promoting values of tolerance among nations, release studies on root cause of fundamentalism and sectarianism. It will further work towards instilling principles of tolerance among youth and nurture ground for tolerance and countering extremism. To demonstrate its commitment to tolerance for the whole world, the year 2017 has been dedicated as the year of tolerance by the UAE government. UAE stands 1st in Arab world for tolerance and 3rd in world.

7.7.3 Dubai Smart Cities and Government Excellence Forum

Dubai Smart Cities Forum [65] is a series of thought leadership sessions organized by Mohammed Bin Rashid School of Government (MBRSG) in cooperation with Dubai Smart Cities Initiative. The initiative convenes experts, decision-makers and leaders from both public and private institutions in a creative and stimulating environment to share experiences and best practices, debate challenges and work collaboratively towards achieving Dubai's smart city ambitions. The forum's sessions highlight local and global smart city trends in sectors such as management, manpower, urban development, services, infrastructure and planning. Since inception in 2014, the forum has become a point of reference to public and private organizations in the UAE and beyond in relation to smart cities

The **Dubai Government Excellence Forum** (DGEF) [66] from Dubai Executive council is the apex agency for tracking Government Best Practices and to promote the concepts and applications of excellence, innovation, quality, and knowledge transfer. The DGEF organizes a number of programmes enabling youth to undertake their active national role as well as support them as a key foundation for the future. In October 2016, the DGEP organized the 9th Dubai Forum for Government Best Practice under the theme Government of Youth, Government of the Future.

The UAE cabinet in 2015 launched a **National Programme for Government Communication** [66] to address society issues and national priorities based on the results of study of Knowledge, Attitudes and Practices (KAP) in the UAE. This programme aims to realise positive changes in the people's behaviour and lifestyles. The NPGC, to be implemented over seven years till the Golden Jubilee of the UAE in 2021, will consist of seven pillars. Healthy Children, Cohesive Family, Quality Education, Green Ideas, Sustainable Food, Prosperous Future and Diabetes-free UAE.

7.7.4 Dubai Open Data Initiative

Real-time data is driver of IoT and smart city innovations. The Dubai Data Law, which has been in force since December 2015 aims to improve efficiency between and facilitate the integration of services provided by federal and local governmental entities. It envisages the use of available data held by 'data providers' are utilized to help inform government decision making, strategic initiatives, and policies. The law also aims to support innovation, economic development, foster competition, and help Dubai achieve its vision of becoming a "smart city".

In the Digital Economy, how organizations use real-time data is the main driver for business competitiveness. Data can help cars from crashing into one another, governments to collaborate better, make factory floors safer and more efficient, and help people and healthcare providers to analyse their personal health instantly.

The Dubai Open Data Law sets out the terms for increased sharing of non-confidential information between government agencies and the public. The new law will make the data accessible to researchers, investors and service developers via an integrated platform, and create opportunities for collaboration between the public and private sectors.

The new law will unify Dubai's data and remove the last legal obstacles for those interested in investing in the digital economy. It will help authorities to prepare policies, implement plans and initiatives efficiently and effectively. Most importantly, the data law will also enable the emirate to achieve its vision of making Dubai a city that can manage data according to a clear and specific methodology consistent with international best practices.

Led by the Dubai Data Establishment, the Dubai Data initiatives has set 11 strategic objectives [67] towards creating a culture of data sharing and innovation, aligned with three core principles for data sharing and publication; data use and re-use; and privacy, confidential information and intellectual property:

(a) Enable Dubai to achieve its smart city vision of becoming the happiest city
(b) Managed data based on clear guidelines aligned with international best practices
(c) Achieve integration and harmony between the services provided by federal government agencies and local government bodies
(d) Optimise use of data for data providers
(e) Promote transparency and establish governance rules on the dissemination and exchange of data
(f) Increase efficiency of services provided by federal and local government entities, in terms of: level of quality, speed of delivery, simplification of procedures, and reduced operational costs
(g) Increase competitiveness of data providers and raise UAE standing on international competitiveness indices
(h) Improve decision making at federal and local levels to enable agencies to effectively process data, prepare policies, and implement strategic initiatives
(i) Encourage and nurture a culture of innovation that will lead to better quality of life for residents and visitors
(j) Balance the dissemination and exchange of data with preservation of data confidentiality and privacy
(k) Provide the necessary data for non-governmental actors in order to support the economic and development plans of the emirate.

Under the Open data law, the federal statistics are published [68] in the federal competitiveness and statistics Authority.

7.7.5 People Engagement

This is at the heart of E-democracy. With 200+ nationalities and a targeted 30 million tourist arrivals by 2020, scores of languages, a multitude of cultures, different levels of literacy and majority of blue collar workers poses a real challenge for the government to engage, Citizens, residents and visitors in the spirit of what a smart city is expected of. This section looks at the number of E-democracy initiatives that Dubai has carried out.

(a) **From Majlis to Hashtag:** [69] The UAE cabinet had a tradition of holding cabinet retreats since 2007 for brainstorming key issues among all stakeholders before the cabinet meeting. In December 2013 the Prime Minister launched a national brainstorming session through twitter urging both citizens and residents to send innovative fresh ideas on education and health that will be considered in the cabinet. This initiative generated 82,000 crowdsourced ideas categorized into 5 most vital issues faced in education and health sector. This demonstrates UAE government's commitment to E-democracy and engaging citizens in governance through social media.

(b) **Smart Majlis** [70] has always been a national cornerstone. People gathered to share ideas and create solutions. Today, in the era of the smart city and the growing community of Dubai, an e-smart majlis has been established to continue receiving ideas and comments through electronic means. Mohammed bin Rashid smart majlis is a platform that allows everyone to participate in building the future of Dubai and to submit ideas directly to the ruler.

(c) **Online Chat and Crowd sourcing**: [71] Most of the government websites and apps has the facility to do an online chat to seek answers to an issue or to submit ideas and suggestions on government service. Prominent among them is the Ministry of interior and Energy and Water Authority's crowdsourcing platform called Ebtikari to encourage inventors and designers from around the world to present their innovations.

(d) **Happiness Meter**: To measure people's satisfaction and level of happiness on the service delivery, Dubai installed happiness meters in several touch points starting from a service counter to an app and the website. People after availing a service has to simply push a button to indicate satisfied, neutral or dissatisfied. The data thus generated is used to create a happiness index. Evaluation of 9 months data consisting of 2 million customers of 38 public entities shows 89% happiness.

(e) **Start Rating System**: In order to bring a culture of excellence in government, Dubai has implemented a 6-star rating system for all its departments through which service delivery is rated, to provide exceptional experience for customer happiness across all service channels.

(f) **Have Your Say** [72] launched by Dubai Municipality since 2012 is an e-complaint and e-suggestion system that makes room for its employees to share creative ideas and innovative thoughts. There is an annual award for best employee engagement program constituted by International Quality & Productivity

Centre Middle East. This platform provides a unified system for suggestions and complaints of Dubai Government.

(g) **Smart Palm** [73] is palm tree-shaped, solar-powered tech hub provided by Dubai Municipality to offer free Wi-Fi service along with charging points for smart gadgets, the first of its kind initiative in the Middle East. This is a big boon for residents and visitors due to the relative high data bandwidth costs in the UAE. There are eight charging points on each Smart Palm, offering extreme recharging capability, two and a half times faster than a regular plug. Smart Palms also issue weather warnings in Dubai beaches.

(h) **Free Wi-Fi and Li-Fi** is an initiative launched by Smart Dubai and Du telecom operator to provide free Wi-Fi across the emirate by the end of 2017 under the Wi-Fi UAE plan. Light Fidelity (Li-Fi), a visible Wi-Fi connection is also being explored and rolled out in phases to meet the exponential growth of wireless data traffic by 2025.

(i) **2016 declared as Year of Reading** [74] to create a generation of book lovers and consolidate UAE's position as a global capital for culture and knowledge. To mark the year of reading, Mohammed bin Rashid Library is being built designed to be the largest cultural center in the Middle East and North Africa (MENA) region with 2600 seats and smart applications for co-creation.

(j) **Library Kiosks** are provided by Dubai Municipality in beaches and in various points where people can borrow books.

(k) **Future Council**: [75] The UAE government is set to establish a world council to implement technologies of the fourth industrial revolution. This council will draw 5000 experts from all over the world and collaborate with government, private and civil public society to establish a framework to focus on renewable energy, water and food security.

7.7.6 Community Development Authority (CDA)

CDA was Established in 2008 [76] to develop frameworks for social development in Dubai. It aims to enhance social development and national identity in cooperation with concerned bodies and strengthen citizens and residents' roles in the community. CDA takes care of people with disabilities, elderly, family, children and youth and other vulnerable groups. CDA has established a number of smart city initiatives for an inclusive engagement with all target groups.

A Local Majlis is established within the community, as a direct line of communication with local residents. CDA works in coordination with official authorities to find solutions to different social issues raise awareness and promote societal development through consistent lectures and meetings, to preserve national identity and bridge the gap between generations.

CDA provides an APP to access all community services including Social Benefits, Human Rights Complaints, Social Studies and Research, Media Centre, elderly services and CDA Events.

7.7.7 Community Inclusive Engagement

Smart Dubai has launched a campaign "A City for Everyone" [77] in order to make Dubai a disability-friendly city by the year 2020 through projects and initiatives that promote the participation and inclusion of persons with disabilities in the community, eliminating all obstacles that may stand in the way of them positively engaging in their environments, as capable individuals of society.

The campaign aims to promote equal opportunities, maintaining social cohesion, Building social capital and Minimizing and eventually eliminating social exclusion. Dubai enacted the disability law in 2014 which is in line with the UAE disability Act with a higher Committee for the Rights of Persons with Disabilities in the Emirate of Dubai in 2016.

Four Apps specifically aimed for differentially challenged persons. Communication service App called Sanad Relay for enabling people with hearing disability or speech difficulties to communicate with persons or entities in the community to receive the service they need in their daily life. CDA Dubai App to access all services for community services, CDA Sanad RV is the third app for track CDA's Remote Vehicles moving across Dubai that offers Speech & Language Therapy, Occupational Therapy, Physical Therapy and general consultation. The 4th App is Dubai Volunteers who can register and offer their services for CDA activities.

7.7.8 Dubai Now

Dubai Now [78] is the first unified service platform that brings through a simple mobile app, more than 55 Smart services by 24 government and private entities, creating a seamless and efficient user experience for the citizens, residents and visitors of the emirate. The aim is to combine most of the daily government needs on a single login for everyone to have an equal access to all apps and services.

The apps in Dubai Now have 1 dashboard, 3 login boards offered by 22 government entities grouped under 11 categories. The apps under the 11 categories are listed below:

7.7.8.1 Utilities and Bills

Pay DEWA, Etisalat, du, Dubai Municipality bills, Dubai Police traffic fines, pay for salik road tolls, bus and metro card and Dubai Customs Accounts. There are also apps to donate to Dubai Cares and Al Jalila Foundation.

7.7.8.2 Driving

The apps under this category allows renewing vehicle registration, top up parking time, see road speed limits, view road accident notifications and traffic, find nearest petrol station and Tasjeel (vehicle registration and testing centers).

7.7.8.3 Public Transport

Track Dubai Airport flight departures and arrivals, call a Taxi, View the Road and Transport Authority (RTA) metro map, journey planning to and from any point and view best routes.

7.7.8.4 Weather, ATM, Post and Address

View Dubai Calendar and Dubai's daily weather conditions, find nearest ATM, Track Emirates Post (Empost) shipments and view shipping rates, send feedback or report complaints regarding any government department service in Dubai, access Makani (10 digit building number) to locate specific buildings and addresses in Dubai. Municipality also provide Alif app for pets and Taheel app for digital device recycling.

7.7.8.5 Islam

View daily prayer timings, find nearest mosque and select the fastest route to reach.

7.7.8.6 Health

Find all doctors, clinics and hospitals registered with Dubai Health Authority (DHA), create a vaccination plan for child and track it, Find nearest pharmacy and view 24 h pharmacies in Dubai.

7.7.8.7 Residency Visas

Track status of the General Directorate of Residency and Foreigners Affairs (GDRFA) visa applications and entry permits for residents and visitors, view all residency visas and entry permits of dependents.

7.7.8.8 Housing

View the Rent Increase Calculator from RERA (Dubai Land Department - DLD), Citizens access to Mohammed Bin Rashid Housing Establishment (MRHE) loan calculator and application status, request maintenance, View Dubai Land Department (DLD) daily and monthly transactions, view electricity and water consumption details and view status of your Ejari (rent) contract.

7.7.8.9 Security and Justice

Report a violation to Dubai Police, Find nearest Dubai Police Station, Inquire about the status of Dubai Court's cases, Call emergency numbers like Dubai Police, Ambulance, Fire Department, and DEWA.

7.7.8.10 Education

Find KHDA registered schools and universities, find their rating, annual fees, curriculum, level, location and other details.

7.7.8.11 Business and Employment

Search for trade activity with the Department of Economic Development (DED), Track and renew trade licenses registered with the Department of Economic Development (DED), Reserve a trade name with the Department of Economic Development (DED).

7.7.8.12 Guard App

CityGuard is a mobile application offered for free to all Abu Dhabi residents, be it UAE nationals, working expatriates, or visitors to the Emirate. CityGuard allows the public to report incidents and submit complaints related to the Abu Dhabi Emirate directly to the government. With its slogan "Your City, Your Community, Your App" CityGuard aims to increase civic participation and collaboration between the public and the government in order to improve the Emirate at all levels.

The aim of the app is to improve government services and the city image with the help of the public. Therefore, several incident types such as public safety, consumer protection and environmental issues have been identified. Participating government entities are the three Municipalities of the Abu Dhabi Emirate, Abu Dhabi Police, Department of Transport, Abu Dhabi Food Control Authority and Health Authority—Abu Dhabi amongst others. CityGuard is another cornerstone in the Abu Dhabi Government's drive to engage with its customers across multiple channels in order to maximise their feedback and collaboration in the service delivery enhancement process.

Through Abu Dhabi City Guard, the customer can instantly report any incidents around the Emirate by taking a photo, movie or audio of the reported incident and use an embedded interactive map to locate the exact geographic position of the incident. The application automatically creates a case with the Abu Dhabi Government Contact Centre which promptly assigns the case to the appropriate government entity for resolution. From open to closed, the case can be monitored by the customer within Abu Dhabi CityGuard or online on abudhabi.ae or by contacting the Abu Dhabi Government Contact Centre. The customer can call the Abu Dhabi Government Contact Centre to report any technical issue that is encountered during the usage of CityGuard, or to provide any feedback regarding the application, its features and potential enhancements.

7.8 Summary of Findings

Dubai City has the heart and soul open to welcome people from all over the world, provide world class facilities for a happy living and foster a culture of reading, innovation and excellence. Since 2015, Dubai leadership is going in full gear on innovation hosting anyone who has an idea on its focus areas, particularly the youth. With a national innovation strategy in place and having established the necessary legal and organization framework to achieve, the run up to the expo 2020 will unfold much advancement in smart city and particularly E-democracy with electronic voting on every government service touch points. The city in its transformation to a knowledge society is aggressively devising a strategy on how to sustain once it achieves the title of smartest and happiest city on earth.

Following list draws summary of key takeaways.

(a) Dubai leadership is able to effectively lay and communicate the clear vision of making Dubai 'the happiest city on earth' to its citizens, residents and the visitors.
(b) Agility, resilience and flexibility of the leadership for restructuring government functions and bodies as needed, if they find something they set up does not work.

(c) Appointing Minister of Future, Tolerance, Youth and Happiness reflects the cabinet's will to promote unity in diversity and to engage people from 200+ countries in its governance.
(d) Established Dubai government accelerator consisting of five core sectors; human resources, environment, security, economy and education. The idea is for 5 ministries to work as a single entity in one location to find suitable solutions that can achieve the aspirations of the people.
(e) Effective utilization of social media for 2-way communication and to solve societal challenges with citizen centred crowd sourcing, co-creation and transparency leading to trust in government.
(f) Time bound goal setting with clear measurement criteria and alignment with global indicators and international standards and measures of evaluation.
(g) Establishing a clear legal framework and mandate for innovation. Apportioning sufficient funding source gave the stakeholders freedom to stretch their imagination on Prototyping and piloting. The innovation fee received on all government transactions ensures that it is well funded.
(h) Dubai has laid the foundation for smart city early on in terms of ICT and a strong IT infrastructure. In certain cases it was a greenfield (ICT in design) approach, but in many cases it was a brownfield that required working with the legacy IT platform. Smart phone penetration of 80% and societal readiness to accept government's smart city initiatives and apps are important success factors.
(i) Break the silos by bringing together all government departments on one common arena and inculcating the idea of 'our city' from 'our department' gave all the stakeholders a common agenda.
(j) Thrust on Innovation led to collaboration, cooperation and co-creation of challenges faced in service delivery.
(k) Empower the youth and provide employment opportunity for them, as 65% of Middle East belong to this age group. The 2016 innovation week hackathon focused on social innovation projects.
(l) Forge public private partnership to nurture talent and host smart city incubation centers for finding solutions to worlds greatest challenges along the themes of opportunity, mobility and sustainability.
(m) Open to ideas from all over the world. Continuous and on-going meetings and discussions with other cities to learn and adapt best international practices and open heart to share experience with other cities.
(n) Make the government open and transparent and capturing people's satisfaction level at every touch points through happiness meters resulted in institutionalizing e-democracy at every touch points. Moving away customer centric 'meeting expectation' to people centric happiness.
(o) Embracing inclusive engagement of citizens, residents and visitors through social media and promoting a culture of tolerance, reading, experimentation and creative innovation through establishing dedicated centers, awards, prizes and societal recognition.

7.9 Conclusion

The government of UAE is engaged in developing a strategy that ensures all sectors' readiness for the future's variability. The ministry is working on employing the relevant tools to shape the future, which helps governments in forecasting opportunities, trends, challenges and future implications, analyzing their impact, developing innovative solutions and providing alternatives. It is currently working on building capacity for the future and long-term planning capabilities through series of workshops for "Shaping Future", to support the government entities efforts in developing innovative models for future services and design pre-emptive flexible strategic plans adjustable to any variability. It also aims to promote the utilization of future trends, foresights and challenges in favour of improving the future of government, and constantly reinventing for customer's expectations fulfilment and satisfaction. The ministry of Cabinet Affairs and the Future is assigned with the post-oil portfolio, and following-up on the programs and policies pertaining to the necessary measures this for a better future for coming generations.

UAE's strategy for future involves building future models for the health, educational, developmental, and environmental sectors, the harmonization of the current governmental policies, in addition to building national capacities in the field of future foresighting, establishing international partnership, laboratories and launching research reports on the future of the various sectors in the country.

The strategy aims at setting governmental systems that make fore sighting the future a part of the strategic planning in the government, and at launching studies and scenarios to forecast the future of all priority sectors in addition to setting plans and policies. Furthermore, it will work on entrenching the culture of the future in government, as well as spreading awareness about its importance and building capacities and reinforcing the position of the UAE as a global destination for the futures.

On the other hand, the UAE Strategy for the future focuses on topics that include the priority sectors of: the future of the human and the youth capital, the future of technology and smart systems, the future of sustainability, the environment and the climate change, the future of the infrastructure and the transportations, the future of health, the future of education, the future of sustainable development, the future of positive and happy life environment, the future of energy, the future of economy and economic and commercial security, the future of financial resources, the future of the government and governmental services, the future of international and political relations, the future of water and food safety and the future of E-Security.

In building the future capacities, the leadership has instructed the establishment of a strong organizational and material infrastructure in order to build the future and invest in national cadres. In this regards, the government will build capacities to achieve sustainability in the field of future foresight through the introduction of future foresight principles in a simplified way in the school curricula, and through national capacity building scholarships to overseas schools and institutions, as well as through launching tools designed specifically for UAE government and teaching future foresight specializations in the national universities and in the public capacity building programs, in order to fulfil the requirements of all sectors.

UAE as a 'Future Destination', the Ministers will transform their ministries into a model of future ministries, as they will represent the country as ambassadors to the future, and they will exhibit its future leadership in the various fields.

Finally, UAE government will work on positioning UAE as a global destination for the sustainable future and through the future oriented initiatives and future platforms, such as the World Government Summit, as well as through hosting new global initiatives like the Annual Meetings of the Global Future Councils. In addition, UAE government will also work on developing strategic partnerships with forums and futures think tanks, experts, platforms and networks, as well as universities inside and outside the country.

References

1. E-Democracy: Strengthening the Participation of Citizens in Government, by Waltraut Kotschy, available @ http://26konferencja.giodo.gov.pl/data/resources/KotschyW_pres.pdf
2. Dubai's Legal System- Creating Legal and Regulatory Framework for a Modern Society, by: Andrew Tarbuck & Chris Lester, Published by Motivate Publishing, available @ http://www.id.gov.ae/assets/kKbkN9NSOGI.pdf.aspx
3. Demography Migration and the Labour Market in the UAE, by Gulf Labour Markets and Migration, year 2015, available @ http://gulfmigration.eu/demography-migration-and-the-labour-market-in-the-uae/
4. Population of UAE, year 2006–2014, by Gulf trading Economics.com, World Bank, available @ http://www.tradingeconomics.com/united-arab-emirates/population
5. Population of Dubai, Dubai Statistical Center, available @ https://www.dsc.gov.ae/en-us
6. Dubai Tourism Statistics and Trends, year 2015, available @ http://www.go-gulf.ae/blog/dubai-tourism-statistics-trends/
7. Population of three Emirates by Nationality, Gulf News, year 2014, available @ http://gulfnews.com/multimedia/infographics/general/demographic-trends-expatriate-population-in-dubai-sharjah-and-ajman-1.1562565
8. Dubai Tourism, Department of Tourism and Commerce Marketing, available @ http://www.visitdubai.com/en/department-of-tourism/about-dtcm
9. Dubai Expophoria 2020: A Message from the Desert on Mobility, Sustainability & Opportunity, by Dr. Nasser Saidi, Publication: PMI Congress EMEA, year 2014, available @ http://nassersaidi.com/wp-content/uploads/2014/05/Dubai-Expo-phoria-2020-May-5-2014-PMI-EMEA-Congress.pdf
10. Highlights of the U.A.E. Government Strategy, Year 2011-2013, available @ http://planipolis.iiep.unesco.org/upload/United%20Arab%20Emirates/United%20Arab%20Emirates_Governement_strategy_2011-2013.pdf
11. Understanding E- Democracy- Government led Initiatives for Democratic Reform, by Julee Freeman and Sharna Quirke, Published by Journal of E- Democracy, year 2013, available @ http://www.jedem.org
12. http://www.wam.ae/en/news/emirates/1395288828473.html
13. http://www.futureagenda.org/
14. http://gulfnews.com/news/uae/government/dh1b-fund-to-shape-future-of-key-sectors-1.1810539
15. https://dubaifutureaccelerators.com/en
16. http://www.smartdubai.ae

17. http://www.smartdubai.ae/foundation_pillars.php
18. http://www.khaleejtimes.com/nation/government/no-room-for-procastination-shaikh-mohammed-says-on-governance
19. http://www.khaleejtimes.com/technology/dubai-smartest-city-in-the-gulf
20. http://www.khaleejtimes.com/business/economy/uae-leads-arab-world-in-un-e-services-index
21. http://www.uaeinnovates.gov.ae/innovation-ecosystem
22. http://www.futureagenda.org/
23. http://www.futureagenda.org/news/whats-on-the-future-agenda-three-business-innovations-for-uae
24. https://www.globalinnovationindex.org/gii-2016-report
25. http://www.uaeinnovates.gov.ae/docs/default-source/pdfs/national-innovation-strategy-en.pdf?sfvrsn=2
26. http://www.uaeinnovates.gov.ae/docs/default-source/pdfs/science-technology-and-innovation-policy.pdf
27. http://motf.ae/
28. http://www.smartdubai.ae/pdfs/PR_SmartDubaiPlatform_EN.pdf
29. http://www.innovationandtech.ae/smart-dubai-platform-rolled/
30. http://www.itp.net/608411-why-do-cities-need-a-smart-platform
31. http://www.khaleejtimes.com/nation/dubai/in-focus-expect-a-hyperloop-in-dubai-under-5-years
32. http://www.khaleejtimes.com/business/real-estate/uae-to-standardise-3d-printed-buildings-with-new-index
33. https://www.adec.ac.ae/en/MediaCenter/News/Pages/The-2016-WRO-UAE-National-Robotics-Challenge.aspx
34. http://mediaoffice.ae/en/media-center/news/17/2/2016/dubai-museum-of-the-future-foundation-announces-launch-of-global-blockchain-council.aspx
35. http://www.khaleejtimes.com/business/local/blockchain-integral-to-regional-smart-cities-digital-economy-booz-allen
36. http://www.khaleejtimes.com/business/local/dubais-global-blockchain-council-welcomes-new-members-adopt-pilot-projects
37. http://www.innovationandtech.ae/visa-opens-regional-innovation-center-dubai/
38. http://middleeast.geblogs.com/en/stories/ge-industrial/aviation-and-travel-tech-incubator-takes-off-in-the-uae/?gclid=CjwKEAjwtNbABRCsqO7J0_uJxWYSJAAiVo5LCpRO2EWtQUSRM-EjrYpFbdiVqu8bDsw_9zkrhG679RoC3svw_wcB
39. http://middleeast.geblogs.com/en/stories/ge-industrial/the-uae-inspiring-everyday-innovation-during-uaes-year-of-innovation/
40. http://m.dubaiprnetwork.com/pr.asp?pr=111686
41. http://www.khaleejtimes.com/un-to-have-centre-of-excellence-in-dubai
42. http://www.enterpriseinnovation.net/article/look-dubais-smart-city-initiatives-1141254266
43. https://www.dewa.gov.ae/en/customer/innovation/smart-initiatives/smart-applications-via-smart-meters-and-grids
44. https://www.dsoa.ae/en/dubai-smart-city/
45. http://smartdubai.ae/partner_story_0524.php
46. http://cloudone.dubai.gov.ae/
47. https://www.rit.edu/dubai/smartcity
48. http://www.khaleejtimes.com/nation/education/a-master-s-degree-in-smart-city-anyone
49. https://www.dsoa.ae/en/news/dubai-silicon-oasis-implements-regions-first-water-saving-subsurface-irrigation-system/
50. http://www.khaleejtimes.com/business/real-estate/dubai-design-district-goes-smarter-with-21-new-initiatives. http://www.dubaidesigndistrict.com/innovation/smart-city-2/
51. http://dubaiholding.com/en/media-hub/press-releases/uae-south-korea-sign-agreement-develop-smartcity-korea/

52. Economic Diversification in the GCC, by: Tim Callen, Reda Cherif, Fuad Hasanov, Amgad Hegazy, and Padamja Khandelwal, Publication: IMF Staff Discussion Note, year 2014, available @ http://www.imf.org/external/np/seminars/eng/2014/mcd/
53. State of Energy Report- Dubai, State of Energy Report, by: UNDP, year 2014, available @ http://www.undp.org/content/dam/rbas/doc/Energy%20and%20Environment/The%20State%20of%20Dubai's%20Energy%20and%20Its%20Path%20to%20Green%20Economy.pdf
54. The UAE Hub of Next Energy Revolution?, by: UAE Analysis, year 2015, available @ https://www.theclimategroup.org/sites/default/files/archive/files/RE100-UAE-brief.pdf
55. Powering Dubai on a Smart Renewable Grid, year 2014, available @ http://www.smartdubai.ae/partner_story_two.php
56. Dubai Solar Park-Mohammad Bin Rashid Al Maktoum Solar Park, By Dubai Electricity and Water Authority, year 2016, available @ https://www.dewa.gov.ae/en/customer/innovation/renewable-energy/solar-park
57. Dubai Traffic Lights going Green, year 2014, available @ http://ems-int.com/blog/dubai-traffic-lights-going-green/
58. Dubai Municipality Uses Green Energy Sources to Implement the Public Parks Projects, by: A Magazine for the Environmental Center for Arab Towns, year 2012, available @ http://en.envirocitiesmag.com/articles/pdf/envirocities-article3.pdf
59. Measuring Health gains from Sustainable development- Sustainable cities • Food • Jobs • Water • Energy • Disaster management, by: Public Health & Environment Department (PHE) Health Security & Environment Cluster (HSE)-World Health Organization (WHO)-Geneva 27, Switzerland, year 2012, available @ http://www.who.int/hia/green_economy/en/index.html
60. http://countrymeters.info/en/Bhutan
61. World Happiness Report 2016| Volume I
62. http://uaecabinet.ae/en/cabinet-history
63. http://gulfnews.com/news/uae/government/shaikha-lubna-tolerance-is-a-value-not-a-service-1.1678601
64. http://www.thenational.ae/uae/20161008/sheikh-mohammed-bin-rashid-launches-award-and-institute-for-tolerance
65. http://www.mbrsg.ae/HOME/Events-and-Lectures/Dubai-Smart-Cities-Forum.aspx
66. http://www.dgep.gov.ae/
67. http://www.smartdubai.ae/dubai_data.php
68. http://opendata.fcsa.gov.ae/
69. From Majlis to Hashtag: Engaging Citizens through Social Media - The UAE National Brainstorming Session. Uae Ministry Of Cabinet Affairs
70. www.MBRMajlis.ae
71. https://www.dewa.gov.ae/en/about-dewa/news-and-media/press-and-news/latest-news/2016/11/dewa-invites-public-to-innovation-week-on-20-24-november-2016
72. https://portal.dm.gov.ae/suggestions/hys/index.html
73. http://gulfnews.com/news/uae/tourism/free-wi-fi-smart-palms-sprouting-in-dubai-1.1496985
74. http://www.uaecabinet.ae/en/details/news/mohammed-bin-rashid-on-presidents-directives-2016-is-uae-reading-year
75. https://www.weforum.org/events/annual-meeting-of-global-future-councils-2016/
76. https://www.cda.gov.ae/en/socialcare/SocialBenefits/Pages/default.aspx
77. http://mycommunitydubai.com/
78. http://ameinfo.com/technology/innovation/dubai-plan-2021-6-smart-city-moves-gitex-2016/

Bibliography

79. Brett G, Lauren D (2013) Beyond transparency: open data and the future of civic innovation. Code for America Press, San Francisco

80. Department of Economic and Social Affairs, United Nations, United Nations E-Government Survey 2016: E-Government in Support of Sustainable Development, www.publicadministration.un.org
81. Dubai Press Club & Mohammed Bin Rashid School of Government (2014) UAE Social media outlook: increasing connectivity between government and citizen. Emirates Media Forum
82. Faisal A, Governance and innovation program, MBRSG, Dubai Smart Cities Forum Series, Session 6: Bringing Smart Cities To Life. Mohammed Bin Rashid School of Government, Smart Dubai
83. Government of Dubai, Dubai Plan 2021, Dubai Pulse 2016 Report
84. John H, Richard L, Jeffrey S, World happiness report 2016, vol I
85. KPMG International, Future State 2030: the global megatrends shaping governments www.kpmg.com/government
86. Law No. (26) of 2015 regulating data dissemination and exchange in the Emirate of Dubai. The Supreme Legislation Committee in the Emirate of Dubai
87. Mohammed Bin Rashid School of Government, Policy Council 6, Building an Inclusive Society: Supporting Youth Employment and Development in the Innovation Economy, www.mbrsg.ae
88. Mohammed Bin Rashid Centre for Government Innovation, Edge of Government: Public innovation from across the globe. Prepared for the World Government Summit by PM's office UAE. www.edge.worldgovernmentsummit.org
89. Neeraj D, Dnyanesh N, Gopalakrishnan H, KPMG International (2015) Dubai—a new paradigm for smart cities, KPMG, KPMG LLP and KPMG Lower Gulf Limited
90. Peter P, Noella E (2016) CeDEM16. Proceedings of the international conference for E-democracy and open government 2016
91. Mohammed Bin Rashid School of Government, The Governance and Innovation Program. From Majlis to Hashtag: Engaging Citizens through Social Media—The UAE National Brainstorming Session. Uae Ministry of Cabinet Affairs
92. Salem F (2016) A smart city for public value: digital transformation through Agile Governance—the case of "Smart Dubai". Dubai: Governance and Innovation Program, Mohammed Bin Rashid School of Government, World Government Summit
93. Soumitra D, Bruno L, Sacha W-V. The global innovation index 2016: winning with global innovation, Cornell University, INSEAD and WIPO
94. The Economist Intelligence Unit Limited 2016, Advanced science and the future of government, World Government Summit Thought Leadership Series, Prepared for the World Government Summit
95. The Executive Council, Government of Dubai. 2021 Dubai Plan. www.dubaiplan2021.ae
96. Tina N, Maxwell School of Citizenship and Public Affairs, Syracuse University. A Manager's Guide to Evaluating Citizen Participation. Fostering Transparency and Democracy Series 2012, IBM Center for the Business of Government
97. UAE Government, Vision 2021 UAE; United in Ambition and Determination. www.vision2021.ae
98. UAE Government, Minister of State for Tolerance, Tolerance: Co-existence, Tolerance, Peace 2016
99. UAE Government, Science, Technology and Innovation Policy
100. Victoria G, Western Kentucky University. Participatory Budgeting: Ten Actions to Engage Citizens via Social Media, IBM Center for the Business of Government. website: www.businessofgovernment.org

Chapter 8
Smart Water Management and eDemocracy in India

Prabh Bedi and Neha Goel Tripathi

Abstract Water management has become a key issue in the 21st century, with increasing population, economic growth and climatic variability. Water is a stressed resource in India. Annual per capita water availability is expected to decline to 1140 m^3 by 2050, from 1545 m^3 in 2011 (Government of India in Water management, [1]). With 45% of population living in urban areas by 2050 (Shukla in How India earns, spends and saves: unmasking the real India. Sage Publications, India, [2]), the availability of water will become a paramount issue. The availability of water is fundamental element to human life, economy and political stability and hence, sustainable water management is critical for existence. Government of India has launched a Smart City Mission for building 100 Smart Cities, in which 24 × 7 availability of water is one of the major emphasis. Traditionally water allocation decisions have been the exclusive domain of technical experts, public officials or political interests, but now there is a shift towards participative involvement of other stakeholders, namely citizens. Adequate water being the essential input in Smart City, calls for an inclusive and participatory management of this resource which extends beyond the city limits. Smart Water management in cities seeks to address challenges in the urban water management through integration of ICT and eDemocracy. Hydro-Informatic approache that is data driven and is technology based can form the basis of analyzing the complexities in water management, particularly in water scarce areas. This chapter looks into inclusive and eDemocratic measures for planning of water smart cities. The need is to develop eDemocracy based collaborative planning efforts to tackle pervading water issues.

Keywords Water · Smart city · HydroInformatics · Smart water management · eDemocracy · India

P. Bedi (✉)
Geospatial Information Technology Consultant, Visiting Faculty,
School of Planning and Architecture, 4, Block-B, I. P. Estate, New Delhi, India
e-mail: prabhb@gmail.com

N.G. Tripathi
School of Planning and Architecture, 4, Block-B, I. P. Estate, New Delhi, India
e-mail: nehagtripathi@gmail.com

T.M. Vinod Kumar (ed.), *E-Democracy for Smart Cities*, Advances in 21st Century Human Settlements, DOI 10.1007/978-981-10-4035-1_8

8.1 Introduction

Water is the lifeline of human civilization. Since time immemorial, access to water has been a defining factor in the location of settlements. Important civilizations have developed on the banks of rivers like Indus, Ganga, Nile, Euphrates, as the rivers have provided water for farming and led to the thriving of cities. Dynasties have been setting up their capitals on the banks of rivers and even today, major cities like New Delhi, Moscow, London and many others are on the banks of rivers and lakes.

As the population of the settlements have grown, the need for management of water has been realised. Water has been ported through channels and disposed off through drains since the times of Indus Valley Civilization. The means have been provided by the various rulers for the consumption of water for various domestic and non-domestic uses. The emphasis has been a lot on its conservation as well. In the modern context, water became a prime concern in 1972 with it being identified in United Nations Conference on the Human Environment as one of the natural resources that must be safeguarded. In 1977, in Argentina, a conference was exclusively devoted to discuss emerging water resources problem, where, there was a growing recognition that technology and infrastructure alone were not sufficient to address persistent water management concerns. Discourse about water governance began to emerge focusing on deliberative democracy, environmental governance and multilateralism, during the Dublin Conference in 1992 [3].

Water management is the activity of planning, developing, distributing and optimum use of water resources under defined water policies and regulations. It includes management of water resources, including for irrigation, flood protection, water treatment of drinking water, industrial waste and sewage and for maintaining the water table.

The ideology of water management has evolved from provisioning water for use in an urban area to include sewerage and drainage to integrating it like a water cycle to the present-day inclusion of democratic governance technology. Water management includes management of water resources systems, which are both natural and man-made. While the natural water resource systems include watersheds, river channels, groundwater floodplains and sub-surface water, man-made systems include conveyance channels, dams and reservoirs, storage facilities, wells and tanks, treatment plants and pumping stations.

When addressed specific to urban areas, water management involves water supply for domestic, commercial and industrial uses, drainage, wastewater treatment and sludge handling. Urban water refers to all the water that occurs in an urban area, either naturally, from atmosphere or on surface through rivers, streams, wetlands and natural aquifers or below the surface as underground water or through human intervention through piped networks and storage facilities.

The objective of urban water management is to build resilience in the urban centers such that they become livable and sustainable. With the advancement in technology and changing ideology, the need has arisen to manage the earth's resources in an integrated manner, leading to Integrated Water Resource Management, which promotes the coordinated development and management of water, land and related resources to maximize economic and social welfare in an equitable manner without compromising the sustainability of vital ecosystem [4].

The chapter traces the water management practices in India since earliest times to the water policies and schemes of the present day to bring to light the changes that came about in management practices. The post-modern development like hydroinformatics and smart water management that are being adopted for the betterment of the cities, citizens and the environment have been discussed and learnings from historical management practices have been highlighted, which when clubbed with the post-modern technological innovations may provide lasting solution to the water problems.

8.2 Historical Perspective

Despite being crisscrossed by a complex network of rivers, some stretches of India have neither river nor lake to depend on. The history of India tells us that floods, droughts or both have been a perennial occurrence. Depending on the agro-climatic conditions, over time and with experience, people developed varied water management practices. Water management practices in India from historical perspective were largely in rural context, as those were the settlement types that existed mostly in the country.

The most important aspect of water management was harvesting of water, which depended largely on the ecological region inhabited by the people. In arid and semi-arid areas of Rajasthan, people developed covered wells for drinking water purposes. In the highlands, comprising of western Himalayas systems as *Zings*[1] and *Kuls*[2] were built for carrying water from the glaciers to the villages and for storing water melted water collected from the glaciers in small tanks. In the northern plains, *Kuis*,[3] *Johad*[4] and *Baandh*[5] were constructed for harvesting the water to recharge the groundwater. These water conservation structures were constructed in the urban settlements as well like stepwells and water soak pits.

---

[1]*Zing* is a small storage tank for collecting water melted from glaciers.

[2]Kul is channels for carrying water from glaciers to villages.

[3]*Kui* is deep tanks 10-12 meters in depth used for collection of water.

[4]*Johad* is an earthen checkdam used in the states of Haryana and Rajasthan, India, that collects and stores water throughout the year, to be used for the purpose of drinking by humans and cattle.

[5]*Baandh* is a dam to contain the flow of water.

Water management as it existed in pre-British period received patronage from the rulers in the form of construction of storage facilities for water supply to the capital cities and important towns. The citizens were also encouraged to build independent as well as community facilities to meet the domestic needs as in the form of wells and storage tanks. All the aspects of management, from identification of a suitable site to construction of the facility was based on experiential knowledge of the people based on the necessities of the people's daily livelihoods [5]. Embree [6] mentions that a thing of value was regarded as being part of aggregate rather than belonging to a single person and water was considered as a valuable resource, common to all for use and a basis of life. The close-knit structure of the communities enabled in managing labor and materials needed for such facilities as water was a common resource and basis of life. As stated by Bhaduri and Singh, pre-British institutions in India were created on the principles of duties and obligations and the rights were overlapping in nature with a strong local fervor in management of resources. There existed a high rate of participation by the people in activities linked to the management of the community resources common to all. It was obligatory for the citizens to perform one's duties towards the maintenance and upkeep of the common resources

During the pre-British period, all over the country traditional practices of managing water were practiced and was decentralized [7]. Having a monarchial system of governance throughout its history, the rulers apparently gave full support to the people and allowed the local communities to nurture the local knowledge base. The use of common properties for developing water facilities was the practice as people rarely owned any property in individual capacities.

Water has been traditionally perceived as a common resource and is the collective responsibility of the community for management. This entails conceptualization of the design based on the requirement of each fraction of the community based on sharing of knowledge and experience of the older people of the community. As highlighted by Singh [8], the decisions by the members of the community are based on collective choices made by mutual discussions under the leadership of senior men of the community, thereby informally following the concept of community participation at the grassroots within the framework of democracy.

Traditional water management practices declined during the British period due to its imperialistic rule. Lack of local ecological knowledge and systems introduced by the British led to disruption of the existing social structure in the country [5]. The traditional water management (harvesting) structures were de-legitimized by the British. To make management easier for themselves and have a complete control over the basic needs of the people, the British made laws to govern the water bodies under the ownership of state. This change gravely impacted the age-old water management practices as land, water and forest resources were addressed in an integrated manner traditionally.

8.3 Post Independence

The Constitution of India does not explicitly address the citizen's right to water. The Supreme Court of India has recognised the 'right to water' as part of the 'right to life' through Article 21 of the Constitution [9] through interpretation of the fundamental right to life in a series of cases making right to safe drinking water as a fundamental right. Further, during its review of the Constitution in 2002, National Commission recommended insertion of a new Article 30D as '*Every person shall have the right—(a) to safe drinking water…*'

As per India's constitutional setup, the authority to formulate, legislate and implement policies on water that entails water supplies, irrigation, canals, drainage, water storage and water power lies at the state level [10, 11]. Water being a state subject, it is the responsibility of the states to formulate policies, administer and implement various schemes pertaining to water and this includes the development and maintenance of urban water supplies as well. At the national level, Ministry of Urban Development and to some extent Ministry of Urban Employment and Poverty Alleviation are responsible for formulation of broad policies, their role being limited to advisory rather than regulatory.

8.3.1 Water Management Through National Five Year Plans

The earliest efforts of the government were to increase accessibility to water, as per the recommendations of the Bhore Committee in 1946 and Environmental Hygiene Committee in 1949, which was achieved by building large physical infrastructure like dams and reservoirs to increase water supply to urban areas amongst its other uses. At the time of independence, most of the towns lacked piped water supply services. To augment the situation systematic approach to provision of water supply commenced in 1950 with the Five Year Plans. However, until the end of the Fourth Five Plan drinking water supply programs were not given required precedence in the national planning process. One of the first policy initiative was in 1954 in the form of 'National Water Supply and Sanitation Policy' under the Central Ministry of Health with the main intention to assist the states in the execution of the water supply schemes. Later, the Fourth Five-Year Plan (1969–1974) stated that in urban areas, water supply schemes should be looked upon as a service which had to be paid for by the direct beneficiaries through capital contributions and water charges [12].

Until the Eighth Five Year Plan, the approach of the urban water schemes was to increase the coverage and access to water supply and sanitation services, having unrealistic targets, like 100% coverage at the end of the plan period, having achieved only 84%. The plan exemplified the problems mainly in the implementation of the schemes as related to operation and maintenance of installations and inadequate sewerage systems polluting the ground water, rivers and other water sources. In the eighth Five Year Plan (1992–1997) as far as drinking water supply was concerned, envisaged 3.85 percentage of total plan outlay on water supply, enlarging the coverage of drinking water to about 94 percentage of the urban population. However, the targets could not be achieved. The Planning Commission at the Centre rarely infringed on the constitutional right of the states with respect to water sector and merely set targets for the states to achieve not making an effort to articulate institutional or political difficulties, if any [13].

Ninth Five-Year Plan (1997–2002) recognised the integral link between rapid economic growth and quality of life. It identified the key urban concern as the growing gap between demand and supply of basic services like drinking water. It stated drinking water supply and sanitation facilities as being important and crucial for achieving goal of health for all. In the Tenth Five-Year Plan (2002–2007) the deteriorating level of urban infrastructure such as water supply and the low capacity on the part of city level administration to maintain it was recognized as a major area of concern. The tenth plan put special emphasis on urban sector reforms and bringing private sector expertise and capital through Public Private Partnerships (PPP) for improving efficiency and better water service delivery.

In these plans, the focus had been on the role of Urban Local Bodies (ULBs) in improving the development of water supply infrastructure, which changed in subsequent plans to demanding accountability from state governments for the deficiencies of development schemes. This national level shift has been due to the constitutional legitimacy attained by the ULBs under the 74th Constitutional Amendment Act (CAA), 1992. Urban planning, water supply and sanitation infrastructure and public health are among the eighteen items of the 12th Schedule that, the 74th CAA prescribes to transfer to the ULBs [14]. Moreover, the shift was from the mere statistical achievement to achieving assurance of water quality, high standards of operation and maintenance and accountability to customers. Reform based missions and schemes like Jawaharlal Nehru National Urban Renewal Mission (JnNURM) attempted to bridge the gap in physical infrastructure as well as governance.

The Twelfth Plan (2012–2017) emphasis, as important as the quantum of water to be supplied, is the problem of its management and equitable supply to all. In most cities, water supply is sourced from long distances. In this system of bringing water from far and in distributing it within the city, the length of the pipeline increases, as does the cost of infrastructure and its maintenance [15].

It needs to be highlighted that despite major programs and policies in India, today access to drinking water in remains a challenge despite massive outlays for drinking water even after six decades of planning and development.

8.3.2 Policies and Laws Governing Water Management

India adopted National Water Policy (NWP) in 1987 based on the recommendations of National Water Resource Council constituted in 1983. The policy aimed at optimum utilization of water resources, with the objective of sustainable development in harmony with the environment. The Policy gave foremost priority to the provision of safe drinking water to every human being followed by irrigation facilities, hydro-power, navigation and water for industrial use.

Subsequently, in 2002 National Water Policy was modified and reintroduced in response to the rapidly changing scenario in water sector to address the then emerging issues and provided critical policy inputs. NWP 2002 laid emphasis for the first time on ecological and environmental aspects of water allocation.

The policy was relooked into and in 2012 suggested an environmentally sound common integrated perspective to govern the planning and management of water resources within the local, regional, and national contexts. The Policy [16] clearly stated that water needs to be managed as a common pool community resource that is held by the State under the public trust doctrine to ensure equitable and sustainable development for all. NWP 2012 has done away with water allocation prioritization mentioned in NWP 1987 and 2002, but has emphasized on treating water, over and above the pre-emptive need for safe drinking water and sanitation, as an economic good. NWP 2012 also emphasized the fact that the service provider role of the State must gradually shifted to that of a regulator of services and facilitator for strengthening the relevant institutions.

The NWP 2012 recommended on aspects like enhancing water availability, water demand management through efficient water use practices, water pricing, conservation of river corridors, water bodies, and infrastructure, project planning and implementation, management of floods and droughts, water supply and sanitation, institutional arrangements, trans-boundary rivers, database and information system, capacity building, adapting to climate change and preparation of a plan of action by the National Water Board based on the National Water Policy.

The major guiding principles stated in the NWP 2012 [16] include:

- The principle of equity and social justice must inform the use and allocation of water resources.
- Planning, development, and management of water resources need to be governed by common integrated perspectives considering local, regional, and national context, having an environmentally sound basis, keeping in view the human, social, and economic needs.
- Safe drinking water and water for sanitation should be give utmost priority followed other basic domestic needs, agricultural needs for ensuring food security and minimum ecosystem needs. After meeting the above-mentioned needs, NWP mandates that water should be allocated in a manner to promote its conservation and efficient use.

- The focus in future needs to be on demand management particularly in view of impending impact of climate change by enhancing water use efficiency and minimizing water wastage.
- Water based activities need to be regulated in consonance with the local geo-climatic and hydrological situations.

Need has been felt for a model law as stated by Kumar [17] which may function as a guideline to the states for water governance. The Draft National Water Framework Bill, 2016 mandates the right to sufficient quantity of safe water for life within easy reach of the household making it the state's responsibility to provision water services even if delegated to private agencies [18]. The implementation of the National Water Framework if done in earnestness may bring in parity over space and time. The proposed law is not intended to centralize water management but, as clarified by the Sub-group on National Water Framework Law [19] set up by NITI Aayog, is an umbrella statement of general principles governing the exercise of legislative and/or executive (or devolved) powers by the Centre, the States and the local governance institutions.

In the recent years, many states have formulated State Water Policy documents and endorsed reforms in the urban water sector. The reforms have been either institutional/governance or financial/economic. States like Karnataka and Goa have formulated state policy documents specifically addressing urban water supply and sanitation. The institutional reforms pan the restructuring the roles and functions of the institutions involved, strengthening the ULBs, seeking participation of private agencies [20], separating the functions of regulation from execution and transferring regulatory functions to newly created independent regulatory bodies [21]. The economic reforms focused on cost recovery through rationalization of tariffs and maintaining accounts and budgets separate from those of other operations of ULBs to strengthen the agencies financially. However, the implementation efforts are not in tandem and ground realities have shown few sporadic and unconnected efforts to implement reform measures [22].

As stated by Ahluwalia [23], Groundwater use in India is currently governed by the framework of British common law sanctified by the Indian Easement Act of 1882. The Act has empowered the landowner with absolute right to draw any amount of ground water from under the land owned by him. The attempt at legislative reform in the past focused mostly on allocation and setting up a public regulatory authority for groundwater regulation and management empowering the state government to take the final decision. The government of India is currently working on a national water framework bill and a model groundwater bill, which may be adapted by the state governments as per their requirement and context. The issues being addressed are equitable access and aquifer protection. The focus on diluting the link between land ownership and control over groundwater. The intent is to treat groundwater as a common pool resource to be exploited only for public good [23].

Water laws in India are largely state based, due to the constitutional scheme, which since the Government of India Act, 1935 has in principle given power to the states to legislate in this sector. There are numerous water laws in the country [24], but water issues are addressed in response to emerging crises (not a proactive approach), resolving rights and settling disputes. No rational scientific framework is available to reconcile ambitious goals of economic prosperity and competitive claims for an increasingly scarce resource by different segments of the society.

8.3.3 Role Players

At the Central level, NITI Aayog (erstwhile Planning Commission) is responsible for allocation of financial resources to states and the ministry for various programs and schemes. Ministry of Water Resources is responsible for laying down policy guidelines and programs for development and regulation of country's water resources. Other than these, the Central Pollution Control Board under the Ministry of Environment, Forest and Climate Change (MoEFCC) deals with water quality monitoring and preparation and implementation of action plans to solve pollution problems. The Ministry of Urban Development is responsible for execution of schemes launched by the government spatially falling in the urban space.

Even though, water is the legislative responsibility of state governments, in the recent years the process of water service delivery has been marked by top-down approach from the center, where the schemes and missions (like JnNRUM) were are launched by the center, to the states and from the state governments to the municipalities or the local government. All the programs and the policies are decided at the central or the state level and passed on to the local level agencies for implementation.

The state level parastatals like Public Health Engineering Department (PHED) are responsible for water supply to urban areas despite the devolution of powers to the ULBs under the 74th CAA in 1992. In some cases, the responsibility of urban water supply is fragmented between different agencies. Besides the PHED at the state level, Water Supply and Sewerage or Drainage Boards at state levels and local government levels, share the responsibilities. Multiplicity in responsibility sharing is one reason for non-accountability in performance.

Water supply in urban areas which is at local level in the hierarchy of governance, is managed by state authorities or parastatal agencies. With increasing mismanagement and huge inefficiencies, the task of water management has now been off-shored to private organisations, making water into a commodity.

8.3.4 Smart City Initiative

Government of India's most recent initiative is of SMART Cities which defines a Smart City as one that provides core infrastructure and give a decent quality of life to its citizens, a clean and sustainable environment and application of 'Smart' Solutions [25]. It is suggested that through Smart City solutions, the government should bring in smart water management through smart urban planning using smart Information and Communication Technology (ICT). As stated by Leinmiller and O'Mara [26] smart water system is designed to gather meaningful and actionable data about the flow, pressure and distribution of a city's water. Smart water system not only considers reducing the losses and managing the distribution but addresses energy consumption issues and suggests measures for its optimal use. Water loss management is becoming increasingly important as water supply agencies are stressed due to ever increasing population growth in urban areas clubbed with water scarcity. Incorporating smart water technologies allows water providers to minimize Non-Revenue Water (NRW) by identifying leakages using technology. Such systems help in predicting possible leakages in the near future based on models and simulations using real-time Supervisory Control And Data Acquisition (SCADA) data.

Some Indian cities which are on path to becoming Smart Cities have adopted smart systems to reduce water losses and for efficient management of water supply. Bengaluru,[6] which had maintained an average of 20 h of supply per day in the 1980s, was able to sustain an average of just 2.5 h in 2000s. Bangalore Water Supply and Sewerage Board (BWSSB), the agency that supplies water to Bengaluru city, launched a program designed to test the feasibility of providing metered individual water and sanitation connections to unauthorized slum households in the city, departing from its traditional practice of servicing such communities via free and shared public taps. For this purpose, it waived its requirement that only households with legal tenure could avail of an individual connection. As a result of this initiative, BWSSB succeeded in mobilizing 46 poor communities by early 2005, accounting for 10% of the city's slums [27].

Thiruvananthapuram city is using IBM Analytics and Mobility solutions to analyze, monitor and manage water distribution [28]. The utility grid is installed with smart sensors working in conjunction with the IBM Intelligent Operations Water software which enables Kerala Water Authority (KWA) workers to receive alerts via smart devices about water leakage cases in near real time. In addition, the sensors installed throughout the treatment process allow KWA to measure water turbidity, salinity, conductivity, PH and chlorine levels in real time. This smart system helps in monitoring and management of both quality and quantity. The smart system has helped to curb NRW losses due to leaking infrastructure and unauthorized use of water as well. The smart water system has helped KWA in tracking its 210,000 water meters across the city on consumption, thereby reducing billing anomalies and improving revenue collection by more than 10% [28].

Surat city, another budding Smart City, has taken control of its high-value and bulk consumers to check for water consumption especially the industrial users where each user has been metered using electromagnetic instruments and water consumption is carefully monitored. The city has adopted initiatives like imposing a leakage charge of 5% on industry. All newly developed areas are metered while in

[6]Bangalore city's name was changed to Bengaluru in November 2014 (http://www.karnataka.com/govt/karnataka-city-name-changes/).

the rest of the city, water bills are charged as a component of the property tax. The city has been mapped for leakage and old city area has been found to be the worst in this regard for which the agency is in the process of taking remedial steps to improve piping. With just one percentage of its area metered, the city has a cost recovery of 92% and its efficiency in collection of water charges is 94% [29].

Cities in India have started adopting technological solutions with the aim of providing better services to the citizens, though the initiative is not as widespread This can be attributed to the issues in water supply management which range from hardcore technical water supply and management problems to those linked to financial constraints and governance issues as discussed below.

8.3.5 Emergent Issues

Urban India has been experiencing limited, irregular, unreliable, inequitable and polluted water supply. The authorities responsible for supply of water are not responsive to the maladies faced by the citizens. With growing urbanisation, the demand for water is increasing as is the generation of wastewater. These issues can be broadly categorized as those linked to quantity and quality.

There is a large spatial and temporal variation in the availability of freshwater. Per capita water supply grossly varies from 40 L per capita per day (lpcd) to 200 lpcd. Only 62% of urban households have access to treated tap water and about 50% are directly connected to a piped network. The average connected household receives water only for approximately 2 h per day [30].

In the near future, urban concentration is likely to be in small and medium sized towns which are ill-equipped in their infrastructure for water and sanitation. Moreover, these towns do not have the required mechanism for generating revenue to cope up with increasing infrastructure demands. Unplanned peri-urban growth too has a low pace of infrastructure development [31].

The issue of adequate quantity, quality and the distance of the nearest available water source is emerging to be the most contentious issue about urban water supply. As the traditional water sources are all drying up, transporting water from far off places (100–400 km) is becoming a common feature and is thus becoming an expensive affair.

Leakage and inefficiencies in the water supply system causes wastage of nearly 50% of usable water. The ground water level is declining at the rate of 10 cm per year. Over 70% of surface water and ground water resources are contaminated [17]. In the absence of strong regulation, industrial and domestic waste water is discharged in rivers, canals and underground water sources. About 70% of underground and surface water resources in India have been contaminated [17]. All this is leading towards a water scarce situation in most urban areas of the country.

Table 8.1 Per capita average annual availability of water in India during 2010, 2025 and 2050

Average annual water resources potential	Estimated population (million)			Estimated per capita average annual water availability M^3		
	2010	2025	2050	2010	2025	2050
1869.3	1162.31	1394.02	1640.00	1608.26	1340.94	1139.82

Source B.P. Directorate, CWC. $: Reassessment of water resources potential of India March 1993, CWC. #: Report of the Standing sub-committee for assessment of availability and requirement of water for diverse uses in the country, August 2000 [32]

As per the report of the Standing Sub-Committee for assessment of availability of water as the population of the country will increase, the per capita water availability will reduce as shown in Table 8.1.

Another serious problem is the lack of maintenance of water supply and sanitation infrastructure, leading to contamination of ground water sources in most of the cities and towns today. The main cause of contamination of groundwater and surface water is the improper discharge of sewage. More than half of the cities have no sewage treatment facilities. The much-needed water quality monitoring and surveillance is a rare phenomenon, except in the case of few privileged metropolitan cities.

Meagre finances with the urban water supply authorities is one of the biggest challenges in Indian urban water supply system. In India, most of urban local bodies are dependent on the state government for investment in water supply system. There are not enough sources for revenue generation to maintain water supply infrastructure. In India, water supply is responsibility of both state government and urban local bodies. Most municipal governments continue to rely heavily on state and central financing, not only for capital investments but also for a significant part of operational expenditures. As a result the Municipal Governments do not feel fully empowered and responsible for service delivery, and continue to 'look up' to state governments for financial and technical support, feeling little need to involve, consult or inform end users about proposed schemes [27]. Unequal distribution of water is a big challenge where poor people do not even have access to potable water system.

Water utilities are not legally bound to report on their performance to customers, and they are only required to report to higher tiers of government on budgets and expenditures. Further, many municipal water departments have no definitive measures and mechanical tracking systems by which to monitor and record water flows, and hence are unable to assess the extent to which poor metering, theft, or breaks and leakages to estimate water losses [33].

No clear-cut distribution of responsibilities to different urban agencies having overlapping roles and responsibilities causes delay in most water projects. Lack of accountability and regulatory framework for core and peri-urban areas is not proper. Lack of initiative taken for public participation in decision making to improve water supply system is another major hindrance for its improvement.

The inability to provide citizens with the necessary infrastructure has caused problems, including the growth of the informal supply of drinking water and improper wastewater collection and disposal systems. These informal systems operate largely unregulated, posing major health risks to the population.

The urban water supply and sanitation sector which is suffering from inadequate levels of service, an increasing demand-supply gap, poor sanitary conditions and deteriorating financial and technical performance needs to be addressed on highest priority. There is a growing realization that government alone cannot solve and manage the water scarcity challenges faced by the country. The approach in water management in urban areas in the near future is likely to focus on water conservation and efficiency, distributed storm water management which captures and uses rainfall, water reclamation and reuse and water treatment and reuse with the involvement and participation of citizens.

8.4 HydroInformatics

The current urban water sector scenario seeks just for technical quick-fixes that may enslave the communities to technology. The high expenses of such technologies may bind the communities to funding agencies eventually alienating the local population from its own resources. The solution to be sought needs to be a combination of technology and people's involvement.

The solution to these problems lies in hydroinformatics based water eDemocracy, where the ultimate objective should be to increase the water table through judicious use of water and conservation practices with emphasis on electronically enabled citizen participation to contribute to decision-making and take ownership to any changes that may be sought in the water regime in their area. eDemocracy is the support and enhancement of democracy, democratic institutions and democratic processes by means of technology [34].

There is an essential need for shift from water 'management' to water 'governance'. Whereas, water management is rooted in centralized decision-making with the government as the prime stakeholder, water governance is the political, social, economic and administrative systems that influence water's use and management. The key factor in water governance is citizen's participation and should be essentially based on informatics approach.

Hydroinformatics is information about hydro, a Greek word meaning water, hence information about water. Hydroinformatics can be defined as a branch of informatics which concentrates on the application of information and communications technologies (ICTs) in addressing the increasingly serious problems of the equitable and efficient use of water for many different purposes. As stated by Price and Vojinovic [35] hydroinformatics which emerged in 1980s integrates the social and technical knowledge to create solutions for the water problems of an area. Social dimension is important in hydroinformatics for understanding the impact of technological changes brought about by ICT. The objective of hydroinformatics is to bring in complete transparency in the decision-making process by involving all the stakeholders. Some kind of information about water has existed since the time man started using water for fulfilling his needs. This information or knowledge has been passed on generation to generation and has been shared between communities. Hydroinformatics, as is understood today, can be said to have emerged largely due to the advancements in the field of information technology.

Hydroinformatics has evolved from a stage where methods used by manual calculation were programmed to run on a machine [36] followed by a period when software codes began to be written initially in mainframes using paper tape and punch cards and later using programming languages as PASCAL and FORTRAN. This period of writing codes for specific application developed into coding for more generic applications in 1970s [36] and grew into software modelling systems. During 1980s commercial modelling packages for personal computers began to be marketed like MOUSE and MIKE 11 by Danish Hydraulic Institute and WALLRUS and WASSP from Wallingford Software. These software empowered the engineers with the ability to manage the water systems even without the core knowledge of computational hydraulics. Modelling become the core of the software for water management and gradually became available to a larger group of people other than water engineers and water management experts. In all the advancement that took place in the water modelling and management software, the parallel developments in computer hardware, software, programming, database management, Geographic Information System and Remote Sensing technologies have played an important role.

Since 1980s till date, as stated by Price and Vojinovic [35], the advancement in hydroinformatics transformed the 'performers of projects' into 'providers of products' and the society too transformed from that of 'knowers' to societies of 'consumers of knowledge'. Internet has played a vital role in this transformation. It has not only facilitated the communication of information but has facilitated access to data and information stored in servers and websites by a large number of people through a wide range of devices from a computer to a mobile phone connected through Internet anywhere in the world. With the information being available online, stakeholders can participate in the decision-making process. In the future, the option of accessing the information is likely to increase tremendously, which will bring in more transparency and bring a change in the society in social terms. As Abbott and Vojinovic [37] state, the purpose of stakeholder participation is to induce a change in the built environment that bring a positive change in the social environment. The societal changes caused by hydroinformatics may lead to more stakeholders participating in decision-making in not a reactive manner to the decisions of the government but in a positive manner [35].

The biggest benefit of hydroinformatics in water management is in the planning, design, construction, installation and operation of water assets [35]. It is necessary to understand that hydroinformatics is not just automatic collection of data on water consumption but should allow for remote access of data on water flow, consumption, quality, pressure at a required resolution, so that the data details can be used to improve the operational decisions. Hence, hydroinformatics is not just collection of Big Data but essentially includes the intelligent processing of the same such that right decisions for the customers can be taken.

Hydroinformatics, which delivers smart water management, uses ICT to address water issues with extensive automation to increase the response time. It helps in capturing data in near real-time and has the ability to transmit the data to remote locations for data processing, interpretation and modelling by the stakeholders [38]. The smart water management tools can be categorized into data acquisition and integration, where network sensors, smart pipes and smart meters are used; data dissemination, where Internet and radio transmitters are used; data processing, storage, modelling and analytics, where GIS and specialized software are put to use; management and control where SCADA and optimization tools are used; and visualization and decision support where web-based communication is used.

There exist some challenges in the adoption of hydroinformatics as a smart water solution. Lack of standardization in the relevant technologies like sensor networks and Internet of Things. Standardization is essential for the success of the solution as it brings in integrity. Organisation and agencies following the standards reduce their risks as there is compatibility and interoperability in the systems. Coherent cross-sector policies and a multi stakeholder formulated water resources management strategy are essential to the success and sustainability of smart water management tools [39], which needs to be supported with appropriate ICT governance as lack of investments in ICT can hamper the planning and implementation of the smart solution. As stated by Boyle et al. [40], a number of factors need to be considered in implementation of smart water management system, amongst which customer satisfaction needs to be the key. The water agency in India is in government's domain and the focus on customer satisfaction is the least. Adoption of hydroinformatics is a re-engineering effort and not just adoption of technology. The problems of water supply and management that are faced by the urban areas can be overcome only by better understanding the customer problems which can be achieved by installing smart water solutions. By the installation of the smart solutions, the water managers will be able to understand the water consumption patterns and trends for various uses and enable in better urban water planning. As stated by Beal and Stewart [41], designs from environment, engineering and economic perspectives can be efficiently designed based on up-to-date and accurate data collected through the smart solution. It will help the managers in proactively reduce water losses by reacting to leakages and alarms [42]. Smart management can lead to development of customized tariff systems as well which can further influence the consumption pattern of the consumers.

The installation of a smart solution like SCADA by Pimpri Chinchwad, near Pune, Maharashtra, in 2008, has helped the municipal corporation is equitable water supply in the city. The system provides information on raw water lifted from the Pavana river, treated at the water treatment plants (WTPs), sent via the pipelines, and received at 85 elevated storage reservoirs. The department has set up benchmarks for the flow of water through the system and if there is any deviation, the official concerned is sent a message on his cell phone, after which he determines the reasons for the deviation and takes corrective action. The municipal corporation embarked on a pilot project to implement 24 × 7 water supply in a specified area which has reduced the water demand as now there is no need to store water with corresponding wastage. Further electricity bills of residents have reduced as there is no need to pump water as the pressure is being maintained in the system [43].

Singapore embarked on a Smart Water project to overcome the challenges of leakages and pipe bursts due to ageing infrastructure in collaboration with the water utilities company, academia, ICT institutes and private agencies called WaterWise. The project has been successful in aiding event detection like leaks and bursts, provides real-time monitoring and decision support [39]. Two important components of WaterWise are the Integrated Data and Electronic Alerts System (IDEAS), and the Decision Support Tools Module (DSTM). Whereas, IDEAS is responsible for data stream management, analytics and alerts, DSTM uses the data aggregated by IDEAS to provide decision support tools on a demand-zone basis. The system can be integrated to the existing GIS platform to display the data collected in real-time spatially. The project has expanded from an eight-node network of integrated smart tools to a 50 nodes network spread over 80 Km^2.

City of Seoul is another example of implementation of smart water solution through which the city has been able to increase its revenue water to 94.3% in 2013 from 64.1% in 1998 [44] This has been possible by maintaining a stable water pressure despite the hilly terrain of the urban area, construction of reservoirs at the correct height, measurement of minimum night flow, leak detection and reorganization of pipes not in use. This has been possible due to use of appropriate smart water management tools clubbed with local knowledge.

Water security ensures the reliability and availability of water in terms of quantity and quality. In present times security means protecting the water supply system from contamination as well, which may come in the form of pollution, accidents, and terrorist attacks. Ensuring the security of water has been the concern for utility managers [39]. United States Environmental Protection Agency (EPA) is entrusted with the task of Water Security, which comprises of online water quality monitoring, consumer complaint surveillance, public health surveillance, routine sampling and analysis and enhanced security monitoring. Developed on ICT, the smart solution is capable of providing timely detection of potential contamination. The system consists of Event Detection Systems (EDS) which is designed to monitor water quality data in real time and produce an alert if water quality is deemed anomalous [45].

Hydroinformatics supported smart water management helps in identifying and rapidly addressing the leakages and reducing the water losses in the distribution network, understanding the consumption time during the day by residential and commercial users to enhance water supply designs and management, enhance customer satisfaction by providing targeted services as leak alerts and taking feedback from customers through web portals and mobile applications to know about changing consumption patterns so as to deliver better services.

8.5 eDemocracy an Essential for Smart Water Management

As stated above, the true benefit of hydroinformatics is when not only are the stakeholders, which includes citizens, are benefited by the equitable distribution of water, reduction in water losses and by capping of the loopholes in appropriate tariff collection by the water utilities, but when the stakeholders are able to participate in the decision-making process linked to water supply in their area.

Thus, such a decision-making process is needed in which water managers inform the initial decision and participate in planning the appropriate responses, interacting with the principal actors, particularly citizens and with the managers of other sectors. Thus increasingly, collective governance 'beyond governments' is part of the solution, with state and non-state actors working together. In this rapidly changing world, it is best to adapt in time to prevent crises rather than adapting in response to them. In the end achieving the goal is a political challenge. It involves determining the distribution of power and resources within a given community well as the interrelationship(s) between communities.

Canadian provinces have amended their laws and introduced new policies to promote delegated governance, which involves delegating decision-making over water management to the local level. Initially, citizen participation in environmental governance tended to emphasize citizen consultation, centered on project-review processes of mega-projects which developed into integration of citizens into decision-making through either consensus-building consultation processes or shared decision-making processes [46]. The Netherlands is another example of the changing policy paradigm for water management towards more integration and adaptiveness. Dutch governance system on water management emphasizes on new governance paradigm that is oriented towards more close collaboration between public, private and societal actors in interactive processes [47, 48].

During the pre-British period, water was managed in the country in a decentralized manner, where water was not considered a commodity, but a community resource. Tasks related to water were handled at a community level and were highly sanctified, which made the people follow the unsaid rules for judicious use and conservation of water. Considering the present-day situation of water crisis in nearly all urban centers in the country, the need is to take on the learning from the historic times by involving the people in the decision-making process. Making people aware about the issues through dissemination workshops, holding relevant and age-appropriate seminars and workshops in schools so as to educate the youth of tomorrow about the essentials of water management and conservation may go a long way in establishing a democratic water regime. Similar educative content needs to be readily available through web portals, where the citizens should have an option of giving suggestions and possible solutions to their problems. The final decision may rest with the relevant agency, but with citizen involvement, there will be better acceptability and improvement in the water situation. With the adoption of hydroinformatics by the agencies, the benefit will incur not only in terms of preventing losses and better management. The benefits would accrue to and from the citizens when the information generated through hydroInformatics is shared and made accessible to the citizens and other stakeholders. Once the information is crowd-sourced, there will be many options and solutions that may be provided by the citizens, again the adoption of which would finally rest with the water supply and management agency. However, that would entail eDemocracy in the true sense.

The democratic model of water governance entails delegation by relevant authority of the government to a lower level, involvement of a wide variety of non-state actors, use of a hydrographic boundary as watershed rather than political boundaries, collaborative decision-making processes, often emphasizing consensus and trust-building, science-based decision-making, often requiring extensive fact-finding and use of hydroinformatics for decision making.

ICT based good governance and community participation in water management are the essential building blocks for water sector reform. Only through a partnership between people who are the users of water and water agencies who believe in democratic functioning can safe, equitable and adequate water distribution be ensured. Such systems and set-ups can understand the need for conservation of resources and ensuring sustainable water systems.

8.6 Conclusions

Traditionally, water resources have been managed by the community as water has been considered as a common resource and not a commodity or private property. The perception about water as a resource changed during the British rule with the gradual privatization of land and removal of water as a common resource. This brought the change in water management practices as well from collectively deciding and building water infrastructure based on experiential knowledge of the older generation to a science based approach which in many cases was not based on local knowledge.

The management of water became the responsibility of the government, which mainly included provisioning and maintenance services for domestic and non-domestic uses. This changed approach continued after independence. With increase in population and urbanisation, the stress on the water sector too increased. Centralised management of water has resulted in weakening of the water infrastructure, depletion of water quality and quantity of water supplied. Disparities have emerged in distribution of water supply. All these issues are primarily due to mismanagement and centralized approach of the government.

As growing scarcity of water is going to put more and more pressure on the water availability, the efficient and democratic governance of water is going to be needed more than ever. Smart cities need to meet their water requirement through successful water management which in turn would depend on co-operation, the active involvement of all water users in a city. There is an essential need of learning from the historic water management practices of involving all stakeholders concerned in the management of the water issues. There needs to be adoption of eDemocratic approach in governance based on hydroinformatics. While eDemocracy will ensure accountability and transparency, hydroinformatics will ensure that decisions taken are well informed. Since eDemocracy is based on electronically enabled citizen's participation, it will entail inclusion of local knowledge and the availability of database from ICT will lead to amalgamation of the localized expertise with scientific knowledge for a lasting solution to transform the urban areas as water sensitive centers.

References

1. Government of India (2016) Water management. Accessed http://sscgj.in/water-management/, 12 Sept 2016
2. Shukla Rajesh (2010) How India earns, spends and saves: unmasking the real India. Sage Publications, India
3. World Meteorological Organisation (n.d.) The Dublin statement on water and sustainable development. Accessed http://www.wmo.int/pages/prog/hwrp/documents/english/icwedece.html on 03 Oct 2016
4. Global Water Partnership (2012). Accessed http://www.gwp.org/the-challenge/what-is-iwrm/ on 03 Oct 2016
5. Bhaduri S, Singh A (n.d.) Decline of traditional water harvesting systems during British india: exploring the issues of 'knowledge incompatibility', 'breaking down of commons' and 'free ridership', Center for Studies in Science Policy, JNU
6. Embree AT (1969) Landholding in India and British Institutions. In: Frykenberg RE (ed) Land control and social structure in Indian history. The University of Wisconsin Press, London
7. Vani MS (2009) Community engagement in water governance. In: Iyer RR (ed) Water and the laws in India. SAGE Publications, New Delhi
8. Singh, Nandita (2004). Water management traditions in rural India: Valuing the unvalued. KTH Royal Institute of Technology. Accessed at https://www.researchgate.net/publication/228435348_Water_management_traditions_in_rural_India_Valuing_the_unvalued
9. IDFC (2011) Water: policy and performance for sustainable development, India infrastructure report-2011, Oxford, India
10. Narain V (2000) India's water crisis: challenges of governance. Water Policy 2(6):433–444
11. Government of India (1950) Constitution of India, seventh schedule, article 246. http://lawmin.nic.in/olwing/coi/coi-english/Const.Pock%202Pg.Rom8Fsss(35).pdf. Accessed 01 Dec 2016
12. Planning Commission (n. d.) Regional development, housing and water supply in fourth five year plan. http://planningcommission.nic.in/plans/planrel/fiveyr/index4.html. Accessed on 10 Oct 2016
13. Government of India (2007) Tenth five year plan. Planning Commission, New Delhi
14. Government of India (1950) Constitution of India, seventh schedule, Article 246. http://lawmin.nic.in/olwing/coi/coi-english/Const.Pock%202Pg.Rom8Fsss(35).pdf. Accessed 01 Dec 2016
15. Sahu S (2009) Politics of access to drinking water in urban areas in India: state and market interventions a case study of Hyderabad, pp 46–80
16. Government of India (2012) National water policy-2012. http://wrmin.nic.in/writereaddata/NationalWaterPolicy/NWP2012Eng6495132651.pdf. Accessed 28 Nov 2016

17. Kumar SV, Bharat GK (2014) Perspectives in water resources policy for India. TERI, New Delhi
18. Government of India (2016) Draft national water framework bill 2016, Ministry of water resources, river development and ganga rejuvenation. http://wrmin.nic.in/writereaddata/Water_Framework_May_2016.pdf. Accessed on 10 Nov 2016
19. Planning Commission (n.d.) The draft national water framework act: an explanatory note. http://www.planningcommission.nic.in/aboutus/committee/wrkgrp12/wr/wg_wtr_frame.pdf. Accessed on 09 Nov 2016
20. World Bank (2006) Indian water supply and sanitation: bridging the gap between infrastructure and service, background paper. World Bank, New Delhi
21. Ehrhardt D, Groom E, Halpern J, Seini C (2007) Economic regulation of urban water sector: some practical lessons. Water sector board discussion series, paper no. 9, The World Bank Group, Washington, DC
22. Wagle et al. (2011) India infrastructure report 2011: policy and performance for sustainable development IDFC, New Delhi
23. Ahluwalia IJ (2016) Cities at crossroads: a looming crisis http://www.financialexpress.com/opinion/cities-at-crossroads-a-looming-crisis/329941/. Accessed 06 Dec 2016
24. Iyer RR (2007) Towards water wisdom. Sage Publications India
25. Government of India (2015) Smart cities: mission statement & guidelines. http://smartcities.gov.in/writereaddata/SmartCityGuidelines.pdf. Accessed on 15 Nov 2016
26. Leinmiller M, O'Mara M. (n.d.) Smart water: a key building block of the smart city of the future. http://www.waterworld.com/articles/print/volume-29/issue-12/water-utility-management/smart-water-a-key-building-block-of-the-smart-city-of-the-future.html. Accessed on 29 Nov 2016
27. Malick B (2004–05) Engaging with citizens to improve services: overview and key findings. The water and sanitation program funded research
28. Financial Express (2015) How analytic tools are helping in fixing water woes in Bengaluru, Kerala. http://www.financialexpress.com/economy/fixing-water-woes-with-data-analytics/169143/. Accessed on 15 Nov 2016
29. Shah M (2016) Urban water systems in India: a way forward, working paper 323: Indian council for research on international economic relations
30. Anonymous (n.d.) http://shodhganga.inflibnet.ac.in/bitstream/10603/8807/9/09_chapter%201.pdf. Accessed on 15 Nov 2016
31. UN Habitat (2012) State of the world's cities 2012/2013: prosperity of cities. https://sustainabledevelopment.un.org/content/documents/745habitat.pdf. Accessed on 24 Nov 2016
32. Central Water Commission (2000) Report of the standing sub-committee for assessment of availability and requirement of water for diverse uses in the country
33. McIntosh AC (2014) Urban water supply and sanitation in Southeast Asia: a guide to good practice, Asian Development Bank, Philippines
34. eParticipation.eu. (2016) E-Democracy. http://eparticipation.eu/information/e-democracy/. Accessed on 23 Dec 2016
35. Price Roland K, Vojinovic Zoran (2011) Urban hydroinformatics: data models and decision support for integrated urban water management. IWA Publishing, London
36. Abbott MB (1988) Contirbutions of computational hydraulics to the foundation of computational hydrology. In: Recent advances in the modelling of hydrological systems, vol 345. NATO ASI Series C. Kluwer Academic Pub, Sintra, Portugal
37. Abbott MB, Vojinovic Z (2010) Realising social justice in the water sector: 1. J Hydroinform 12(1)
38. International Telecommunication Union (2014) Smart water management in cities, United Nations. https://www.itu.int/…/web-fg-ssc-0122-r7-smart_water_management_in_cities.docx. Accessed on 15 Dec 2016
39. International Telecommunication Union (2014) Partnering for solutions: ICTs in smart water management, UNESCO

40. Boyle T, Giurco D, Mukheibir P, Liu A, Moy C, White S, Stewart R (2013) Intelligent metering for urban water: a review. Water 5:1052–1081
41. Beal C, Stewart RA (2013) Identifying residential water end uses underpinning peak day and hour demand. ASCE J Water Res Plan Manage. doi:10.1061/(ASCE)WR.1943-5452.0000357
42. Britton C, Stewart RA, O'Halloran KR (2013) Smart metering: enabler for rapid and effective post meter leakage identification and water loss management. J Clean Prod 54:166–176
43. India Water Portal (2011) PLC & SCADA based water quality, treatment and distribution management system for 428 MLD PCMC water supply system. http://www.indiawaterportal.org/articles/plc-scada-based-water-quality-treatment-distribution-management-system-428-mld-pcmc-water. Accessed on 16 Nov 2016
44. Choi YJ, Ahn JC, Im HT, Koo A (2014) Best management practices for water loss control in Seoul. In: 16th conference on water distribution system analysis, WDSA
45. EPA (2013) Water quality event detection system challenge: methodology and findings. Office of water (MC-140) EPA 817-R-13-002 April 2013. Retrieved from: http://water.epa.gov/infrastructure/watersecurity/lawsregs/upload/epa817r13002.pdf. Accessed on 30 Dec 2016
46. Paper for the BC Water Governance Project, a partnership of the Fraser Basin Council, Ministry of Environment, Fraser Salmon and Watershed Program, Georgia Basin Living Rivers Program and Fisheries and Oceans Canad, Linda Nowlan and Karen Bakker, November 2007
47. Tortajada C (2010) Water governance: a research agenda. Int J Water Res Dev 26(2):309–316. doi:10.1080/07900621003683322. (To link to this Article)
48. Biswas A, Tortajada C (2010) Future water governance: problems and perspectives. Int J Water Res 26(2):129–139

Chapter 9
Preparation of Town Planning Schemes—An E-Democracy Framework for Citizen Centric Planning

R.D. Desai, J.E.M. Macwan, Krupesh A. Chauhan and Partha Tripathy

Abstract Cities are for citizens and also '***by** & **of***' its citizens. However, citizens' participation which is one of the key agenda of Good Governance is generally in conflict when it comes to urban planning process which is more technocratic with a top down approach. Citizens generally are more aware and concerned about private interest rather than city as whole when they participate about the Development plans and neighborhood when we consulted on the Town Planning Schemes (TPS). In the context of Urban Planning when the plans are laid out, regulations are drafted which is meant **for** the citizens ought to ensure the factor that these documents are "by" & "of" the citizens indicating proper consultation in the process. Gujarat has been one of the progressive states of India so far as rate of urbanization is concerned and Town Planning schemes which have been practiced since early twentieth century has been a model for other cities. Practitioners state that the process for preparing a TPS as prescribed in GTPUDA (Gujarat Town Planning and Urban Development Act, [3]), and its rules, seem to be a complex and long process involving over 50 steps that can take a maximum of four years and one month as provisions of the GTPUDA, but in practice takes much longer. Administrators of central government are leading to educate Indian citizens to practice the excellent platform of E-Governance, which shall enable the citizens to express their thoughts

---

R.D. Desai
Gujarat Regional Chapter of ITPI, GCPIA, Surat, India
e-mail: gcpia.vnsgu@gmail.com

R.D. Desai
Surat Centre, GCPIA, Surat, India

J.E.M. Macwan
Head Civil Engineering Department, SVNIT, Surat, India
e-mail: jemm@ced.svnit.ac.in

K.A. Chauhan
Urban Planning Section, Civil Engineering Department, SVNIT, Surat, India
e-mail: kac@ced.svnit.ac.in

P. Tripathy (✉)
Urban Planning, Civil Engineering Department, SVNIT, Surat, India
e-mail: partha.tripathy@gmail.com

T.M. Vinod Kumar (ed.), *E-Democracy for Smart Cities*, Advances in 21st Century Human Settlements, DOI 10.1007/978-981-10-4035-1_9

and willingness in a more free way. The intention of this chapter is to throw light on ICT, specifically the planning process adopted in a fast growing state of Gujarat. Apart from that, issues regarding E-Democracy initiatives are also addressed. Though the process of E-Governance is in its infant state from user's point of view, but this will surely gain momentum in the days to come. This practice will instil high-democratic values and plan implementation within desired time frame.

Keywords E democracy · Urban planning · Town planning schemes · Citizen consultation

9.1 Introduction

In this era, increased urbanization poses challenges before any local governments to provide sustainable local services such as urban transport systems, waste management systems, and urban water supply and sanitation. To meet these challenges, local governments often benefit if they are able to improve and strengthen democratic urban governance, develop a more holistic planning system built on integrated environmental solutions, inclusive planning, social sustainability and economic growth. While the center of every activity by any Local government is well being of the all its citizens, but while planning activities are undertaken, they adopt a pseudo consultative approach, collating the statistics of the habitat and inhabitants, like area, typology, age, socio economic characteristics etc. Urban spatial planning has been in the doldrums for many years. It has typically been regarded as old-fashioned, technocratic, and bureaucratic, stifling development by wrapping it up with red tape [7].

There has been a transformation in the approach of City planning. The most visible change has been the expansion of the urban political system from 'government' to 'governance', which in developed countries represents a response to the growing complexity of governing in a globalizing and multilevel context, as well as the involvement of a range of non-state actors in the process of governing. In developing countries, the concept of governance has been promoted as a policy measure, along with decentralization and democratization.

One of the important messages which transpire from literature, even though urban planning has changed relatively little in most countries since its emergence about 100 years ago, a number of countries have adopted some innovative approaches in recent decades [8]. The approaches include, strategic spatial planning, use of spatial planning to integrate public-sector functions, new land regularization and management approaches, participatory processes and partnerships at the neighborhood level, etc. To ensure that participation is meaningful, socially inclusive and contributes to improving urban planning, certain conditions need to be satisfied, including: a political system that allows and encourages active citizen participation; a legal basis for local politics and planning that specifies how the outcomes of participatory processes will influence plan preparation and decision-making; and mechanisms for socially marginalized groups to have a voice in both representative politics and participatory planning processes. Citizen participation is now an integral component of ideas like "smart growth" and "New Urbanism" in the field of urban planning.

Urban governance with its emphasis on stakeholder participation is often thought to be in conflict with the traditional urban planning which is technocratic and top down in its approach [4].

It is worth to understand the evolution of approaches to urban planning. There have been four broad phases in the evolution of approaches to urban planning and management:

- Master planning,
- Strategic or area planning,
- Urban management and
- Urban governance.

The advent of new paradigms of governance called for reinvention of urban planning. New planning practices around the world have incorporated participatory approaches as it has come to be recognized that the expert- driven decision making processes of the past were quite unsuitable for effective planning. But the elements of new urban planning are not yet fully formed [7].

Land being dealt and managed at the State level, with emergence of Planning, States during 1970s adopted to enact legislations to enable physical planning of the urban forms. Most of the States in India adopted legislations formats which provisioned for creation of Development Authorities, preparation of Macro level and Micro level plans in time bound manner and also clout to regulate the development through land use planning and height and bulk regulations for macro level planning as well as micro level planning. The legislation also provisioned for citizen participation through objections and suggestions. The legislations have also provisioned for the process to be followed including the manner in which the citizen engagement should be carried out.

Innovation in technology has brought about the changes in the processes and ICT (Information Communication Technology) has played a robust role in any process re-engineering. In the last few decades, availability of satellite images and development in the software tools have enabled planners to use these information, data and modern tools to put up the urban plans. Accordingly, professionals and executive bodies have procured new technologies, undergone capacity building programs and at certain instances also legislations have warranted for improvisation in the processes. For instance the maximum time period provisioned in preparation of TP Schemes in the GTPUDA have undergone amendments and has been shortened substantially.

Still the questions remain unanswered, if there been a conscious effort by the policy makers to improvise the processes and procedures for public participation and moreover if the City administrators, planners improved upon their practices to maximize citizen participation in the process to ensure an enhanced democratic framework.

Participation and public–private partnerships have become important elements in all of the innovative planning approaches. Participatory planning empowers communities and builds social capital, and also leads to a better design of urban projects and can allow for participants' concerns to be incorporated within strategies. Successful participation is, however, dependent upon certain preconditions relating to the political context (a political system that encourages active citizenship and that is committed to equity and redress), the legal basis for participation (processes and outcomes are legally specified) and available resources (skilled and committed professionals, well-resourced and empowered local governments, and informed and organized communities and stakeholders) [8].

In many countries of the developed world, master planning has now been replaced by strategic spatial planning which emphasizes *only on inclusive stakeholder participation processes* and key strategic elements of urban growth.

As Cities are known as *engines of growth*, urbanization offers significant opportunities to reduce poverty and gender inequality, as well as to promote sustainable development. Post-independence, one of the focused key agenda for the Government of India has been to build planned cities and strengthen the urban infrastructure, however, administrators have struggled to handle the menace of slums, in-equality and economic disparity which is increasing and reasons could be information asymmetry within the citizens and regulatory capture by selected stakeholders.

How can cities avoid calamity and make the most of their opportunities? Increasingly, it is hoped that improved urban governance will be the answer. The term "urban governance", formerly equated with urban management, has come to be understood as both government responsibility and civic engagement. Generally, it refers to the processes by which local urban governments—in partnership with other public agencies and different segments of civil society—respond effectively to local needs in a participatory, transparent and accountable manner.

India is a democratic republic and the philosophy of justice, equality, liberty and fraternity are enshrined in its constitution. The democratic principles of the country flow from the Preamble of the Constitution itself. Governance was process and procedure centric and generally a top down approach was used in policy making. In addition, the country, given its vast size, federal structure of governance with over 240,000+ local governance institutions and large population coupled with its other complexities viz. multi-cultural, multi-ethnic, multi-religious and multi-lingual society, did not make itself amenable to large scale public consultation in policy making.

Town planning in modern era is famous since end of 20th century. Since then, Development Plan concept is more or less been used by many cities. Development plan is a process of developing a city as per the proposed guideline within specified time limit. E-Democracy is something which relates ICT with political participation of stakeholders in decision making. There are many misconceptions regarding e-governance and e-democracy. E-Governance is specifically related to give efficient information of government activities to public (like voting counts information, water quality information, online application of birth certificate etc.). E-Democracy is related with distributing control and decision making process.

E-Governance is about Transparency and E-Democracy is about Directness

In planning, government interventions include preparation of Development plan, Zonal plans, Town Planning schemes etc. All activities were done with citizen participation before ICT era also. But what else can ICT improve in Democratic process is major question of 21st century. Government centricity in E-Government policies and practices neglect civic inclusion in decision making. Citizens are generally excluded from the list of decision making that these all policies are going to be implemented on and such decisions are generally through representative democratic.

E-Democracy is '**vertical expression of horizontal policies**' (Meaning: policies are made by only a handful of people can be improvised by adding more groups of citizens in a column). There are three kind of liberty (or say, process) given in any local bodies, Information, Consultation and Participation.

'**Information**' is a one-way relationship where governments produce and distribute information to citizens, such as occurs through websites and e-newsletters. This includes active attempts by governments to increase information dissemination on particular issues and arbitrary citizen access to information available through digital means upon demand.

'**Consultation**' involves a limited two-way process through which citizens can provide feedback to government; for instance, via online surveys and petitions. This requires that citizens are provided with the information necessary to make informed decisions, but feedback is restricted to topics predetermined by governments, which means civic input has a limited capacity to shape political agendas and discourse.

'**Participation**' concerns the development of stronger relationships between citizens and governments, in which citizens are viewed as partners. It includes active involvement of citizens in the policy-making process and may take place through the use of, for example, digital discussions and wikis, where citizens can propose policy options and shape the direction of political dialogue.

9.2 Genesis of Town Planning Schemes in India

First excellent effort of developing today's metro city Mumbai was in 1948. In 1948, "Modak-Mayor Master Plan" was proposed for Mumbai's overall development. History of development plan starts from immediately after independence of the nation India. In 1954, Bombay Town Planning Act was enacted which laid down the concept of Town Planning Scheme. After Bombay was divided in two different states (Maharashtra and Gujarat) in 1960, Gujarat legislated its own Act for town planning in 1976, The Gujarat Town Planning and Urban Development

Act [3], which has a well laid out detailed process of preparation and implementation of Town planning schemes.

Development plan for any city is generally for the period of 20–25 years which outlays the broad land-use, physical infrastructure and also development regulation and guidelines. This plan is periodically reviewed and revised (generally at interval of 5 years). According to development plan proposed, it is planned through the means of sub-sectoral or micro level by TP schemes. Development plan of any city will show zoning large area of a city into various types such as, Residential land, Institutional land, Industrial land, Agricultural land, Public purpose, transportation, logistic parks, Education zones, SEWZ etc., while TPS shall further detail an area in the Development Plan with all necessary details.

Preparation of Development Plan, it is mostly bureaucratic process instead of democratic. It is also evident from a recent petition made by the group villages which opposed the notification of the Development Plan prepared by Surat Urban Development Authority in 2016. Could the reason be lack of early engagement of citizens in the process at the primary stage? Could it be an outcome of the Top-Down Approach? For certainly, the breakup between would delay the process.

In the last decade, there has been reflects of efficiency so far as timelines are concerned in preparation of TP Schemes and Development Plans are concerned due to adoption of innovative methods, technology and acquisition of the soft skills by the Authorities. But still, the time period mentioned in the guidelines for Town Planning schemes has not been met in almost every Town Planning schemes in Gujarat. It is observed that, there is no change in the time consumed in the process of preparing a TP Scheme from conception to award by Government, during pre-1990 and post 2000. Prior to use of novel ICT and soft technologies, delay in Town Planning Scheme finalization was primarily in preparing the scheme which was overcome later due to available of updated land records online, digitization of revenue maps by the concerned department, accurate surveying techniques, availability of satellite imageries, and tools like GIS/Autocad etc. However, it is evident that the Authorities have not been able to engage with citizens to the desired extent and manage the consultation process within the time limits. Issues could be either the target citizens size has increased whereas resources handling the process have not, or citizens' and stakeholders have diverse expectations from the Authorities (Table 9.1).

Table 9.1 Standard town planning scheme preparation time as per guideline [3]

Particulars	Period (months)
Making and publishing draft TP scheme	9
Amendment in draft TP scheme	3
Sanction of draft scheme by state government	6
Appointment of TPO	1
Making of preliminary and final scheme by TPO	12
Scheme to come into force	1
Total	32

Source Gujarat Town Planning and Urban Development Act [3]

9.3 Town Planning Scheme Process

The Bombay Town Planning Act of 1915 first introduced Town Planning Schemes in India. This Colonial British Act incorporated not only the 'betterment recovery' principle from the English Act but also the stratagem of 'reconstitution' of plots was borrowed from the 'Lex Adics' procedure from a German Act [5].

The State of Gujarat which was a part of the Bombay State followed the Bombay Town Planning Act of 1915. Local planning authorities adopted the town planning schemes approach for faster development of the City. As T P Schemes were typically micro level plans, they were employed in parts of the city without considering the larger implications of these actions. To ensure planning at the macro scale as well the subsequent Bombay Town Planning Act, 1954 came into force enabling local authorities to prepare Development Plans for cities and towns. This brought in the DP–TP (Development Plan followed by Town Planning Scheme) mechanism which is how it is commonly referred to today. A spurt in use of this mechanism occurred with several local agencies taking up this macro and micro planning exercise (TPVD). In 1960, the erstwhile State of Bombay was bifurcated to form the States of Maharashtra and Gujarat. Both States hence continued to use the same Act. The State of Gujarat eventually, contextualized the Act for their developmental needs and enacted the Gujarat Town Planning & Urban Development Act, 1976. Subsequently, further amendments over time in the Maharashtra Act and later on in the Gujarat Act ensured that TP Schemes remained relevant and grew in implementation efficiency.

TP Scheme model has been widely appreciated as one of successful citizen centric participative model in urban planning at micro level. TP Scheme has been efficient to manage the new growth implies that the agricultural land at the periphery of the cities and towns or smaller settlements that are not yet "urban" are transformed to be made suitable for urban or non-agricultural uses. The TP Scheme planning ensures that the irregular landholdings and plots are given regular shapes; each plot has access; infrastructure services such as water supply and drainage are provided; land is appropriated for providing roads, parks, social amenities, and low income housing. Development controls are prescribed to result in a good quality-built form and development or betterment charges are levied to offset the cost of developing the physical and social infrastructure. But most importantly, all of this must happen in a timely manner and such that it is acceptable to the "landowners" to avoid conflict in the growth management process [1].

The Gujarat Town Planning and Urban Development Act, 1976 is the Act (along with its Rules) that relates to the making and execution of Development Plans and Town Planning Schemes in the State of Gujarat. As discussed earlier, the land use oriented Development Plan serves as the macro plan post which TP Schemes are taken up to enable implementation at a micro level. Details related to the TP

Scheme process are described below and have been divided into 5 distinct stages from Intention to prepare till the final TP Scheme is notified and Final TP Scheme is notified by Government.

9.3.1 Stage-I: Inform and Engage with Citizens About the Intent

Declaration of Intention to make a Town Planning Scheme (Sec 41 and 42). The first step of the process of preparation of the Town Planning Scheme (TPS) is to declare intent for making a TPS. The authority in consultation with the Chief Town Planner may, by resolution, declare its intention to make such a scheme in respect of such area as identified by the authority. Within 21 days from the declaration of intention to make a scheme, the authority shall publish it in the prescribed manner and shall dispatch a copy of the same along with a plan showing the area which it proposes to include in the TPS to the State Government. The publication and the manner in which citizens are informed and engaged is stated in the GTPUD Rules.

Publication of declaration under section 41: (GTPUD, Rules 1979 section 16)

1. The declaration under sub-section (1) of section 41 shall be published in the Official Gazette and shall also be published by means of an *advertisement in one or more Local Language newspapers* circulating within the jurisdiction of the appropriate authority.
2. Every advertisement published under sub-rule (1) shall contain the resolution of the appropriate authority in respect of the declaration under section 41IDH_41 and shall announce that a copy of the plan of the area proposed to be included in the town planning scheme and the surrounding *lands is kept open for inspection of the public at the head office of the appropriate authority during office hours.*

9.3.2 Stage II—Preparation and Publication of the Draft Scheme (Sec 48-1)

In accordance with the GTPUD Act, 1976 within Nine months from the declaration of intention to prepare the scheme the authority shall prepare a draft scheme of the area and publish the same in the Official Gazette along with the draft regulations. Invite suggestions on the draft scheme. **After considering the suggestions/objections received within a month of publication, draft scheme is sent to Govt. for sanction u/s 48(1) within three months (these three months are required to incorporate the suggestions and objections of citizens)**.

The GTPUD Rules 1979 prescribes the manner of public participation at the initial stage of Draft TP scheme.

Meeting of owners and framing of tentative proposals (GTPUD Rules, 1979, section 17) For the purpose of making the draft scheme under section 42 the appropriate authority shall call a meeting or meetings of the owners of the lands included in a town planning scheme by ***a public notice*** as well as ***by individual notice*** to every owner ***whose address is known*** to the appropriate authority and explain in such meeting the tentative proposals of the draft scheme for eliciting ***public opinion and suggestions on the said proposals***. The appropriate authority ***may*** take into consideration all such suggestions made and objections raised on the proposals for making the draft scheme under section 42.

9.3.3 Stage III—Sanction of Draft TP Scheme

The Draft scheme is sanctioned by the state Government under section 48(2) of the Gujarat Town and Country Planning Act, 1976. As per Act the Government may within six months, time sanction the Draft TP Scheme through a Government Notification. The Act prohibits any development in the area falling in the Draft TP Scheme area, by any person unless; the person has obtained necessary permission from the Authority. The State Government takes steps to appoint Town Planning Officer (TPO) for preparing the Preliminary and Final TP schemes.

9.3.4 Stage IV—Preparation of Preliminary TP Scheme by Government

Within one month of the sanction of the Draft Town Planning Scheme the Town Planning Officer who is a quasi-judicial officer is appointed under section 50 of the GTPUD Act, 1976. The function of the TPO is to prepare the preliminary and the final town planning schemes. The TPO's task is to deal with each landowner both on the physical planning proposal—the shape and location of the Final Plot (FP) and the financial proposal—the compensation and betterment issues, and eventually demarcate the FP on ground and hand it over to the owner. The sanctioned draft TPS is divided in two parts to facilitate functioning of TPO: a preliminary TPS to deal with the physical planning proposal and a final TPS to deal with the financial proposal. The Act provides for 12 months to prepare the Preliminary TP Scheme by the TPO. At this stage, citizen consultation is of utmost importance as the decision on Final Plots, sizes, valuation of the plots etc. are undertaken. In this section the features and manner in which the consultation is ought to be carried out as per the provisions of GTPUDA and its Rules has been analyzed and presented.

9.3.4.1 One to One Hearings to Each Landowner on the Preliminary TPS

The TPO gives individual hearings to each landowner and revises the preliminary TPS if required. Inputs from the State Government and local authority and development authority are sought. **The preliminary TPS may be modified if required, and the TPO gives a second round of hearings**.

Consultation with Citizens by the TPO prior to his activities The Town Planning Officer shall, before proceeding to preparation of Preliminary TP Scheme, publish a notice in the Official Gazette and in one or *more Local newspapers* circulating within the area of the appropriate authority. Such notice shall specify the matters which are proposed to be decided by the Town Planning Officer and State that all persons who are interested in the plots or are affected by any of the matters specified in the notice shall communicate in writing their objections to the Town Planning Officer ***within a period of twenty days*** from the publication of notice in the Official Gazette. Such notice shall also be pasted at the office of the Town Planning Officer and of the appropriate authority and the substance of such notice shall be pasted at convenient places in the said locality.

Finalization of the Preliminary TPS and Its Approval

The preliminary TPS is again modified based on the second hearing. At this stage, demarcation of FPs commences. The TPO finalizes the preliminary TPS by writing his or her decisions with regard to every plot. This is referred to as the award of the preliminary TPS and is published in the local newspapers. At this stage, the preliminary TPS is sent to the State Government for sanction. The GTPUDA provisions for a timeframe of two months, for the Government to sanction the Preliminary Scheme. The preliminary TPS comes into effect from the date of sanction and land for roads and plots appropriated for public purpose vest with the local authority or development authority.

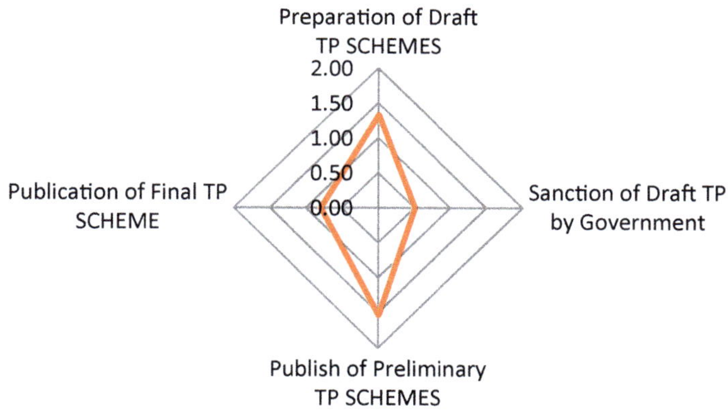

Fig. 9.1 Time periods in different stages for town planning scheme finalization

9.3.5 Stage V—Final TP Scheme

At this stage, the TPO prepares the financial proposals which are taken up with each owner. All the owners are intimated about their compensation to be received or incremental contribution to be paid by them. These pertain to the compensation and incremental contribution. The TPS at this stage is called the final TPS. Once the hearings are done, the financial proposals may be modified and sent to the State Government for opinion. There may be some modifications. The TPO then finalizes the TPS and publishes it in the local newspapers. This is referred to as the "Award of the Final TPS". A Board of Appeals for further issues on valuation is constituted. Once all appeals are resolved and the final TPS is modified, it is sent to the State Government for approval. The State Government is required to sanction it within three months.

The process for preparing a TPS is prescribed in GTPUDA, 1976, and its Rules. It is quite a complex and long process involving over 50 steps that can take a period of four years and one month as provided by the act, but in practice takes much longer [1].

As shown in Fig. 9.1, Spider web diagram shown below will help to understand time period elapsed in various phases of town planning scheme finalization. It shows the total time taken as guided in Gujarat Town Planning and Urban Development Act [3]. That sum comes up as Three years.

In current practice of preparing final town planning scheme, total time taken to finalize it is around 4.2 years with all extension periods in each stage. The process undertaken and the important stages for preparation TP Scheme as per the GTPUD Act 1976 and Rules, 1979 is presented in Fig. 9.2. The stages of preparation have been put in the boxes outlined rectangular boxes with dotted lines whereas the consultation is presented in callouts format.

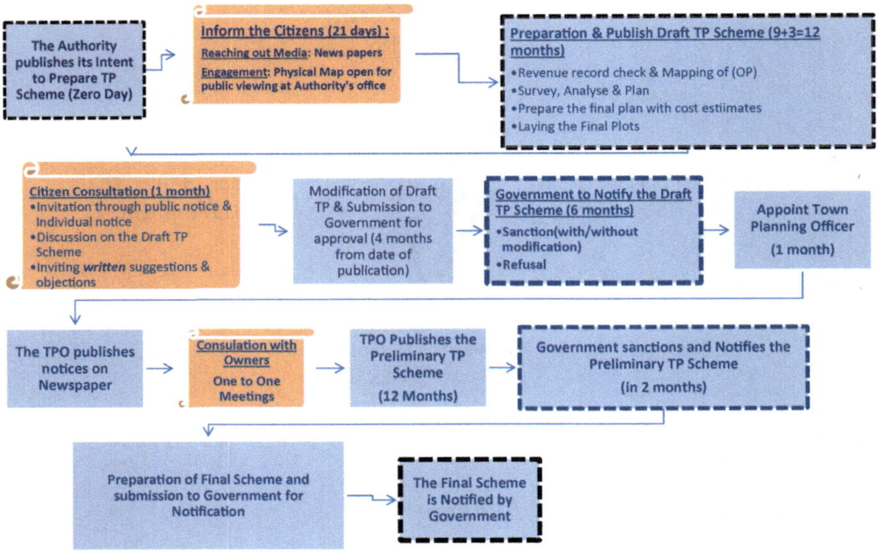

Fig. 9.2 The TP scheme process as per GTPUD act, 1976

Fig. 9.3 Area of Surat under TP scheme

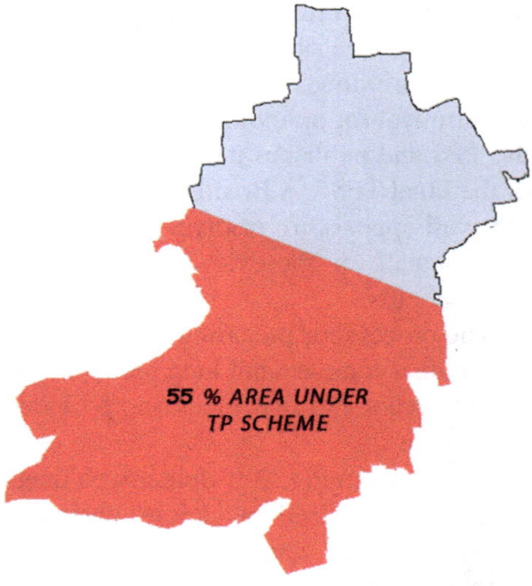

9.4 TP Schemes Implementation in Surat

Surat is a city located on the western part of India in the state of Gujarat. It is one of the most dynamic cities of India with one of the fastest growth rate due to immigration from various part of Gujarat and other states of India. Fig. 9.3 shows the area of Surat Municipal Corporation area developed through TP Schemes.

Surat is one of the cleanest city of India and is also known by several other names like "THE SILK CITY", "THE DIAMOND CITY", "THE GREEN CITY", etc.

The City accounts for:

- 42% of the world's total rough diamond cutting and polishing,
- 70% of the nation's total rough diamond cutting and polishing,
- 40% of the nation's total diamond exports,
- 40% of the nation's total man made fabric production,
- 28% of the nation's total manmade fiber production,
- 18% of the nation's total manmade fiber export, and
- 12% of the nation's total fabric production.

Out of 325 Km^2 of area, total area of Surat city planned under TP scheme is around 177.40 Km^2. As per this data, half of Surat city area is planned under TP scheme. The first TP scheme of Surat was initiated on 26/02/51 (Source: https://suratmunicipal.gov.in/Departments/TownPlanningTPDetails). As per the latest records, 128 TPS have been taken up by the Government Authorities.

On the time consumed in preparation of TP Schemes, an analysis was made of all the TP Schemes and the results are shown in the figure below. It is self-explanatory that, maximum time period has gone in Preliminary and Final town planning scheme. These two steps include the major share of public participation. It is clear that before the period of Internet, the total time taken for overall Town Time taken in preparation of planning scheme preparation and finalization has been decreased after Internet and communication came i.e. after the year 2000.

9.5 TP Scheme Is a Process Which Is Dynamic—Changes Are Inevitable

Most of the literature suggests that, TP Scheme has been a successful framework for implementation of planning proposals of any cities, which otherwise have struggled at early stages in procurement of Land. TP Scheme as a planning tool and more of as a process has been tested, implemented, reviewed and improved in various cities in Gujarat. It has been adopted as a Mantra for implementation of Development Plans in almost all large cities in Gujarat. The process has a long history—it was introduced in 1915 by legislation, and since then the legislation has been continuously improved to make the process more responsive to the changing context of development [1]. There have been amendments in the GTPUDA, 1976, almost five times i.e. in 1986, 1995, 1999, 2000 and 2002, and the processes have been modified. Almost after 20 years of implementation of the Act, in the year 1995 the certain amendments were made which included modification on process time stipulated for preparation and of Draft Plan and submission by the Authority which was reduced from 22 to 15 months including extension and modification post citizen consultation and also the time stipulated by Government for sanction was reduced from 6 to 3 months. Therefore, for preparation of the Draft TP Scheme, the overall reduction has been from 28 to 18 months which is almost 35%, however, there has been no amendments in the time period stipulated for preparation of Preliminary TP Scheme by Town Planning officer which is more than 12 months. During 1990s, availability of Satellite maps, capabilities to use the GIS platforms to prepare maps, total station surveys, digitisation of revenue maps, could be the push factors for bring in efficiency in the process and reducing the timelines. The TP Scheme processes have been re-engineered to enhance the efficiency level in terms of reducing the time consumed in preparing the TP Scheme and its approval process.

The literature suggests that the TP Scheme process as followed in the State of Gujarat has number of advantages, and one of them is that it is "Democratic and Participatory" and some of the limitations of the process is "it is time consuming", and more of a "centralized process—dependent on State Government clearances and approvals at all the Stage", "lack of transparency—encourages corruption and biasness in decision making", "resource crunch—lack of adequate manpower".

Balleny [1] suggests that "These time frames were determined when computerization was non-existent, communication technologies were cumbersome, and there was virtually no private sector in place to assist the public sector. In view of changes in technology, the timelines can be easily reduced".

This paper analyses TP Scheme built in the process and suggests measure to:

1. Enhance citizen participation
2. Make it democratic so far as common interests are concerned
3. Reduce the time consumed.

9.6 Proposal for Better Democratic Process in Town Planning Scheme

Since rise of 21st century, use of mobile telephony has exponentially increased. After 2010 c. smart phones have become common in every sectors of daily life. Some local governing bodies of cities have also liked to do their governance work and some municipal work (like, bill payment, birth certificate etc.) with help of mobile applications.

Implementation of E Governance has been undertaken by various Government Entities in many states and some of the remarkable achievements have been in the reforms in the Land registration details. Similarly initiatives have been taken to digitize the revenue maps also. These two interventions by other departments have played a major role in improving on the timelines stipulated for preparation of Base Plans and collating ownership details. Similarly, good quality Satellite imageries have been available to assess the ground conditions including existing land use of the area, and GPS enabled advanced surveying equipment have also contributed to reduce the time for physical surveys. Table 9.2 shows the process and rules before

Table 9.2 Details of Surat TP schemes

Sr. No.	TP scheme details	No. of schemes	Total area under schemes (Ha)
1	Sanctioned final schemes	39	4575.44
2	Sanctioned preliminary schemes	28	2964.98
3	Sanctioned draft TP schemes	58	9654.96
4	Draft TP scheme submitted to Govt. for sanction	03	542.83
Total		128	17,738.21

Table 9.3 Surat City profile

Area	326.515 km^2
Population	4,466,826 (census 2011)
Density	13,680 persons/km^2 (census 2011)
Location	Latitude: 21.112 °N Longitude: 72.814 °E
Municipality established	1852 AD
Corporation established	1966 AD

Information Technology and after Information Technology. However, so far as citizen consultation processes, analysis of representations has been being done in a conventional process. There has been no improvisation made in the processes which could establish "***to be democratic***" so far (Tables 9.3 and 9.4).

It can be observed that, while technology has been used and adopted in many of the activities but the citizen engagement process, consultation mode, and approval mechanism still remains unchanged. With increase in population and shortage in manpower to handle TP schemes, the efficiency of the consultation process, collation of all the representations, analysis, and incorporating the suggestions in Draft Plan has been time consuming and a key area where delays have been unavoidable. Decisions are democratic or not is generally not established, as the process followed does not reveal such information.

An efficient citizen engagement is proposed which would improve the areas of concern for the Authority and also enhance the efficiency of the manpower resources deployed to manage the process be it from Authority preparing the Draft TP or office of Town Planning Officer responsible for preparing Preliminary TP Schemes and Final TP Schemes.

Entire town planning scheme stages can be inserted in mobile app of Municipal Corporation. There can be different stages in application like; Draft Town Planning Scheme suggestions, Preliminaries and final Town planning schemes suggestion etc.

Families which are covered under the geographic area of proposed Town Planning schemes are informed beforehand with help of municipal corporation application. They can get application installed on their smart phones or access the site through the computers. Now families under any particular Town Planning

Table 9.4 Rules and process: before and after ICT

Activities as per GTPUDA	Before	Now
Informing stakeholders	By news paper and post	By mobile message and post
Survey of the area	Conventional surveying equipment	Total station survey
Establishment of land ownership	Collection of data from revenue department	Data are available online (www.bhulekh.in) maintained by National Informatics Centre, Govt. of India
Digitization of revenue maps	All maps were required to be digitized	Digitized maps are available with state revenue department/department of science and technology
Preparation of ownership details with land area (original land) and price	Was done manually	Certain TP schemes have done using excel tool for calculation. Could be improvised with a data management software
Preparation of TP schemes (road layout, public facilities, final plot)	Manual preparation earlier later autocad was used	Normally GIS is used
Consultation notice	Newspaper advertisement and letters	No change
Consultation	One to one and group consultation at Authority/TPO office	No change
Display of scheme for public viewing and comments	The draft schemes are placed at office of authority's office/zonal office for public viewing	No change
Submission of suggestions/objections	Letter addressed to authority	No change
Analysis of suggestions and objections	Collated and done manually	No change
Approval mechanism	Physical plans are sent to government for sanction and approval	No change

scheme can login with the cell number. They will get One Time Password (OTP) on their mobile and they can login. As soon as they login, they will be able to see the list of town planning schemes proposed in municipal corporation boundary area. A person can select his/her town planning scheme number by simply clicking it and can participate and can give his or her view on various criteria suggested (planned town planning scheme). The stakeholders can give their queries on criteria like

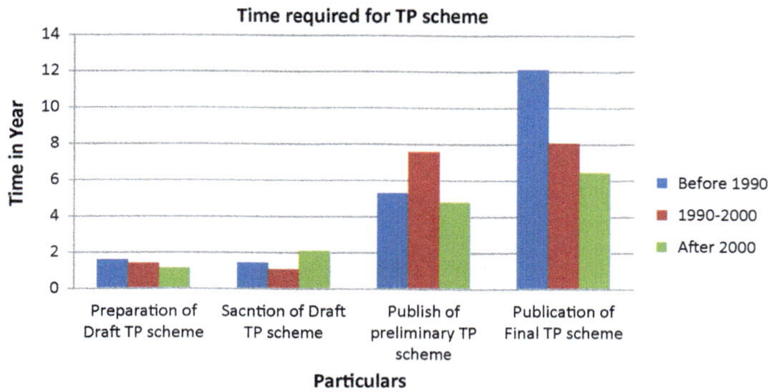

Fig. 9.4 Time required for TP scheme in various periods

roads, size of property (final plot size) and any other problems. For every such participatory stage, municipality can fix some time period (say a fortnight). It can get maximum participation within this short period. So the families which are having any queries on particular criteria can only be invited for communication with Town Planning Officer. People can also select the time period within which they are willing to meet Town Planning Officer. It can be provided with a set of time span and people can select their comfortable time span and interact with the Town planning Officer.

After listing problems of all the participants, TPO can revise the scheme and can propose a new town planning scheme (preliminary town planning scheme). This scheme can be sent to state government for its approval and can be approved within time period mentioned in guidelines. Now further notifications can be obtained by mobile application installed in stakeholders' phones in form of pop-up notifications shown in Fig. 9.4.

Now, with this process of consultation in a town planning scheme by mobile app; it can reduce time of preparation to 15–18 months. As it can be seen from Fig. 9.2 that maximum time is taken in publication of preliminary town planning scheme which includes public participation. But public participation period can be reduced to 60 days instead of a very long period taken currently. Due to this step, preliminary town planning scheme publishing period can be reduced to 5–6 months instead of 18 months currently. Entire process of mobile participation for Town Planning scheme can be divided into six stages as shown in Fig. 9.4. These stages includes all the public participation provisions as per the GTPUD Act. (Figs. 9.5 and 9.6).

The second major concern which consumes substantial time is in preparation of draft town planning scheme which is 16 months. Due to readily available of geographical data and base map (GIS technology), this can be optimised and shortened. Preparation of draft town planning scheme can be reduced to 3–6 months instead of 16 months.

Publication of final town planning scheme takes 9 months can be reduced to 3 months because communication has been improved very well nowadays. So

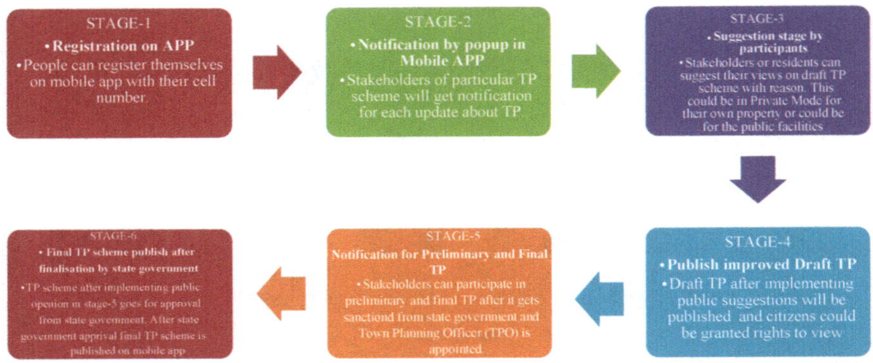

Fig. 9.5 Stages of mobile application for TP scheme finalisation with public participation

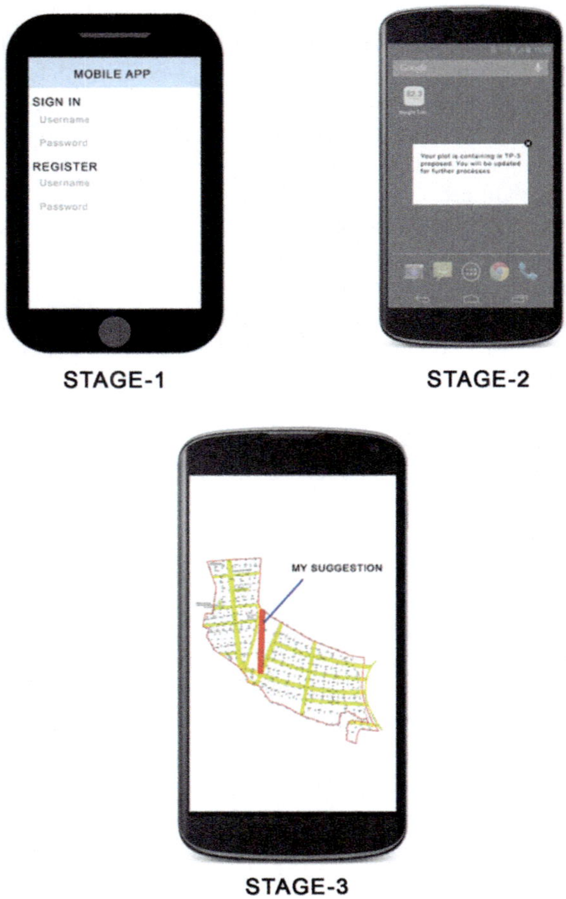

Fig. 9.6 Mobile application process views

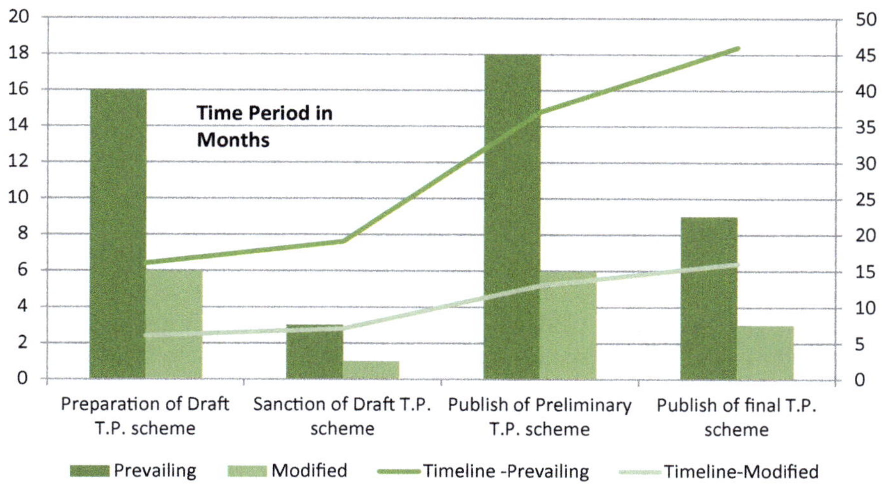

Fig. 9.7 Time reduction in process of TP scheme by proposal given

finalization after some modifications suggested by state government can take a much smaller period.

The period taken by state government for finalization of the TP Scheme is too high which does not involve any preparation other than receiving objections from the public. This time period can be reduced to 1 month. Because due to good amount of public participation, all major concerns of public, Authority are incorporated at the preparation of final town planning scheme stage. So very limited scope for amendment of town final town planning scheme remains (Fig. 9.7).

9.7 Conclusion

As the Urban population is growing at a faster pace, so are the challenges for City Administrators to match supply of developed land and infrastructure. Administrators are also finding it difficult to manage the growth of slums as wells the rising inequalities within the economic class of the Indian Cities. One of the reasons for this could be information asymmetry and regulatory capture. And these are outcomes of limited or poor citizen participations. A new approach to urban planning has emerged with the incorporation of the principles of urban planning.

With the support of full proof technical experts along with administrative will coupled with political support and enlightened public participation will definitely help in preparing and implementing Town Planning Schemes resulting in better environment with best infrastructure services and social facilities for all citizens to live best quality of life.

Gujarat has been one of the progressive states of India and has a led in field of Urban Planning. And Town Planning schemes has been one undisputed, well proven model which has been tested, modified and improved with a process re-engineering approach. Still against stipulated time frame of 36–40 months, completion timeframe has been more than 70 months in the recent times even more at earlier. When literature states of it's participative and democratic approach, processes mentioned fails to assess these attributes or even they are not a part of assessment either. This study of TP Schemes implemented by Surat Municipal Corporation and Surat Urban Development Authority depicts that while Schemes implemented prior to 1990 took more time to prepare draft TP Schemes, at later stages there has been substantial reduction probably due to use of new technology, but still the overall time consumed in approval of TP Schemes more or less remains unchanged as there has been no improvisation in the engagement of citizens including consultation and grievance handling.

As cities have now launched e-governance and m-governance mode to engage with citizens, a new ICT framework for TP scheme process has been proposed which could be integrated on the existing ICT frame of the Cities and connect with the citizens. Secondly, a backend analytical tool may be developed to assess the participation, suggestions, objections and guide decision making following the democratic methodology. The overarching GTPUD Act 1976 warrants for various statutory approvals including that of State Government, a suggestion is being made if such approvals could be made online in a secured network. This would improve accountability and transparency in the process which shall help in building trust. Similar approaches could also be made at a larger scale to engage with citizens for preparing Development Plans for the City.

Acknowlegement The Authors acknowledge the valuable efforts of Mr. Devang Kapadia, Research Student, Urban Planning, Civil Engineering Department, SVNIT, Surat, Gujarat, India for his contribution in analysis of TP Schemes process and suggestions on process re-engineering.

References

1. Ballaney S (2008) Town planning mechanism in Gujarat, India by Shirley Ballaney. World Bank Institute. Retrieved from http://hcp.co.in/file_manager/publications/Town-Planning-of-Gujarat_Research-Paper.pdf
2. Freeman J, Quirke S (2013) Understanding E-democracy. JeDEM 5(2):141–154
3. Gujarat Town Planning and Urban Development Act-1976
4. Mahendra B, Harikrishnan K, Krishne G (2010) Urban governance and master plan of Bangalore City, pp 1–18
5. Panchal K, Salgaonkar J, Vyas DP, Mathews R (2016) Scaling up land readjustment for urban and industrial development through the town planning scheme in Dholera, Gujarat, India
6. Promises and Problems of E-Democracy, OECD-2003
7. Taylor P (2004) Planning for a better future. Habitat Debate 4(4):4
8. UN Habitat. (2009). Planning sustainable cities (2009th ed.). UN Habitat

Web

9. https://www.researchgate.net/publication/221096885
10. http://www.internetlivestats.com/internet-users/
11. http://researchassessment.in/mumbai-before-development-plan-2/
12. http://edemocracy.weebly.com/what-is-edemocracy.html
13. https://suratmunicipal.gov.in/
14. http://townplanning.gujarat.gov.in/dp-tps-information/town-planning-scheme.aspx

Chapter 10
Smart City Project Selection, Prioritization, Customization and Implementation—Voice of Smart Citizens Case Study of Bhubaneswar and Surat

Partha Tripathy, Krupesh A. Chauhan and Anjali K. Khambete

Abstract The Smart City Mission launched by Government of India focuses on the comprehensive development of physical, social, institutional and economic infrastructure to enhance the quality of life, attract people and investments to a City, setting growth and development in motion and all these happening with a citizen centric approach. And moreover, the selection process of smart city mission trained the City Authorities to adopt a citizen participative approach. The cities which scored high in the challenge were those which had adopted citizens identifying the problems and democratic process for selection of solutions. And what was expected to be smart, was the role of ICT in the entire consultation process be it the social media, mobile telephony or the web based applications, citizens participation was overwhelming. Now the question is do we need to make of citizens smart to make a city smart or is it other way round, a smart city shall make the citizens smart? It is felt that the former approach would yield constructive and sustainable results, where a citizen graduates to a netizen in a smart city. While Indian citizens are vocal and being a democratic set up each citizen have aspiration to be represented in the decision of the Muhalla (vernacular term which means a place where a community stays together), Ward, Zone, City, State or the Nation as the case may be. Smart City Mission guided ULBs to increase participation of citizens to all its functions citing them as City's Eyes and Ears. The National e-Governance plan of

P. Tripathy (✉)
Urban Planning, Civil Engineering Department, SVNIT, Surat, India
e-mail: partha.tripathy@gmail.com

K.A. Chauhan · A.K. Khambete
Civil Engineering Department, SVNIT, Surat, India
e-mail: kac@ced.svnit.ac.in

A.K. Khambete
e-mail: akk@ced.svnit.ac.in

T.M. Vinod Kumar (ed.), *E-Democracy for Smart Cities*, Advances in 21st Century Human Settlements, DOI 10.1007/978-981-10-4035-1_10

Government of India, states that the citizen engagement has various stages starting from inform, consult, involve, collaborate and empower. Under the mandatory reforms prescribed in earlier schemes under the aegis of Central government or the State government, Cities Authorities had undertaken certain e-governance and m-governance initiatives which were to the level of inform the citizens. After being tagged as Smart Cities, many Indian Cities have strived to enhance their citizen engagement activities but to what extent in the engagement ladder is something requires assessment. More importantly, does the system put in place by the Authorities for engagement, provisions for citizens participation making decisions democratic is what this paper has assessed for two Indian Cities, Bhubaneswar and Surat. It also suggests how these tools could be utilized further in engaging citizens from identification of issues, project conceptualization, budgeting, and implementation.

Keywords e-Democracy · Participatory project selection · Citizen centric approach and smart city

10.1 Introduction

India is a democratic republic and the principles of democracy flow from the preamble of the Indian Constitution itself. Democracy is said to be the Government of the people, by the people and for the people. This expresses that the Government is elected by the people, it is responsible and accountable to the people and it's also of the people. The most effective way to do that is to actively engage with the public while making policy decisions that impact them directly. However, the public participation in policy making is mostly through representative.

Governance has been a process and procedure centric and a top down approach in policy making. India has over 600,000 local governments and a very large population of around 1.25 billion coupled with its other complexities under "Unity in Diversity" i.e. cultures, ethnicities, religions and languages which has not made it easy for large scale public participation in policy decisions.

Post liberalization in 1990s, introduction of ICTs, internet technologies and Good Governance centric approach with increased participation of non-government entities in delivery of public services has made a paradigm shift in Governance with potential to connect every stake holder in real time. This was conceived as E-Governance which catered to citizen centric approach rather than process-centric approach through increased ICT enabled public participation. Government of India has taken the leap for the theme and has included them in schemes e.g. NeGP (National e-Governance Plan), JnNURM (Jawaharlal Nehru National Urban Renewal Mission), AMRUT (Atal Mission for Rejuvenation and Urban Transformation), by incentivizing the implementation of e-Governance by means of mandatory reforms at the Urban Local Bodies (ULB). The focus on public participation has increased in the country and various initiatives have been taken by central as well as local governments to engage people with ultimate goal of better service delivery, transparency and accountability. Recently launched Smart City Mission (SCM) by Government of India, mandated ULBs to increase participation of citizens in all its functions citing them as City's eyes and ears. The cities had to compete and the selection process, mandated the City Authorities to adopt a citizen participative approach. The cities which scored high in the challenge were those which had adopted citizens identifying the problems and democratic process for selection of solutions regarding the comprehensive development of physical, social, institutional and economic infrastructure to enhance the quality of life, attract people and investments to a City, setting growth & development in motion. The smart part in this was the role of ICT in the entire consultation process be it the social media, mobile telephony or the web based applications, citizens participation was overwhelming in such a short time.

This shows that public participation and engagement are the rising steps towards for good governance which generally Government have adopted as a policy decision. If in the same way the citizens can, not only participate and engage but also be empowered towards decision making, the democracy in real sense could be established. It can be a critical shift in delivery of public and essential services and to take policy decisions up to some extent using ICT based interfaces. Citizen can actually participate not only in the suggestions or feedback stage but from Project Selection in terms of setting up priority of the implementation of the project; Customization in terms of budget allocation, location selection, technology selection; accountability identification and implementation in terms of continuous monitoring and feedback.

Smart City Mission launched by GOI was a real step forward towards E-democracy. The mission was announced in June, 2015 and state governments nominated cities by 31st July, 2015. The final list of 98 cities to submit a smart city proposal (SCP) was announced on 27th August, 2015 and the last date to submit SCP was 15th December which meant around 90 days were provided to prepare the proposal. The major aspect of the selection criteria was that the extent of the engagement of people were given certain weightage. Extent up to which public opinion is considered in the preparation of the final proposal was crucial for any City to rise in the competition. The results announced clearly showed that the cities that garnered most participation by citizens. It represents a culmination of democracy in a right way.

The extent of participation each cities could manage in just the short duration of 90 days during the Smart City Challenge, was mainly due smart use of ICTs to reach out to people. Cities generally utilized the Social Media e.g. Facebook, Twitter, Mobile Apps., websites, GOI created MyGov.in website etc. Competitions were organized to get the results/proposal from within the people of the cities. Obviously the offline efforts such as face to face interviews, meetings etc. were also organized but to achieve this level of participation via conventional method was just not possible. These mediums were used to identify the key issues and drivers for the cities and proposal were also invited which were thoroughly examined by the authorities and experts.

The success of E-Democracy in Smart City Mission warrants for further identification of avenues in the existing democratic setup of ULBs to improve upon. If smart city projects could be identified via public participation, can this be ventured in decision making process of ULBs? Do we need to make of citizens smart to make a city smart or is it other way round, a smart city shall make the citizens smart? It is felt that the former approach would yield better and constructive and sustainable results, where a citizen graduates to a netizen in a smart city. If anything, SCM has profoundly shown that the citizen are and want to be informed and if given an acceptable platform to engage, the current system can be improved and can be made more democratic. Unlike to the conventional way the E-Democratic set-up is quicker, more easily accessible and is not bound by time.

Various Cities have taken up the opportunity to involve the citizen using ICTs via the social media, mobile telephony or the web based applications under the SCM. e.g. Bhubaneswar Smart City has My City My Pride application, Surat has Citizen Connect SMC application, Hyderabad has My GHMC (Greater Hyderabad Municipal Corporation) application, Pune has Pune Connect App etc. This study will focus on up to which extent the engagement has risen in the existing scenario in two major city apps i.e. Bhubaneswar and Surat ULB apps and their services. The services will be compared against spectrum of citizen engagement.

To achieve the smart citizen based E-Democratic setup, creating awareness, training of and continuous engagement with both the service provider as well as the service seeker to use new tools for better service and more effective interaction with each other; a framework is proposed to identify the extent up to which citizens can participate and be empowered, objectives of the services, opportunities, techniques and outcomes to realize the true sense of democracy using ICTs.

10.2 Existing Framework for E-Governance

According to framework for citizen engagement in E-Governance, Citizen Engagement is an interactive two way process that encourages participation, exchange of ideas and flow of conversation. It reflects willingness on part of government to share information and make citizens a partner in decision making [1].

In an ideal scenario, citizen engagement requires governments to:

- Permit participation in agenda-setting, and
- Ensure that policy or project proposals that are generated as a result of this engagement are taken into account while making a final decision.

Citizen engagement may be undertaken at all stages of the policy or project development process and is an iterative process that continually infuses citizens' priorities in policy making/project implementation.

In processes of citizen engagement, citizens may be represented themselves as individuals and sometimes through interest groups such as civil society organizations. Engagement has been understood and explained in a variety of ways. Engagement can be understood by following three aspects individually or collaboratively.

- Contributor
- Organization builder
- Empowering process

However, mostly, citizens are often considered either as beneficiaries of government welfare programmes or in Commercial terminology referred to as customers, neither of which truly reflects the government-citizen relationship.

Active engagement gives the right to hold others accountable, and accountability is the process of engaging in participation. It seeks greater accountability from the service providers through increased dialogue, consultation and by monitoring and assessing performance externally and mutually.

Citizen engagement goes beyond conventional public consultation by enabling citizens to do more than simply voice an opinion—it also allows them to participate in the deliberation process leading to decisions.

10.2.1 Ways of Citizen Engagement

- Information sharing
- Consultation
- Joint assessment
- Shared decision-making and collaboration

10.2.2 Core Values for the Practice of Citizen Engagement

It is important to note that if care is not taken while identifying the groups and mechanisms for interaction, citizen involvement can lead to a sense of disempowerment and a reduced sense of agency, and participation can be perceived as meaningless, tokenistic or manipulated. Therefore, following must be kept in mind while undertaking citizen engagement [1]:

- Public participation is based on the belief that those who are affected by a decision have a right to be involved in the decision-making process.
- Public participation includes the promise that the public's contribution will influence the decision.
- Public participation promotes sustainable decisions by recognizing and communicating the needs and interests of all participants, including decision makers.

- Public participation seeks out and facilitates the involvement of those potentially affected by or interested in a decision.
- Public participation seeks input from participants in designing how they participate.
- Public participation provides participants with the information they need to participate in a meaningful way.
- Public participation communicates to participants how their input affected the decision.

10.2.3 Challenges in Citizen Engagement

- Limited Trust in Government
- Political Reluctance
- Limited capacity to engage
- Lack of Commitment
- Exclusion

10.2.4 Citizen Engagement Framework

In order to ensure a meaningful engagement with citizens, all interactions must be undertaken in a well thought out and planned manner, wherein all stakeholders must be able to voice their inputs/concerns, due consideration to all must be given and a proper feedback mechanism must be put in place to inform all those participated about the decisions and the reasons thereof (Fig. 10.1).

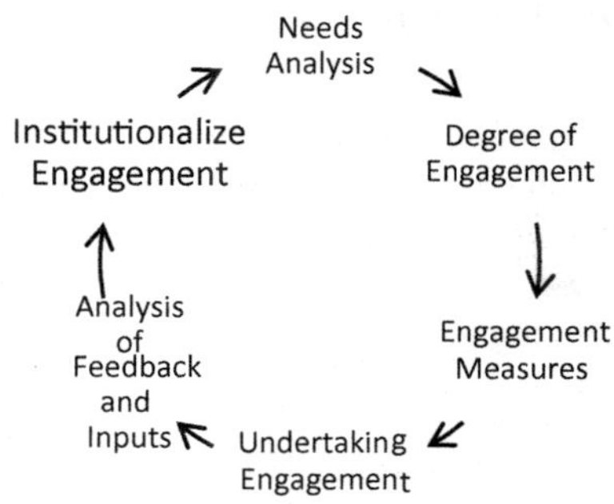

Fig. 10.1 Framework of citizen engagement

- **Needs Analysis**

It is easy to identify need for engagement for a new project. However, for ongoing projects, it is often difficult to determine such points of interventions. However, opportunities to engage with citizens exist at all stages of the project. For the purposes of this Framework, a project life cycle has been broadly demarcated into 4 stages namely Conceptualization and DPR preparation, Pilot and Roll Out, Post Implementation and Project Enhancement.

- **Degree of Engagement**

In an ideal scenario, the citizens may collaborate from conceptualization to implementation of the project and may even be empowered to reject or alter the project design at a later stage of the project. The spectrum of citizen engagement can be represented visually as under (Fig. 10.2).

In Present Scenario. The extent of Engagement is limited up to involvement of citizen only, which is good but to achieve E-Democracy whole spectrum needs to be followed. Citizen should actually be able to make decision in a democratic way for all stages.

- **Engagement Measures**

To effectively engage citizen a framework should be followed in which required measures are to be taken for Active Engagement. Those measures can be universal or project to project basis. The major aspects can be identified as (Fig. 10.3)

- Identifying the internal and external stake holders
- Developing a detailed project information
- Publicizing the information through various media including ICT
- Designing benchmarks and criteria for evaluation
- Selecting tools for citizen participation
- Reporting the outcomes of the process
- Making recommendations based on the outcomes

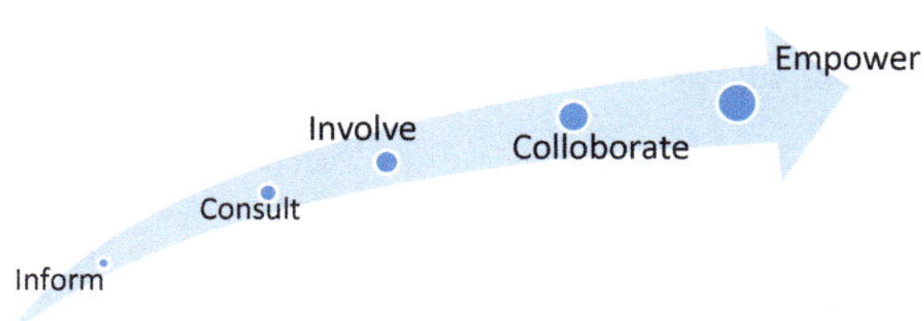

Fig. 10.2 Spectrum of citizen engagement

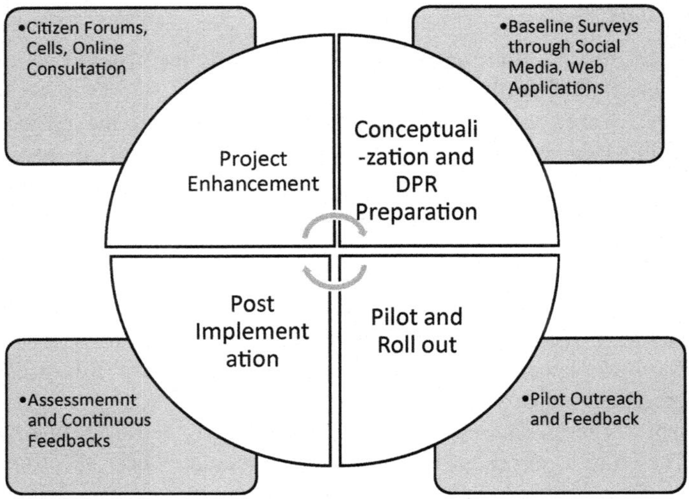

Fig. 10.3 Citizen engagement methods vis-a-vis project stage

10.3 My City My Pride (Bhubaneswar Smart City) Application

Bhubaneswar Municipal Corporation has taken a positive step towards citizen engagement and created "My City My Pride App" a simple mobile app. It's a platform to empower the citizens to bring about transformation, mobilize the public, and connect with the government. This solution helps citizens to post complaints along with images of the issues and get desired results within stipulated time [2].

10.3.1 Functionality of MCMP

MCMP allows citizens to register certain complaints relating to civic amenities such as

- sanitation,
- street lighting
- drinking water, and
- sewerage

To take corrective actions for the same by the concerned officials while enabling the aggrieved parties to track their grievances and complaints (Fig. 10.4).

Citizens/Netizens can report a complaint, the moment they see one through any of these three modes: mobile app, web portal or through the call center. Once

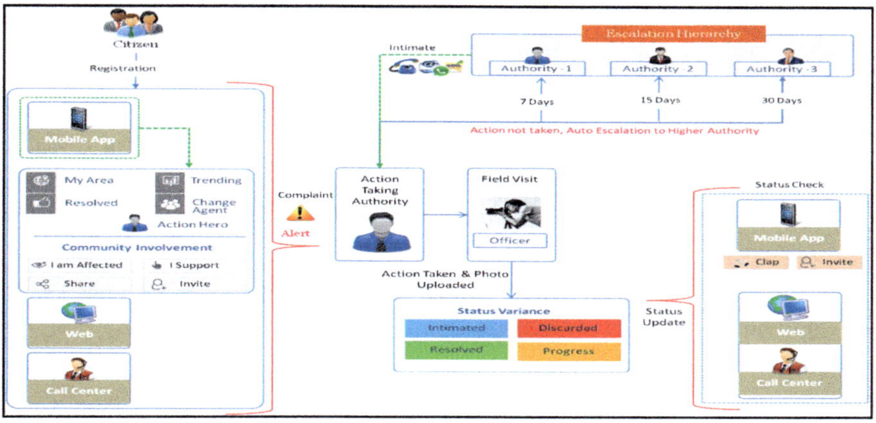

Fig. 10.4 Flowchart of MCMP app

citizens register complaints they turn into the Change Agents. Their complaint is then escalated to the right officials who will work on fixing the issue.

MCMP is powered by m-office, a mobile app powered with a backend portal. M-Office is used by the officials who will take action. They are called Action Heroes. MCMP acts as a platform for the citizens to interact with the officials, whereas m-office allows the admin to set permission, map hierarchy, configure location and perform all administrative activities. Then the complaint is forwarded to the Action Taking Authorities with specific escalation procedures for ensuring appropriate and quick remedial actions. Then the complainant is issued with a unique complaint ID to help the party track the complaint. Once the complaint is resolved, the status is uploaded immediately as Intimated/Resolved/Discarded/In Progress via any of the media. And they start to become the "Action Heroes". MCMP escalates unresolved complaints to the next higher official in the hierarchy [2].

10.3.2 Components of MCMP

- **Call on a toll-free number (1800-345-0061)**
- **Register on the web portal** (www.bmcsampark.in/ www.mycitymypride.org)

Citizens can also directly register their complaints through the web portal i.e. www.bmcsampark.in/ www.mycitymypride.org. Here citizens can login in three different ways:

- Facebook
- Scan from your My City My Pride mobile app
- Register through the user log in id.

The functionalities of MCMP are:

- **Share**: Citizen can post photos, experiences, comment and participate in trending talks, discuss and support issues on just about anything in neighborhood.
- **Affected**: Network for civic action- mobilizes local civic community, connect with like-minded citizens, get to know the government and engage with them.
- **Clap**: This feature shows the appreciation for the participation and success of any efforts via MCMP.

For each participation effort citizen receives points and periodically they can be Champion of Change.

10.3.3 Modules

MCMP is the total solution to all civic issues which is achieved by structured and modular components.

Trending Screen—This screen shows all the complaints along with the images and status posted by various citizen users in or outside the city. This page not only allows viewing complaints but also allows you to share, support or invite others to view, support, or vote your complaints (Fig. 10.5).

- Post a complaint—Citizens can here post a complaint by choosing the complaint type and upload the image to show the issue.
- My area—This screen allows you to see the complaints of your respective areas on providing the user ward no and can choose I'm Affected, I Support or share.
- Resolved Screen—This screen shows the details of the complaints that are already resolved by the department.

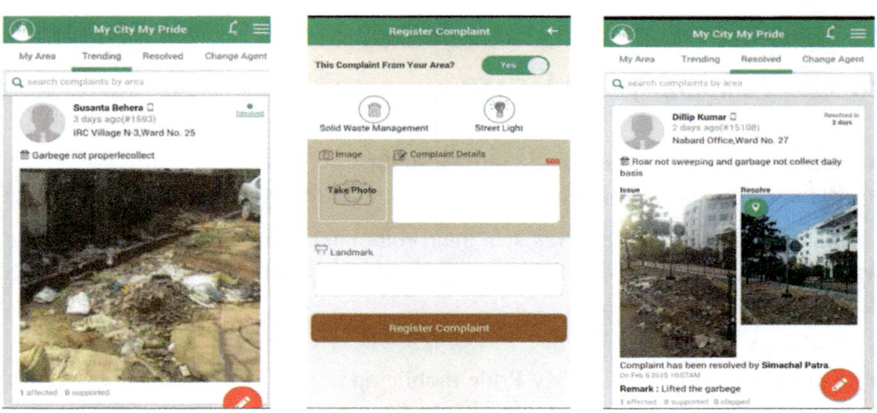

Fig. 10.5 Various screens of MCMP App—1

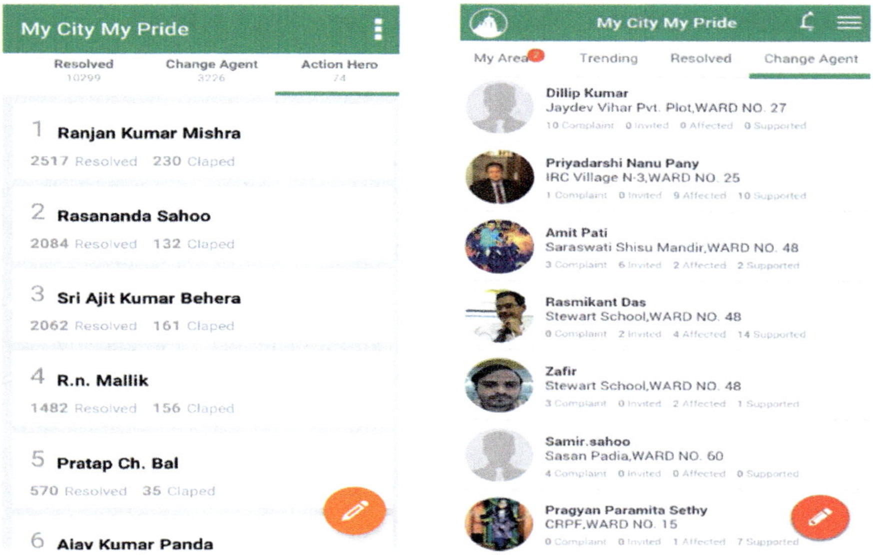

Fig. 10.6 Various screens of MCMP App—2

- Change agent—This screen is all about the citizens belonging to different wards and areas depending on the number of points earned. This change agent option encourages citizen to lodge their complaints by posting photos and the corporation will try to fix the problems within a fixed time period and make a safer smarter city for today's citizenry (Fig. 10.6).

MCMP app has garnered a lot of advantages by being a multi-faceted citizen touch points for greater engagement and experience to provide timely redressal of complaints and pro-active information dissemination and has helped to build transparent and pro-citizen image of urban agencies.

It empowers citizens to be sensors in the public space and allows government and other ULB's to use the mobile tools to track down the issues in the field. The digitized approach of MCMP to problem addressing and resolving, is fool-proof, error-free, effective and quick. The MCMP ICT to bring citizen and ULBs closer together and make neighborhoods better places to live and work.

MCMP app is a step in right direction towards a full fledge E-democracy which actually includes the whole spectrum of Citizen Engagement and it can evolve by increasing the extent of participation for project selection and budget allotment.

10.4 Citizen Connect (Surat Municipal Corporation) Application

Surat Municipal Corporation had created Citizen Connect App before the launch of the Smart City Mission, hence it focuses more on the Governance part rather than citizen engagement. The objective of the development of the app was to make the most sought after information and services available to the citizens of Surat through the Citizen's Connect—SMC Mobile App. It was thought to develop and offer a comprehensive mobile app that enabled availing of information and services anytime and anywhere. To create awareness regarding the service, various communication channels such as, print and electronic media including radio and TV were used. Social Media like Facebook, Twitter were also used to reach out the citizens.

10.4.1 Functionalities of Citizen Connect App

- To check Property Tax details with outstanding amount information.
- To check Professional Tax (EC) details with outstanding amount information.
- To check Water Meter connection details with outstanding amount information.
- To check Birth Registration details.
- To check Death Registration details.
- To check Shops and Establishment Registration Certificate details.
- Register Complaints.
- To share Feedback regarding services.
- Rainfall information (details of dam level, discharge, etc. during monsoon).
- Location tracking using GPS technology which can find nearest zone or city civic center to access services (Fig. 10.7).

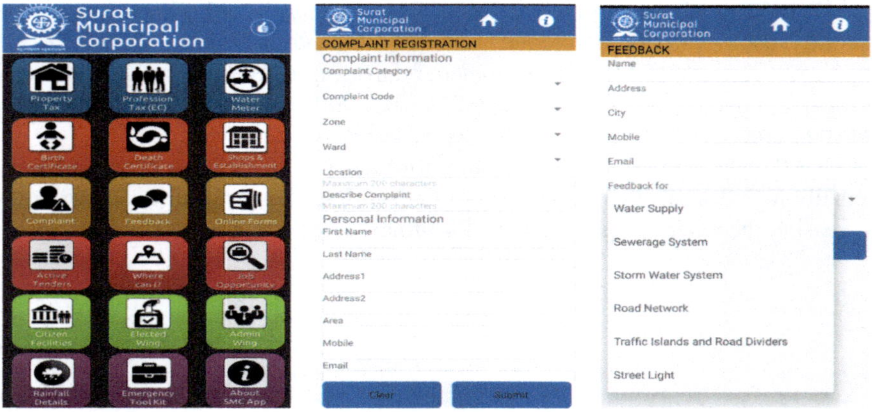

Fig. 10.7 Various screens of citizen's connect: SMC App

Though the application provides interaction opportunities, the major function is to provide services by E-governance or M-governance mode. The extent of engagement is only up to inform and consult. There has been a good response to Citizens Connect SMC Mobile App. Citizens can now obtain various services which were earlier rendered only through physical Civic Centre during working hours. These services are now accessible at citizens home/office located anywhere within the city, outside city from anywhere in country and out of country at their convenient time. The SMC Mobile App has offered an alternate channel of service which is very convenient and completely free having no additional cost. But to evolve the E-governance to E-Democracy the whole spectrum of citizen engagement is required to be implemented [7].

10.5 Comparison of Both the Initiatives

See Table 10.1.

Table 10.1 Extent of MCMP and citizen connect on the spectrum of citizen engagement

Sr. No.	Purpose (on spectrum)	MCMP, Bhubaneswar	Citizen Connect, Surat
1	Inform	Informs about the issues in the neighbourhood, no functions to broadcast information, news regarding city	Informative app, detailed information regularly updated regarding services, weather, water level in dams, taxes etc.
2	Consult	Facilitates discussions on specific issues	No function to consult, only information exchange
3	Involve	Citizen can post issues and give feedback for services and reach out to accountable entity in the system, a transparent system	Citizen can post issues and provide feedback regarding various services. No interaction via app. No update. Further process is offline and requires continuous efforts. No accountability identified via app
4	Collaborate	Citizen can up-vote the issue to identify the severity of the issue. Communities/Focus groups can be formed	No function to collaborate
5	Empower	Citizen can identify the accountable entity for the issue. Hence they can directly reach out for solution. The more an issue is supported, priority is given to the issue. Citizen can actually prioritize the action sequence with different issues by actively participating	No Function to empower citizen through application

10.6 Initiatives by Surat Municipal Corporation for Citizen Engagement

After the Smart City Challenge in December 2015, Surat Municipal Corporation, took up an initiative for Citizen's engagement for opinion *Beautification of Surat City* through web based application during April 2016. The poll was online for a month, was also hosted in the www.mygov.in website where the Smart city proposals were hosted. However, the participation level were recorded abysmally low which is less than 0.05% (294 opinions) of the total households in Surat. This poses a question, while the participation level in the Smart City Challenge was substantially higher, was may be due to adequate marketing & promotion done by the authority and there was a component of competition with other cities. The Smart City Proposal was also prepared in consultation with many stakeholders and promoted through Facebook, Twitter and Whatsapp groups and word of mouth.

The low level of participation could be the platform of engagement, citizens are more hooked to mobile internet than computer based internet, secondly, Surat being a substantially populated by migrant from other states, language barrier may require attention while designing the applications and most importantly, as this has not been a routine practice, adequate promotion and marketing to reach out to citizens to participate shall also be an factor of concern for all the cities in the initial years.

Has the citizen participation up to adequate level prior to accepting and implementing a Policy been a conscious Operating Procedure needs to be understood and internalized by the city authorities? Recently, Surat Municipal Corporation is preparing Parking Policy for the Surat City through a consulting firm. However, if we look at the terms of reference of the assignment, stakeholder consultation has been considered to be one of the routine consulting done to assess the issues. Such consultation target interviews of academic institution, government functionaries, & business associations and conduct Stakeholder workshop of selected groups to identify issues. There has been neither one to one consultation proposed during preparation stage or nor post preparation of draft parking policy.

10.7 e-Democracy Framework—Project Selection to Commissioning

See Fig. 10.8.

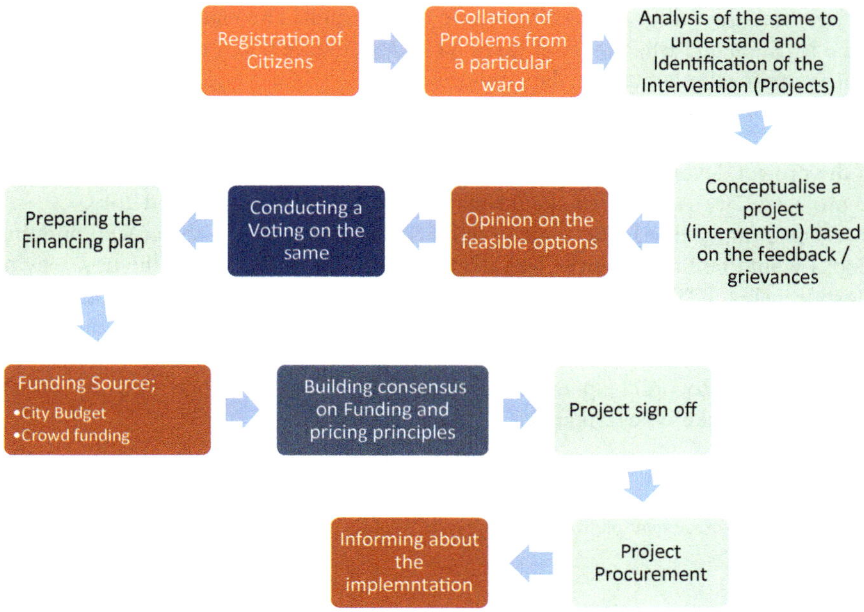

Fig. 10.8 E-democracy framework—project selection to commissioning

10.8 Conclusion

Different tiers of government obviously have different responsibilities and consequently different relationships with citizens. National governments, responsible for the welfare of a whole country, are unlikely to understand the needs of communities in the same way as local government. Therefore many e-democracy researchers see the local government level as the most natural place for government to engage with citizens.

Democracy in the context of Project Selection or Policy making when the projects or policies as the case may be are conceptualized and are implemented, which are meant for the citizens ought to ensure the factor that these projects are "by" & "of" the citizens indicating proper consultation in the process. Most of the projects those are implemented without consultation, have failed due to many reasons one of them being ownership of Citizens for whom it has been built. With this perspective, while cities have adopted e governance and m governance to reach out to citizens for sharing information and receiving complaints, the systems could be made more comprehensive to transform these problems to create public welfare projects. Such an engagement would also educate citizens about the likely investment, source of funding, likely cost and pricing mechanism. This consensus building would help in decision making by the Authorities and also more importantly bring in social cohesion and ownership by the citizens. Success rate of such project from the utility perspective would also be high. The analysis of Apps put in place by Surat and Bhubaneswar gives an understanding about the degree of citizen centric approach adopted by the Authorities.

City Authorities need to take an informed decision to include, citizen participation as a Standard Operating Procedure for any developmental work in the cities. In the project development activities right at the beginning of preparation of policy or projects, City Authorities should make citizen participation mandatory.

When country like India strive to build Smart Cities, one of the crucial building blocks would be Smart People who adapt to the change and accept the new way of lifestyle using ICT, be it travelling in metros with cash cards, be it undertaking online classes, or be it surviving in a cashless economy, its people who would make it a success. Exposure of citizens to such a transformed set up shall require City Administrators to build up engage and expose citizens to such systems. Smart People shall make cities Smart.

Acknowledgements The Authors acknowledge the valuable efforts of Mr. Harsh Desai, Research Student, Urban Planning, Civil Engineering Department, SVNIT, Surat, Gujarat, India for his analysis of the engagement platforms of cities.

References

1. Concept Note on My City My Pride (MCMP) (2016) Concept note on my city my pride (MCMP). CSM Technologies, Bhubaneswar, Oodisha, India
2. Department of Electronics & Information Technology, M. o (2012) A framework for citizen engagement in e-governance. Government of India, New Delhi
3. Peixoto T (2008) e-participatory budgeting: e-democracy from theory to success?. World Bank—Governance Global Practice, Washington
4. Tero Päivärinta ØS (2006) Models of e-democracy. Commun Assoc Inf Syst 17:818–839 (Article 37)

Websites

5. https://bmc.gov.in/
6. https://www.suratmunicipal.gov.in/
7. https://socialapphub.com
8. https://negp.gov.in/
9. https://nextcity.org
10. http://wikipedia.org

Chapter 11
Smart Grid: Energy Backbone of Smart City and e-Democracy

Jignesh G. Bhatt and Omkar K. Jani

Abstract In broad context, e-Governance is a functional subsystem under e-Democracy, since Governance and hence, entire Government itself is an important section of democracy itself. Hence, smart city development is in fact, the transformation of approach in their futuristic contexts towards responsibilities, interests and rights of all its stakeholders; such as citizens, government, administration, services, utilities, etc. Globally, consistent focus for financial growth and quality of life has ultimately resulted into unprecedented rise in consumption of electrical energy. Electrical power management in India has become a critical issue with rising population, increasing life expectancy, economic growth and, more importantly, due to vibrantly changing and difficult to predict weather. Electrical power supply has always been a resource in deep scarcity in India, so 24 * 7 electricity is an issue causing concern. Efficient decision-making at different levels rely upon reliable availability of electricity, which has been an integral element of citizens' life, therefore stable and sustainable power management, especially supply reliability is a need of today. Government of India has launched a Smart City Mission for building 100 Smart Cities, in which 24 × 7 availability of electricity has been one of the major focus. Historically, electricity generation, transmission and distribution decisions have been the exclusive domain of utility companies, domain experts, bureaucrats or sometimes political interventions, but now a paradigm shift could be observed towards citizens' participation and active involvement. Sufficient electricity being the mandatory element of a smart city necessitates an inclusive and participatory management of this resource. Smart electricity management in upcoming smart cities encounters challenges in the urban electricity

J.G. Bhatt (✉)
Department of Instrumentation and Control Engineering, Faculty of Technology, Dharmsinh Desai University, Nadiad 387 001, Gujarat, India
e-mail: jigneshgbhatt@gmail.com

O.K. Jani
Solar Energy Division, Gujarat Energy Research and Management Institute (GERMI), Pandit Deendayal Petroleum University (PDPU) Campus, Gandhinagar 382 007, Gujarat, India
e-mail: omkar.j@germi.res.in

T.M. Vinod Kumar (ed.), *E-Democracy for Smart Cities*, Advances in 21st Century Human Settlements, DOI 10.1007/978-981-10-4035-1_11

management through smart grid via integration of ICT and e-Democracy. Smart Grid as an energy backbone of smart city is immensely vital and serving at the core of Smart City realization. Evolving e-Democracy, smart grid includes highly interactive participation of citizens in energy consumption domain, based on humanitarian and customer centric approach. Different types of prosumers (producers + consumers), their different energy requirements at different timings, different types of energy resources and their switching feasibilities considering different aspects have been integrated. Under National Smart Grid Mission, Government of India has launched Restructured Accelerated Power Development and Reforms Programme for funding Smart Grid initiatives and Gujarat has been one of the leading states of India as far as self-reliance and reforms like solar city, smart grid, etc. are concerned. Renewable Power Plants and Smart Grid Pilot Projects implemented in Gujarat have been proved models and case studies for other locations. Moving ahead from our earlier contributed chapter "E-Governance for Photovoltaic Powergrid: Solar City Gandhinagar, Gujarat, India", in the book *E-Governance of Smart Cities*, in this chapter, we now present citizen-centric approach of design-implementation of smart grid with case study of pilot project at Naroda, Ahmedabad.

Keywords e-Democracy · Electricity · Smart city · Smart grid · Smart electricity management · Collaborative planning · Electricity scarcity · Information and communication technologies (ICT) · Internet of things (IoT)

Abbreviations

2G	2nd generation
3G	3rd generation
4G	4th generation
AMI	Advanced metering infrastructure
BAS	Building automation system
CT	Communication technology
DDU	Dharmsinh Desai University, Nadiad, Gujarat, India
DISCOM/DisCom	DIStribution COMpany/Distribution Company
DR	Demand response
EA 03	Electricity Act 2003
GEB	Gujarat Electricity Board
GEDA	Gujarat Energy Development Authority
GENCOM/GenCom	GENeration COMpany/Generation Company
GenCos	Generation Companies
GERC	Gujarat Electricity Regulatory Commission
GERMI	Gujarat Energy Research and Management Institute, Gandhinagar, Gujarat, India
GIS	Geographic information system
GoG	Government of Gujarat
GoI	Government of India

GPS	Global positioning system
GPRS	General packet radio service
GSM	Global system for mobile communication
GUI	Graphical user interface
HAN	Home area network/home automation network
HTLS	High temperature low sag
ICT	Information and communication technology
IOT/IoT	Internet Of Things/Internet of Things
IPDS	Integrated power development scheme
ISGF	India Smart Grid Forum
ISGTF	India Smart Grid Task Force
IT	Information technology
JNNSM	Jawaharlal Nehru National Solar Mission
LTE	Long-term evolution
MNRE	Ministry of New and Renewable Energy, Government of India
MoP	Ministry of Power, Government of India
NLDC	National Load Dispatch Center
NSGM	National Smart Grid Mission
PGCIL	Power Grid Corporation of India Limited
PLC	Programmable logic controller
PLCC	Power line carrier communication
PMU	Phasor measurement unit
Prosumer	Producer + consumer
PV	Photo voltaic
PXIL	Power Exchange India Limited
RAPDRP/R-APDRP	Restructured accelerated power development and reforms programme
RE	Renewable energy
REMC	Renewable Energy Monitoring Center
RET	Renewable energy technology
RLDC	Regional Load Dispatch Center
SCADA	Supervisory control and data acquisition
SG	Smart grid
SGCT	Smart Grid Communication Technology
SGKC	Smart Grid Knowledge Center
SoS	System of systems
SLDC	State Load Dispatch Centre
SLPMU	State level project monitoring unit
UGVCL	Uttar Gujarat Vij Company Limited
UI	Unscheduled interchange
VAR	Volt-ampere reactive
WAMS	Wide area monitoring system
WBAS	Wireless building automation system

11.1 Introduction [1–5]

One of the major and most significant changes observed globally in world development is 'Urbanization'. Adil and Ko [1] have indicated that the growth of Decentralized Energy Systems (DES) have been directing towards a new frontier in urban energy planning and design of local energy systems. As affordability of RET, particularly solar PVs and solar thermals, have been rising, cities and urban regions are transforming into venues, not only for energy consumption, but also for energy generation and distribution. This is clearly establishing a need for systemic and paradigmatic change in local energy infrastructure.

Focusing upon the importance of high quality, reliable, efficient and uninterrupted electrical power and its importance for energizing core services of smart cities, Al-Ali [2] have outlined the need for IoT and cloud computing for interconnecting and synchronizing basic application services of smart cities which have been now transforming from conceptual models to developmental stages. While observing growing developments in AMI, Arasteh et al. [3] have noted that smart cities have now been getting equipped with IoT based electronic devices in order to increase their smartness further.

Bansal et al. [4] have estimated that approximately 500 million people would have been living in cities by 2030, which is around 60% of the world's population. Cities contribute to economic growth by turnover of financial resources, workforce, etc. and have potential to provide higher opportunities of wealth creation, improved health-medical supports, quality education and much better standard of living with enhanced infrastructure facilities. High-level integration of existing technologies to deliver a smart energy network, enhanced electricity transmission, energy efficient transportation, and low carbon building footprints, will make it easier to manage the unfolding urbanization, and could have much positive impact on energy use and consumption. Policy interventions and government investments are important determining tools to its success.

Bhatt et al. [5] presented descriptions on "E-Governance for Photovoltaic Powergrid: Solar City Gandhinagar, Gujarat, India", in the book *E-Governance of Smart Cities*. In this chapter, the authors presented technological development and e-Governance aspects of solar PV based electrical powergrid in the upcoming smart city-solar city Gandhinagar, Gujarat state, India.

Proceeding further from earlier contribution, in this book chapter, the authors have presented interesting descriptions on 'Smart Grid'—the technological development serving as energy backbone of the upcoming smart city. Beginning with introductory details including historical perspective and present situation, need analysis has been included. Subsequent section describes the relationship among electricity, urbanization and human settlement with useful literature cited. In the next section, by including conceptual details and salient features and merits of smart grid, smart grid development in India has been discussed; wherein various government initiatives and other information have also been included. In the same section, smart grid developments in state of Gujarat have been discussed along with

UGVCL's approach (as methodology) and case study of UGVCL's smart grid pilot project at Naroda, Ahmedabad has been included. Next section is based on e-Democracy and citizens' participation consisting subsections like internet based e-tools and mobile based M-tools, followed by separate sections on experiences and lessons learnt. Various relevant issues and challenges have been identified and mentioned in the subsequent section. Finally, sections like vision and roadmap, technological inputs and useful conclusions have been included to mark end of the chapter with acknowledgement and list of cited references.

11.1.1 Historical Perspective [6]

In India, electricity reforms started with the re-evaluation of Electricity Supply Act, 1948 and the Indian Electricity Act, 1910, which led to the Electricity Act, 2003. The EA 2003 has been brought about to facilitate private sector participation and to help cash strapped state electricity boards to meet electricity demand. EA 2003 envisages competition in electricity market, protection of consumer interests and provision of power for all. The act recommends the provision for national electricity policy, rural electrification, open access in transmission, phased open access in distribution, mandatory SERCs, license free generation and distribution, power trading, mandatory metering, and stringent penalties for theft of electricity. One more welcome step that the Indian electricity market has taken is the implementation of ABT, which brought about the effective day-ahead scheduling and frequency sensitive charges for the deviation from the schedule for efficient real-time balancing and grid discipline.

To promote power trading in a free power market, CERC approved the setting up of Indian Energy Exchange (IEX) which is the first power exchange in India, in June 2008. IEX has been modeled based on the experience of one of the most successful international power exchanges, Nordpool. At present, two power exchanges are operating in India, namely, IEX and Power Exchange India Limited (PXIL). These exchanges have been developed as market based institutions for providing price discovery and price risk management to the electricity generators, distribution licensees, electricity traders, consumers and other stakeholders. The participation in the exchange operations is voluntary. At present, exchanges offer day-ahead operations whose time line is set in accordance with the operations of RLDCs. Power exchanges coordinate with the NLDC/RLDCs and SLDCs for scheduling of traded contracts' to get up-to-date network conditions. Currently, about 96% of the market transactions in India are in the form of bilateral (long and short terms) contracts. The rest is dealt by two power exchanges.

The Indian power sector has come a long way since the introductions of second wave of reforms vide the EA 03. This gave a major boost to the power sector by creating conducive environment for enhanced private sector participation, which resulted in a large increase in generation capacity as well as investments in T&D as well as the introduction of a power market and power trading activities. In addition to this, the focus was shifted on reforms aimed at providing affordable, good quality and reliable power to consumers, making the electricity industry commercially viable and promoting efficiency improvements.

11.1.1.1 Generation

Power Generation has gained the most due to the entry of private players. The magnitude of capacity being added each year has increased manifold when compared to previous planning periods. Also, with the use of new and more advanced technologies, efficiencies of thermal power plants have been improving and emission levels falling. Operational requirements related to scheduling and dispatches are driving the implementation of automation across the power system and for the generators. All new plants now have sophisticated operational IT systems and the existing generation fleet is slowly upgrading to match. RE based electricity generation has gained prominence over the years. Several fiscal and policy measures have been introduced to promote RE. On an average, over 3000 MW of RE installed capacity has been added every year with major contribution from the wind and solar energy segments. Solar energy is gaining momentum through the union government's JNNSM and similar state level policies. Given the economics of coal and gas, fuel security issues and environmental concerns that are being faced, generation from RE is increasingly assuming a central role in the entire power system.

11.1.1.2 Transmission

The transmission sector in India is moving towards higher voltage levels of 1200 kV and is introducing a higher level of automation and grid intelligence. PGCIL has already installed PMUs for WAMS on a pilot basis in select regions and is now pursuing a plan to install PMUs nationwide. Significant technological advancements such as increasing the capacity of transmission corridors through the use of Static VAR compensation and re-conductoring of lines using HTLS wires are also being taken up. Power system operation is also under evaluation as a result of the disturbance in July 2012 and it is expected that policy reform will lead to more system control being given to the load dispatch centers and the phase out of the current UI mechanism designed to discourage DisComs and GenCos from deviating from published schedules. The UI mechanism is expected to be replaced by an ancillary services market, which would be managed by the power exchanges, thus further liberalizing power markets and providing greater transparency on costs and prices of services. Managing these systems will require real-time monitoring and control only possible with robust state-of-the-art information and communication systems.

11.1.1.3 Distribution

The electricity distribution sector in India is currently in the worst shape, plagued by high network and financial losses in almost all states. There is an urgent need to bring in new technologies and systems to arrest these leaks. R-APDRP introduced by the GoI has been aimed at reducing the network losses to 15%. Part-A of the program is aimed at creating IT Infrastructure and automation systems within utility operations, which until its introduction was largely missing in most of the distribution utilities in the country. Part B is aimed at strengthening the physical network. The R-APDRP is still under implementation and completion is expected during the 12th Five Year Plan. Once completely implemented, the program would provide a strong foundation for evolution to smart grids in the power distribution segment.

For the distribution sector, smart grids will mean the introduction of DR programs, managing the expected introduction of electric vehicles and integrating distributed energy resources in a way that can help the DisComs balance local supply and demand and reduce peak time consumption. Building to Grid (B2G) or development of "Green Buildings" which could be incentivized to manage their consumption and even distributed energy resources to match grids conditions will also play their part in helping DisComs to manage supply and demand. For realizing all these changes, AMI as well as reliable communication technology infrastructure is essential.

11.1.2 Present Situation [7–9]

India is the 3rd largest producer of electricity in the world. Indian power system is today one of the largest synchronous grids in the world with approximately 250 GW capacity, >3 million km^2 area covered and serving ~240 million consumers. It is largely dominated by government owned utilities (central and states—29 states and 7 union territories own majority of power utilities); private sector role is about 27% in generation, <1% in transmission and about 5% in distribution. Transmission Grid in India is one of the largest in the world, comprising ~1,03,000 ckms lines of 765 kV/400 kV and ~132,000 ckms lines of 220 kV capacity, with 1200 kV AC and 800 kV HVDC networks being under development (Fig. 11.1).

India operates with very high T&D losses—about 26.5% nationally (>40% in many states!). ~400 million + people have no access to power. Large parts of the country experiences power cuts for several hours every day—customers keep storage (invertors)/standby generation facilities. Power quality being poor, consumers require voltage stabilizers, UPS, inverters, etc. The world loses ~$89.3 billion, while India loses ~US$ 10–15 billion annually in power theft.

Over population in cities always create problems for the authorities such as electricity shortage, supply breakdowns, etc. As per the studies, approximately

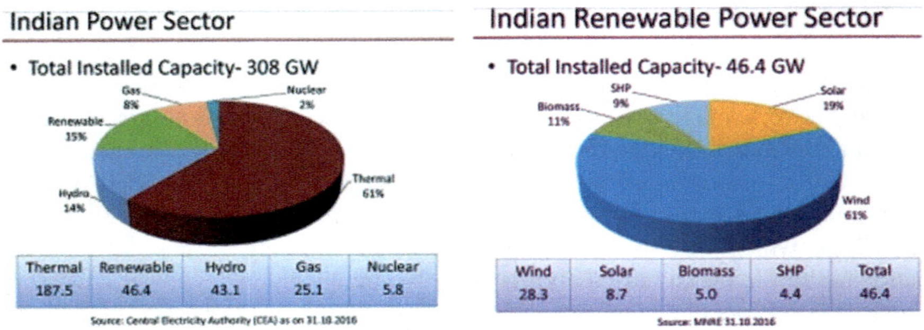

Fig. 11.1 Scenario of Indian power sector [8]

2.5 billion Indians have either unreliable or no access to electricity. By 2035, energy consumption will increase by 35% (World Energy Council 2010).

Not only availability of energy, but the challenge of converting from its existing form to useful form in the form of electricity needs to be properly tackled, if India wishes to turn her dream "power for all" in a reality. Majority of Indians still live in rural areas utilizing traditional ways for cooking and live below poverty line. Solar, wind and hydro power resources are available and applicable; however, suitable and efficient technologies are required to make such small installations to serve local striving due to insufficient and/or no power supply situations. While electrical power generation from renewables keep fluctuating due to various reasons including vibrant climatic situations, intermittent nature as well as difficulties to predict power production, the existing conventional electrical grid find it difficult to handle such variability due to unbalance and instability in electrical networks. Therefore, a new grid which shall resolve such renewable interconnection issues as well as ensure reliability, safety, security, profitability and public-private participation is required. Smart grid is emerging as a novel concept to provide insight to policies maker, grid operators and end users for sustainable energy efficiency.

11.1.3 Need Analysis [6, 10, 11]

The legacy conventional unidirectional power-only flowing electrical grid served well for over a century, now there are stringent requirements to upgrade, due to aging infrastructure as well as on account of upcoming challenges on technological, environmental and societal fronts. Therefore, not only Indian, but all national governments and relevant stakeholders have been planning substantial and significant developments for futuristic smart grids.

The urgency for Smart Grids in India emerges from the key challenges being currently faced. India operates the 3rd largest transmission and distribution network in the world, yet faces a number of challenges such as: inadequate access to electricity, supply shortfalls (peak and energy), huge network losses, poor quality and reliability and rampant, theft. The evolution towards Smart Grid would address these issues and transform the existing grid into a more efficient, safe, reliable and less constrained grid that would help provide access to electricity to all.

According NIST, the following are anticipated benefits and requirements of smart grids:

(i) Improving power reliability and quality.
(ii) Optimizing facility utilization and averting construction of back-up (peak load) power plants.
(iii) Enhancing capacity and efficiency of existing electric power networks.
(iv) Improving resilience to disruption.
(v) Enabling predictive maintenance and self-healing responses to system disturbances.

Smart Grid Drivers

(i) Increase in electricity demand and supply shortfall: India's increasing electricity demand especially during peak hours continues to outpace India's power supply. So, managing growth and ensuring supply is a major driver.
(ii) Loss reduction: In India, the AT&C losses are around 32% of installed generating capacity. Smart grid can make a substantial contribution in reducing these losses.
(iii) Increase in unit cost of electricity: With the increase in unit cost of electricity, there is a need for utilities to replace and renew aging transmission and distribution infrastructure with a pressure of using the assets wisely.
(iv) Managing human element: Human errors and deliberate errors can be lowered by using smart instruments like smart meters.
(v) Reliability: The smart grid can improve outage management performance by responding faster to repair equipment before it fails unexpectedly.
(vi) Efficiency: The smart grid can improve load factors and reduce system losses.
(vii) Renewable energy integration: In order to integrate renewable energy projects into the grid.
(viii) Grid improvement: Electricity demand in India is growing much faster than the transmission system in tremendous strain. The smart grid will improve the grid's resilience and robustness.
(ix) Technological advances: The advances in computing and telecommunications during the last half century have affected almost every facet of life. One reason the smart grid is taken seriously because advance computing and telecommunications made it possible.

11.2 Urbanization, Electricity and Human Settlement [1, 4, 12, 13]

Fistola and La Rocca [12] have outlined interesting relationship of technology and human settlement. The authors have stated that technology represents a basic element in the process of building up the urban artifact. Today the term "technology" is used to refer to ICT tools available. City and technology have very close relationship, since the first human action could have been made to resist to natural events or produce a human settlement. Technology has to be considered like a basic and fundamental factor inside the evolution process of the human beings, which can extend human senses in order to have a best perception of the environment and to develop capabilities to understand natural events, to carry on scientific development and to build common space for more people. Today, technology has a new acceleration, which can separate it from the human evolution process. The smart city has to consider the adoption of the ICT inside its process of development and not the addition of the technological equipment (like sensors) to the physical system of the city.

Joshi et al. [13] have described "smart city" as a futuristic approach to alleviate obstacles triggered by ever-increasing population and fast urbanization, to benefit the governments as well as the citizens. Smart cities are an endeavor to make cities more efficient, sustainable and livable by delivery of vital elements like quality of life and socio-economic development, etc. In other words, a smart city is a city that can monitor, integrate and control various functionalities of all critical infrastructures and can help in optimizing the resources maintaining security issues as well. The authors have identified six significant pillars for developing the framework as: Social, Management, Economic, Legal, Technology and Sustainability (SMELTS) to indicate how these factors can make the smart city initiative a successful project.

Bansal et al. [4] have mentioned that urbanization propelled by economic reforms is putting cities under perpetual pressure of population concentration and energy intensive growth model. The cities are often confronted with a multitude of key problems like high urban densities, traffic congestion, energy inadequacy, unplanned development and lack of basic services. Due to high land values, migrants often have no choice but to settle in shantytowns and slums, where they lack access to decent housing and sanitation, health care and education; thus adding to urban poverty. Urbanization is also contributing significantly to climate change as 20 largest cities consume 80% of the world's energy and urban areas generate 80% of greenhouse gas emissions worldwide. The challenges of rapid urbanization are to deal with the social, economic and environment development through more effective and comprehensive land administration functions, supported by efficient per capita infrastructure supply, resolving issues such as climate change, disaster management, insecurity, energy scarcity, environmental pollution, and extreme poverty. Urbanization must be able to support urban planning to achieve sustainable development in order to meet the growing energy and housing demands, reliable public transportation systems and be able to meet essential urban services without putting pressure on resources. Therefore, it needs to support innovative urban planning policies and strategies beyond traditional urban planning paradigms. Urbanization on the positive side provides an unparalleled urban planning opportunity to pre-address social and environmental problems, including reduction of greenhouse gas emissions combined with the retrofitting and upgrading of facilities and networks in existing urban centers, as well as smart urban planning of cities can provide better education, healthcare and high-quality energy services more efficiently and with less emissions because of their advantages of scale, proximity and lower geographic footprints. Thus "Smart Urbanization" is the key to safer cities of tomorrow. Building cities sustainably using smart growth principles, compact development planning form, using eco-city concepts, concept of low carbon electricity ecosystem etc., provides an opportunity to avoid future sources of greenhouse emissions, while developing more livable and efficient urban centers. It could also alleviate population pressure on natural habitats and biodiversity thus reducing the risks to natural disasters.

Bansal et al. [4] have mentioned that electricity is one the major energy consuming sectors and need to be stratified for smart urbanization. To make electricity efficient, larger-scale use of smart grids and superconductors are needed for transmission and distribution of electricity in dense urban settings. This would reduce their overall carbon footprints. Smart, information-rich energy network that uses superconductors for enhanced electricity transmission capacity and allows transportation needs to be met by multiple approaches not reliant on private vehicles. Widespread adoption of such technologies will make it easier to manage the unfolding urbanization, and could have much positive impact on energy use and consumption.

Adil and Ko [1] have suggested that centralized approaches to energy production, delivery and consumption constituted the original framework for provision of modern energy services. With the invention of high voltage AC power in late 19th century, electricity could to be transmitted over large distances through electric cables. Mass production, delivery and consumption of energy were expanded, stretched across the country through one nation-wide grid, designed as if to convey electricity perpetually to the energy-hungry masses of residential, commercial and industrial consumers. In addition, increasing resource conflicts tied to fossil fuels, depletion of the fossil fuel reserves, and anthropogenic climate change offer ample reasons for designing an alternate energy system that enables weaning off of carbon-intensive fossil fuels in meeting the world's current and future energy needs. The alternate approach is decentralized; it is predicated on recognition of the finiteness of fossil fuels and their carbon-intensive character, deteriorating energy infrastructure and growing demand for energy. DES suggests a paradigm shift in the way energy is produced, delivered and consumed. DES conceived on the basis of RET offer a clean and inherently resilient approach towards reaching sustainable development goals. There are four major advantages of such DES over centralized energy systems: (i) the ability to offer low to zero-carbon emissions, (ii) offset capital-intensive investments for network upgrades, (iii) impart local energy independence and network security, and (iv) motivate social capital and cohesion.

As RET becomes more affordable and penetrate the energy markets, cities become the place for celebrating these benefits of DES.

11.3 The Smart Grid [9, 10]

A smart grid is an electrical grid with automation, communication and IT systems that can monitor power flows from points of generation to points of consumption (sometimes even down to the appliances level) and control the power flow or curtail the load to match generation on real-time basis. It involves two-way communication among the generating units, the control centers of distribution utilities and the consumers. The smart grid enables increased, predictability and control of generation and demand through consumer involvement, thus bringing flexibility in both generation and consumption, enabling the utility to better integrate intermittent renewable generation and reducing costs of peak power. A smart grid is cost-effective, responsive, and engineered for reliability of operations.

Interesting comparison of basic distinctive features of conventional and smart grid can be referred in Table 11.1. It could be observed from this comparative analysis that unlike conventional unidirectional electricity-only flowing conventional grid, smart grid is high speed, more accurate-efficient and customer-oriented bidirectional electricity + information flowing grid, which has potential to serve need based customized requirements of variety of customers and with useful intelligent features.

Murthy Balijepalli et al. [9] have described smart grid as a concept for transforming an electric power grid by using advanced communications, automated controls and other forms of information technology. It integrates new innovative tools and technologies from generation, transmission and distribution all the way to consumer appliances and equipment. This concept, or vision, integrates energy infrastructure, processes, devices, information and markets into a coordinated and collaborative process that allows energy to be generated, distributed and consumed more effectively and efficiently, with focus on sustainable options for customers. Smart grid enables devices at all levels within the grid (from utility to customer) to

Table 11.1 Comparison: conventional grid versus smart grid [10]

Conventional grid	Smart grid
Electromechanical	Digital
One way communication	Two way communication
Centralized generation	Distributed generation
Few sensors	Sensors through out
Manual monitoring	Self monitoring
Manual restoration	Self healing
Failures and blackouts	Adaptive and islanding
Limited control	Pervasive control
Few customer choices	Many customer choices

independently sense, anticipate and respond to real-time conditions by accessing, sharing and acting on real-time information.

Referring to IEEE P2030, Bhatt et al. [10] have described smart grid as power, communications, and information technologies for an improved electric power infrastructure serving loads while providing for an ongoing evolution of end-use applications. IEEEP2030 also defines Smart Grid as "System of Systems" (SoS) as the smart grid is a complex system made up of interrelated systems.

As mentioned in Bhatt et al. [10], SG facilitates for active participation of consumers with timely access and control to their energy usage. Consumers can bid their energy resources at the electric market. SG supports real-time power quality monitoring and active diagnostics to respond power quality deficiencies and reduces loss to customers due to insufficient quality of power. SG possesses the capability to anticipate and respond to system disturbances by continuous self-assessment to take corrective action. Constantly changing customer choices-behavior, increased integration of renewables, varieties of DR programs, etc. all are likely to increase the fluctuations in the ratio of produced and consumed power. Hence, utilities must make strong efforts to deal with the increasing volatility and vibrancy in classification, affordability, feasibility and final choices of power production, demand, distribution and consumption. Important features, such as Real Time Pricing (RTP), require intensive monitoring of the consumers' power consumption patterns along with close real-time asset monitoring and timely provision of control actions. This necessitates data prioritization and delay-responsiveness using communication links with sufficient reliability, data rates and latency.

Futuristic conceptual architectural diagram of Fig. 11.2 provides interesting illustration of how a smart city would evolve with smart grid. It could be observed from the diagram that the smart grid provides level playing field to different types of

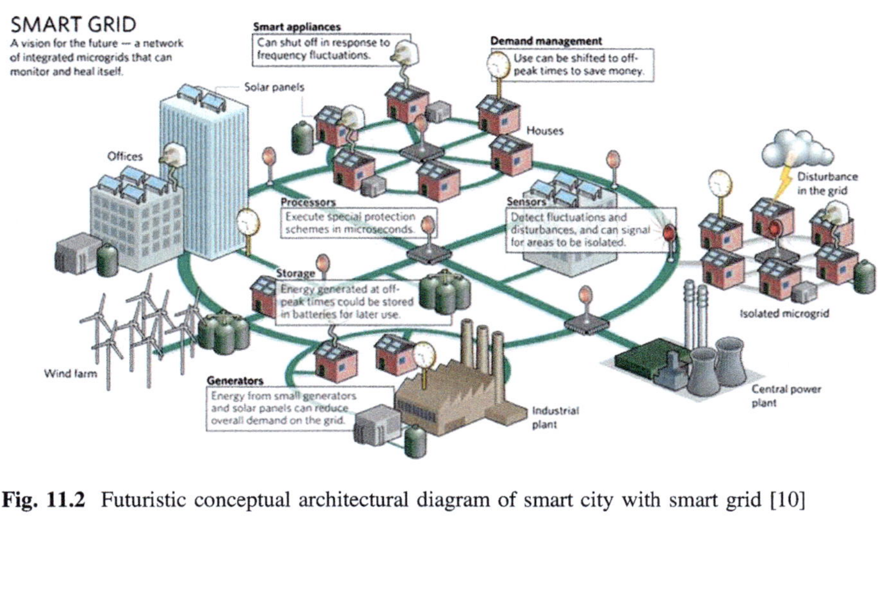

Fig. 11.2 Futuristic conceptual architectural diagram of smart city with smart grid [10]

energy sources as well as different types of prosumers, as in this novel setup, a consumer would also be a producer of electricity as well.

Figure 11.3 presents the SG architecture with associated systems and sub-systems. Overall SG architecture could be split into three major systems—smart infrastructure system, smart management system and smart protection system.

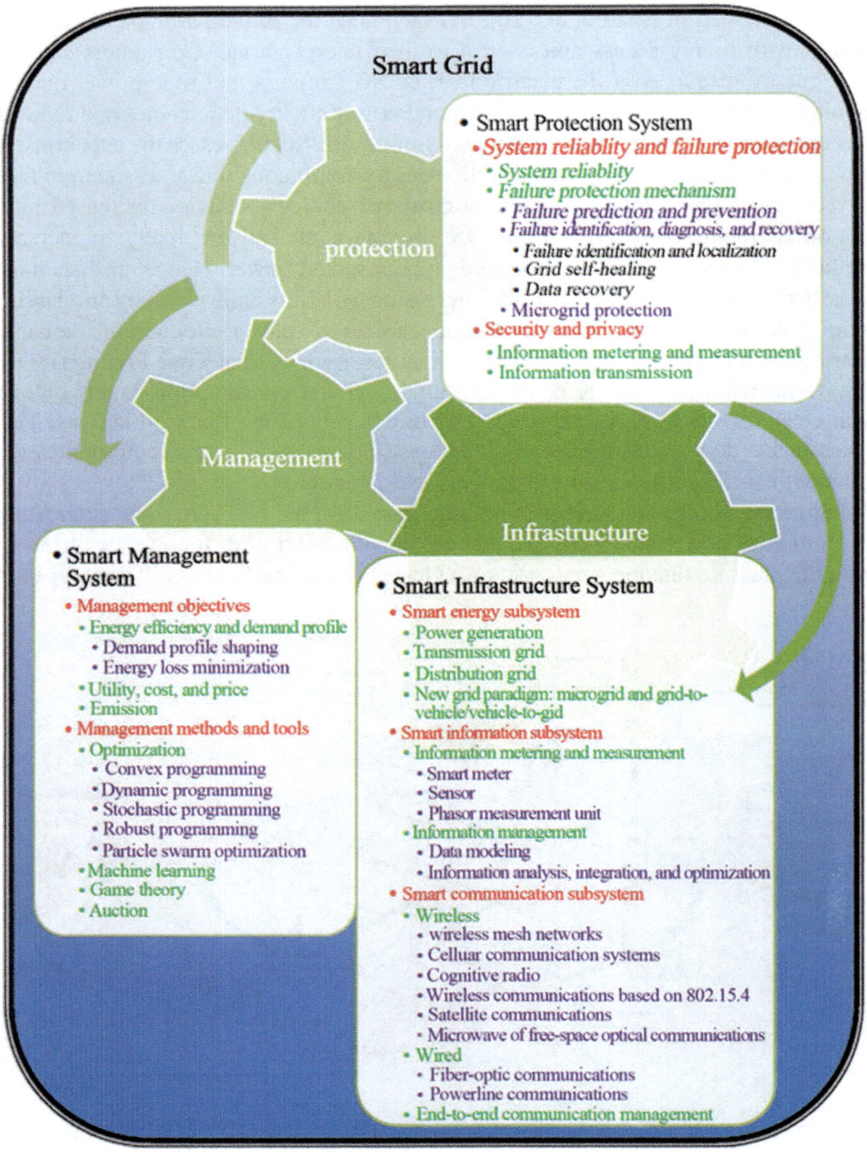

Fig. 11.3 Technical smart grid architecture with associated systems and sub-systems [10]

(a) *Smart infrastructure system* is the energy, information, and communication infrastructure underlying of the SG that supports (i) advanced electricity generation, delivery, and consumption, (ii) advanced information metering, monitoring, and management and (iii) advanced communication technologies. The smart infrastructure system has been further split into three subsystems: smart energy subsystem, smart information subsystem, and smart communication subsystem.
(b) *Smart management system* is the subsystem in SG that provides advanced management and control services. The smart management system has also been split further into two subsystems: management objectives, and management methods and tools.
(c) *Smart protection system* is the subsystem in SG that provides advanced grid reliability analysis, failure protection, and security and privacy protection services. The smart protection system has also been split further into two subsystems: system reliability and failure protection; and security and privacy.

11.3.1 Key Characteristics and Benefits [9, 11]

Key Characteristics:

(i) *Superior citizen involvement in e-Governance*:
Involves consumers by engaging them as active prosumer (producer + consumer) participants in the electricity market
Provides timely information and control options to consumers
Real time monitoring, automated outage management and faster restoration enables consumers to enjoy improved reliability
In-house displays, IoT based equipments, programmable control thermostats, portals and energy information tools like mobile apps enable consumers to track and manage their energy usage and identify opportunities to reduce and conserve electricity

(ii) *Improved power quality*:
Provides high quality of power and reduces the occurrence of distortions of power supply
Respond to local and system-wide inputs and have much more information about broader system problems

(iii) *Enable demand response*:
Dynamic pricing mechanisms incentivize consumers to alter their usage during different times of day based on pricing signals, enabling them to optimize their electricity bills through better energy management
Extends within home, so that consumer appliances, devices, etc. can be controlled remotely allowing for demand response

(iv) *Market empowerment*:
Provides greater transparency and availability of energy market information
Enables more efficient, automated management of market parameters
Provides level playing opportunities to new market products

(v) *Enhanced automation*:
Incorporate extensive measurements, rapid communication, centralized advanced diagnostics, and feedback control that quickly return the system to a stable state after interruptions or disturbances

(vi) *Clean and green*:
The energy conservation and improvements in end-use efficiency enabled by the smart grid reduce half of the emissions
Deploy and integrate distributed resources and generation including renewable resources
Facilitates distributed generation, especially the roof top solar generation, by allowing movement and measurement of energy in both directions

(vii) *Self-healing and resilient*:
Performs real time self-assessment to detect, analyze and respond to sub-normal grid conditions
Detects and addresses emerging problems on the system before they affect service

(viii) *Asset optimization and operational efficiency*:
Enables better asset utilization from generation to the consumer end points

(ix) *Integration of advance and low carbon technologies*:
Enables 'Plug and Play" scalable and interoperable capabilities
Permits higher transmission and distribution system penetration of renewable generation, distributed generation and energy storage
Integrates "smart" appliances (IoT based) and consumer devices like hybrid cars, electric vehicles, etc.

Benefits:

Benefits extended by smart grid to its various stakeholders have been included in Table 11.2. It could be observed that smart grid provides win-win situation for all the stakeholders without many tradeoffs. Overall financial implications could be reduced by generation of renewable energy by consumers and thereby reaching the final break-even point easily.

11.3.2 Smart Grid in India [14–19]

Indian power sector is expanding at a fast pace. India has a power sector characterized by deficient generation and high distribution losses. The first power exchange of India was introduced in June 2008. Smart grid activity has been reported in some distribution pockets mostly by private players. Many utilities in

Table 11.2 Benefits of smart grid [11]

Stakeholders	Benefits
Residential and small commercial consumers	(i) Improved system reliability (ii) Individual control over energy used and monthly bills (iii) Customers are transformed from passive ratepayers to active, engaged participants in electricity markets (iv) A more reliable grid will limit the risk of outages
Large industrial and commercial customers	(i) A smart grid will provide additional benefits from more detailed information and better reliability (ii) A smart grid will allow large customers to integrate their production, shortage and efficiency investments easily into wholesale market operations
State governments	(i) State governments can benefit from higher reliability and lower duration of outages (ii) Greater information and control over distribution system will also allow grid operators to assist with emergency situations, such as fires and storms, by turning off power selectively or by restoring power faster and more efficiently (iii) State governments are also consumers of electricity and can take advantage of consumer-related benefits of smart grids (iv) Environmental benefits
Utility grid operators	(i) Grid operators will benefit from direct cost reductions, enhanced system reliability, and higher customer satisfaction (ii) More efficient deployment of field staff as a result of better information on grid conditions (iii) Improvement in efficiency of billing, customer connections, and many other utility processes (iv) Reductions in working capital needs and bad debt expenses (v) Reduction in theft and energy losses (vi) Improved and more efficient customer service, more efficient planning and maintenance of the system, and more efficient use of back office resources

power sector have now smart grid activity with high priority in their research and development agenda.

The efforts for the development and deployment of smart grids in India are presently being carried out through India Smart Grid Task Force (ISGTF) and India Smart Grid Forum (ISGF) under the aegis of Ministry of Power (MoP). 14 smart grid pilot projects have been carried out by different state utilities.

11.3.2.1 Government Initiatives [11, 14–16]

(a) *Smart Cities Mission* [14]
Smart Cities Mission is an urban renewal and retrofitting program by the Government of India with a mission to develop 100 smart cities all over the country making them citizen friendly and sustainable. Ministry of Urban Development is responsible for implementing the mission in collaboration with the state governments of the respective cities. The government of India has a vision of developing 100 smart cities as satellite towns of larger cities and by modernizing the existing mid-sized cities. Smart cities are projected to be equipped with basic infrastructure to offer good quality of life through smart solutions. Assured water and power supply, sanitation and solid waste management, efficient urban mobility and public transport, robust IT connectivity, e-Governance and citizen participation along with safety-security of its citizens are some of the likely attributes of these smart cities.
Smart Cities Mission of the Government is a bold, new initiative. It is meant to set examples that can be replicated both within and outside the Smart City, catalyzing the creation of similar Smart Cities in various regions and parts of the country. The core infrastructure elements in a smart city would include:

(i) adequate water supply
(ii) assured electricity supply
(iii) sanitation, including solid waste management
(iv) efficient urban mobility and public transport
(v) affordable housing, especially for the poor
(vi) robust IT connectivity and digitalization
(vii) good governance, especially e-Governance and citizen participation
(viii) sustainable environment
(ix) safety and security of citizens, particularly women, children and the elderly
(x) health and education

Solar PV and Smart Grid serve as essential subsystems of any smart city in form of "Reliable Energy Backbone" to supply continuous power and seamless connectivity for all smart applications conceived under a smart city.

(b) *National Smart Grid Mission (NSGM)* [15]
Government has approved the National Smart Grid Mission (NSGM), an institutional mechanism for planning, monitoring and implementation of policies and programs related to Smart Grid activities. Major activities envisaged under NSGM are development of smart grid, development of micro grids, citizen engagements, training and capacity building, etc. NSGM entails implementation of a smart electrical grid based on state-of-the art technology in the fields of automation, communication and IT systems to monitor and control power flows from points of generation to points of consumption.

A smart Grid Vision and Roadmap for India was approved by the Ministry of Power in August 2013 which also envisaged the launch of National Smart Grid Mission (NSGM) having its own resources, authority, functional and financial authority to plan and monitor implementation of the policies and programmes prescribed in the roadmap.

(c) *India Smart Grid Task Force (ISGTF)* [11]

The Government of India formed the India Smart Grid Task Force in 2010 as an inter-ministerial group and will serve as the government focal point. It is a body composed of officials from different government departments and is primarily meant for understanding and advocating policies in smart grid technologies. Major functions of the ISGTF are:

- (i) To ensure awareness, coordination, and integration of diverse activities related to smart grid technologies
- (ii) To promote practices and services for R&D of smart grids
- (iii) To coordinate and integrate other relevant intergovernmental activities
- (iv) To collaborate on an interoperability framework
- (v) To review and validate recommendations from the ISGF

Corresponding to NSGM, each of the states also has a State Level Mission which is chaired by the Power Secretary of the state. The administrative/operation and maintenance expenses in this regard have been borne by respective states. NSGM provides support for training and capacity building to State Level Project Monitoring Units (SLPMUs) for smart grid activities.

(d) *India Smart Grid Forum (ISGF)* [11]

The Government of India also formulated the India Smart Grid Forum in 2010 as a non-profit, voluntary consortium of public and private stakeholders with the prime objective of accelerating development of smart grid technologies in the Indian power sector. The ISGF has roles and responsibilities complementary to the ISGTF.

The goal of the forum is to help the Indian power sector to deploy Smart Grid technologies in an efficient, cost-effective, innovative and scalable manner by bringing together all the key stakeholders and enabling technologies.

The India Smart Grid Forum will coordinate and cooperate with relevant global and Indian bodies to leverage global experience and standards where ever available or helpful, and will highlight any gaps in the same from an Indian perspective.

India Smart Grid Forum (ISGF) is a public-private partnership initiative of the Ministry of Power (MoP), Government of India for accelerated development and deployment of smart grid technologies in the Indian power sector.

11.3.2.2 DRUM, R-APDRP and Smart Grid Pilot Projects in India [11, 16–20]

(a) Distribution Reform, Upgrades and Management (DRUM) [11]
The Ministry of Power, Government of India, and the U.S. Agency for International Development (USAID)—India jointly designed the Distribution Reform, Upgrades and Management (DRUM) Project with the purpose of demonstrating "the best commercial and technological practices to improve the quality and reliability of 'last mile' power distribution in selected urban and rural distribution circles in the country." The project is in synch with the Indian Government's policy on power sector reforms, the Electricity Act of 2003, and the Re-Structured Accelerated Power Development and Reforms Program (R-APDRP) scheme.
The overall programmatic goal of the DRUM Project is to demonstrate commercially viable electricity distribution systems that provide reliable power of sufficient quality to consumers and to establish a commercial framework and a replicable methodology adopted by India's financial institutions for providing non-recourse financing of DRUM activities and programs.

(b) *Restructured Accelerated Power Development and Reforms Programme (R-APDRP)* [11, 17]
Ministry of Power, Govt. of India, as a part of Reforms in the Power Sector, has launched the RAPDRP in the XI Five year Plan. The focused objectives of the program have been on the actual demonstrable performance in terms of AT&C loss reduction, establishment of the reliable and automated sustainable systems for collection of base line data, adoption of information technology in the areas of electricity accounting, customer care and strengthening of distribution network of state power utilities. To achieve these objectives, it is proposed to strengthen and upgrade distribution network and adopt Information Technology systems and services. Various IT systems are planned to be implemented in the distribution companies across the nation under first phase of the programme. The basic infrastructure is being created under RAPDRP/IPDS in large towns, over which the smart grid development can be superimposed. IT implementation and SCADA being carried out will be leveraged for smart grid.
R-APDRP is one of the largest IT initiatives by electric utilities anywhere in the world—in one integrated project, all state owned distribution utilities in India are building IT Infrastructure. The Govt. of India has proposed to continue R-APDRP during the XI Plan with revised terms and conditions as a Central Sector Scheme.
Proposed Coverage and Scheme
R-APDRP has been proposed to cover urban areas—towns and cities with population of more than 30,000 (10,000 in case of special category states). In addition, in certain high-load density rural areas with significant loads, works of separation of agricultural feeders from domestic and industrial ones, and of High Voltage Distribution System (11 kV) will also be taken up.

Projects under the scheme shall be taken up in Two Parts. Part-A shall include the projects for establishment of baseline data and IT applications for energy accounting/auditing and IT based consumer service centers. Part-B shall include regular distribution strengthening projects. The activities to be covered under each part are as follows:

(c) *Smart grid pilot projects in India* [18–20]
Ministry of Power (MoP) injunction with India Smart Grid Task Force had shortlisted 14 smart grid pilot projects and one smart city R&D platform at under mentioned geographical locations: (Ref. Fig. 11.4)

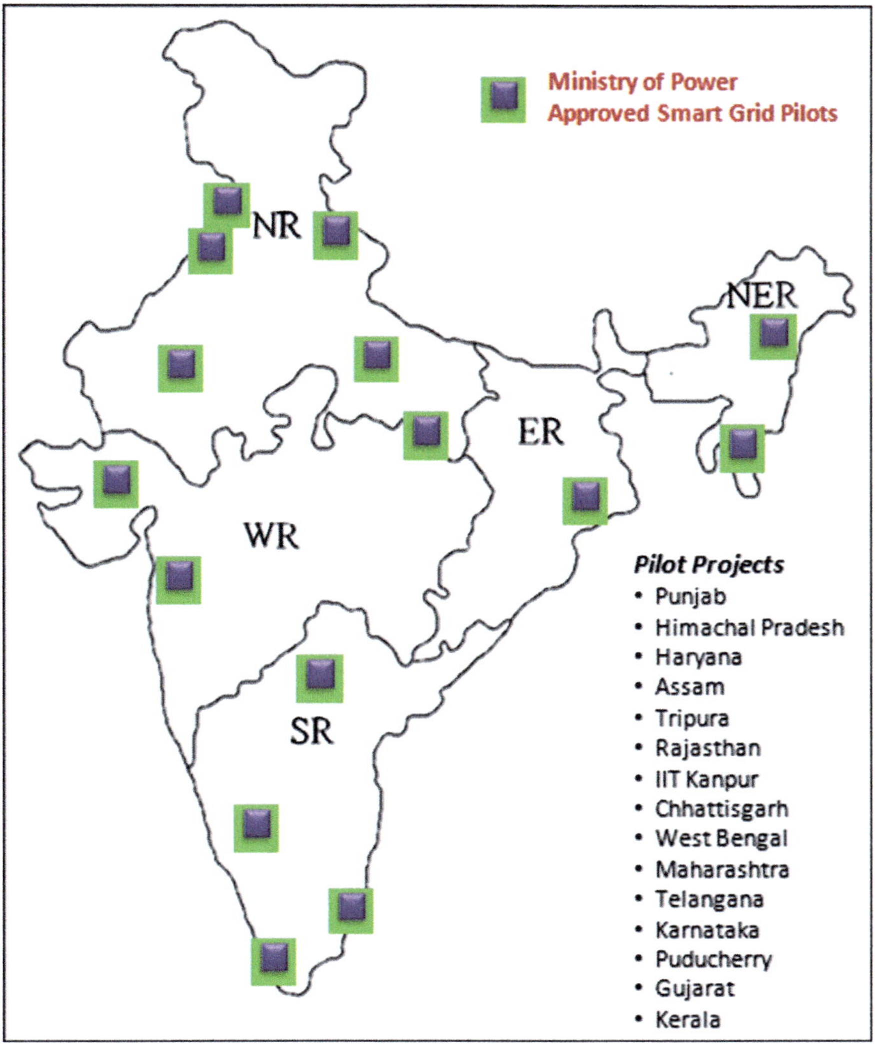

Fig. 11.4 Approved smart grid pilot projects in India [18–20]

(i) APDCL, Assam
(ii) CESC, Mysore
(iii) CSPDCL, Chhattisgarh
(iv) HPSEB, Himachal Pradesh
(v) JVVNL, Rajasthan
(vi) KSEB, Kerala
(vii) MSEDCL, Maharashtra
(viii) PED, Pondicherry
(ix) PSPCL, Punjab
(x) TSECL, Tripura
(xi) TSSPDCL, Telangana
(xii) UHBVN, Haryana
(xiii) UGVCL, Gujarat
(xiv) WBSEDCL, West Bengal
(xv) IIT Kanpur

Most of these projects have been part funded by MoP (50% project cost as grant from GoI) with combined costs of approximately about US$ 80 million. Most of these projects have been either executed or on the verge of completion with active involvement/participation of 20,000 or more citizens.

11.3.3 Smart Grid in Gujarat [20–22]

With duly implemented IT systems under 'E-Urja' program to manage their business functions, Gujarat DisComs have been leading among all utilities across the nation in terms of reduction of losses and IT systems implementation. To utilize the opportunity provided by Govt. of India under R-APDRP scheme, Gujarat DisComs have been further strengthening their distribution network and IT systems and integrating them with existing 'E-Urja' systems Fig. 11.5.

11.3.3.1 Methodology: UGVCL's Approach [23, 24]

Incorporated in 2003, as a part of power sector reforms pursuant to the unbundling of erstwhile Gujarat Electricity Board (GEB), Uttar Gujarat Vij Company Limited (UGVCL) commenced its commercial operations in 2005, pursuant to notifications of the Government of Gujarat as an independent distribution licensee engaged in distribution and retail supply of electricity. As shown in Fig. 11.6, northern districts Banaskantha, Patan, Mehsana, Sabarkantha and central districts of Ahmedabad and Gandhinagar are covered by UGVCL electrical supply.

Fig. 11.5 Renewable energy map of Gujarat, India [22]

11.3.4 Case Study: UGVCL's Smart Grid Pilot Project at Naroda, Ahmedabad [25–27]

UGVCL's smart grid pilot project is located at Naroda of Sabarmati circle in Ahmedabad city, Gujarat, India. The pilot project covers approximately 20,524 customers of industrial, commercial and residential type and accounting for input energy of around 1700 MU. The functionalities of Peak load management, Outage Management, Power Quality Management are being planned by implementing Automated Metering Infrastructure (AMI). Some additional functionality like load forecasting, asset management and integration of renewables have also been underway. The approved project cost is around Rs. 82.70 Crore, out of which GoI is funding Rs. 41.35 Crore. Powergrid is approved consultant for the project.

Benefits Envisaged

(i) Peak load management/DSM
(ii) Reduction in AT&C losses
(iii) Savings in remote connect/disconnect and peak power purchase cost by reduction of peak load
(iv) Reduction in transformer failure rate

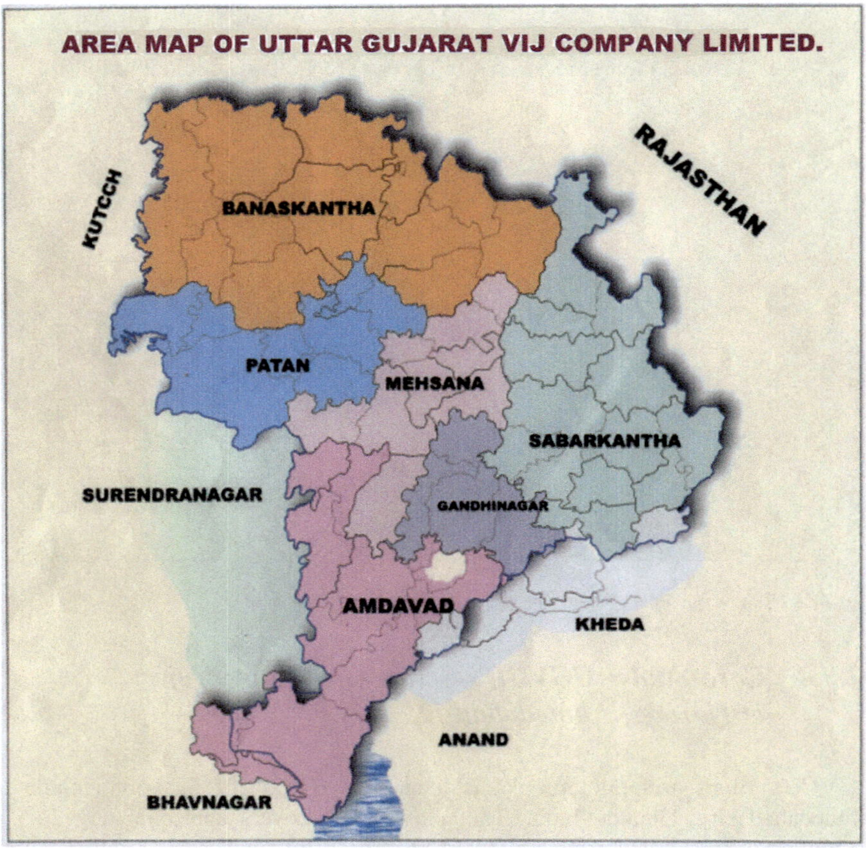

Fig. 11.6 Power area map of UGVCL, Gujarat state, India [24]

(v) Reduction in number of outages
(vi) Reduction in cost of meter reading, cost of payment collection, etc.

In Figs. 11.7 and 11.8, transit and hybrid maps of the UGVCL's smart grid pilot project have been shown. The pilot project site is in the heart of the Ahmedabad city and well connected by road and train transports. The project site is also quite near to river Sabarmati and domestic as well as international terminals of Ahmedabad airport. It could also be observed that the variety in types of consumers offers good opportunity for trail based pilot run of this novel technological-humanitarian experiment.

Fig. 11.7 Transit map of UGVCL's smart grid pilot project, Naroda, Ahmedabad city, Gujarat state, India [26]

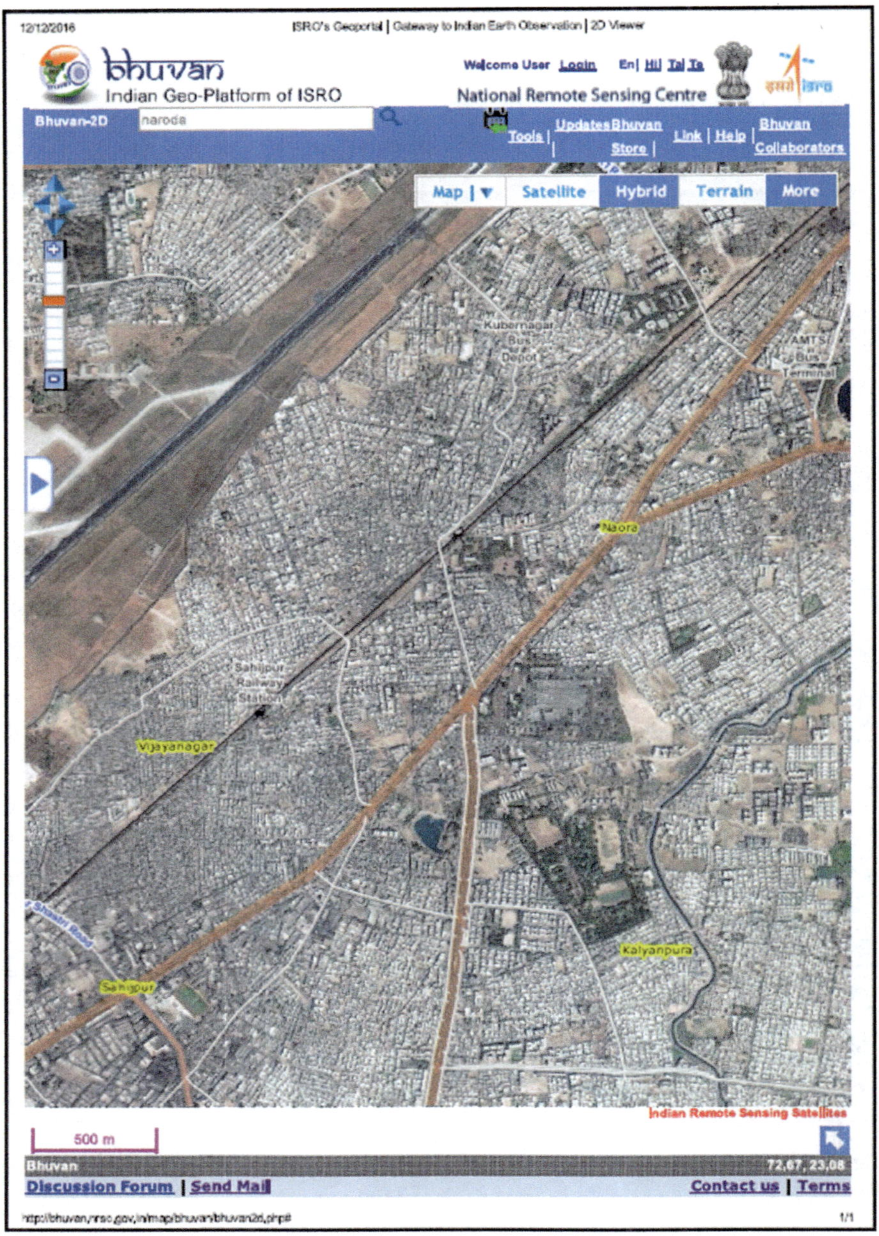

Fig. 11.8 Hybrid map of UGVCL's smart grid pilot project, Naroda, Ahmedabad city, Gujarat, India [27]

11.4 e-Democracy and Citizens' Participation

11.4.1 Internet Based e-Tools [23, 28]

11.4.1.1 UGVCL Consumer Support Website [28]

(a) Consumer support services
In addition to official website http://www.ugvcl.com, UGVCL has launched a distinct website http://ugvcl.in for providing consumer support.
This new secured customer support website is designed and maintained by Tata Consultancy Services (TCS) Limited since year 2010. All the new updates like billing rates, updated contact information, tenders, online bill payment, online complaint booking-follow-up, etc. have been availed through this website. Snapshot of registration/login page of this new customer support has been shown in Fig. 11.9. Snapshot shown in Fig. 11.10 has been displaying zone-wise UGVCL offices with their addresses nearby registered consumer address with toll-free 24 * 7 helpline number.

(b) *Complaint related services* [23, 28]
Snapshot of online complaint booking screen has been shown in Fig. 11.11 along with facility to track status and follow-up past unresolved complaints, if any. Snapshot of online energy theft reporting screen has been shown in

Fig. 11.9 UGVCL consumer support website—registration and login [28]

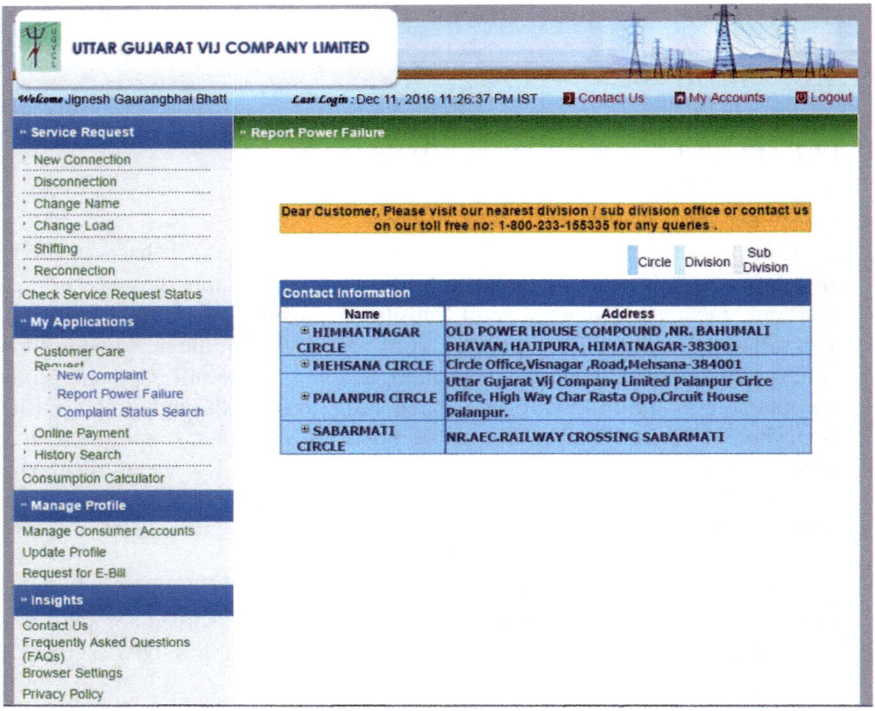

Fig. 11.10 UGVCL consumer support website—contact information display for different circles covered [28]

Fig. 11.12, wherein consumers can report unlawful stealing of electricity to prevent unauthorized unbilled usage.

(c) *Bill related services* [23]
Online facilities to estimate the billing amount have been provided by UGVCL to its consumers, which is available as "Bill Calculator" in old website and "Consumption Calculator" in new website as displayed in snapshots shown in Figs. 11.13 and 11.14 respectively.

(d) Online bill payment facility
Online bill payment facility has been provided in new website by UGVCL to its customers as displayed in snapshot shown in Fig. 11.15. UGVCL has availed common payment gateways to help customers for e-payments.

Fig. 11.11 UGVCL consumer support website—complaint section [28]

11.4.2 Mobile Based M-Tools [24, 29, 30]

11.4.2.1 UGVCL App [29]

In recent times, internet access and payment transactions are more preferred via mobile as compared to desktop and laptop computers due to obvious reasons including convenience, etc. UGVCL has also availed all major services for customer supports using android based mobile app which could be downloaded from Google Play Store. Snapshots of UGVCL mobile app have been included in Figs. 11.16 and 11.17a, b, which enables the basic facilities of bill payment and complaint to the consumers.

11.4.2.2 Solar PV Potential App by GERMI [24]

GERMI has also developed an app for estimation of production of electricity generation by solar PV for given geographic location. This app could be

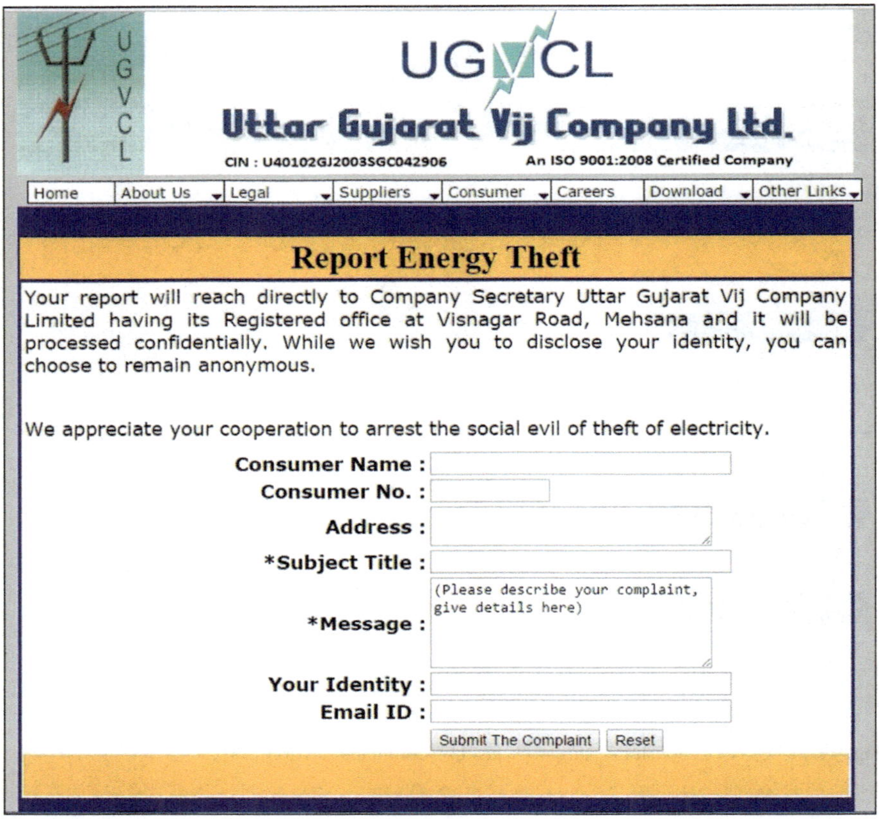

Fig. 11.12 UGVCL consumer support website—reporting energy theft [23]

downloaded from Google Play Store. Snapshots of GERMI mobile app have been included in Figs. 11.18, 11.19a, b and 11.20a, b.

11.5 Challenges [7, 11, 31, 32]

(i) "*To provide energy storage solutions, which would allow excess power to be used in less favorable weather conditions*" has been identified as a major challenge by Pillai and Forum [7].

(ii) "*To generate and sustain high rate of growth in power sector*" to support economic growth and employment generation.
Estimated demand in India has been estimated to rise up to 900 GW by 2032, almost quadrupling the existing capacity with per capita consumption is one-fourth of the world average at present. 79 million households are yet

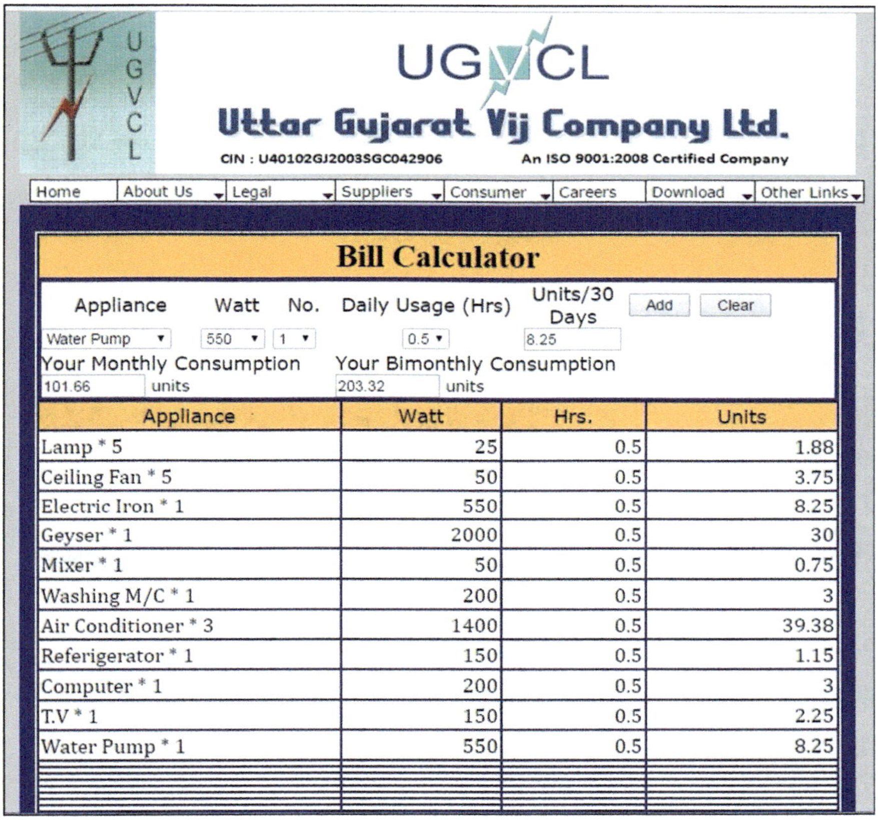

Fig. 11.13 UGVCL consumer support website—bill calculator [23]

to be electrified as per 2011 census. To address the above needs, the Indian power system should grow ~8–10% per annum for next several decades. Managing such a rapidly growing power system of this much gigantic size is a challenge in itself.

(iii) "*To integrate renewables and EVs along with maintaining grid stability*" India is pursuing one of world's largest grid connected renewable energy programs, sustainable integration of such renewable resources to such large fast growing grid would be a difficult task. India launched the National Electric Mobility Mission with a target of 6 million EVs by 2020; successful rollout of EVs along with maintenance of grid stability shall be quite complex problem.

(iv) "*Reduction of T&D losses*" continues to be top priority of both Government and utilities.

Fig. 11.14 UGVCL consumer support website—consumption calculator [28]

(v) "*Evolution of policies and regulations*"
As no defined standards and guidelines exist for the regulation of smart grid initiatives in India, and presently available policy and regulatory frameworks have been typically designed to deal with the existing networks and utilities, while moving towards smart grid, new policy and regulatory frameworks must evolve to encourage incentives for investment. The new framework should be such that it should match the interests of the consumers with the interests of the utilities and suppliers, ensuring that the societal goals are achieved at the lowest cost to the consumers.

(vi) "*Cost minimization*"
Cost is clearly one of the biggest hurdles in implementing smart grids and to make them a good sensible business case. Replacement of non-compatible equipments is expensive and difficult as it would result in pre-mature retirement of such equipments. High initial costs of implementing smart grids should be justified in terms of higher equipment availability, better quality of supply and greater use of green energy. Further, careful societal cost-benefit analysis, beyond return of investment calculations, would also be needed for such justification.

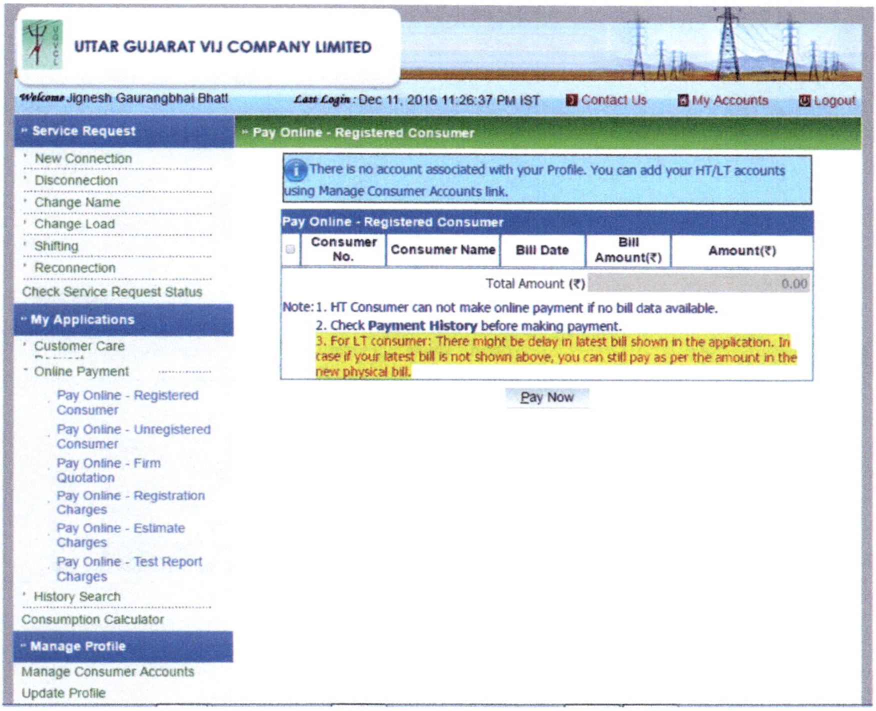

Fig. 11.15 UGVCL consumer support website—online bill payment section [28]

(vii) *"Lack of awareness"*
Sufficient spreading of awareness shall be required among prosumers regarding operational and economic aspects while ushering into this new low carbon environment. Similar and relevant awareness should also be spread into all the stakeholders including utilities, government, etc.

(viii) *"Cyber security and data privacy"*
Keeping the smart grid network secured from malicious hacking attacks shall be one of the most critical challenges. As the consumption data can provide sufficient insight regarding citizens' behavior and choices, invasion of privacy and security of such data shall be a serious concern. Resolution of issues relevant to such matters should be in transparent manner minimizing any negative impact on a customer's perception. Fan et al. [32] have identified various such challenges that must be addressed in order to have fully robust, secure and functional smart grid network.

(ix) *"Last mile connectivity"*
Last mile connectivity is the major challenge in smart grid applications and these field trials to determine the most appropriate communication solutions. Especially, in remote-rural areas of villages and difficult to reach or

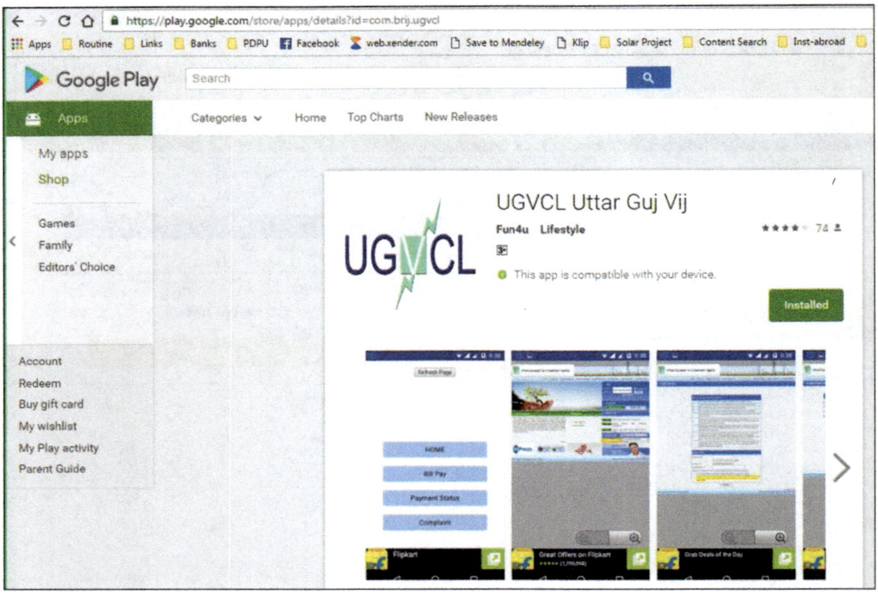

Fig. 11.16 UGVCL android app at Google play store [29]

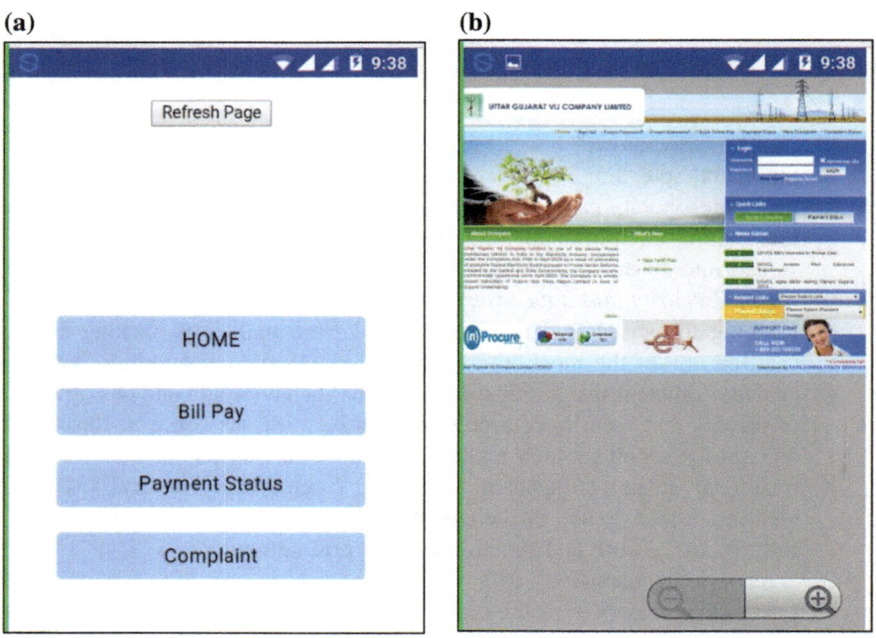

Fig. 11.17 a, b UGVCL android app snapshots [29]

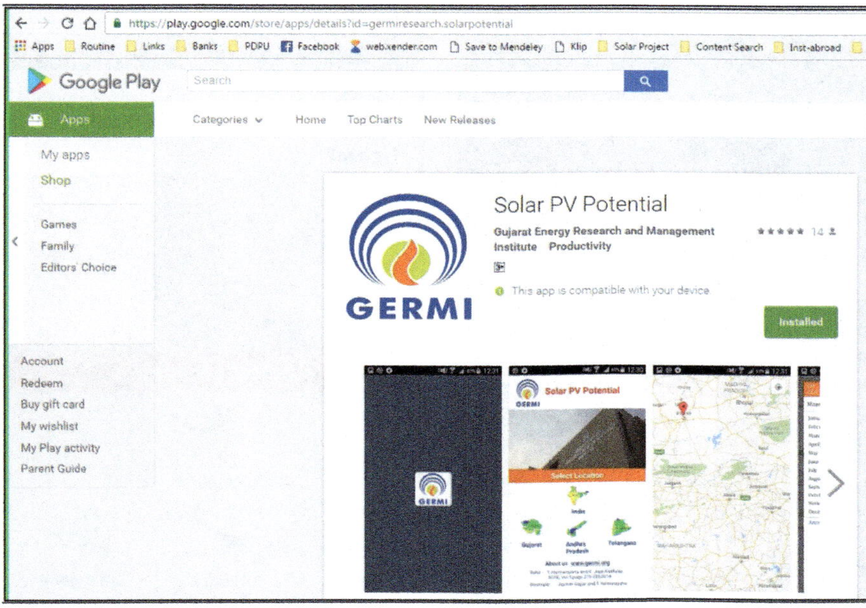

Fig. 11.18 GERMI's Solar PV Potential android app at Google play store [30]

inaccessible locations of cities in India, providing reliable and seamless connectivity for data is still remains a dream to realize!!

In (Babel), the authors have identified and summarized similar challenges especially in Indian context, which could be learnt in Fig. 11.21.

Barriers identified in [11] are similar to challenges identified in [31] in Fig. 11.21 and summarized in Table 11.3.

11.6 Vision, Way Forward and Roadmap [7, 11]

Vision

To transform the Indian power sector into a secure, adaptive, sustainable and digitally enabled ecosystem that provides reliable and quality energy for all with active participation of stakeholders.

Way forward

(i) Regulatory changes to support dynamic pricing
(ii) Time of Use pricing mechanism to be worked upon and finalized
(iii) Replacement of ageing assets

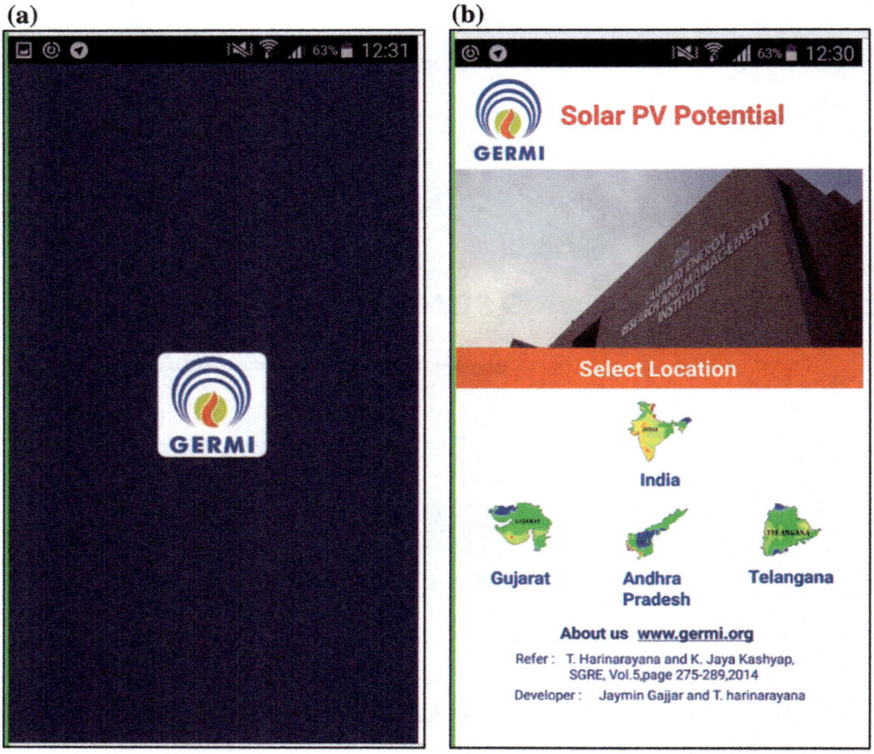

Fig. 11.19 **a**, **b** GERMI's Solar PV potential app snapshots [30]

(iv) To encourage consumer engagement both in communicating effectively with the consumer and in delivering high quality implementation
(v) To develop a communication infrastructure supporting Smart Grid activities at sub-transmission and distribution level required
(vi) Challenges around data privacy, cyber security and inter-operability need to be dealt by creating and adopting appropriate standards in the respective fields
(vii) Assessing and prioritization of benefits from AMI (smart meter) through use-cases.

Roadmap
Distribution

(i) Appropriate policies and programs to provide access for electricity for all, Electrification of 100% households by 2017, 24 × 7 quality supply on demand to all citizens by 2027

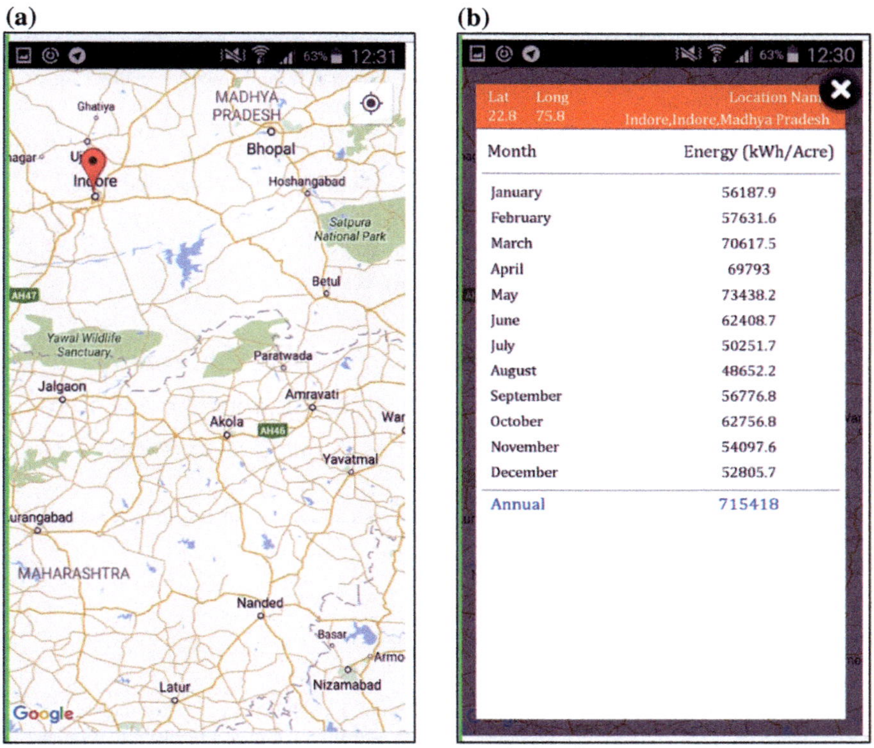

Fig. 11.20 GERMI's Solar PV Potential app: **a** geographical location and **b** solar power generation potential [30]

(ii) Based on outcome of smart grid pilots, full rollout of smart grids in pilot project areas by 2017; in major urban areas by 2022 and nationwide by 2027

(iii) Completion of existing complementary or building block projects such as R-APDRP. Planning for integration of such systems into future smart grid deployments

(iv) AMI roll out for all customers in a phased manner based on size of connection, geography and utility business case. Starting with consumers with load >20 KW by 2017, 3-phase connections by 2022 and all consumers by 2027. Development of innovative and sustainable financing/business models for smart meter roll outs

(v) Working with other stakeholders, building of National Optical Fiber Network by connecting all the 2,50,000 g Panchayats in the country by Optical Fiber Cable and including the telecom link at the nearest 33/11 kV substation to support smart grid in distribution by 2017

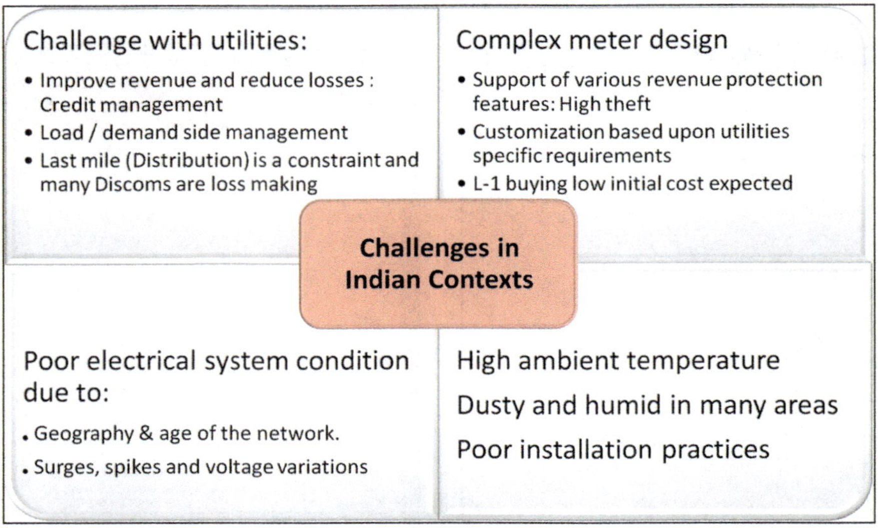

Fig. 11.21 Challenges in Indian Contexts [31]

Table 11.3 Barriers to implement smart grids in India [11]

Barriers	Details
Political and regulatory	In many cases utilities don't get business case for the smart grid as there are regulatory and policy barriers in place that either create reverse incentives or fail to create sufficient positive incentives for private sector investment
Technology maturity and delivery risk	Technologies have significant technology risks associated with them because agreed standards have not emerged. In addition, there are only a handful of examples of large scale implementation of more than 50,000 premises and therefore there continues to be significant delivery risks priced into the estimates
Business case	Where policy makers and utility executives are aware of the role that smart grids can play, they are often unable to make the business case for the smart grid investments
Lack of awareness	There is lack of awareness among stakeholders about role of smart grids in enabling a low-carbon future
Poor financial health of utilities	In India, debt burden utilities find it difficult to invest in smart grid initiatives
Skills and knowledge	In the longer term, a shortfall is expected in critical skills that will be required to architect and build smart grids
Cyber security	Digital communication networks and more granular and frequent information on consumption patterns raise concerns in some quarters of cyber in-security and potential for misuse of private data

(vi) Enabling programs and projects in distribution utilities to reduce AT&C losses. Below 15% by 2017, below 12% by 2022, and below 10% by 2027
(vii) Conversion of existing EHV sub stations in all urban areas and sub transmission and medium voltage substations in metro cities to advanced (such as) Gas Insulated Substations (GIS) in a phased manner through innovative financing models
(viii) Development of Microgrids, storage options, virtual power plants (VPP), vehicle to grid (V2G), solar to grid (PV2G), and building to grid (B2G) technologies in order to manage peak demand, optimally use installed capacity and eliminate load shedding and black-outs
(ix) Push for mandated roof top solar power generation for large establishments with connected load >20 kW
(x) EV charging facilities to be created in all parking lots, institutional buildings, apartment blocks etc.; and quick/fast charging facilities to be built in fuel stations and at strategic locations on highways
(xi) Microgrids in 1000 villages/industrial parks/commercial hubs by 2017 and 10,000 villages/industrial parks/commercial hubs by 2022. Islanding facility from main grids during peak hours to be availed
(xii) Optimally balancing different sources of generation through efficient scheduling and dispatch of distributed energy resources (including captive plants in the near term) with the goal of long term energy sustainability
(xiii) Improvement in power quality and quantum across the board

Transmission

(i) Development of a reliable, secure and resilient grid supported by a strong communication infrastructure that enables greater visibility and control of efficient power flow between all sources of production and consumption by 2027
(ii) Implementation of WAMS, using PMUs for the entire transmission system. Installation of a larger number of PMUs on the transmission network by 2017 or sooner, as guided by the results of initial deployments. Indigenization of WAMS technology and PMU development and development of custom made analytics for synchrophasor data by 2017
(iii) Setting up of Renewable Energy Monitoring Centers (REMCs) and Energy Storage Systems to facilitate grid integration of renewable generation
(iv) Installation of 50,000 km of optical fiber ground wire (OPGW) over transmission lines by the year 2017 to support implementation of smart grid technologies
(v) Enabling programs and projects in transmission utilities to reduce transmission losses to below 3.5% by 2017 and below 2.5% by 2022

(vi) Implement power system enhancements to facilitate evacuation and integration of 30 GW renewable capacity by 2017, 80 GW by 2022, and 130 GW by 2027, or as mutually agreed between MoP and MNRE

Policy, standards and regulations

(i) Formulation of effective customer outreach and communication programs
(ii) Development of state/utility specific strategic roadmap(s) for Smart Grid deployments. Required business process reengineering, change management and capacity building programs
(iii) Policies for grid-interconnection of consumer generation facilities (including renewable) where feasible. Policies for roof-top solar, net-metering/feed-in tariff as well as peaking power
(iv) Policies supporting improved tariffs such as dynamic tariffs, variable tariffs, etc., including demand response programs. Bulk consumers; extending to all 3-phase (or otherwise defined) consumers by 2017
(v) Policies for implementing energy efficiency in public infrastructure and EV charging facilities starting and Demand Response ready appliances by 2017.
(vi) Finalization of frameworks for cyber security assessment, audit and certification of utilities
(vii) Development of business models to create alternate revenue streams by leveraging the Smart Grid infrastructure to offer other services (security solutions, water metering, traffic solutions etc.) to municipalities, state governments and other agencies.
(viii) Build upon the results of smart grid pilot projects and recommend appropriate changes conducive to smart grid development in Acts/Plans/etc.
(ix) Development of Indian Smart Grid Standards. Active involvement of Indian experts in international SG development bodies

Other initiatives

(i) Tariff mechanisms, new energy products, energy options and programs to encourage participation of customers in the energy markets that make them "***prosumers***"—producers and consumers—by 2017
(ii) Create an effective information exchange platform that can be shared by all market participants, including prosumers, in real time which will lead to the development of new and enhanced energy markets.
(iii) Investment in research and development, training and capacity building programs for creation of adequate resource pools for developing and implementing smart grid technologies in India—can also become a global leader and exporter of smart grid knowhow, products and services

11.7 Technological Inputs

11.7.1 Hybrid Communication Technologies [32–35]

Fan et al. [32] have opined that optimization of energy consumption in future intelligent energy networks (or Smart Grids) will be based on grid-integrated near-real-time communications between various grid elements in generation, transmission, distribution and loads.

Wired networking technologies such as RS485, Modbus, PLCC, BACnet, LonWorks, EIB/KNX, etc. and wireless networking technologies such as ZigBee, Wi-Fi, WiMax, GSM, GPRS, 3G-4G, LTE, etc. have been utilized for development of communication and integration of various smart grid applications.

Wired networking technologies are popular due to their superior reliability, security and latency performances. However, frequent rewiring of existing facilities may be difficult and expensive, further it also affects the aesthetic of the building; hence in such cases power line and wireless networks have been preferred to enhance coverage and improve scalability. Bhatt [33] have suggested that wireless technologies are demonstrating their relative strength as compared to wired technologies on account of their salient features and obvious merits.

In Roy et al. [34], the authors could observe sufficient potential in pure wireless as well as hybrid technology solutions (combined of wired and wireless) to serve critical need based applications of smart grid and therefore, those of a smart city.

Overall, need based customizable, mixed or hybrid technology solutions which are effectively combining wired and/or wireless both type of technologies based on actual requirements could evolve the best and optimum solution(s) for given smart application(s).

Güngör et al. [35] have addressed critical issues on smart grid technologies primarily in terms of information and communication technology (ICT) issues and opportunities. Main objective of the authors is to provide a contemporary look at the current state of the art in smart grid communications as well as to discuss the still-open research issues in this field. The paper provides a better understanding of the technologies, potential advantages and research challenges of the smart grid to provoke future research interests.

11.7.2 IoT, HAN and BAS [2, 3, 5, 36−44]

As described by Arasteh et al. [3], an IoT prototype is subject to smart and self-configuring objects that are connected to each other through a global network infrastructure. IoT is mostly considered as real objects, broadly scattered, with low storage capability and processing capacity, with the target of improving reliability, performance and security of the smart city and its infrastructures. In the IoT context, devices can be integrated based on the geographic location and evaluated by using an analyzing system. An IoT infrastructure could be utilized for the monitoring of intended domains using sensor services for the collection of particular data. IoT is not only likely to affect the various aspects of the smart city citizens' life like health, security, and transportation, it could play an important role at the national level for making policy decisions in important aspects such as energy saving policies, pollution control, economic considerations, reliability enhancement, etc.

Al-Ali [2] have suggested that the concept of Smart Cities is becoming a reality as it evolves from conceptual models to developmental stages. Resilient, reliable, efficient and seamless energy and electrical power flow are essential parts to energize and power the services of smart cities such as smart hospitals, smart buildings, smart factories, smart traffic and transportations. All of these smart services are expected to run without interruptions by the use of smart energy and electrical power grids which are considered among the most important pillars for such cities. To keep the services of smart cities interconnected and in sync, the Internet of Things (IoT) and cloud computing are key in such transfers.

The World Wide Web (WWW) is evolving from a traditional host that contains text, images, audios and videos to a physical host that enables users to control physical objects. Home appliances, remote CCTV cameras and factory floors can be monitored and controlled using Internet of Things (IoT) using communication media. The physical web concept is emerging nowadays. An IoT communication network is utilized in energy generations and consumptions in residential areas.

The world is becoming an integrated global community through multiple technologies and numerous areas of applications and services. IoT concepts are leading to a world where real, digital and virtual things are converging to make our cities smarter and more intelligent. Nowadays, traditional web technologies are empowered by IoT to connect physical objects (things) such as home appliances and smart grid devices with a unique address for each thing. This has been made possible with the help of IPV6 protocol which has 2^{128} unique IP addresses compared to 2^{32} addresses used by IPV4. Using IPV6, billions of objects can be connected, monitored and controlled at the same time. In terms of communication networks, the consumption domain is divided into three different networks: Home Area Network (HAN), Business Area Network (BAN) and Industrial Area Network (IAN).

Many communication protocols are utilized in these networks, such as ZigBee, PLCC, Z-wave, Wi-Fi, WiMax, 3G/GSM and LET. Each renewable energy resource is considered as an object and it is assigned a unique IP address. Using bidirectional communication, it becomes possible to monitor each object as control is carried out via its unique IP address. Such IoT networks are scalable and more devices and appliances such as local batteries storage, home appliances, smart meters, etc. can also be included as objects to join the same network with functionalities of remote monitoring and control.

Shrihariprasath and Rathinasabapathy [36] have presented an effective implementation of an intelligent remote monitoring system for solar Photovoltaic (PV) Power Conditioning Unit (PCU) for using in a greenhouse environment. The proposed smart remote monitoring system is based on internet of things and its

GPRS based design could be installed to remotely monitor solar PV PCU in real time and to solve management problems, maintenance and shortens MTTR.

Perera et al. [37] presented and interesting survey of one hundred IoT smart solutions in the marketplace and examine them closely in order to identify the technologies used, functionalities, and applications. Based on the application domain, the authors have classified and discussed those solutions under have different categories such as (i) smart wearable; (ii) smart home; (iii) smart city; (iv) smart environment; and (v) smart enterprise. This survey is intended to serve as a guideline and a conceptual framework for future research in the IoT and to motivate and inspire further developments. It also provides a systematic exploration of existing research and suggests a number of potentially significant research directions.

Domingues et al. [38] have presented that a BAS consists of a system installed in buildings that controls and monitors building services responsible for heating, cooling, ventilation, air conditioning, lighting, shading, life safety, alarm security systems, and many more. A BAS aims at automating tasks in technologically-enabled environments, coordinating a number of electrical and mechanical devices interconnected in a distributed manner by means of underlying control networks. These systems may be deployed in industrial infrastructures such as factories, in enterprise buildings and malls, or even in the domestic domain. Building automation has been receiving greater attention due to its potential for reducing energy consumption and facilitating building operation, monitoring and maintenance, while improving occupants' satisfaction.

Bhatt and Verma [39] have presented simple, user-friendly, re-usable and customizable design and implementation of wired security system and wired building automation system. Indicating clear trend of legacy stand-alone security systems transforming into intelligent-computerized-network based building automation systems, the authors have developed web-based GUI for live run-time monitoring and control of a residential building as major outcome.

Bhatt and Verma [40] have presented the design of a three-layer BAS for a residential building using PC based controller. GUIs for live remote monitoring via CCTV camera, critical events alarm annunciation through SMS and distribution of audio-video entertainment using intranet/internet are some of the important features of the design.

In Patel et al. [41, 42] have presented an alternative design of Programmable Logic Controller (PLC) originally aimed at industrial safety and emergency shutdown-trip systems, however its present commercially available modern and cost-effective versions could also be used in BAS as well as HAN with suitable customizations.

Bhatt [43], Bhatt and Verma [39] have described that building automation system (BAS) is a data acquisition and control system that incorporates various functionalities provided by the control system of a building. In general, the BAS are also known by other names such as: energy management control systems (EMCS), building management systems (BMS), building energy management systems (BEMS), facility management systems (FMS) and so on. Modern BAS is a computerized, intelligent network of electronic devices, designed to monitor and control the lighting, internal climate and other systems in a building resulting in optimized energy usage, safety, security, information, communication and entertainment facilities. BAS maintains the internal climate of building within a specified range by regulating temperature and humidity regulates lighting based on parameters like occupancy, ambient light and timing schedule, monitors system performance and device failures and generates audio-visual-email and/or text notifications to building operation and maintenance (O&M) staff. The BAS reduces building energy consumption and, thereby, reduces O&M costs as compared to an uncontrolled building. A building controlled by a BAS is often referred to as an Intelligent Building. Home automation or domotics is possibly the system having the highest level of functional similarity to building automation. But the former has significantly different characteristics. While the BAS focuses on financial benefits, the domotics focuses on comfort and mental peace. Domotics are of comparatively quite smaller scale, with low costs of equipment as well as commissioning and maintenance.

Bhatt and Verma [39] have described that Home area network (HAN) is the network within the premises of a house or building enabling devices and loads to communicate with each other and dynamically respond to externally sent signals (i.e. price). This type of network, characterized by a low data rate requirement, provides the communication infrastructure behind the meter. As a result, smart grid applications are extended into the home premises and energy management systems (EMS) capabilities are enabled. Further benefits derived from the realization of such HANs are anticipated for customers, utilities and society as a whole. Overall, Energy savings and user satisfaction are two major design considerations for modern HAN systems. Interesting descriptions on useful developements of remote monitoring and control applications of solar PV could be found in Bhatt et al. [5].

Ejaz et al. [44] have presented that the drastic increase in urbanization over the past few years requires sustainable, efficient, and smart solutions for transportation, governance, environment, quality of life, and so on. The Internet of Things offers many sophisticated and ubiquitous applications for smart cities. The energy demand of IoT applications is increased, while IoT devices continue to grow in both numbers and requirements. Therefore, smart city solutions must have the ability to efficiently utilize energy and handle the associated challenges. Energy management is considered as a key paradigm for the realization of complex energy systems in smart cities. Along with a brief overview of energy management and challenges in smart cities, a unifying framework for energy-efficient optimization and scheduling of IoT-based smart cities has been presented in the article which is illustrated in Fig. 11.22.

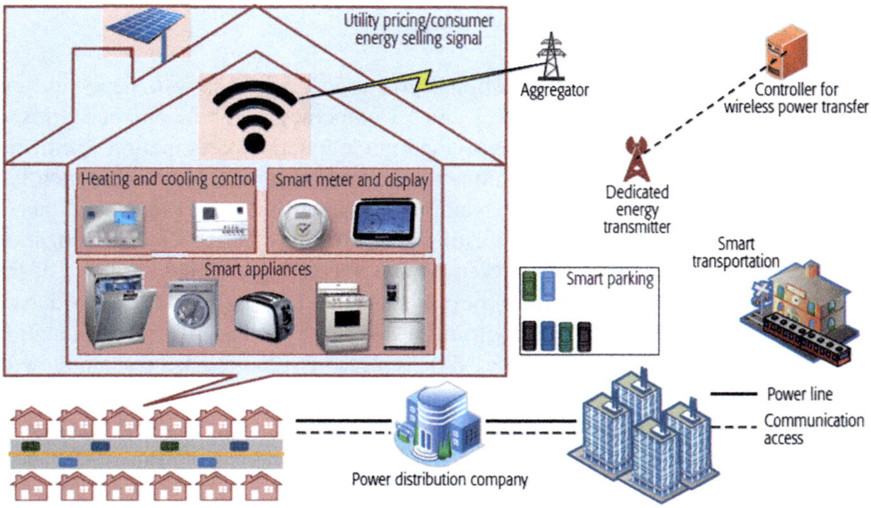

Fig. 11.22 IoT-based illustration of smart cities focused on smart homes [44]

11.8 Experiences and Lessons Learnt

Cities are prime movers of global economy and platforms for providing opportunities in employment, goods and services, commerce, culture and enhanced quality of convenient life with better services, health-education facilities, etc. They have some mega issues like poverty, poor health, social issues and environmental degradation. The cities are often confronted with a multitude of key problems like high urban densities, traffic congestion, energy inadequacy, lack of basic services, water-soil-air pollution, aging infrastructure and environmental impact on climate change coupled with poor urban governance.

Reliability of critical power infrastructures has been the area of major focus today, wherein smart grids are expected to play game changing role. As energy backbones of smart cities, smart grids are deployed with intelligence, automation, citizens' active participation. Smart grids have started attracting attention of global research community and demonstrating excellent growth potential.

Distributed energy sources and demand response programs contribute to environmental, economic and social aspects of urban sustainability; they are now viewed from different perspectives. Most of the available studies and literature on smart grid subject mainly takes only technocratic approach, while novel upcoming approaches have been now found based on their social impacts, including policy regulations, e-Democracy, etc.

11.9 Conclusions

Transformation from legacy conventional grid to smart grid is now necessity and reality by effective applications of ICT and e-Democracy. The role of smart grids as energy backbones of smart cities, involving high interactive participation of citizens in energy management, based on humanitarian and customer centric approach has been discussed. After presenting the historical perspectives and present situations of electrical power sector in India, interesting relationships of electricity, urbanization and human settlement have been presented. An interesting case study of one of the most successful Indian smart grid pilot project by UGVCL at Naroda, Ahmedabad, Gujarat, India has been included with brief details regarding citizen participation tools based on internet and mobile. Experiences, issues, challenges along with vision and future roadmap have been shared to mark the end of the chapter.

Rapidly rising urbanization has led to many critical issues like scarcity of electricity. Developing countries like India have been aggressively invest in both strengthening their existing legacy electricity grids and also by building smart grids and renewables.

Comprehensive study of renewables' socio-technical co-assessment, their responses and combined impacts upon urban planning and energy policies—have been indeed interesting domains of research. Transition to smart grids necessitates detailed investigation and comprehensive evaluation regarding co-evolution of renewables and social responses together in cities and further how they could be related to urban planning-policies and citizen-centric e-Governance.

Acknowledgements The authors thankfully acknowledges the support received with appreciations to all the faculty members, Management and office bearers of Gujarat Energy Research and Management Institute (GERMI), Gandhinagar—382 007, Gujarat, India, Pandit Deendayal Petroleum University (PDPU), Gandhinagar—382 007, Gujarat, India, and Dharmsinh Desai University (DDU), Nadiad—387 001, Gujarat, India for their kind cooperation, active support and timely help in development of this book chapter. Useful contributions and cooperation received from UGVCL, GEDA, Govt. of Gujarat, Govt. of India and all the cited sources of references are also hereby thankfully acknowledged.

The authors also take this opportunity to express their sincere thanks to Editor Prof. TM Vinod Kumar and his entire team including Springer staff for their valuable guidance, excellent cooperation and timely help extended.

References

1. Adil AM, Ko Y (2016) Socio−technical evolution of decentralized energy systems: a critical review and implications for urban planning and policy. Renew Sustain Energy Rev 57:1025−1037. doi:10.1016/j.rser.2015.12.079
2. Al-Ali AR (2016) Internet of things role in the renewable energy resources. Energy Procedia 100:34−38. doi:10.1016/j.egypro.2016.10.144

3. Arasteh H, Hosseinnezhad V, Loia V et al (2016) Iot-based smart cities: a survey. EEEIC 2016–Int conf environ electr eng. doi:10.1109/EEEIC.2016.7555867
4. Bansal N, Shrivastava V, Singh J (2015) Smart urbanization–key to sustainable cities. Real Corp 2015 2:551–560
5. Bhatt J, Jani O (2014) E–governance for photo voltaic powergrid: solar city Gandhinagar, Gujarat, India. In: E-governance of smart cities
6. ISGF (2013) India smart grid day. India Electr 2013:48
7. Pillai RK, Forum G (2015) Introduction to India smart grid forum, smart grid vision and roadmap for India. Smart Grid Projects and India Smart Grid Week 2015 Overview of Indian Power Sector Smart Grids–Challenges for India, pp 1–13
8. Urvish Dave (2016) Year in review for renewable energy sector across India–a new dawn in Renewable Energy! https://urvishdave.wordpress.com/2016/12/19/2016-year-in-review-for-renewable-energy-sector-across-india-a-new-dawn-in-renewable-energy/. Accessed 23 Dec 2016
9. Murthy Balijepalli VSK, Khaparde SA, Gupta RP, Pradeep Y (2010) SmartGrid initiatives and power market in India. IEEE PES Gen Meet PES 2010:1–7. doi:10.1109/PES.2010.5589947
10. Bhatt J, Shah V, Jani O (2014) An instrumentation engineer's review on smart grid: critical applications and parameters. Renew Sustain Energy Rev 40:1217–1239
11. Smart Grid http://indianpowersector.com/wp-content/uploads/2013/04/Smart-Grid.pdf. Accessed on 09/01/2016
12. Fistola R, La Rocca RA (2014) New technologies for sustainable energy in the smart city: the WET theory. TeMA J L Use, Mobil Environ 7:29–42. doi:10.1007/s13398-014-0173-7.2
13. Joshi S, Saxena S, Godbole T, Shreya (2016) Developing smart cities: an integrated framework. Procedia Comput Sci 93:902–909. doi:10.1016/j.procs.2016.07.258
14. What is Smart City? http://smartcities.gov.in/writereaddata/What%20is%20Smart%20City.pdf. Accessed 12 Jan 2017
15. National Smart Grid Mission (NSGM). http://www.nsgm.gov.in/content/nsgm.php. Accessed 23 Dec 2016
16. India Smart Grid Forum (ISGF)–Smart City Project, UGVCL, Gujarat. http://www.indiasmartgrid.org/ProjectsDesc.php?id=56. Accessed 23 Dec 2016
17. Restructured Accelerated Power Development & Reforms Programme (R–APDRP). http://ipds.co.in/Forms/Know_More.aspx. Accessed 23 Dec 2016
18. Smart Grid Pilots in India. http://www.nsgm.gov.in/content/sg_pilots.php. Accessed 23 Dec 2016
19. Smart Grid Pilots–Snapshots. http://www.nsgm.gov.in/content/sg_snapshot.php. Accessed 23 Dec 2016
20. Updates on execution of Smart Grid Pilot Projects in India–IV. http://www.desismartgrid.com/2014/06/updates-execution-smart-grid-pilot-projects-india-iv. Accessed 23 Dec 2016
21. Gujarat Energy Development Authority (GEDA) website. http://geda.gujarat.gov.in/index.php. Accessed 12 Jan 2017
22. Renewable energy map of Gujarat. http://geda.gujarat.gov.in/state_renewable_energy_maps.php. Accessed on 12 Jan 2017
23. UGVCL website. http://www.ugvcl.com. Accessed 23 Dec 2016
24. Power area map of UGVCL. http://www.ugvcl.com/cprofile/pau.htm. Accessed 12 Jan 2017

25. Uttar Gujarat Vij Company Limited Smart Grid Pilot Project–PoC studies 23rd May 2016. http://www.ugvcl.com. Accessed 12 Jan 2017
26. Transit map of Naroda area, Ahmedabad city, Gujarat state, India. http://googlemap.com. Accessed 12 Jan 2017
27. Google map of Naroda area, Ahmedabad city, Gujarat state, India. http://googlemap.com. Accessed 12 Jan 2017
28. UGVCL Customer support website. http://www.ugvcl.in. Accessed 23 Dec 2016
29. UGVCL app in Google Play Store. https://play.google.com/store/apps/details?id=com.brij.ugvcl. Accessed 23 Dec 2016
30. Solar PV Potential app by GERMI in Google Play Store. https://play.google.com/store/apps/details?id=germiresearch.solarpotential. Accessed 23 Dec 2016
31. Babel BS, Smart metering : to help Indian utilities address the revenue challenge Ind'a's energy scenario, Accessed 02 Sept 2016
32. Fan Z, Kulkarni P, Gormus S et al (2013a) Smart grid communications: overview of research challenges, solutions, and standardization activities. IEEE Commun Surv Tutorials 15:21–38. doi:10.1109/SURV.2011.122211.00021
33. Bhatt J (2010) “Wireless networking technologies for automation in oil and gas sector”. In: Yadav S (ed) Int Conf on Manag Pet Sect. Pandit Deendayal Petroleum University, Gandhinagar, India, p 25
34. Roy S, Jani O, Bhatt J (2013) A record note of workshop on smart meter and communication technology solutions. Gandhinagar, India
35. Güngör VC, Sahin D, Kocak T et al. (2011) Smart grid technologies: communication technologies and standards 7:529–539
36. Shrihariprasath B, Rathinasabapathy V (2016) A smart IoT system for monitoring solar PV power conditioning unit. In: IEEE WCTFTR 2016–Proc 2016 world conf futur trends res innov soc welf. doi:10.1109/STARTUP.2016.7583930
37. Perera C, Liu CH, Jayawardena S (2015) The emerging internet of things marketplace from an industrial perspective: a survey. IEEE Trans Emerg Top Comput 3:585–598. doi:10.1109/TETC.2015.2390034
38. Domingues P, Carreira P, Vieira R, Kastner W (2016) Building automation systems: concepts and technology review. Comput Stand Interfaces 45:1–12. doi:10.1016/j.csi.2015.11.005
39. Bhatt J, Verma H (2015) Design and development of wired building automation systems. Energy Build. doi:10.1016/j.enbuild.2015.02.054
40. Bhatt J, Verma H (2010) RS–485/MODBUS based intelligent building automation system using LabVIEW. In: 4th Int Conf Comput Appl Electr Eng Adv IIT Roorkee, Roorkee, India, p 5
41. Patel D, Bhatt J, Trivedi S (2015) Programmable logic controller performance enhancement by field programmable gate array based design. ISA Trans. doi:10.1016/j.isatra.2014.08.019
42. Patel D, Trivedi S, Bhatt J (2011) Design and implementation of field programmable gate array based programmable logic controller. In: Soni HB (ed) 2nd Int Conf Signals, System Autom, Universal Publishers, USA–G.H. Patel College of Engineering & Technology, Vallabh Vidyanagar, India, pp 332–336
43. Bhatt J (2013) Building automation system: Design and Development. Ind Autom 16:18
44. Ejaz W, Naeem M, Shahid A et al (2016) Efficient energy management for internet of things in smart cities efficient energy management for internet of things in smart cities

Part III
E-Democracy Tools and Issues

Chapter 12
Integration of Existing Information Systems as a Spatial eDemocracy Tool for Smart Cities in India

Mahavir and Prabh Bedi

Abstract During the past couple of decades, advances in digital technology have led to establishment of information systems at sectoral levels like urban, environment, forest, demography (census) and more. The information system for urban areas in India exists in the form of National Urban Information System (NUIS) with the objective to facilitate planning, maintenance and governance of municipal areas. The use of NUIS for planning has been initiated in some urban centers by utilizing the data for preparation of master plans. There are other sectors that play an important role in an urban milieu. These sectors have their own existing systems in place based on information technology. The call of the day is to use these systems in tandem with each other so as to build a comprehensive plan for urban area which is a step towards building smart cities. The authors advocate the integrated use of the information systems that exist in isolation. The approach being exhorted is to particularly use the data in NUIS and data in ENVironmental Information System (ENVIS) as eDemocracy tools towards building smart cities. The available data in both the information systems can be integrated with the perspective of preparation of a master plan as per the Urban and Regional Development Plans Formulation and Implementation (URDPFI) Guidelines. An emphasis is being laid out to incorporate the environmental aspect in mainstream master planning through a suggested framework so as to come up with environmentally sensitive master plans to build sustainable smart cities for posterity. The integration of the two systems will bring in the spatial component into information based democracy and lead to a spatial eDemocratic society.

Keywords Spatial eDemocracy · NUIS · ENVIS · Master planning · Smart cities

Mahavir
School of Planning and Architecture, 4, Block-B, I. P. Estate, New Delhi, India
e-mail: mahavir57@yahoo.com

P. Bedi (✉)
Geospatial Information Technology Consultant, Visiting Faculty, School of Planning and Architecture, 4, Block-B, I. P. Estate, New Delhi, India
e-mail: prabhb@gmail.com

T.M. Vinod Kumar (ed.), *E-Democracy for Smart Cities*, Advances in 21st Century Human Settlements, DOI 10.1007/978-981-10-4035-1_12

12.1 Introduction

Two important Information Systems for urban areas in India are National Urban Information System (NUIS) and ENVironmental Information System (ENVIS). National Urban Information System is a scheme for developing urban database to build an information system to enable planning, management and de-centralized governance in the context of provision of scientific planning and implementation of the 74th CAA [1]. Though NUIS is a contributor to National Spatial Data Infrastructure (NSDI), Bedi [2] and hence is based on standards and procedures defined at the national level, ENVIS is not so. ENVIS is a decentralized system with a web-based distributed network of environment specific databases [3]. The basis of NUIS is Geographic Information System. Government of India has recently embarked on building SMART Cities for its citizens. Conforming to the level of development of the urban areas in the country, the level of smartness that is to be achieved varies when compared to the smart cities of more advanced countries. However, the prime basis remains the same that is to improve the quality of life of the citizens, thereby bringing in the concern for environment of the urban areas. The means of achieving smartness too has been cloned with Information Systems being the platform.

Two important Information Systems of the urban areas, however, exist in isolation to each other. It is imperative that these be integrated to build smart cities. This chapter discusses the framework that can bring together the two information systems and lead to building SMART Cities in the country by functioning as a spatial eDemocratic tool.

The chapter is broadly segregated into three sections. The first section discusses the urban sector in the country ranging from ideological developments to the functionaries that have come to be. The second section focuses on Information Systems and its evolution in India. The focus gradually moves to NUIS and ENVIS and concludes with a suggested framework to integrate the two systems and using it as a spatial eDemocratic tool for better building smart cities.

12.2 Urban Domain in India

Census of India classifies those areas as urban that have a Municipality or Corporation or Cantonment or Notified Town Area and satisfies the criteria of a minimum population of 5000, has more than 75% of male working population in non-agricultural activities and a population density of at least 400 persons per km^2 [4]. Urban means those land areas which are characteristic of or belong to a town or a city [5].

Land, as per the Constitution of India, falls under the legislative competence of state legislature [6], with exception in some sectors like defense and railways etc., where the main task is judicious use of land. The use of land in urban context is planned and managed by organisations like Town and Country Planning

Organisation/Departments (TCPO/Ds), Development Authorities (DAs), Urban Local Bodies (ULBs) and Parastatals. Planning and management of the urban areas, whether physical or urban or regional or all-encompassing spatial, fall within the purview of the State. However, there is some bit of urban planning and management that is under the national level authorities like Ministry of Urban Development (MoUD). The National Land Utilization Policy 2013 (Draft) clearly states that land is required for development of essential infrastructure and for urbanisation, while at the same time there is a need to protect land under environmentally sensitive zones and land which provides ecosystem services.

12.2.1 Authoritarian to eDemocratic

Planning in India has been evolving from its pure physical state to spatial form. In India, it has not been evolutionary; planning has been either replicated from the western world or is 'scientisation of politics' [7] and [8]. With no national policy for urban areas and urban being a state subject, planning has remained economic and sectoral at higher levels of hierarchy lacking the spatial components much needed to diffuse its impact on the local levels. At the same time, at the local level, it has lacked the economic approach. The emphasis is now shifting towards spatial planning which is all inclusive.

The advances in science and technology in 1960s and the advent of computers opened unlimited avenues of creating data which was being done manually till recent years. The data converted into information could now be used for creating multiple scenarios and in making prediction. The ever-growing complexity in the relations between nature, man and his creations, and the increasingly emergent and glaring imbalances could now be handled with sophisticated tools provided by computers like systems analysis and operations research. With all these inputs, in the recent years the shift has been towards governance from the traditional government. Governance is the process of horizontal and vertical coordination and communication between public and private sectors.

For the success of the urban spatial plans, the process needs to be collaborative. A carefully designed participation program with an open exchange of information and ideas between the general public, decision makers, and planners needs to be incorporated into the planning and decision-making process.

Public participation, 'voice of the people', has become important in planning decision-making process. The idea of public participation gained form in the United States of America from the advocacy planning movement during the 1960s. It has continued to be redefined by planners, politicians, developers, and citizens over the past few decades. In its various forms, participatory planning has been instigated by different institutions and within the context of many different agendas. These agendas range from political manipulation to consultation and ultimately to redistribution of power to marginalized communities.

The emerging concept of spatial planning is inclusivity, with an increased role of certain sections of the society. It is divergent from traditional planning as it is perceived as an ongoing process, having an integrated approach to governance as against the 'silo' and centralized approach with an increasing thrust on deliverance of services and meeting the defined targets.

Democracy means that people participate in making decisions either directly or through their elected representatives. The differences exist in the levels of democratization which is deduced from the level of citizen involvement in decision-making. As the society matures it moves up the ladder of citizen participation and democracy from non-participation where the citizens are manipulated to being informed and consulted about the happenings in the society. With greater maturity in the community, partnerships develop between the government and the community leading to powers being delegated to some representatives giving way to complete citizen control [9] (refer Fig. 12.1). With regard to India, it can be said that the Indian community is in various stages of 'tokenism' from planning perspective, where the people are informed about the actions and plans the authorities are intending to make to the betterment of the society. In some instance, through meeting and deliberation inputs are sought from the representatives of the people in the plan-making process or some subject matter experts are consulted to provide relevant inputs during planning process.

Over the last few years, the process of planning has been evolving into a democratic system with increasing public participation, hence, people being able to influence the decision-making process. Planning and its process have been much influenced by various policies and guidelines that have been formulated over the years since Independence, which have taken learning from best practices world over. With the advent and popularization of technology, the shift has started

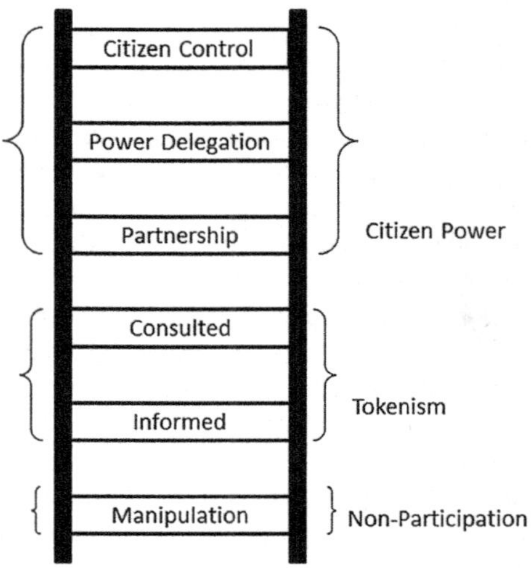

Fig. 12.1 Levels of citizen's participation indicating the democratic maturity of a community. *Source* Adapted from Arnstien (1969) the ladder of citizen participation. http://www.informaworld.com reprinted by permission of The American Planning Association, www.planning.org and the publisher Taylor & Francis Ltd, http://www.tandfonline.com

towards eDemocracy, where Information Systems and communication technology is being used to support democratic decision-making process.

eDemocracy is not merely self-generation of utility bills and making payments online or checking availability of parking spaces in real time. It is not just limited to online registration and generation of birth and property certificates. eDemocracy is also not merely developing Apps where citizens can report a choking sewer line or overflowing waste dump. There are transactional activities that have been incorporated as part of good governance adapting from emerging advancements in eCommerce and eBusiness solutions. eDemocracy is essentially the participation of the citizens in the decision-making process through an electronic medium, in present day age of web based technologies, through online services and applications. eDemocracy has to graduate to spatial eDemocracy, where the citizens can participate and make their voice heard in the spatial planning and monitoring process. With the coming in of spatial eDemocracy, the citizen will be able to participate in decision-making through spatially enabled tool, in simpler terms through interactive maps available in the government's enhanced geospatial portals like Bharat Maps and BHUVAN and platform like data.gov.in which is largely non-spatial.

At the current stage eDemocracy is aspatial. It entails access to the information from the government in the form of reports, tables and questionnaires, which are all non-spatial. The access in spatial terms from master planning perspective is rare and if it exists is to some static maps either in print or on websites like on some DA websites. Government of India has guaranteed right to information through Right to Information (RTI) Act, 2005 to bring in accountability and transparency in governance [10], perceived by the authors as a step towards eDemocracy. The provisions of the Act empower the citizens to know the number of trees that may have been cut. As of date this information is available to the citizen in non-spatial form. However, the Act does not abstain from accessing to spatial information, thus, ensuring spatial eDemocracy. Further, free and open access to spatial data would enhance the e-participation [11].

Things in spatial connotation are based on geospatial technology of which Geographic Information System (GIS) is an important component. Geographic Information System consists of hardware, software, data, users and procedures and geo-spatial information is a derivate of GIS. With large amounts of data being produced in spatial context and available in public domain, Geospatial Information and data, too is infrastructure [12] and a public good [13]. Spatial eDemocracy as well as eDemocracy may be in danger considering the Geospatial Information Bill (Draft) 2016 [14] and may not be achievable.

eDemocracy takes place at different levels, ranging from local to regional to state and national levels. The form of eDemocracy varies with the level at which it exists taking different forms of citizen's participation. Information System has the ability to greatly enhance the level of societal decision-making. The role of the planner is no longer top-down and centralized but has become that of one who attempts consensus building between the stakeholders, based on most viable options. Democracy and to a large extent eDemocracy is crucial to the success of smart cities. Transforming the cities into smart entities has been taken on a mission mode by the government and now forms an important part of the urban milieu.

12.2.2 Smart Cities

Technologically, the concept of Smart Cities has grown from technopolis or science parks to Networked Cities to Digital Cities or Virtual Cities to Intelligent Cities to the present day Smart City. At an ideological level, the concept can be traced back to the concept of Eco City of 1970s which grew into sustainable city to Green City and later to Inclusive City to the present-day concept of Smart City to tackle the issues of urbanisation [15].

In the Indian context, Government of India focuses on 'sustainable and inclusive development in compact areas' basing the Smart City concept on core infrastructure elements. These elements are adequate water supply, assured electricity supply, sanitation including solid waste management, efficient urban mobility and public transport, affordable housing especially for the poor, robust IT connectivity and digitization, good governance, sustainable environment, safety and security of citizens, health and education.

The smartness in the urban areas is to be brought about through smart solutions such that employment is generated, incomes are enhanced and overall quality of life is improved especially of the poor and disadvantaged people so as to build inclusive cities [16]. The identified Smart Solutions address e-Governance and citizen services, waste management, water management, energy management, urban mobility and other segments like health, education and trade. Such cities should essentially be developed based on citizen-friendly governance. The government categorically stated the use of geospatial technologies for building of Smart Cities so as to best capture real-time data. However, the proposed Geospatial Information Regulation Bill (Draft) 2016, if enacted in its present form, may hugely and negatively affect the mega initiatives of the Government as it would not only hamper the development of the SMART Cities that are based on technology and Information Systems but at the ideological level too, its implementation may not be possible as it would a deterrent to the basic democratic rights of the people and would further digress from eDemocracy.

12.2.3 Urban Functionaries

Key activities of the urban areas planning, service provisioning and maintenance are delivered by the planning authorities, ULBs and parastatals, which mainly exist at the state and local levels. Among these, planning is largely guided by Urban and Regional Development Plans Formulation and Implementation (URDPFI) Guidelines, 2014.

Planning that takes place at the national level is limited to formulating policies, guidelines, model laws and fund outlays for urban development and maintaining centrally aided urban projects and research and training.

NITI Aayog (erstwhile Planning Commission) makes national five year plans, which have an economic approach. The Aayog addresses urban development as a component of its economic plans having the main function of ensuring a balanced and optimal utilization of country's resources.

Other authorities at the national level directly responsible for activities in the urban areas are MoUD and Ministry of Housing and Urban Poverty Alleviation (MoHUPA). Other ministries too deal with the urban areas but have a sectoral approach, like Ministry of Commerce and Industry, which is responsible for industrial development in the country, which may be in urban or rural areas. Such authorities are thus, indirectly involved in the development of the urban arears. Moreover, at the ministerial level, the approach is generally aspatial and economic in nature.

Among the two ministries, MoUD at the center is the nodal authority for urban planning and management having a constitutional and legal control over the urban development activities in Delhi and Union Territories. It plays a decisive role only for Delhi and the Union Territories. It plays a major role in territories cutting across many states, like in the case of National Capital Region (NCR), through a Board. For, rest of the States, it has an advisory role in policy and guideline formulation regarding development, environment, transportation, land, water supply and sanitation in urban areas. It is responsible for directing the growth of urban development in the country, which becomes a guideline for the states. Though sometimes, it is forced upon indirectly through controls associated with funding to intervene in urban development projects in the States.

Land, by and large, being a state subject, urban planning and management, falls under the purview of the state governments. The responsibilities are discharged through different agencies at state, regional, district and local levels. The State Planning Board formulates the state five year plans and annual plans within the framework of the national five year plans. The focus of the Board is economic planning [17] and like NITI Aayog, lacks spatial approach to planning.

In most of the states, a ministry in-charge of urban and regional development formulates the policies, strategies, priorities and programs, which are very similar in its role to that of Ministry of Urban Development except that the jurisdiction of the state ministry is not beyond its respective states. There is no similarity or standardization in the structuring of the ministry across the states [17]. Hence, in some states the urban ministry is separate, whereas in some it is clubbed with other ministries, particularly in smaller states, like in Mizoram reflecting the level of emphasis given to the sector. The state ministry is supported by Town and Country Planning Directorate/Department and other specialized agencies for meeting their objectives.

At the district level, urban planning is limited to few states like Gujarat, Maharashtra, Kerala and Goa and there too the approach is economic in nature. The planning at this level has regional connotation and falls within the purview of Metropolitan Planning Committee (MPC) and District Planning Committees

(DPC) catering to urban and rural areas under the 74th Constitutional Amendment Act (CAA). The MPC and DPC are empowered with the authority to prepare District Development Plan and Metropolitan Development Plan. Both need to incorporate the common interests of the plans prepared by the ULBs and panchayats, including spatio-economic planning [17].

ULBs in India exist as Municipal Corporation, Municipal Council, Nagar Panchayat, Town Improvement Trust, Notified Area Committee, Town Area Committee, Cantonment Board and others. The variation among these is in terms of their function, power, status and area of jurisdiction. Similarity among these is broadly in their near common purpose of provision and maintenance of facilities and amenities.

The core functions of the ULBs are providing water supply for domestic, industrial and commercial purposes, drainage and sewerage, solid waste management, preparation of plans for social and economic development and social justice, communication system including construction and maintenance of roads, footpaths, pedestrian pathways, transportation terminals, both for passengers and goods, bridges, over-bridges, subways, ferries and inland water transport system, transport system accessories including traffic engineering schemes, street furniture, street lighting, parking areas and bus stops, community health and protection of environment including planting and caring of trees on road sides and elsewhere, markets and slaughterhouses, promotion of educational, sports and cultural activities, record of births and deaths and aesthetic environment.

As the size of urban areas increased due to influx of population, the government set up multiple agencies to address the growing needs of such areas. The Urban Development Authorities (UDAs) mainly address the planning function while the parastatal agencies largely address the infrastructure related aspects in the urban areas and many a times have overlapping functions with the ULBs.

Moreover, the existence of UDAs came about to handle the incompetence of the existing setups rather than enhancing the capabilities of those setups [18]. The functions of UDAs are to promote and secure the development of the areas defined for such purpose in the plan document. To fulfil the function, the DAs prepare the master plan for the concerned area. For the development purpose the DA has the power to acquire, manage and dispose off land and to carry out building, engineering and other operations, to execute work in connection of supply of water and electricity, to dispose sewage and to provide other services and amenities, undertake maintenance and generally to do anything necessary to expedite for purpose of such development.

The UDAs are backed by respective state town and country planning acts. Over time, these authorities took over the task of maintaining of the newly developed urban areas, like Haryana Urban Development Authority in Haryana. In some states like Madhya Pradesh, there are Special Area Development Authorities (SADA) which perform the functions of planning and development in their jurisdiction.

Like in the case of ULBs, there is no uniformity in the setting up of these organisations across the states in the country. In some states these exist at the state level like in Haryana and Punjab. In other states, these bodies have been constituted at the city level, like Ahmedabad Development Authority, Kolkata Metropolitan Development Authority and many more. As noted by Bedi [18] there has been no set criterion for constituting these, say on the basis on population or area. There are some states that have the state level UDA as well as city/town level development authorities, like in Punjab, where at the state level, Punjab Urban Planning Development Authority exists and there exist six authorities in the cities of S. A. S. Nagar (Mohali), Amritsar, Bhatinda, Jalandhar, Ludhiana and Patiala [19]. The development authority in Delhi, known as Delhi Development Authority, is an exception as it exists under MoUD, a central government ministry. These authorities have overlapping functions with TCPDs and ULBs. Even though the jurisdiction with ULBs may differ but functions remain the same and there is a delayed transfer of the functions and jurisdiction to the ULBs. The duality in function and jurisdiction has led to redundancy and duplicity in tasks including data collection and creation within the individual IT setups of these organisations.

The 74th CAA has transferred the task of urban planning and development at the local level to the ULBs. At present, from urban spatial planning perspective the involvement of the ULBs is more in the implementation of the town planning schemes and in some instances in the enforcement of the master plans. The 74th Constitutional Amendment Act (1992) emphasised the role of ULBs in spatial and economic planning. It called for the integration of the town and city plans in the annual plans of the state and national level. Urban planning including town planning has been declared as the urban local body function and ULBs as the planning and development agencies.

As of date the delivery of the functions of urban spatial planning is primarily done by the TCPDs and by the development authorities, where these exist. The other function of urban planning is preparation of the town planning schemes; implementation of the plan and monitoring the development is done by the DAs and in some instances by the ULBs. However, there is no interstate or intrastate consistency in these either. It must be noted that the implementation, monitoring and enforcement are a part of the planning process as well and form a weak link in the spatial planning process. The process of spatial planning, from planning decision to monitoring has been split and assigned for operational purposes between different organisations. The split in the functions has led to the divergence of digital activities undertaken by these organisations as well as is made clear in sections below. This further iterates the need for a unified information system for the complete urban segment in the country.

12.3 Information Systems

Valacich and Schneider [20] define Information Systems based on components as combinations of hardware, software, and telecommunications networks that people build and use to collect, create, and distribute useful data, typically in organizational settings. Whereas, Laudon and Traver [21] have defined Information Systems based on role as interrelated components working together to collect, process, store, and disseminate information to support decision making, coordination, control, analysis, and visualization. Information Systems in present day world is networked hardware and software that people and organizations use based on some standards and defined procedures to create, collate, process and disseminate data. The terms Information Systems and Information Technology are used interchangeably, though Information System is an umbrella term which includes Information Technology within it. Information Technology comprises of essentially all the hardware, networks and software and excludes the people who are the users and the procedures that the people follow.

The information systems have expanded significantly over the years from ‘data processing’ in 1950s to 1960s, wherein the role of Information Systems was simple: transaction processing, record keeping, accounting, and other electronic data processing applications to ‘management reporting’ in 1960s to 1970s which consisted of processing of all the data into useful information reports, to help the manager in the decision-making. From 1980s to 1990s, new phenomenon appeared called ‘end user computing’, and it came as result of many factors such as the rapid development of microcomputer processing power, application software packages, and telecommunication networks. From 1990s to 2000s, rapid growth of Internet, Intranet, Extranet, and other interconnected global networks dramatically changed the capabilities of information systems. Internet-based and Web-enabled enterprise and global electronic business (eBusiness) and commerce systems have become commonplace in the operations and management of today’s business enterprises. Now eGovernance and eDemocracy is taking over the traditional practices of governance and democracy.

Till very recent times ‘Information Systems’ was understood more in terms of management of tabular information using computers, wherein the inputs were made in an out-of-the-box or customized dialog boxes and the outcomes or the results were in the form of either tabular reports, popularly known as the Management Information System (MIS). Information System in urban planning and management initiated with MIS. Though every aspect of the management and planning of urban area has a spatial dimension, it has been largely lacking in day-to-day activities. With the change in perspective of the System from just tabular to a more holistic one to include the location aspect, GIS is becoming an integral part of the Information System.

Evolution is an integral part of Information System. It is not a one-time activity. All the components of an Information System, be it hardware, software, networks, procedures, standards or users themselves need to be upgraded and updated with the changes in technology. As the hardware and software would need updation over time, so would the people who are the users in one form or other would need to update their skills through trainings and workshops. Hence, an Information System once set up needs to be looked at as an infrastructure [12] like a road network which is repaired and fixed and upgraded from time to time rather than discarding it and building a new one from scratch. This act is in opposition to the premise of building an Information System as it brings in redundancy.

In the urban context, based on the objectives being fulfilled, Information Systems can be categorized into Land Information System, Planning Information Systems, Cadastral Information Systems, Property Information System, Disaster Information System and many more that are domain specific. Similarly based on the kind of data being used as a prime input, information system can be categorized into MIS and GIS as well. Whereas GIS's prime concern is with spatial data having non-spatial data linked to it in form of attributes, MIS concerns only with non-spatial data. These can be broadly considered as Spatial Information Systems as well. Other than these, in recent decades some such systems have developed that are combinations of the information systems that have evolved since inception. These are the ones that incorporate information technology as the base, follow prescribed standards, are built on vast data, both spatial and non-spatial, incorporate applications for decision-making and are data dissemination nodes as well. Such systems and phenomena are called Spatial Database Infrastructure (SDI).

With the advancement in technology, many new concepts have come into play. The key among these that have come to be vital in the field of planning are crowd sourcing and cloud computing, Big Data and Internet of Things (IoT). Big Data is very large quantity data which measures in units over an Exabyte and may soon go over units like a zettabyte and yottabyte. As succinctly defined by Batty [22], Big Data is any data that cannot fit into an Excel spreadsheet. A human perspective over time especially in planning terms ranges over long term up to 25–30 years. The plannable human sense spread over space is rarely in minutes or hours. Whereas the data collected in today's world of everything and everybody connected and networked, falling under the regime of Big Data is measured and recorded in smaller measures of time. There is no doubt that the outputs from Big Data and IoT will enable in setting up smart cities, as like these newly emergent technological concepts smart city concept too is based on functioning of urban systems in short time [22]. In order for the outputs of IoT and Big Data to be used within the existing planning framework, time would play an important role, so that patterns over a year or more at least can be obtained so as to base planning decisions on which may become an input towards GIS based master planning for Atal Mission for Rejuvenation and Urban Transformation (AMRUT) Cities, which is aimed at providing basic services to households and build amenities in cities to improve the quality of life, especially the poor and the disadvantaged. AMRUT focus is on ensuring water supply, sewerage management, storm water drainage, urban transport and availability of green and open spaces, reform management and support, and capacity building [23].

12.4 Information Systems for Urban Areas

Information System is the use of systems, especially computers and telecommunication for gathering, retrieving, storing and sending information using some set procedures and standards with or without human intervention. The development of

Information System in the country has been influenced by the political climate and government policies. Computer age in India began in 1955 with the installation of the computer system in Indian Statistical Institute, Calcutta (now Kolkata) [24]. In the urban sector, the use of computers began mainly in two segments, namely in service delivery and management by the ULBs like in billing and MIS applications and for land use interpretation from aerial photographs and satellite images by the planning authorities, other than the routine tasks.

As stated by Rai and Kumar [25] aerial photographs have been long used for urban analysis. The photographs have been used for base map generation and creating thematic maps, however were found to be expensive and restricted to use in smaller urban areas. The dependence on aerial photographs lessened after the launch of remote sensing satellite in 1988, which can be considered a landmark year. The restrictions that the use of aerial photographs had like in metropolitan regions were overcome with the coming in of the satellite images. The application of satellite images in urban and regional planning gained momentum. The images were used for broad land use, land cover and urban sprawl mapping of large urban areas as well. Over the years with technological advancements, as the cost reduced and resolution of the satellite images improved from 90 m to sub meter, the use of the images spread to other divisions of planning, municipal and parastatal agencies.

The use and deployment of information technology be it computers, software, hardware peripherals, network or satellite images initiated as per the need clubbed with availability of funds and willingness to adopt technology as against the traditional planning and management practices. Hence, the development in adoption of information technology was departmental. With the liberalisation policies of 1991 and opening up of the private sector in this field, various proprietary technologies were adopted by numerous planning authorities, ULBs and UDAs. Even within the same municipal agency various departments and divisions made use of different technologies making it difficult for the departments to share their data.

The inclusion of information technology in other national level programs and missions like Jawaharlal Nehru National Urban Renewal Mission (JNNRUM), Rajiv Awas Yojana (RAY) was emphasized and promoted. However, the use and application of same was separate from the NUIS Scheme.

In parallel, country had embarked on building NSDI and National GIS. NSDI came into being through a cabinet legislation in June 2006 with the goal of being all-inclusive. Existing spatial data producers pledged to contribute towards the creation of national SDI [26]. NSDI addresses the critical need for acquiring, processing, storing, distribution and improving utilization of spatial data generated by various agencies of the Government of India. This structure of Indian NSDI has given it a sectoral character, very distinct from NSDIs in other countries, but very much in line with sectoral and segmented approach of national five year plans.

Geographic Information System is perceived as an effecting tool that can be integrated in the national level developmental plans and programs and hence decision-making [27] by the then Planning Commission, which has been the basis of the emergence of National GIS (NGIS). Planning Commission further iterated that GIS applications and decision support system addressing all the sectors is the key to the integration of NGIS within the framework and routine working of the government machinery. National GIS is envisaged as a multi-participant system available for government, public and business use. National GIS is essentially having a seamless GIS database, infrastructure, decision-support system with relevant applications and a single point access portal for the socio-economic development of the country so as to achieve inclusive growth. It is hence envisaged to enhance the decision-making capabilities of administrators [28]. In the most recent initiatives of the government, like SMART Cities Mission and AMRUT, the use of GIS based information systems is mandatory.

12.5 National Urban Information Systems

Internationally, Urban Information System does not exist at national level, but the inputs towards planning and management of the urban areas is provided by the national level set ups like NSDI. NUIS scheme contributes data to a trans-disciplinary and multi-organizational national level initiative of Government of India—NSDI. It is a unique scheme which can become the backbone of the urban setup of the country [18].

12.5.1 Genesis of National Urban Information System

The realization for an integrated system in the form of an information system in urban and regional planning was felt as early as 1976 during the annual conference of Chief Town Planners of states and union territories. Same was iterated in 1977 with a study and recommendation by United Nations (UN) experts on Information Systems who visited India to study the feasibility of Urban and Regional Information System (URIS). The experts recommended the implementation of computer-based spatial information system.

In pursuance to the UN experts recommendations, in a national seminar organized by TCPO and School of Planning and Architecture (SPA), New Delhi in 1979 deliberated upon the conceptual framework of URIS, information requirements and organizational framework. It was decided to set up a steering group on URIS to oversee its progress and set up URIS in TCPO, New Delhi in collaboration with state Town and Country Planning Departments (TCPD). In 1981 and 1983, the steering group suggested in the form of two reports, Urban Information Systems and Regional Information Systems, a minimum critical data to initiate URIS and suggested action on Urban Information System (UIS). Three broad categories of data were suggested, namely, physical, socio-economic and basic infrastructure, services and amenities. This was for the first time that urban information system was a separate entity from the regional aspect. Emphasizing on the urban aspects, the group recommended actions on Urban Information Systems, urban indicators and city data banks.

In 1984, National Natural Resources Management System (NNRMS) set up a task force on Urban and Rural Studies to study the existing system of information collection and management decision making in the area, ways to improve the existing system using conventional techniques, extent to which modern Remote Sensing (RS) techniques can improve the existing system and evolve balanced and integrated information and management system with an optimal mix of conventional and RS techniques. The task force identified the elements amenable to Remote Sensing and the areas where operational systems need to be put on priority basis.

However, none of these committees led to the setting up of an information system for urban or regional areas. From time to time deliberations and recommendations were made to promote the use of computers, remote sensing and GIS technologies. In 1988, National Commission on Urbanisation reiterated the need for URIS using GIS and Remote Sensing technologies. It laid importance on not just the routine collection of data and analysis but stressed on the use of georeferenced database for spatial planning. The Commission recognized that for developing such an information system, basic data had to be organized at the national level along with the data on environment for urban areas. The Commission stated that the designing of appropriate information system be using GIS and remote sensing technology must receive high priority.

In early 1990s, the parliamentary Committee observed the inadequacy of the up-to-date base maps for large number of urban areas and suggested the need for a program to generate base maps for every urban center in the country. Subsequent to this, in the eighth Five Year Plan (1992–97), Urban Mapping Scheme was taken up to map 50 urban centres in two phases from different states on a scale of 1: 2,000 and 1:10,000 in TCPO, New Delhi. The objective was also to enhance the capacity of the TCPO and TCPDs. This was the first initiative that was a translation from mere recommendations to an actual working scheme.

The URDPFI Guidelines (2015) talk about the techniques for the preparation of the base maps and the scales of mapping for various plans at various planning area levels. As a suggested action, the Guidelines spell out the need for URIS, with Town and Country Planning Organisation, New Delhi to be designated as the nodal agency.

As stated earlier, URDPFI Guidelines act as a foghorn for the planning authorities. The key deliverable of planning is a master plan. The Guidelines prescribe the data content, methodology and techniques to be used in the preparation of the master plans. The emphasis is laid on environmental aspects as well, particularly energy efficiency, climate change, green buildings and city's biodiversity status. It is mandated that the plans be prepared within the binding framework of environmental laws like EIA Notification, 2016, Environmental Protection Act, 1986, Forest Conservation Act, 1980 and Environmental Guidelines for Industries, Eco-Fragile Zones including coastal zones and water bodies in urban areas. The Guidelines provide specifications on data and procedures at macro level (regional) as well as micro (zonal) levels to include waste management, sewage treatment, solid waste management and rain water harvesting.

In 2006, Ministry of Urban Development, Government of India's launched National Urban Information Systems (NUIS) scheme for developing urban database to build an information system to enable planning, management and de-centralized governance in the context of provision of scientific planning and implementation of the 74th CAA. NUIS is a contributor of data to NSDI.

12.5.2 National Urban Information System

At the time of launch, the scheme envisaged covering all the 5161 (as per 2001 census) urban agglomerations and towns in the country. Apart from the mapping and attribute data creation of the urban centers, the scheme also addresses the prerequisites of data creation that is design standards (for image, mapping, GIS database, output and quality), hardware and software, capacity enhancement, procedures and identification of the urban areas [1].

The Scheme is being implemented in phases, primarily three. In the first phase select towns (152) have been taken up, for which the spatial as well as attribute data (GIS and Urban Data Bank) is being created. For 24 urban areas, detailed maps are being created from utility mapping perspective.

The scheme comprises of two components, Urban Spatial Information System (USIS) and National Urban Databank and Indicators (NUDB&I). USIS consists of development of Geographic Information System (GIS) with application tools to support master/zonal plan preparation and for ULB administration and utilities management. NUDB&I include designing and establishing a comprehensive data bank and its integration to support planning and monitoring the health of urban settlements and hence, caters to the attribute part of the spatial data. Both USIS and NUDB&I are built of a set of standards, procedures and parameters defined for GIS database, thematic mapping, applications, hardware and software, quality control and capacity enhancement. These standards find their basis in those of National Spatial Database Infrastructure (NSDI).

NUIS Scheme has been projected as an "integrated" tool that will support various planning levels/scales for urban areas. NUIS Scheme is much more than base maps and includes thematic maps (spatial), non-spatial data and should be developed as a "one-stop" resource for urban planning and management for each urban area. The scheme details the standards to be used in mapping and data collation for both USIS and NUDB&I. The layers required for mapping and its associated attributes have also been detailed out in the guidelines. The mapping is being done on 1:10,000 for the urban center and its demarcated periphery and 1: 2,000 scales for the urban core for all the centres. Since NUIS is based on the URDPFI Guidelines, it is built to incorporate the legal aspects and by-laws as well. The key use that NUIS has been put to is for master planning.

In the most recent scheme of government, Atal Mission for Rejuvenation and Urban Transformation (AMRUT), through deliberations, it was concluded that 1: 10,000 scale database content and accuracy generated for master planning was inadequate for the purpose. It was suggested that for the generation of comprehensive GIS database for Master Plan formulation, very large scale maps on a scale of 1: 4,000 should be prepared. This would entail revision of the design and standards of the NUIS Scheme for generation of comprehensive GIS database to meet the requirements of existing schemes as AMRUT [23], SMART Cities, Heritage City Development and Augmentation Yojana (HRIDAY) and other state level urban development projects.

NUIS is an Information System based on defined standards and procedures and not just a scheme. NUIS is an evolutionary concept and is not static. It was conceptualized on the concept of periodic updates and evolution. Hence, the data gaps that may be apparent in preparation of the master plan and in the management of the urban areas can be overcome by timely updates. The system is dynamic for inclusion of existing and new legal aspects as well.

12.6 ENVIS

As an information system exists for urban areas, environmental data exists at the national level in the form of ENVironmental Information System, popularly known as ENVIS. ENVIS Scheme was initiated by the central government in December 1982 (Sixth Plan) as a decentralized system with a web-based distributed network of subject specific databases under Ministry of Environment, Forest & Climate Change (MoEF&CC) [3]. ENVIS is governed by its guideline document that details out its objectives and framework. It specifies the guidelines for establishment, management and monitoring of the centers. However, there is no specification on the spatial component. The purpose of ENVIS is collection, collation, storage, retrieval and dissemination of information relating to a variety of themes associated with environment sector. The centers are mandated with providing referral service where information is not readily available.

The network presently consists of sixty-nine centres, of which twenty-nine address "State of the Environment and Related Issues" and are hosted by the Environment/Forest Department of State Governments/UT Administrations. Remaining forty centres, hosted by environment-related governmental and non-governmental organisations/institutes of professional excellence, have a thematic mandate. These include toxic chemicals, human settlements, coastal states of India, endemic and threatened plants, wildlife species, green planner, coastal districts of India, environment in Indian Parliament, Bibliography on forests, sacred caves, mining environment, renewable energy, microorganisms and environmental management, hazardous waste, biodiversity of Western Ghats and environmental news. Amongst these thematic specializations, the one most relevant to urban area is human settlements [3].

The key objectives of ENVIS is to build up a repository and dissemination centers in Environmental Science and Engineering to provide national environmental information services relevant to present as well as future needs and to support and promote research, development and innovation in environmental information technology. The State ENVIS centres are tasked with collecting data from line departments and other sources as required. ENVIS produces value-added information products relating to environment, forest, wildlife and climate change which is in the form of reports, apps, database and thematic maps.

Within the framework of ENVIS is Indian State-Level Basic Environmental Information Database (ISBEID) [3]. ISBEID helps the State level centers to collect, compile and disseminate environmental information on a centralized server, particularly covering the gaps on environmental related parameters like pollution, biodiversity and waste management amongst others. Its purpose is to provide internet-based GIS application having interactive maps and querying ability other than being a one-stop information source to users. The environmental data is collated for parameters like general administrative profile, ecology, infrastructure, energy, agriculture, industries, tourism and heritage, natural resources, forest resources, water resources, ground water resources, water pollution, air pollution, biodiversity, disaster, waste and sanitation. Nearly all the parameters, except agriculture, forest resources and biodiversity, are directly linked to urban areas and there is a necessity for their inclusion in the planning process of the city depending on its environmental condition.

The centers and ISBEID, however, do not have any specific standards and defined procedures for data creation or manipulation. As observed from the various ENVIS websites, the data is largely available in form of reports, tables and static maps, making the possibility of integration with other Information System a gargantuan task. The format in which the data is available refutes the purpose of creating an Information System as it leads to repetitiveness in data creation by the user and hence build on redundancy. This adds to the hindrance in creating an eDemocratic society which is an essential for Smart Cities.

12.7 Framework for Integration of Information Systems

Both NUIS and ENVIS are independent systems. Over time both will be updated and upgraded depending on the ideological and technological advancements in the fields of Information System as well as the domains of urban and environment.

ENVIS is an infrastructure base for the environmental aspects like NUIS is for urban sector. It must be noted that both the systems have been created with a well-defined purpose and objectives. The purpose of NUIS is to provide data towards master plan preparation among other planning and management activities of urban areas by urban authorities. Since the focus of urban planning and management is ensuring proper use of land and service delivery, NUIS creates data to meet these requirements especially in terms of scale, detail and quality. ENVIS on the other hand caters to the whole gamut of environmental issues and creates data ranging from terra, air to water; on, below over the land, be it natural or man-made.

As the need has been arising in the urban areas, the emphasis of the environmental aspects has been increasing. The environmental condition in the urban areas has been steadily deteriorating with increasing pressures of urbanisation. This may be attributed to not so efficient spatial, economic and social planning, plan implementation and monitoring. With the global concerns for climate change and increasing concern about its impact on humans, the need has arisen to look at the man-made environmental conditions at local levels. These levels are many times at sub-urban scale. In such cases the data gathered at the sub-urban micro levels needs to be aggregated at urban levels to be integrated within the planning framework.

With increasing environmental deterioration, it is imperative to include environmental aspects across the hierarchy of urban plans like the Coastal Regulation Zone (CRZ) lines within the NUIS database. There have been studies proposing an interface between NUIS and ENVIS [29], to overcome the gaps in both the systems from the perspective of building climate change responsive cities.

The framework proposed here is a conceptual structure to serve as a support for building an environmentally sensitive master plan. The prime objective of proposing a framework for integration of the two systems is to enable the citizens in taking informed decisions about their urban areas entailing not only the physical, social and economic aspects but environmental as well. The integration would contribute in achieving an environmentally sensitive spatial eDemocratic Smart Cities.

For the integration of the two systems, the existing design, methods and procedures of both the systems need to be assessed for conflict resolution. The assessment may lead to some fuzzy areas for which conflict resolution may have some underlying difficulties. In such the scenarios as suggested by Resnick et al. [30] and Breese et al. [31], pragmatic perspective may be used for simplifying the design rather than adopting a normative approach. A change that may be sought within the ENVIS would be adoption of the standards laid down under NSDI, such that its integration with other systems becomes possible. On a spatial level, this would consist of bringing parity in data for integration. For example, the environmental data may be collected at point source, which should be available with required details for integration on a spatial dimension, like essentially may consist of location information not just consisting of the name of the location but its latitude and longitude information other than other environmental data about the location. The framework for the integration of the two systems considers, other than the standards and specification for data creation, the differential ability of the users in the two existing systems and their existing methods as well as the urban functionaries who would be the prime users of the two systems with an intent to build environmentally sensitive cities, based on the principle of high degree and level of citizen's partnership.

Integration would be possible through the bringing together of the value-added data produced by the various ENVIS centers with the data produced under NUIS. In order for the users to be able to use the deliverables from both the Information

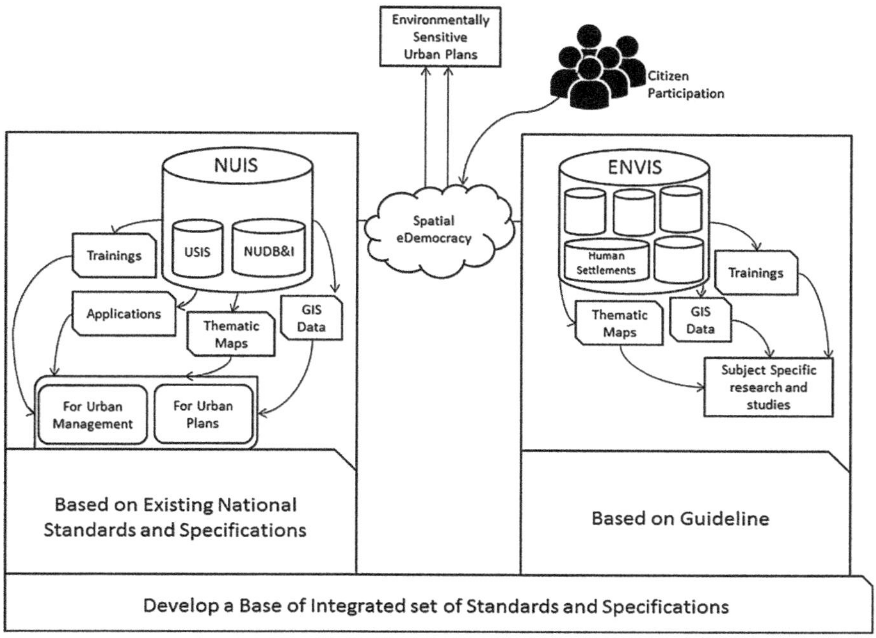

Fig. 12.2 Framework for integration of NUIS and ENVIS for building environmentally sensitive spatial eDemocratic smart cities

Systems, it is essential that the structure and platform of both the systems be same. Hence, it is proposed that similar standards for data creation, GIS database and thematic mapping be used as have been defined and updated for NUIS at the national level. The standards can be developed from the base provided by NSDI and NGIS (Fig. 12.2).

This would make possible the extrapolation and interpolation of the environmental data within the urban areas and further integration with NUIS database seamless and meaningful and can be further extended to GIS based planning for AMRUT cities.

It would be essential for one organisation to take up the onus of integration. Since the area under consideration here is urban, it could be an organisation within the urban ministry that could take on the responsibility. With the integration of the two systems, the organisation can ensure the inclusion of not only environmental aspects in master planning and building smart cities, but also e-participation of the people and hence eDemocracy. eDemocracy depends on citizen's ability to access and generate information relevant to governance. With spatial component becoming available, through the integration of NUIS and ENVIS, to the citizens through web portals, the organisation would be allowing the citizens to influence the policy decisions and bring in more transparency in spatial dimension.

The ulterior goal of the information system could be achieved by negating repetition and overcoming redundancy. This would help towards the national mission of SMART Cities by providing integrated systems, which can be used for building environmentally sensitive master plans through a spatial eDemocratic approach. This would be possible as the citizens would be able to access spatially enabled planning derivatives to voice their opinion about their cities and meet the criteria of the geospatial data being a public good as suggested by Rao [9].

12.8 Conclusions

NUIS and ENVIS are two important information systems in the context of urban area, particularly pertaining to planning and environment. Since now, information systems has come to be understood as part of country's infrastructure and is a public good, its updation and upgradation is imperative. This would entail inclusion of new schemes like AMRUT. The master planning has progressed from being a centralized activity to being participative.

Whereas, NUIS along with its newer *avatar*, AMRUT, is essentially spatial in nature, having one of the objectives of easing the task of master plan preparation by providing GIS based updated database of the urban areas based on the national standards, ENVIS exists as an aspatial system, where data is collected by various Centers and disseminated through the Web mainly in the form of reports.

As our urban areas are progressing to become Smart Cities in the near future, the prime focus is making these areas environmentally, socially and economically viable by providing a good quality of life. To achieve this, it is imperative that spatial eDemocracy be made the basis of attaining the objectives of Smart City. The case of spatial eDemocracy is attainable when the platform is Information Systems, specially NUIS and ENVIS, the two existing systems relevant to urban areas. The two systems that exist in isolation to each other need to be integrated for which the framework has been suggested.

Currently, though the progress has been from government to governance and eGovernance and from manipulative democracy to eDemocracy, both are still non-spatial. Integration of the systems will help transform the combined information systems into a spatial entity, as the non-spatial information would complement in the form of attribute data. Further open and free access to spatial information would enable citizens to e-participate on a spatial dimension, thus promoting spatial eDemocracy.

Once the standardization has been built into these existing systems, its integration with others database that may be a part of NSDI and NGIS or may exist in isolation like Health Management Information System, could be achieved as a future step.

References

1. Town and Country Planning Organization (2006) National urban information systems—design and standards. Ministry of Urban Development, New Delhi
2. Bedi P, Mahavir (2013) Indian national urban information system as a decision making tool for urban planning. Geospatial World Forum, Rotterdam. Accessed 07 Dec 2016 geospatialworldforum.org/2013/presentation/Prabh%20Bedi.pps
3. Ministry of Environment and Forest (2013) Guidelines for environmental information system scheme. Government of India, New Delhi
4. Census of India—Census Terms. Census of India. http://censusindia.gov.in/Data_Products/Library/Indian_perceptive_link/Census_Terms_link/censusterms.html. Accessed 30 Sept 2016
5. Kulshrestha SK (2006) Dictionary of urban and regional planning. Kalpaz Publications, Delhi
6. Government of India (1950) Constitution of India, Seventh Schedule, Article 246. http://lawmin.nic.in/olwing/coi/coi-english/Const.Pock%202Pg.Rom8Fsss(35).pdf. Accessed 13 Sept 2016
7. Habermas J (1971) Towards a rational society. Heinemann, London
8. Mcloughlin BJ (1985) The systems approach to planning. A critique, Victoria
9. Arnstein SR (1969) A ladder of citizen participation. J Am Plan Assoc 35(4):216–224
10. Government of India (2011) Right to Information Act, 2005. http://righttoinformation.gov.in/rti-act.pdf. Accessed 15 Oct 2016
11. Gonzalez-Jaramillo VH (2015) Use of geographic information systems with open source solutions, an approach to access edemocracy and egovernment. 2015 Second International Conference on eDemocracy & eGovernment (ICEDEG)
12. Mahavir, Bedi P (2012) A case for financing the 'data' as part of infrastructure', Coordinates July 2012
13. Rao M (2016) "Public Good" of geographic information. Mycoordinates.org. doi:06.2016
14. Mahavir (2016) The geospatial information regulation bill: the national development missions would directly suffer setback. Mycoordinates.org. doi:06.2016
15. Bedi P, Mahavir (2016) Smart cities and not so smart geospatial information regulation bill 2016, in SPANDREL, Issue 12, December 2016
16. Ministry of Urban Development (2015) Smart city: mission statement and guidelines. Government of India, New Delhi
17. Kulshrestha SK (2012) Urban and regional planning in India: a handbook for professional practice. Sage Publications, New Delhi
18. Bedi P (2013) 'Spatial technological initiatives in key central government urban development initiatives' spatio-economic development record 20(1):16–21
19. Punjab Urban Planning and Development Authority 2011, Authorities, Government of Punjab http://puda.nic.in/. Accessed 13 Sept 2016
20. Valacich J, Schneider C (2009) Information systems today—managing in the digital world, 4th ed. Prentice-Hall, Englewood Cliffs
21. Laudon KC, Traver CG (2012) Management information systems, 12th ed. Prentice-Hall, Englewood Cliffs
22. Batty M (2013) Big data, smart cities and city planning. Dialogues in human geography. Sage Publications, New Delhi. doi:10.1177/2043820613513390
23. Town and Country Planning Organization (2016) Formulation of GIS based master plan for AMRUT cities. Ministry of Urban Development, New Delhi
24. Rajaraman. Computer Age. http://www.cbi.umn.edu/hostedpublications/pdf/Rajaraman_HistComputingIndia.pdf. Accessed 29 Sept 2016
25. Kumar VK, Rai PK (2011) Role of geoinformatics in urban planning. J Sci Res 55:11–24
26. Kumar RS (ed) (2009) Indian NSDI: a passionate saga. NSDI, New Delhi
27. Planning Commission (2011) Establishment of 'National GIS' under Indian National GIS Organisation (INGO), Government of India, New Delhi

28. Nayak S, Rao M (2011) 'National GIS: Shaping India', Coordinates, vol. VII, issue 7, July 2011. Available from: http://mycoordinates.org/national-gis-shaping-india-2/. 20 Sept 2016
29. Bandyopadhyay P, Mahavir (2016) Climate change responsive city planning: a case for developing interface between ENVIS and NUIS. SPACE, vol 19. School of Planning and Architecture, New Delhi
30. Resnick O, Iacovou I et al (1994) GroupLens: an open architecture for collaborative filtering of netnews. ACM Conference on Computer Supported Cooperative Work, pp 175–186
31. Breese J, Heckerman D et al (1998) Empirical analysis of predictive algorithms for collaborative filtering. In: Proceedings of the fourteenth conference on uncertainty in artificial intelligence, Morgan Kaufmann. http://www.research.microsoft.com/users/breese/cfalgs.html. Accessed 12 Nov 2016

Chapter 13
Geo-enabled e-Democracy Tools and Services for Smart Cities

Pramod Kumar, Kshama Gupta, Harish Chandra Karnatak, Asfa Siddiqui and A. Senthil Kumar

Abstract In the recent past, an overwhelming growth in geo-enabled open source data and tools through web services and data repositories is witnessed. Internet technology has significantly enhanced the utility of geo-enabled data and applications by making them more accessible to a wider range of users, planners and decision makers through geoportals, mobile Apps and Cloud GIS. The Cloud Computing Architecture (CCA), Internet of Things (IoT) and Service Oriented Architecture (SOA) represent new technological development which allow them to send and receive data without requiring user interaction and enhance interoperability in data and information services. The geospatial information available through geoportals and online data repositories have immense scope for its utilisation in smart city planning with many success stories world over. Geo-enabled data and tools can go a long way in creating a range of smart city applications where citizen participation is one of the key objectives. These tools and services have immense application potential for public participation, grievance management and to address many more aspects of e-democracy and e-governance such as Tourism GIS, Municipal GIS and so on. These citizen-centric Apps and web services facilitate faster information dissemination and improve the efficiency and

P. Kumar · K. Gupta (✉) · H.C. Karnatak · A. Siddiqui · A. Senthil Kumar
Indian Institute of Remote Sensing, 4, Kalidas Road, 248001 Dehradun, India
e-mail: kshama@iirs.gov.in

P. Kumar
e-mail: pramod@iirs.gov.in

H.C. Karnatak
e-mail: harish@iirs.gov.in

A. Siddiqui
e-mail: asfa@iirs.gov.in

A. Senthil Kumar
e-mail: senthil@iirs.gov.in

T.M. Vinod Kumar (ed.), *E-Democracy for Smart Cities*, Advances in 21st Century Human Settlements, DOI 10.1007/978-981-10-4035-1_13

management of infrastructure, which is essential to enhance the quality of life of urban dwellers and one of the key objectives of the smart city movement. In India, the "*Bhuvan*" geoportal developed by the Indian Space Research Organisation (ISRO) provides a milieu of data sets which can be used for building smart city applications. *Bhuvan* portal hosts high-resolution data (~ 1 m resolution) of more than 350 Indian cities till date and planning to cover other cities in near future. It also offers thematic maps useful for Master Plan formulation for 152 towns prepared under National Urban Information System (NUIS). Effort is on to use high-resolution satellite images for the overlay and fine-tuning of Urban Framework Survey. It also hosts many other data sets, e.g., land use/land cover, road network, soil, geomorphology, etc. which can be used to plan and manage the smart cities effectively.

Keywords Geospatial · Geoportals · Mobile apps · *Bhuvan* · Smart cities

13.1 Introduction

Globally, more than 54% of population now live in urban areas [1]. The urbanisation is witnessed across the world but Asia and Africa continents are getting urbanized more rapidly than other parts of the world. Asia has a low level of urbanisation as compared to developed world yet comprise of 53% of the world's urban population. India is also one of the countries, where urbanisation has happened slowly compared to other developing countries, yet the size of growing urban population in terms of absolute numbers is more due to its sheer population size. Besides, it is expected that in near future India will witness urbanisation at a speed which has never been seen before [2]. The growing urbanisation is expected to bring its own challenges in terms of urban population living in substandard housing, overcrowding, congestion, deficient infrastructure, degraded environment, climate change and so on.

To address the issues faced by today's urban areas and to make the cities liveable and sustainable, the smart city concept is advocated. There is no universally accepted definition of smart city, however, smart city is expected to improve quality of life of its citizen and to provide core infrastructure efficiently. The term 'Smart City' has different connotations in different parts of the world. In general, planners always aspire to develop the entire urban ecosystem, which can be represented by the four pillars of sustainable development—physical, social, economic and institutional infrastructure. This is definitely a long term goal and smart city can add a layer of smartness to achieve these goals (http://smartcities.gov.in/writereaddata/smartcityguidelines.pdf).

Webster's defines democracy as "a government in which the supreme power is vested in the people and exercised by them directly or indirectly through a system of representation." The core concept of democracy is to actively engage an informed citizen in decision making process. Therefore, to achieve the goal of comprehensive smart city development, it is desirable that citizen ought to be made part of planning process and decision making to empower the democratic reforms. E-Democracy is simply the utilisation of information technology to facilitate the goal of achieving full democracy.

Geo-enablement of E-Democracy tools is a concept that adds location to existing information, thereby disseminating knowledge to stakeholders more efficiently in an interactive manner and assume an essential part in comprehension and tending to the challenges that is confronted in an inexorably intricate and interconnected world. Geo-enablement of citizens are a need of the hour in order to have the right tools and information in hand to take the right decisions [3]. The information provision is the first step for enhancing citizen participation in e-democracy, where geo-enabled e-Democracy can play a major role. Geo-enabled e-Democracy is facilitated by the Geospatial technologies which relates to the collection and processing of data that is associated with geographic location. It is a fusion of multiple technologies and majorly comprises of Remote Sensing, Geographic Information System (GIS) and Global Navigation Satellite System (GNSS). Remote Sensing is a technology which is concerned with acquisition of information and data about an object without coming in direct contact. The example includes photographs taken from GPS enabled smart phones, satellite images etc. which portray earth features with embedded location information. GNSS provide the satellite based geo-location and is increasingly being used in day-to-day life for navigation and geo-location. The best example can be given in the form of navigation app that helps a common man to locate the unknown address with precision. The remote sensing data and GNSS information is integrated with GIS to produce meaningful information and analysis of data for host of applications in present scenario. The advancements in this field from Desktop GIS to Geoweb 2.0 platforms have enabled hosting of geoportals for information dissemination and public participation. The popular geoportals such as Bing maps, Google maps, *Bhuvan*, OpenStreetMap, etc. are being used by international community and governments for providing datasets and information for facilitating collection of geographic information volunteered by people and for enhancing people participation in decision making. The *Bhuvan* geoportal hosted and developed by the Indian Space Research Organization (ISRO) is providing an integrated platform to decision makers and planners for accessing multitude of geospatial datasets which can be utilized for the development of a range of smart city applications. Thus, Geo-enabled E-Democracy tools can provide necessary datasets, tools and mechanism for facilitating the citizen participation in improving the liveability of urban areas. They provide better visualization of location specific data and improved dissemination of information and ideas.

13.2 E-Democracy

E-democracy is a combination of the words—electronic and democracy, is also known as digital democracy and it incorporates information and communication technology to promote democracy (https://en.wikipedia.org/wiki/E-democracy). In this form of government, it is believed that all citizens are equal and they can participate equally in the process of decision making. The technological advancements in this century has enabled governments to connect with masses through internet media which provides a massive digital network with open standards, universal and inexpensive access.

The United Nations Economic Commission for Europe (UNECE) Convention on Access to Information, Public Participation in Decision making and Access to Justice in Environmental Matters (Aarhus Convention) outlines three mainstays of E-democracy as access to information, public participation and access to Justice [4]. Access to information is the first step to ensure accountability, transparency and involvement of people, which is highly desirable for democratic systems. It enables people to be an active participant in decision-making process at all levels, which is essential to uphold democratic organizations. Similarly, E-democracy is intended to increase the public participation in decision making. Effective public participation in decision making enables public to express their views, be a part of development process and also allows decision makers to address the opinions and concerns of the target groups, which are relevant to the issues in hand [5]. The process yields far better results as it minimizes the later unrest and can speed up the implementation process.

Gaining access to information and public participation in decision making is one of the hallmark of smart city concept. A crucial objective of smart city concept is to empower urban citizens and groups by offering them necessary information and skills to bring changes in their own community. This places a new challenge in front of planners and policy makers to bridge the digital divide and coping with Information and Communications Technology (ICT) illiteracy for effective participation. In this direction, geo-enabled tools and skills can be highly valuable as it helps users relate to their immediate surroundings with locational dimension and visualize the information in hand. It assist them in better comprehension of the information as well as to understand the corollaries of decisions taken.

13.3 The Need for Geo-enablement

Geo-enabled information has biggest advantage that it carries the location information. Location is considered as the fourth driver in decision-making other than the social, economic and environmental factors. Geo-enablement is not just about developing and using GIS. It engages client to dissect data in its locational setting, which furthermore empowers them to comprehend different components having an effect in decision making process. Directly or indirectly, the spatial information is used in every aspect of day to day life, be it identifying an address or a facility or a service. For smart city to be truly smart, the planners need to analyse a host of information pertaining to multifarious aspects in spatial context. In many Indian cities, the phenomenon of urban sprawl and haphazard unplanned growth is making a colossal issue.

Many cities in the world are grappling with the issues of overpopulation, inefficient infrastructure and transport systems, pollution, congestions and environmental degradation. During the recent times, the pollution levels in Kanpur and Delhi in India, Xingtai in China, Al Jubail in Saudi Arabia, etc. has reached to its highest level and that has caused serious health issues to the urban dwellers. Among the various reasons, poor availability of Transit Oriented Development (TOD) and mismanagement of available resources have caused the precarious situation. The TOD requires spatial distribution of land uses as one of the primary information. Hence, availability of comprehensive geo-information is the basis for outlining the solutions to the multiple problems and the welfare of urban populace.

Mitigation and adaptation strategies for any disaster management calls for multitude of spatial information for delineating the hazardous zones, be it flood, earthquake or a fire disaster. Geospatial technologies especially remote sensing is highly capable of providing geo-enabled information regarding slope, land use, building information, earthquake prone zones, etc., which is useful to identify vulnerable areas. Even in post-disaster scenario, damages to property and infrastructure could be assessed, if geo-enabled information is available and accessible to stakeholders. Provision of uniform and well-distributed urban green spaces to cater to the recreational and environmental needs of the population is envisaged in Smart City mission [6]. Geo-enabled spatial data assist hugely in the assessment of distribution of Urban Green Spaces [7]. Similarly, there are many needs and challenges in front of planners which require geo-enabled spatial data. Although, geo-enabled data alone cannot provide the solutions but it is most crucial for addressing the needs and forms the basis of urban planning and analysis.

13.4 Geospatial Technologies

Geospatial technologies are an amalgamation of various technologies, namely Remote Sensing, GIS and GNSS. It provides unique datasets, techniques and tools for the geo-enablement of spatial information and also has a role to play in every aspect of smart city planning be it smart living, smart environment, smart mobility, smart people, smart economy or smart governance, the pillars of smart city concept covering all facets of sustainable growth [8]. These technologies have been used effectively for addressing the various aspects of urban sustainable development such as urban sprawl and growth monitoring [9, 10], suitability analysis for urban development [11], evaluation and distribution of urban green spaces [8, 12], urban hazard analysis and so on. The geospatial technologies provide multitudes of datasets and has immense capabilities for analysis, visualization and dissemination of spatial information.

13.4.1 Data Sources

The data acquired from remote sensors provides unique information about earth system and also provides innumerable benefits over conventional methods due to its synoptic coverage, repetitive view and maps produced at multiple scales. This technique has advanced notably in last few decades since the first remote sensing image which was acquired from a hot air balloon in 1858 [13]. Today, a host of remote sensing datasets are available ranging from 1 km spatial resolution for global coverage to a very high spatial sub-meter resolution which offers opportunities to study and monitor urban areas in great detail [14]. It does not only provides information in two-dimension (2D) but also in third-dimension (3D). The high resolution satellite stereo imageries and 3D point cloud generated from Light Intensity Detection and Ranging (LiDAR) data is being increasingly used for obtaining information on 3D profile of urban areas. These datasets provide a large volume of information which can be harnessed for providing up-to-date geo-enabled information to the citizens for decision making.

The Town and Country Planning (TCP) Departments in many cities across the world have a few information in spatial or non-spatial framework to strengthen smart city applications. Most of these information in developing countries are in paper form and need to be digitized. As the urban managers, typically lack capacity in developing countries, to digitalize the analogue maps often they hire specialists to create digital database. Generally, the geospatial information for smart city applications could be derived from the following datasets:

- **Satellite remote sensing**: Satellite images of varied resolution are available across the globe, thus mostly obviate the need to freshly procure the data. Fresh data acquisition is necessary only for particular purposes or date of acquisition following some changes that necessitate the investment in remote sensing data. The stereo images can provide height information to define terrain and 3D city models.
- **Aerial photograph**: Traditionally, aerial photographs provided the efficient means of capturing information with greater details and scale of mapping. However, restrictions in some part of the world to capture details through aerial photography imposed limitations and/or lengthy procedures (i.e., defence clearance, etc.) for data acquisition. While traditional (nadir) aerial photography offers the most efficient form of image capture, oblique aerial photography provides images of building facades. This is especially helpful to build 3D city model, a vital element in smart city planning.

- **Airborne LiDAR**: LiDAR mounted on aircraft, can map building, tree, power line, etc. and useful for terrain modeling (even under vegetation) and 3D city modeling. Being aircraft based, it has the same benefits and limitations as outlined for aerial photography. Aerial cameras are often co-mounted with the LiDAR sensors. The airborne LiDAR data is rare and by and large not available for the urban areas in developing countries.
- **UAV**: Unmanned Aerial Vehicles (UAV) offer a viable platform for capturing smaller areas on earth surfaces. Current technology provide good quality 3D photographic solutions. However, the developing world is still groping with the security provisions of UAV data.
- **Terrestrial LiDAR**: Terrestrial LiDAR is useful to generate point clouds from multiple view angles to define building façade and interiors (or Building Information Models—BIMs). Terrestrial LiDAR data can also be captured from a moving vehicle which is well suited to define the complex assets and infrastructure along urban road corridors.
- **Smart Phones**: GPS mounted smart phones can easily engage citizens and help in geo-tagged data and photographs collection. If the data is collected from multiple viewpoints then it can also help to build 3D models based on photogrammetric principles. GPS mounted digital cameras or smart phones capture images with multiple view points and following the image co-registration using tie-points help to build point clouds and 3D models.
- **Field survey**: Field survey remains the traditional method for data collection or for the validation of data collected through secondary means.

13.4.2 Decision Making through GIS

Geographic information system (GIS) or Geospatial Information Systems is an integration of technologies that capture, store, analyse, manage, and present data and information related to location(s). In the simplest terms, GIS is an integration of cartography, statistical analysis, and database system with information technology. During the initial phases of development, GIS has been widely used for digital data conversion known as digitization, storing and generating printed maps with little focus on spatial analysis (http://elearning.iirs.gov.in). Later, GIS emerged as a tool which integrates spatial and non-spatial data to solve the location- specific real world problems through modeling. Modern GIS technologies uses digital information for which various digitized data creation methods are being used. The most common method of data creation is digitization, where a hard copy map or survey plan is transferred into a digital medium through the use of a Computer-Aided Design (CAD) program or GIS software with geo-referencing capabilities. With the availability of ortho-rectified images both from satellite and aerial sources, onscreen digitization is becoming the main avenue through which geographic data is extracted. On-screen digitization involves the tracing of geographic features directly on top of the images by tracing the geographic form on a separate digitizing tablet.

13.4.3 Global Navigation Satellite System

Global Navigation Satellite System (GNSS) is also termed as positioning infrastructure and it has gained widespread use for geodesy, surveying, navigation and for capturing of geo-enabled data. GNSS has three main segments—(a) Space segment which consists of constellation of navigation satellites, (b) Control Segment comprises of ground-based network of master control stations, data uploading stations and monitor stations for monitoring the satellites' signals and status and (c) User segment which consist of equipment that processes received signals from GNSS satellites to derive the location and time information (https://www.gsa.europa.eu/european-gnss/galileo/galileo-european-global-satellite-based-navigation-system). Hence, GNSS is considered to be one of the important technologies for providing geo-enabled location data [15]. Major GNSS systems around the world are as follows:

(a) **Global Positioning System (United States)**—It is the first operational GNSS system, launched by USA in 1970s. It has a constellation of 24 satellites providing global coverage (https://www.nasa.gov/directorates/heo/scan/communications/policy/GPS_History.html).
(b) **GLONASS (Russia)**—It is launched by Russian Government in 1982 for providing weather positioning, velocity measuring and timing services anywhere in the world and has constellation of 24 satellites distributed over three orbital planes (https://www.glonass-iac.ru/en/guide/).
(c) **Galileo (European Space Agency)**—It is a GNSS system launched by ESA named after Italian astronomer Galileo Galilei launched in 2005. The basic aim of the system is to provide high-precision positioning services to European nations, independently. 14 satellites out of 30 satellites have been launched under this system by May 2016 and the system will be fully operational by 2019 (http://www.navipedia.net/index.php/Galileo_Future_and_Evolutions).
(d) **BeiDou (China)**—It is the Chinese Navigation System which envisages to have 35 satellites in its constellation to provide global coverage by 2020 (http://en.beidou.gov.cn/).
(e) **Indian Regional Navigation Satellite System (India)**—It is a regional navigation System, also known as NavIC, being developed by India. It has constellation of 7 fully operational satellites to provide high-precision positioning services in Indian region covering primary service area of 1500 km from its boundary. Its extended service area covers the area between primary service area and area from 30°S to 50°N latitude and 30°E to 130°E longitude (http://www.isro.gov.in/irnss-programme).

GNSS system provide high-precision positioning services with such a precision that could be envied by traditional surveyors. Presently, its application ranges from Terrestrial, Aerial and Marine Navigation, Search and Rescue operations, Disaster Management, Vehicle tracking and fleet management, Integration with mobile phones, Precise Timing, Mapping and Geodetic data capture, Terrestrial navigation aid for hikers and travelers, Visual and voice navigation for drivers (http://www.isro.gov.in/irnss-programme). Its use has been increasing exponentially in day to day life with increasing use of navigation apps and services.

13.4.4 Benefits of Geospatial Technologies

The geospatial technologies have several benefits in smart city applications and they generally fall into five basic categories:

(a) **Time and cost saving**: The remote sensing data provide repetitive and synoptic view of urban areas at relatively greater efficiency. Using these datasets, maps could be prepared with better accuracy and at relatively lesser time and cost. The traditional method of land use/land cover survey typically costs Indian Rupees 100 per sq. km as compared to Indian Rupees 20 per sq. km using remote sensing data [16].
(b) **Data visualization**: GIS-based maps and visualizations enormously help in building scenarios and subsequent planning. It enhances communication between various groups, divisions, disciplines, professionals, associations, and people in general.
(c) **Data catalogue**: GIS enables better data management. The GIS enabled data could be easily retrieved, manipulated and distributed over variety of platforms. It becomes feasible due to data cataloguing.
(d) **Informed decision making**: The geospatial technologies enable the creation of "what-if" scenarios for informed decision making. Common examples include site suitability analysis, route/corridor selection, zoning, planning, conservation, natural resource extraction, etc.

13.5 Evolution of GIS

The advancements in geospatial technology is allied with high computing capabilities and advanced visualization systems using contemporary ICT solutions. GIS has quickly incorporated distributed systems and databases for interoperable solutions for effective decision making whereas microcomputer has allowed GIS to be applied to new fields and has improved GIS education and awareness. Due to advancements in ICT, the data creation and storage mechanism has drastically changed in recent past using online participatory approach known as crowdsourcing or collaborating mapping. Today, the current GIS technology enables the concept of "*map anywhere and serve anywhere*". With recent developments, there is a leap in the development of spatial analysis tools and logical processing methods in online GIS applications and platforms. This enabled the development of numerous spatial algorithms, spatial modeling techniques and better display and visualization of geospatial data for effective planning and decision making.

These advances in geospatial technologies can play a very meaningful role in urban development specially in smart city planning where the citizen gets involved through participatory approach. The geographic distribution of citizen services can provide a value added information and service platform for geo-enabled governance. The coming sections of this chapter provide a brief overview of various technological advancements in geospatial domain which can be effectively used for urban and regional development with a specific emphasis on smart city planning.

13.5.1 Desktop or Single User GIS

The desktop GIS or single user GIS is an entry level, low cost solution that provides basic GIS data access, querying as well as map production and spatial analysis capability to its users. Desktop GIS applications provide excellent platform for geospatial data analysis to GIS professionals by providing the tools and technologies to analyse the patterns and problems for effective land management solutions. Technically, the desktop GIS is a collection of software and data products that runs on standard desktop computers. In a typical desktop GIS configuration, the digital geospatial data storage mechanisms vary from simple file system to database systems. Typically, the geospatial data is created and stored either in raster or vector data format (Fig. 13.1). The raster and vector format of geospatial data storage have certain advantages and disadvantages over each other.

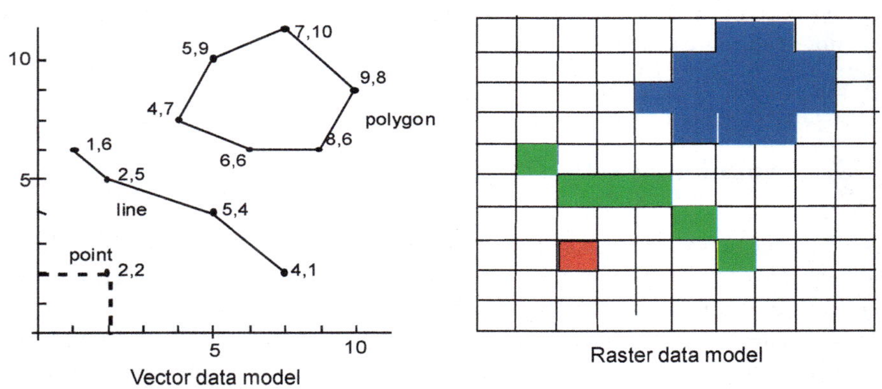

Fig. 13.1 Vector and raster representation of geographic features

The desktop GIS applications are build using specific GIS software and initially, data interoperability was a major issue. The maps produced in initial phases by the local agencies were not made accessible to the public as they need to install GIS software as well as has to have knowledge of GIS. However, with technological advancements, now a host of user friendly interoperable desktop GIS software are available in proprietary commercial format as well as free and open source software [17]. Some of the popular desktop GIS software are listed in Table 13.1.

The desktop GIS can be used in smart city projects for GIS data creation, analysis and also for the preparation of GIS based plans. It fits well in offline mode where the system network is not accessible. The handling of large size data sets will give better performance in desktop based GIS. The major limitation in this GIS platform is synchronization of offline and online data sets and collaborations in multiple stockholders, especially in participatory approach.

Table 13.1 Popular desktop GIS software

Name of the software	Software type	Description
GRASS GIS	Free and open source	Major tools available in the software includes geospatial data management and analysis, image processing, graphics and maps production, spatial modeling, and visualization (http://grass.osgeo.org)
QGIS	Free and open source	QGIS is one of the most popular user friendly and open source GIS software which provides extensive tools for vector data handling. It can be easily integrated with GRASS and R (http://www.qgis.org/en/)
gvSIG	Free and open source	**Gvsig Desktop** (gvSIG from this point forward) is a free GIS software and aims to represent, edit, analyse and manage information from the point of view of spatial relations (http://www.gvsig.com/en)
RGeostats	Free and open source	RGeostats is the geostatistical package developed by the Geostatistical Team of the Geosciences Research Center of MINES ParisTech (http://rgeostats.free.fr/)
ILWIS	Free and Open SOURCE	ILWIS is popular GIS and image processing platform which provides various tools for raster and vector data analysis with special emphasis on water resource applications (http://www.ilwis.org/)
ArcGIS	Commercial	ArcGIS is a popular commercial desktop GIS software which provides vector and raster based GIS data creation, processing and analysis tools with various extensions especially in windows operating system platform. Vector data handling is major strength of ArcGIS (https://www.arcgis.com/)
ERDAS Imagine	Commercial	ERDAS is a popular commercial image processing software which provides various spatial data analysis tools for raster data formats (http://www.hexagongeospatial.com)
ENVI	Commercial	ENVI is another popular image processing software which provides tools for LiDAR, SAR, multispectral or hyperspectral data sets (http://www.exelisvis.co.uk/)

13.5.2 WebGIS and Geospatial Web 2.0

The recent advancements in Internet and related technologies has extended the use of geospatial data and information for variety of applications at different levels. The integration of GIS and Internet technology has emerged as one of the exciting technological advancement in geospatial domain known as Internet or Web based GIS. The web enabled GIS facilitates decision making at the strategic, tactical, and operational levels. It has enhanced the performance of administrative operations and serving as a gateway for decision-makers and general users to access the system conveniently and effectively. Some of the popular web based spatial data and information services are Google Maps, *Bhuvan* geoportal, Bing Map, OpenStreetMap, etc. Today, Internet is emerging as a popular means of GIS data accessing, analyses and transmission and information services for various online applications.

In Internet technology, the World Wide Web (www), File Transfer Protocol (FTP) and Hypertext Transfer Protocol (HTTP) make it convenient to access and transfer data files across the globe. The Internet provides GIS users easy access to acquire GIS data from central server system to diverse data source in distributed environment. It is a means for users to exchange GIS data, conduct analysis and prepare outputs in the form of maps, report and web services. The users of Internet GIS application can use advanced GIS tools for analysing their data without having any specific data or GIS software installed in their computer device [18]. The advent of web 2.0 in Internet technology has opened a new dimension for geospatial data applications by introducing Geoweb 2.0 for GIS data and applications. Geoweb 2.0 allows more interactive and latest GI system as a mashup architecture which is very effective for development of user defined geospatial applications for planning and decision making [19]. The web GIS is more accessible than Desktop GIS to the public and hence is advantageous for viewing real time data with dynamic queries. Geoweb 2.0 has high potential for enhancing public participation as it covers all aspects of participation starting from information dissemination to collaboration to empowerment of citizens in decision making. It empowers users to view, modify and provide alternatives to the solutions without extensive training [20–22].

The web based GIS applications are dynamic in nature, which is very important for its wider utilisation e.g. if any client(s)/user(s) or database administrator updates the data or information at server end, it will be available to all the users concurrently. It can also link real time data and information such as satellite images, weather information, traffic movements and accident information, etc. by integrating online sensors and observatories. A typical GIS application usually includes three essential elements: (a) presentation also known as 'Graphical User Interface' (GUI), (b) logic or processing, and (c) data. The relationship between these three elements is that one element sends the request to other element and the other elements responses to the request through the same interface. This making and fulfilling of request is called client/server-computing model. The element that makes request is called 'client' and the element that fulfills the request is called 'server'.

The client/server is a software defined system architecture of computing. The client/server environment is generally heterogeneous in nature where the hardware platform and operating system of the client and server are not generally the same. In such cases, the communications mechanism may be further extended through a well-defined set of standard i.e. Application Program Interfaces (APIs) and Remote Procedure Calls (RPC) protocols. The client/server computing provides the capability to use the most cost-effective user interface, data storage, connectivity, and application services. The web based GIS implementation requires configuration of various server and client side components. In the coming section of this chapter, various server and client side components of web based GI systems have been explained. In the development of web based GIS, following four major server components are required at host level: (a) Database server, (b) GIS server, (c) Application server, and (d) Web server.

13.5.2.1 Database Server

The database server may have a file based system or Relational Database Management System (RDBMS) or a combination of files and RDBMS. In a typical web GIS application, the spatial data is organized in RDBMS environment which allows better performance, data security, data consistency and many more advantages of RDBMS for GIS data sets. Table 13.2 presents well known and famous database server software solutions with RDBMS support available for geospatial data sets.

13.5.2.2 GIS or Map Server

Map server or GIS server is a software package or program, which is responsible for rendering the GIS data into web browser. The standard web servers and browsers support only standard image and data formats like .jpeg, .gif .txt, .html, .xml, etc. To represent or publish geospatial data in web compatible format, there is a need of intermediate software components called as a map server or GIS server. Today, many of the map server products are available either as commercial product or as an open source/freeware products for map publishing in internet environment [23]. The important GIS/map server products available at present are shown in Table 13.3.

13.5.2.3 Application Server

An application server in a web GIS platform is a software which provides customized software applications with the services like query system, GIS analysis and

Table 13.2 Important GIS database servers with RDBMS support

S. No.	Database server software	Strength	Web address	Status
1	PostgreSQL + POSTGIS	Performance and advanced analysis	http://postgis.refractions.net http://www.postgresql.org	Open source (Freeware)
2	ArcSDE + Selected RDBMS	Technical support	http://www.esri.com	Commercial
3	Oracle Spatial	Support for JAVA and store common spatial data types in a native Oracle environment	http://www.oracle.com/technology/products/spatial/index.html	Commercial
4	My SQL	Compatibility with PHP and other open source s/w	http://www.mysql.com	Open source (Freeware)
5	TerraLib	Time series analysis and supported by many RDBMS	http://www.terralib.org/	Open source (Freeware)
6	SpatiaLite	Spatial extensions for the open source SQLite database	http://www.gaia-gis.it/spatialite	Open source (Freeware)
7	IBM DB2	Strong error handling	http://www-01.ibm.com/software/data/db2	Commercial

processing, report generation, data security and authorizations, etc. The web GIS application servers are customized using API and spatial libraries. Many application servers like JBOSS (java application server) are bundled with GIS servers like ERDAS Apolo server and available as a single package. The development of application server using open API like OpenLayer API, Google API, Yahoo API, etc. are becoming popular in user community due to its easy development and interactive support. The development of server application can be done using Java SDK, .net framework, PHP, Javascript etc. Table 13.4 lists few of the important application development environment for web GIS applications.

13.5.2.4 Web Server

A web server is a computer program which uses the client/server model and the www's HTTP to serve the files that form web pages for web users (whose computers contain HTTP clients that forward their requests). The primary function of a web server is to deliver web pages on the request of clients. This means delivery of HTML documents and any additional content that may be included in a document, such as images, style sheets and scripts. A user agent, commonly a web browser or web crawler, initiates communication by making a request for a specific resource

Table 13.3 Important GIS/map server products and their strengths

S. No.	Map/GIS server software	Strength	Official web address	Status
1	Geoserver	Performance, security, vector support and OGC web services	http://www.geoserver.org	Open source (Freeware)
2	UMN Mapserver	Performance, raster and vector support	http://www.osgeo.org	Open source (Freeware)
3	Arc GIS Server	Technical support and GIS processing	http://www.esri.com	Commercial
4	SkylineGlobe	3D visualization	http://www.skylineglobe.com	Commercial
5	Mapguide	Support and quick customization	http://mapguide.osgeo.org	Open source (Freeware) and commercial
6	Degree	OGC web services	http://www.deegree.org	Open source (Freeware)
7	ERDAS APOLLO Server	Raster support	http://www.erdas.com	Commercial
8	Intergraph Geo web server	Engineering based application	http://www.intergraph.com	Commercial

Table 13.4 Important application development environment for web GIS applications

S. No.	Database server software	Strength	Official web address	Status
1	OpenLayers	AJAX library for accessing geographic data layers of all kinds	http://openlayers.org	Open source (Freeware)
2	GeoBase	Geocoding, navigation and route optimization	http://www.geobase.org	Open source (Freeware)
3	Geomajas	Aggregation and transformation of GIS data sources	http://www.geomajas.org	Open source (Freeware)
4	GeoTools	GIS data creation, editing and processing using JAVA framework	http://www.geotools.org	Open source (Freeware)
5	GDAL/OGR	Compatibility with any development environment	http://www.gdal.org	Open source (Freeware)
6	GEOEXT	Rich GUI using open layer API	http://www.geoext.org	Open source (Freeware)

Table 13.5 Popular web server products

Product	Developer
Apache and Tomcat	Apache
IIS	Microsoft
Nginx	Igor Sysoev
GWS	Google
Resin	Caucho Technology
Lighttpd	Lighttpd
Sun Java System Web Server	Oracle

using HTTP and the server responds with the content of that resource or an error message if unable to do so. Table 13.5 lists popular web server products which facilitates the delivery of web pages to the user.

13.6 Service Oriented Architecture (SOA) and Geo-web Services

The technological trends in information technology has made possible the transformation towards Service Oriented Architectures (SOAs) and distributed computing to achieve high interoperability in data and information. In SOA, the services are intended to be independent building blocks that collectively represent a software applications. These web services have many unique characteristics that allow them to participate as part of a SOA. One of the unique quality is complete autonomy from other services which is important for interoperability and also setting a relationship and linkage between different services. This means that each service is responsible for a specific task and represents a particular domain. The most widely accepted and successful implementation of web services are XML based. XML stands for Extensible Markup Language which is a markup language similar to HTML and used for data definition while HTML is used for data presentation purpose.

The web services architecture has three roles: a provider, a requestor and a broker as shown in Fig. 13.2. The provider creates the web service and makes it available to clients or users who want to use it. A requestor is a client application that consumes the web service at other end. The broker, such as a service registry, provides a way for the provider and the requestor of a web service to interact as a catalogue. The provider, requestor, and broker interact with each other through the three major operations viz. publish, find and bind. A typical example of web service implementation in Internet domain could be online airline ticket booking system through travel agency or third party service providers such as MakeMytrip, Yatra.com, etc. The airline operator publish their web service to the service registry of these ticket booking agencies. The customer search for best price or most suited

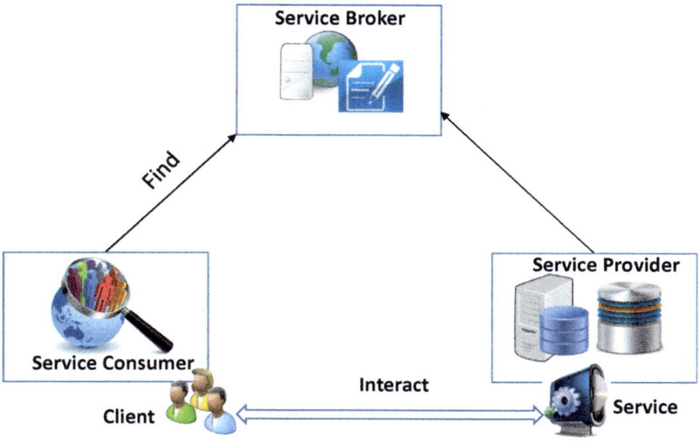

Fig. 13.2 Service oriented architecture

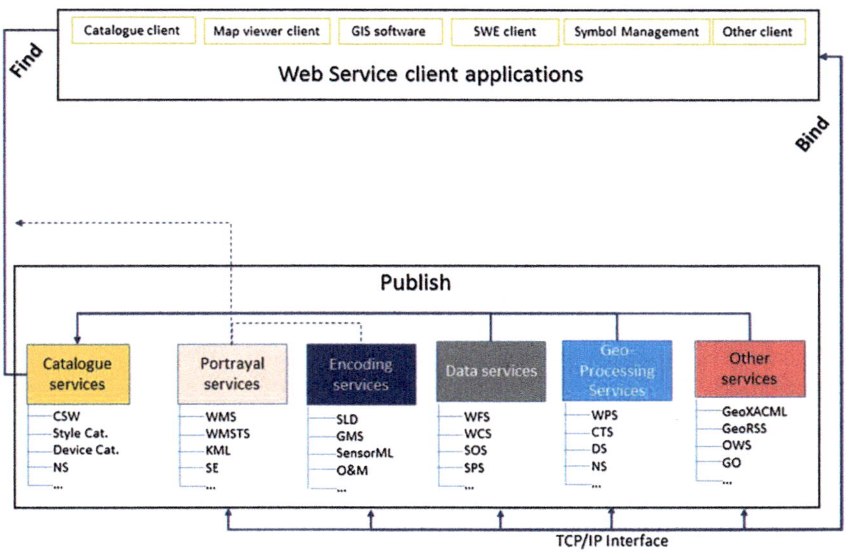

Fig. 13.3 Web service framework of OGC (*Source* http://live.osgeo.org)

timing of any airlines through online web services. These web services works on real time data sets from remote servers hosted by different airline operator using various technological implementations.

In geospatial domain, the web services are very important to achieve interoperability in data and information available with different data providers. Today, the geospatial services available in Internet through various geoportals are increasing rapidly. There is a need of a methodology to locate desired services that provide

Table 13.6 OGC GIS service specifications

S. No.	Type	Use	Example
1	Catalogue services	Allows users and applications to classify, register, describe, search, maintain, and access information about Web Services	CS Core, CS-W ebRIM, CS-W 19115/19119 CS-W ebRIM for EO, etc.
2	Processing services	Provide operations for processing or transforming the data as per user defined parameters	Sensor planning service (SPS), Web Processing Service (WPS), Coordinate Transformation Service (CTS and Web Coverage Processing Service (WCPS), etc.
3	Encoding	Defines symbology encoding, an XML language for styling information that can be applied to digital feature (vector) and coverage data (raster)	Geography Markup Language (GML), CityGM Styled Layer Descriptor (SLD), SWE Common, etc.
4	Data services	Tightly coupled with specific data sets and offer access to full or a portions of geospatial data. Original data access without physical download	Sensor Observation Service (SOS), Web Feature Service (WFS), etc.
5	Portrayal service	Used for simple data visualization like map rendering and cartographic representation of the maps	Web Map Service (WMS) and Web Map Tiling Service
6	Others services	Project/operation specific and other Web 3.0 services	GeoXACML, GeoRSS, Geospatial Objects, OWS Common, etc.

access, data discovery and analysis capabilities for geospatial data. The interoperability of services across organizations and providers is important for seamless integration and sharing of spatial data from a variety of sources. Different organizations and commercial vendors develop their own data standards and storage structures for geospatial data. If GIS services are not interoperable, these data sets cannot interact or get overlaid to each other even though they are in the same organization or they belong to same commercial vendor. To solve the interoperability problems in GIS, the Open Geospatial Consortium (OGC) has introduced data and service standards by publishing specifications for the GIS services. OGC is a not-for-profit, international, voluntary, consensus standards organization founded in 1994. The major objectives of OGC are to lead in the development, promotion and harmonization of open geospatial standards. OGC have around 500+ members from industry, government, research and university across the world. The GIS service specifications developed by the OGC are based on SOA (Fig. 13.3). Such systems unify distributed services through a message-oriented architecture. The OGC GIS services can be grouped into six major categories as shown in Table 13.6.

13.7 Mobile and Participatory GIS

The mobile GIS is a new technological advancement in geospatial domain and it has the strength not only to deliver the geospatial data to the mobile users everywhere and anytime, but also it can be used as an effective mode of geographic data creation and collection using participatory approach. Mobile GIS is an extension of Internet or web based GIS which provides GIS functionalities in a portable devices such as mobile, PDA, tablet etc. Mobile GIS is an integration of various technologies viz. Mobile technology, GIS, GNSS, wireless communications, etc. [24]. Mobile GIS is a cost effective solution for data collection and creation directly from fields. Some of the exciting applications of mobile GIS include Location Based Services (LBS), crowdsourcing/Voluntary Geographic Information (VGI) and participatory GIS. Today, the mobile devices are embedded with GNSS which adds a location tag with the information transmitted by the device. The communication channels (like General Packet Radio Service (GPRS), Code division multiple access (CDMA), Worldwide Interoperability for Microwave Access (WiMAX), etc.) are required to transmit the information from mobile device to the receiver. The second-and-half generation and above (2.5G, 3G and 4G etc.) networks provide the possibilities of a mobile Internet with high-speed data transfer at a rate up to 256 kbps or better. Two of the leading network protocols used in high-speed mobile applications are GPRS and CDMA20001x.

The mobile GIS architecture is a combination of client at mobile and host at server level. The data and information is published at server end for mobile user similar to web based GIS application. The only major configuration required at server end is to tune the data size for low storage and limited computation devices like mobile and PDA. The mobile GIS applications are best suited for smartphone devices where the multimedia components with GPS facility are available for the users. GIS operations in smart mobile phones are through the ICT evolution which has many exciting new applications of GIS. The mobile users are moving in a geographic space and they know their positions and they have access to the widely available geographic data and information from variety of servers. The mobility is "key" for mobile GIS applications. The theme specific GIS applications like mobile GIS for mapping, field data collection, spatial analysis and Location Based Services (LBS) etc. have become very popular among user communities.

The mobile GIS requires GIS applications and software to be installed on the smart mobile devices where the communication part can be online or offline depending upon requirements and availability. The GPS is one of important component of Mobile GIS where the location of the geography feature will be presented using coordinates recorded using GPS. The accuracy of the location varies from kilometer to a meter level.

13.8 Cloud Computing GIS

Cloud computing is one of most recent advancement in technology which has directly influenced the involvement of IT and related technologies in various business sectors. Cloud computing platform offers various services in virtual environment such as Software as Service (SaS), Infrastructure as Service (IaS) and Platform as Services (PaS). Here, Infrastructure refers to IT infrastructure and Platform refers to computer platform which are required to run any software application(s). The cloud computing technology has revolutionized the functioning and operations of IT services in Internet platform. Typically, the cloud computing is an integration of hardware, software, peoples (users), data and applications. GIS has adopted the cloud computing technology to enhance its uses and outreach by shifting various geospatial tools and functionalities to the cloud. Cloud-based tools are accessed for web-based GIS as geo-web services.

The cloud GIS can be considered as next generation GIS which offer 'GIS on Demand' based on user(s') or organization(s') requirements. The GIS software, data (spatial and non-spatial), computation infrastructure and platforms requirements are available for hire with a full scalability options. Various public and private clouds are getting popular such as ArcGIS Online, Cloud GIS (www.giscloud.com), Amazon web service, ThunderMaps, etc. In India, two major initiatives from Government namely, MEGHDOOT from Center for Development of Advanced Computing (CDAC) and National Informatics Centre (NIC) Cloud (http://cen.gisserver.nic.in) are getting popular among the users.

GIS in Cloud computing provide shared resource pooling such as hardware, software, data, networks, etc. among user organizations for effective utilisation and collaboration with common or shared goals. The cloud GIS offers on demand services in GIS such as online mapping, sharing of geospatial data (raster as well as vector), GIS data analysis and processing. It also supports viewing, creating, monitoring, managing, analysing and sharing maps and data with other users. It facilitates inputs, validation and collaboration by a global mobile workforce in real time using mobile GIS technology. Some of the possible applications of Cloud GIS in smart city planning could be citizen and social science services, road infrastructure projects, mobile data collection and integration, traffic management, e-commerce and geo-targeted marketing and awareness, geo-referenced weather services, web collaborative mapping and public safety and emergency response.

13.9 Internet of Things

The Internet of Things (IoT) is a system of interrelated computing devices, mechanical and digital machines, objects, animals or people that are provided with unique identifiers and the ability to transfer data over a network without requiring human-to-human or human-to-computer interaction (http://Internetofthingsagenda.techtarget.com). In the concept of IoT, all the objects (static or movable) will have one IP address (Internet Protocol) with a live interconnectivity with other objects. IoT has evolved from the convergence of wireless technologies, micro-electromechanical systems (MEMS), micro-services and the Internet with IPv6. The inclusion of IP version 6 or IPv6 in Internet domain has increased the Internet address space drastically which is an important factor in the development of IoT. The present form of data and information in Internet is being generated by various means such as typing, pressing a record button, taking a digital picture or scanning a bar code. Sometimes, the data and information is also being generated using some onboard sensors and auto detective devices. Most of the time, the involvement of person is required in all these data and information collection systems and it may have limitations of time, attention and accuracy. This can be addressed with IoT where the computation devices will generate and communicate the data and information required for their operations and maintenance. If the connected devices are having embedded with location sensors such as GPS in IoT then it will be known as GIoT or geo-enabled IoT.

The concept of IoT can play a significant role in smart city planning where the geo-embedded sensors of various interconnected computer devices will generate lot of meaningful data and information for smart planning. Some of the possible applications of IoT in smart city planning could be smart garbage collection system in the city, smart traffic management system, emergency services, public distribution system, etc.

13.10 Open Geo-data Sources and Repositories

Geospatial data sharing and dissemination is made easy with the advent of Internet based geoportals and web services. The WWW and high speed network access has given a new dimension to geospatial domain where larger data sets may be shared and processed, more complex models can be established, computation intensive analysis for decision-making can be performed and better methods of display and visualization for virtual reality can be achieved. Today, many organizations and individuals are hosting their data sets online as an open geo-data through online data repositories. This approach is enhancing the outreach of geospatial data for its better utilisation. These open geo-data sets can be used for various thematic applications either as a geo-web services or as a data product. The coming sections of this chapter presents a status of various online open data repositories which can be used for smart city planning and other urban and regional studies. The online data repositories are presented in two forms i.e. online geoportals and data repositories.

13.10.1 Geoportals

Website or websites are modern ways of sharing information using Internet technology. Typically, the website use Hyper Text Markup Language i.e. HTML to present the information in web browser environment. HTML is a markup language which is used for data presentation purpose. Generally, the HTML based websites are static in nature which present the predefined data and information to its users through web server. If database systems such as DBMS, RDBMS or ORDBMS are used for data storage and management purpose for online web application, then it will be known as web portal. In this approach, the data is accessed and processed using an additional middle tier at server end using any web programming language such as PHP, C#, JAVA, Python, etc. This middle tier is also known as business logic or application server. In case of GIS data, the database servers hosts the raster and vector data in addition to non-spatial data sets. To access geospatial data from database servers, the middle tier needs to be modified/customized to understand the GIS data types from geo-data servers. The modified application server for geo-data is known as GIS or Map server which converts geospatial data in web compatible formats. The web portals with geo-data are known as Geoportal. Some of the popular geoportals are Google Maps, Bing Map, ISRO *Bhuvan*, MapMyIndia, Rediff Map, etc. In Table 13.7, the popular geoportals with their unique capabilities are listed.

The GIS based web portals provides a centralized and uniform interface to access the distributed and heterogeneous resources and data services. Most of the web GIS based portals available in internet are designed for specific theme and are targeted to specific class of users.

13.10.2 Geo-data Repositories

The online geo-data repositories offers variety of data for different application. They differ from geoportals as they allow users to download the data either freely or on payment. URLs and status of some of the popular geo-data repositories are shown in Table 13.8. These datasets are available at multiple scales catering to differing requirements ranging from global scales to urban or local scales (http://freegisdata.rtwilson.com/). However, there is still a long way to go for creating necessary datasets for addressing multiple requirements of urban areas.

The above list is just an indicative list which tries to summarize the plenty of datasets which are now available freely on web for the use by international community and urban planners.

Table 13.7 Popular geoportals and their services

S. No.	Web portal	URL	Salient features
1	Google maps	http://maps.google.com	Global satellite images and maps (2D and 3D) at street level with various applications
2	Bing Map	http://www.bing.com/maps/	Global satellite images and maps (roads and POI)
3	Openstreet Map	http://www.openstreetmap.org	Open and free street level vector data created using collaborative or participatory approach
4	Wikimapia	http://wikimapia.org	Crowdsourcing approach for tagging ground information
5	ISRO *Bhuvan*	http://bhuvan.nrsc.gov.in	Rich contents and seamless availability of multi-temporal, multi-resolution satellite data for entire Indian region. Thematic and disaster services with data download facility
6	MapMyIndia	http://www.maps.mapmyindia.com	Rich POI and detailed maps of India
7	Indian NSDI	https://nsdiindia.gov.in	Metadata catalogue and policy document
8	Biodiversity information system	http://bis.iirs.gov.in	Biodiversity spatial viewer and data download utility. Rich data contents on plant biodiversity of India
9	Indian Bio-resource information network	http://ibin.gov.in	Rich data contents for Indian bio-resource information. System of distributed database
10	India-WRIS	http://india-wris.nrsc.gov.in	Rich data contents for water resources of India
11	NIC GIS	http://gis.nic.in	Village level tagging of information in GIS domain

13.11 Bhuvan

Bhuvan (is a Sanskrit word, meaning Earth) is a geoportal (http://bhuvan.nrsc.gov.in) of Indian Space Research Organisation (ISRO) with a host of services and countrywide coverage of systematic geographic database that include visualization of satellite data; thematic map display, query and analysis; data downloads; near real-time disaster services; apps for crowdsourcing and diverse geospatial applications. The platform has multiple languages interface. It has the seamless, ortho-rectified satellite images of 2.5 m spatial resolution available for entire India for visualization and mapping. The platform allows seamless transitions from

Table 13.8 Online free and open data repositories

S. No.	Product	URL	Coverage	Category
1	Gateway to Indian Earth Observation—*Bhuvan* Geoportal	http://bhuvan.nrsc.gov.in/data/download/index.php	India	General and theme specific
2	Meteorological and Oceanographic Satellite Data Archival Centre—MOSDAC	http://www.mosdac.gov.in	India	General and theme specific
3	Indian Biodiversity Information System	http://bis.iirs.gov.in	India	General and theme specific
4	Indian Bio-resource Information Network IBIN	http://www.ibin.gov.in	India	General and theme specific
5	Visualization of Earth observation Data and Archival System (VEDAS)	http://vedas.sac.gov.in/vedas	India	General and theme specific
6	Natural Earth—Vector	http://www.naturalearthdata.com	Global	General
7	Global Map	https://www.iscgm.org/gmd	Global	General
8	DIVA-GIS Country Data	http://www.diva-gis.org/gdata	Global	General
9	UNEP GEOdata	http://geodata.grid.unep.ch	Global	General
10	Koordinates	https://koordinates.com/about	Global	General
11	GeoNetwork	http://www.fao.org/geonetwork/srv/en/main.home	Global	General
12	GSHHS	http://www.ngdc.noaa.gov/mgg/shorelines/gshhs.html	Global	Land and Ocean Boundaries
13	GDEM	http://asterweb.jpl.nasa.gov/gdem.asp	Global	Elevation
14	SRTM	http://srtm.csi.cgiar.org	Global	Elevation
15	Global Multi-Resolution Topography	http://www.marine-geo.org/portals/gmrt/	Global	Elevation
16	OpenTopography	http://www.opentopography.org/index.php	Global	Elevation
17	WorldClim	http://www.worldclim.org	Global	Weather and climate

(continued)

Table 13.8 (continued)

S. No.	Product	URL	Coverage	Category
18	Climate Analysis Indicators Tool	http://cait.wri.org	Global	Weather and climate
19	HydroSHEDS	http://hydrosheds.cr.usgs.gov/index.php	Global	Hydrology
20	Delineation of major watersheds in the World	http://www.fao.org/geonetwork/srv/en/metadata.show?id=30914	Global	Hydrology
21	Water Isotopes	http://wateriso.utah.edu/waterisotopes/pages/data_access/da_main.html	Global	Hydrology
22	Natural Disaster Hazards	http://www.ldeo.columbia.edu/chrr/research/hotspots/coredata.html	Global	Natural disasters
23	USGS Earthquakes Database	http://earthquake.usgs.gov/earthquakes/eqarchives/epic/kml	Global	Natural disasters
24	Global Seismic Hazard Map	http://www.seismo.ethz.ch/static/GSHAP	Global	Natural disasters
25	NOAA/WDC Historical Tsunami Database	http://www.ngdc.noaa.gov/hazard/tsu_db.shtml	Global	Natural disasters
26	MODIS Fire Detection Data	http://activefiremaps.fs.fed.us/gisdata.php	Global	Natural Disasters
27	Natural Disaster Hotspots	http://sedac.ciesin.columbia.edu/data/collection/ndh	Global	Natural disasters
28	GLOBCOVER	http://due.esrin.esa.int/page_globcover.php	Global	Land cover
29	MODIS Global Land Cover	http://visibleearth.nasa.gov/view.php?id=61004	Global	Land cover
30	EarthEnv Global Consensus Land Cover	http://www.earthenv.org/landcover	Global	Land cover
31	Global Lakes and Wetlands Database	http://www.worldwildlife.org/pages/conservation-science-data-and-tools	Global	Land cover
32	World Soil Information	https://www.soilgrids.org/#/?layer=geonode:taxnwrb_250m	Global	Ecology

(continued)

Table 13.8 (continued)

S. No.	Product	URL	Coverage	Category
33	Harmonized World Soil Database	http://webarchive.iiasa.ac.at/Research/LUC/External-World-soil-database/HTML/index.html?sb=1	Global	Ecology
34	ERS/MetOp Soil Moisture	http://www.ipf.tuwien.ac.at/404/	Global	Ecology
35	Global High Resolution Soil Water Balance	http://www.cgiar-csi.org/data/global-high-resolution-soil-water-balance	Global	Ecology
36	Global Carbon Storage in Soils	http://www.docs_wri.org/page_forest_gh_sc_final.zip	Global	Ecology
37	World Bank Geodata	https://velluminformation.com/2012/01/28/google-earth-world-bank-data-and-kml-files/	Human geography	General
38	Google Maps Gallery	http://www.maps.google.com	Human geography	General
39	History Database of the Global Environment	http://themasites.pbl.nl/tridion/en/themasites/hyde/index.html	Human geography	General
40	Natural Earth	http://www.naturalearthdata.com	Human geography	Administrative boundaries
41	IUCN 2013 Red List	http://www.iucnredlist.org/technical-documents/spatial-data	Human geography	Environmental boundaries
42	Global Land Use Dataset	http://nelson.wisc.edu/sage/data-and-models/global-land-use/grid.php	Human geography	Land use
43	Human Influence and Footprint	http://sedac.ciesin.columbia.edu/data/collection/wildareas-v2	Human geography	Land use
44	Crop Calendar GIS	http://nelson.wisc.edu/sage/data-and-models/crop-calendar-dataset/index.php	Human geography	Land use
45	ESPON Urban Morphological Data	http://database.espon.eu/db2/resource?idCat=43	Human geography	Land use
46	European Urban Morphological Zones	http://www.eea.europa.eu/data-and-maps/data/urban-morphological-zones-2000-1	Human geography	Land use

(continued)

Table 13.8 (continued)

S. No.	Product	URL	Coverage	Category
47	Global Terrorism Database	http://www.start.umd.edu/gtd	Human geography	Wars, conflict and crime
48	Gridded Population of the World	http://sedac.ciesin.columbia.edu/data/collection/gpw-v4	Human geography	Population
49	Global Rural-Urban Mapping Project	http://sedac.ciesin.columbia.edu/data/collection/grump-v1	Human geography	Population
50	WorldPop	http://www.worldpop.org.uk	Human geography	Population
51	Large Urban Areas 1950-2050	https://nordpil.com/resources/world-database-of-large-cities	Human geography	Population
52	Global Urban Extent	http://nelson.wisc.edu/sage/data-and-models/schneider.php	Human geography	Population
53	GeoHive	http://www.geohive.com	Human geography	Population
54	OpenStreetMap	http://www.geofabrik.de/data/download.html	Human geography	Buildings, roads and points of interest
55	OSM Metro Extracts	https://mapzen.com/data/metro-extracts	Human geography	Buildings, roads and points of interest
56	POI Factory	http://www.poi-factory.com	Human geography	Buildings, roads and points of interest
57	SimpleGeo's Places	http://s3.amazonaws.com/simplegeo-public/places_dump_20110628.zip	Human geography	Buildings, roads and points of interest
58	Nuclear Power Station locations	https://fusiontables.google.com/DataSource?dsrcid=579353#rows:id=1	Human geography	Buildings, roads and points of interest
59	Open Flights	http://openflights.org/data.html	Human geography	Transport and communications
60	World Port Index	http://msi.nga.mil/NGAPortal/MSI.portal?_nfpb=true&_pageLabel=msi_portal_page_62&pubCode=0015	Human geography	Transport and communications
61	Global Roads Open Access Data Set	http://www.ciesin.columbia.edu/confluence/display/roads/Global+Roads+Data	Human geography	Transport and communications

(continued)

Table 13.8 (continued)

S. No.	Product	URL	Coverage	Category
62	Undersea Telecommunications Cables	http://www.cablemap.info	Human geography	Transport and communications
63	Geonames Country Information	http://download.geonames.org/export/dump/countryInfo.txt	Human GEOGRAPHY	Gazetteers (place/feature names)
64	GRUMP Settlement Points	http://sedac.ciesin.columbia.edu/data/set/grump-v1-settlement-points	Human Geography	Gazetteers (place/feature names)
65	G-Econ	http://gecon.yale.edu	Human geography	Miscellaneous
66	Internet Map	http://www.netdimes.org/new/?q=node/65	Human Geography	Miscellaneous
67	World Urban Database and Access Portal Tools (WUDAPT)	http://www.wudapt.org/	Urban geography	Urban form and function
68	UN-Habitat Urban Data	http://urbandata.unhabitat.org/	Urban Geography	Urban Indicators
69	Urban Development (World Bank)	http://data.worldbank.org/topic/urban-development	Urban geography	urbanisation, traffic and congestion, and air pollution
70	Urban Data Platform by European Commission for European cities	http://urban.jrc.ec.europa.eu/	Urban Geography	Demography, urban development, economic development, transport and accessibility, environment and climate, resource efficiency and social issues
71	DLR's Global Urban Footprint	https://urban-tep.eo.esa.int/#!	Urban geography	Urban footprint

national to village level while depicting varieties of geospatial datasets. It provides hydrological base for India from basin to watershed, transport network from national highways to city roads and rich location information called as Points of Interest (POI). *Bhuvan* showcases more than 3.5 million location-based information that are created through crowdsourcing mechanism [25]. It has Digital Surface Model (DSM) for entire India that gives excellent depiction of country's topography on the fly. Presently, satellite image data for more than 350 cities are hosted in Bhuvan at 1 m spatial resolution that could help in preparing various plans for town/city development schemes of the Government [25].

Bhuvan, as a platform, is open and being used by diverse users community. The Government agencies use this platform to share and host their data, as per their requirements, enabling specific applications of their choice. It presently hosts 30 state geoportals. A number of collaborative geospatial applications involving 22 ministries of Govt. of India have been integrated into *Bhuvan* platform for specific use related to g-Governance programme. There are more than 6000 services offered by Bhuvan. It provides support to disaster like cyclone, floods, landslides, earthquakes, forest fire and drought, which is useful for various phases of disaster including preparedness and response. The disaster that struck in Uttarakhand, India during 2013, floods and *Hudhud* cyclone during 2014 in India, and Nepal Earthquake in 2015 are some of the recent examples where *Bhuvan* provided unique services in terms of online disaster information update, forecasts and post-disaster scenario. *Bhuvan* is also acting as a clearing house for satellite data and satellite data derived products. They are effectively used for scientific studies as researchers are utilizing the data for applied research as more than 3.5 lakh products have been downloaded by the users in last 3 years. *Bhuvan*-2D visualization interface is designed to provide map view, satellite image view and hybrid of both. In addition, terrain view is also provided in a single integrated viewer [25]. The Table 13.9 presents the various data sets which are available on *Bhuvan* for visualization, use and download.

13.11.1 EKTA

Embedding Knowledge with Technology and Association (EKTA) on *Bhuvan* platform is a universal Mapper (http://bhuvan.nrsc.gov.in/updates/bhuvan_may2014.html). It provides interactive and user-friendly tools for adding, modifying and updating points, lines and polygons with data versioning. The open source libraries at the backend provide a robust system for map making process and data collection. Following applications are realized and in operation using *Bhuvan* Mapper—(a) Crowd sourcing of location (point type) data, curvilinear feature (line) and area (area) data along with the rich customized attributes for each feature, Implemented in Generic *Bhuvan* Mapper for mapping cities and other areas, (b) Mapping the neighbourhood of Uttarakhand to assess the damage, (c) Watershed development activity mapping/monitoring with hierarchical customized attributes in Integrated Watershed Management Programme (IWMP) application, (d) Mapping of forest assets in Karnataka Forest application, (e) Mapping of reservoir/tanks with detail list of attributes for Maharashtra, (f) Water Resources Development System of Maharashtra Engineering Research Institute (MERI), (g) Neighbourhood mapping for School Kids, and (h) Online shapefile creation and downloading with user wise privileges.

Table 13.9 Typical satellite data and thematic GIS data resources available on *Bhuvan*

Satellite data resources			
1	High resolution satellite image	2.5 m spatial resolution natural	Entire India
		Colour composite (NCC)	
2	Very high resolution satellite image	1 m spatial resolution natural	Major cities/towns
		Colour composite (NCC)	
Base and thematic GIS data services			
1	NUIS scheme base and urban	1:10,000	152 towns
	Thematic GIS database		
2	SIS-DP database and land use/land cover	1:10,000	Entire India—state wise
3	Land use/land cover	1:50,000	Entire India
			2005–2006
			2011–2012
4	Geomorphology	1:50,000	Entire India (in collaboration with GSI
5	Geological structures	1:50,000	Entire India (in collaboration with GSI
6	Wastelands	1:50,000	Entire India
7	Ground water prospects	1:50,000	Entire India
8	Land use/land cover	1:250,000	Entire India (9 cycles from 2004 to 2013

Source http://bhuvan.nrsc.gov.in/bhuvan_links.php

13.11.2 Base Layers

Rail and road network data from national highways up to street level is available for display through multi-level geospatial visualization mechanism. This information is of prime interest for any GIS application particularly for urban and regional planning. The rail network database also contains details on station locations and other assets. The scale based rendering of the administrative and infrastructure layers with essential styling and annotations are enabled for ease of interpretation. Figure 13.4 shows the typical base layer information available on *Bhuvan*.

13.11.3 Multi-date Satellite Images

Various applications such as urban sprawl and growth monitoring require temporal satellite data to understand the changes over the period for decision making,

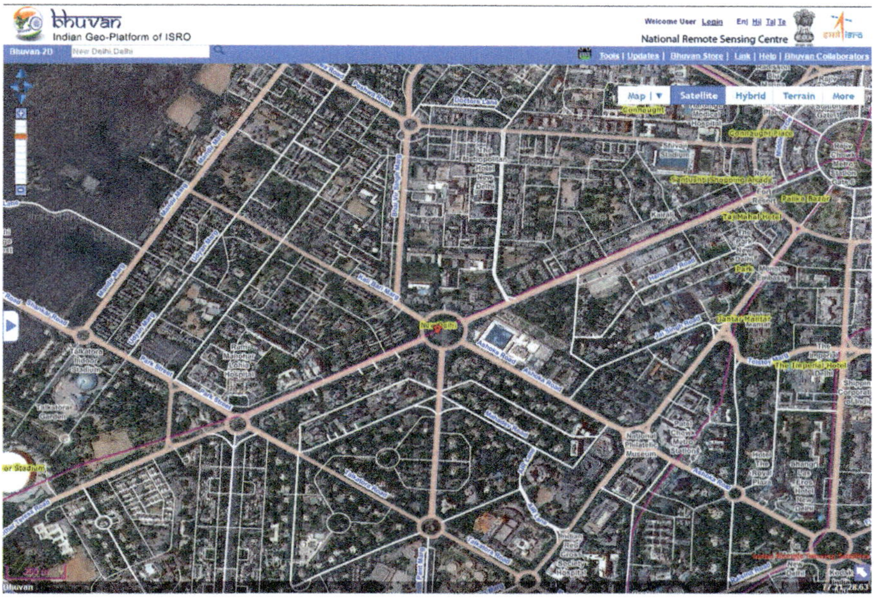

Fig. 13.4 Typical base layer information available on *Bhuvan* (*Source* http://bhuvan.nrsc.gov.in/map/bhuvan/bhuvan2d.php)

monitoring and evaluation. These applications require multi-date images along with software tools to swipe, overlay and analyse changes and these functions are well supported on *Bhuvan*.

13.11.4 Point of Interest (POI)

The Point of Interest (POI) data is an essential requirement for building applications catering to urban planning, tourism, location based services, disaster, etc. These POIs ranges from place names, localities, municipal, tourism information and information derived from various projects on diversified themes. *Bhuvan* contains more than 8 million POI location information for visualization and rich place name search option (http://dhte.puducherry.gov.in/download/rusa/1_Bhuvan_User_Handbook.pdf).

13.11.5 Crowdsourcing

Bhuvan contains various tools and applications for crowdsourcing the data from community. Exclusive mapper application for creating detailed maps, add content option to contribute POI data, customized Android applications for field data collection are available on *Bhuvan*. The primary challenge for the crowdsourced data through community participation is robust architecture to handle large volumes of data received and moderation. *Bhuvan* with its proven platform and architecture contains more than 3.5 million crowdsourced data.

13.11.6 G-Governance Dashboard for Ministries

Geospatial governance dashboard developed for Ministries of Govt. of India allows the users including administrators to directly reach the specific programmes/projects according to their interest in alphabetical order. This framework is also developed using responsive design to access the applications in mobile phones or tabs. It hosts 64 developed applications till date [25].

13.12 Geo-applications on *Bhuvan*

13.12.1 Census GIS

Bhuvan has Census 2011 information integrated at village level for the users to make necessary visual analysis. Users can interactively generate choropleth maps based on equal interval classification and unique values by providing Census 2011 code or name. Point maps can be generated by providing longitude and latitude information. Specific features like village boundaries with Census 2011 attributes are also integrated in a ready-to-use query form (Fig. 13.5).

13.12.2 Municipal GIS

Municipal GIS in *Bhuvan* enables the citizens to know about wards and facilities available, various schemes executed by the government, grievances redressal system besides facilitating administrators/planners to have an online planning tool for better governance. The municipal corporation can utilize the data for urban planning and also in the effective implementation of developmental schemes by providing information on street layout, size and location of buildings. It will help in identifying illegal encroachments and similar other violations, along with to

Fig. 13.5 Census data visualization on *Bhuvan* (*Source* http://bhuvan.nrsc.gov.in/governance/socialjustice_census)

calculate property tax. The database for Ludhiana town on *Bhuvan* has been prepared with the help of Punjab Remote Sensing Centre (PRSC), Ludhiana, Punjab, India for Ludhiana Municipal Corporation (http://main.mcludhiana.gov.in/pdf/MCL_Blocks_gismap.pdf). The process of allotting a unique identity to the property would help city dwellers to avail services from municipal body ranging from water-related issues to property tax. The 18-digit UID number contains city, block, locality and property codes. The UID is 'geo-tagged' with the exact spatial dimensions of every property and has the latest photograph attached with each property. The integration of existing municipal database like water/sewage ID no., house tax property no. with the UID helps to maintain a single number for each property and useful to generate online bills for all the municipal services. With the help of door-to-door survey, approximately 4.5 lakh properties in the city are allotted the UIDs. Similar activity is also scheduled for the area under Municipal Corporation of Bathinda town, Punjab state, India (Fig. 13.6).

13.12.3 Management Plans for Heritage Sites and Monuments

India has a rich cultural heritage and whole country is dotted with architecture and heritage sites. Some of these heritage sites receive enormous global and national attention. However, many of these heritage sites face increasing risk from urbanisation, economic development and implications of unanticipated changes. The conservation of world heritage sites, ancient monuments and archaeological sites is of national importance and also helps in development and promotion of tourism, which is one of the major engines of economic growth. Ministry of Culture, Govt. of India has taken up a national project for inventory and monitoring of

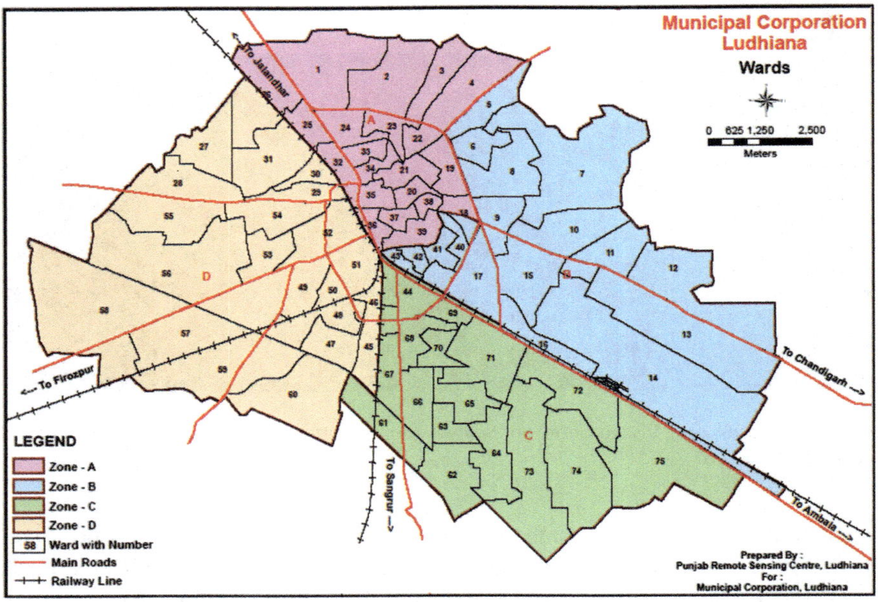

Fig. 13.6 Database prepared under Municipal GIS for Ludhiana city and hosted on *Bhuvan* (http://main.mcludhiana.gov.in/pdf/MCL_Blocks_gismap.pdf)

archaeological sites of the country in collaboration with ISRO. The systematic database of heritage sites and site management plans generated using space technology will help to take appropriate measures in conserving, preserving and monitoring of heritage sites (https://www.geospatialworld.net/news-posts/isro-to-create-3d-visualisations-of-indian-heritage-sites/). A Standard Operating Procedure (SOP) using high resolution satellite data as primary data source and employing state-of-the-art geospatial technology and open source tools has been developed for operational use by Archaeological Survey of India (ASI). The remote sensing data from Cartosat-1, Cartosat-2 and Resourcesat LISS-IV are used for creating the database on heritage sites and monuments. Three management zones (protected, prohibited and regulated) around the heritage site are delineated using GIS tool after locating the site/monument on the satellite image. All the land use features within each zone are precisely mapped. The collateral data e.g., total station survey data, revenue village, cadastral maps and other maps/attributes related to the inventory, plans etc. are geo-referenced and integrated with the satellite data and form part of the geo-database. Other data like ground photographs, different views of the monuments, type of buildings and their heights, street view of the buildings and lanes, open spaces are also part of the database. Smart phone based applications are developed for geo-tagging and uploading by different stake holders and citizens as part of crowd sourcing. The 3D digital models of the sites and monuments are

Fig. 13.7 3D visualization of the heritage site and fly-through (*Source* http://www.isro.gov.in/3d-digital-model-of-gol-gumbaz-%E2%80%93-one-of-largest-dome-structure-world)

generated and overlaid over the satellite data, which provide virtual reality walk-through and visualization (http://www.isro.gov.in/3d-digital-model-of-gol-gumbaz-%E2%80%93-one-of-largest-dome-structure-world) (Fig. 13.7).

13.12.4 Assets Mapping

The asset mapping datasets are hosted on *Bhuvan* geoportal for information dissemination and management. Specialized Android field data collection and mapper applications have been developed which allows for capturing assets with customized categories and attributes. The crowdsourcing application for AP Housing Corporation Ltd. in India has enabled geo-tagging of more than 3.5 million houses with field photographs and important housing attributes (http://bhuvan.nrsc.gov.in/state/AP_housing/#). It has the facility to map the assets as point or line features as per requirement and provisions to overlay administrative boundaries and other thematic datasets, e.g. landuse/landcover, wastelands, etc. The major benefits from the assets mapping are interactive community participation, facility mapping, gap analysis for damage assessment and mitigation strategy (Fig. 13.8).

13.12.5 Tourism

The GIS mapping of Amritsar city in India has been done using high-resolution satellite data and depicts tourist attractions and various amenities required by visitors during their stay in the city. The GIS database has been launched by the Punjab Tourism Department and prepared by the Punjab Remote Sensing Centre, Ludhiana, India. Tourism Web GIS through *Bhuvan* facilitates the users to select the tourism places of their interest and also search nearby places along with routing. It would facilitate tourists to locate hotels, restaurants, ATMs, tourist spots, shortest

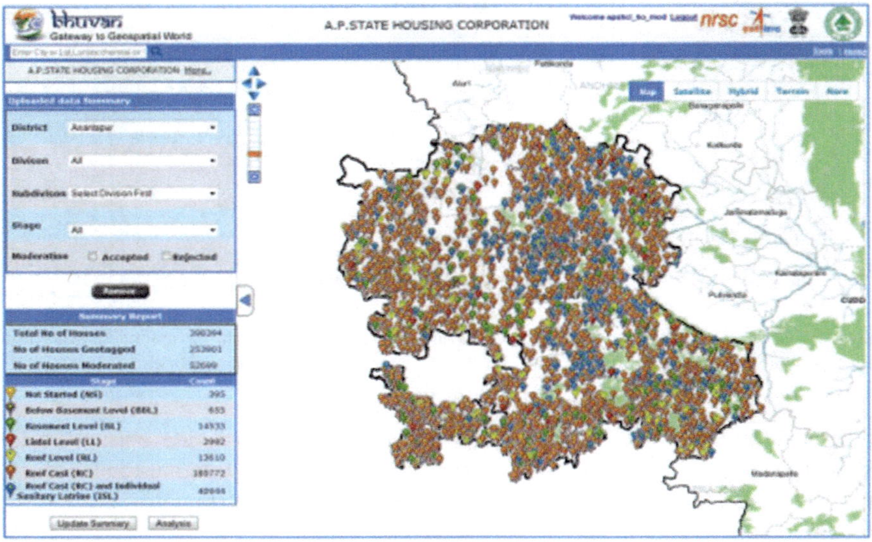

Fig. 13.8 A snapshot showing asset mapping points (over 25 lakhs houses) on *Bhuvan* geoportal (*Source* http://bhuvan.nrsc.gov.in/state/AP_housing/#)

Fig. 13.9 Tourism GIS developed for Amritsar city and hosted on *Bhuvan* (*Source* http://bhuvan.nrsc.gov.in/tourism/tourism.php)

route, name of streets, etc. Considering the 57 lakh tourists that visit the holy city of Amritsar every year, the facility launched will facilitate their stay in the city. Tourists will automatically be guided to their destination using turn-by-turn GPS navigation. The tourism GIS data is also available for Amritsar, Punjab; Hampi and Badami, Karnataka; Nalanda, Bihar and Viyaywada in Andhra Pradesh states of India on *Bhuvan* geoportal (Fig. 13.9).

13.12.6 Urban Growth

Urban growth refers to the extent of urbanisation, which is a global phenomenon mainly driven by population growth and large scale migration. Urban growth characterization and monitoring using temporal and multi-spectral satellite data is available on *Bhuvan* platform for the period from 1991 to 2014 for 82 cities of India. The salient features of this database are incremental view of urban growth and provision for animated view to visualize growth monitoring of urban areas (Fig. 13.10).

13.12.7 Urban Framework Survey

The Ministry of Statistics and Programme Implementation (MoSPI), Govt. of India is the nodal agency for planning and facilitating the integrated development of statistical system in the country. It collects statistics at local level and later aggregates systematically at higher level of governance and planning. In case of data gaps, sample surveys are carried out at national/state level. Two systems are in vogue, the enumeration blocks of Census of India and the urban blocks of the National Sample Survey Office. The Urban framework survey (UFS) provides complete, updated, identifiable, operationally convenient and readily accessible sampling frame for the urban area. Under the UFS, towns are physically surveyed and updated over a five year period. Supplementary information on type of area such as slum area, residential area, industrial area, education, hospital, commercial area, prohibited area etc. are collected during the survey/updation and some auxiliary information/landmarks such as school, post office, bus stand, clinic etc. are

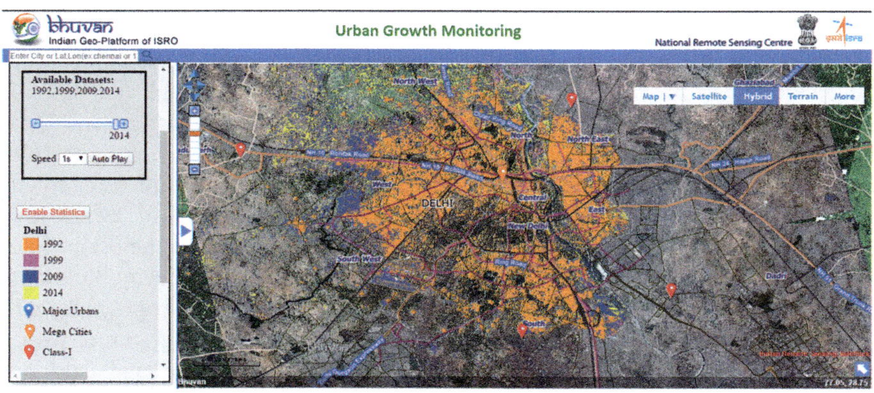

Fig. 13.10 Urban growth monitoring for Delhi city on *Bhuvan* geoportal (*Source* http://bhuvan.nrsc.gov.in/urban/sprawl/urbangrowth.php)

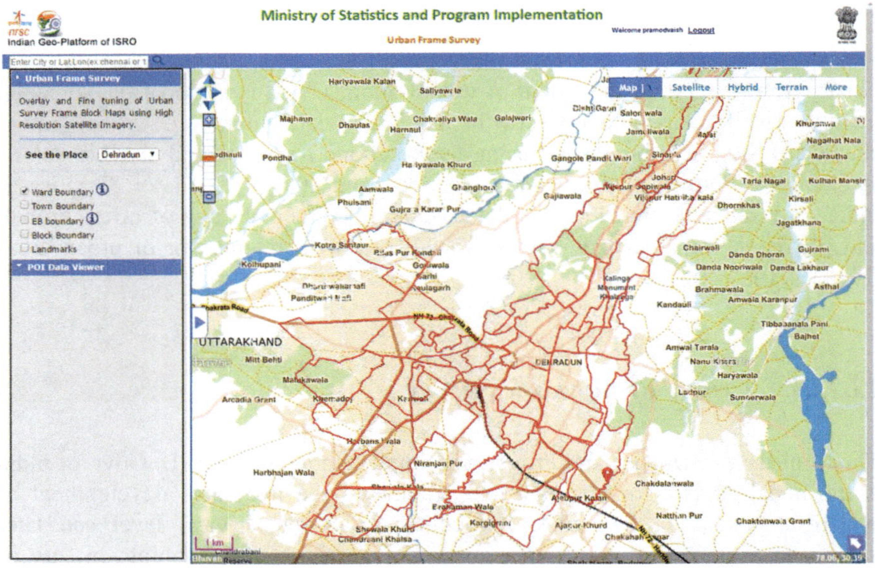

Fig. 13.11 Urban framework survey of Dehradun city (*Source* http://bhuvan.nrsc.gov.in/governance/mospi_urbanframesurvey/#)

recorded for each block. This helps in selection of blocks with certain characteristics for the socio-economic surveys. The UFS data for Bodhan and Dehradun cities of India are available on *Bhuvan* geoportal (Fig. 13.11).

13.12.8 National Urban Information System

The National Urban Information System (NUIS) Scheme of Ministry of Urban Development (MoUD), Govt. of India was launched in 2006 and contains GIS database for 152 towns/cities. The spatial and attribute database thus generated can be useful for the preparation of Master/Development plans, detailed town planning schemes and serve as decision support for e-governance. For these cities and towns, detailed base maps, administrative and infrastructure data is also available at 1:10,000 scale. The *Bhuvan* 2.5D depiction of buildings further helps in better visualization of such large scale maps of towns and cities. The capacity building is one of the major objectives of NUIS Scheme. About 28 training programmes with approximately 20 participants at various levels under each programme, over a period of two years, were organized to train the urban managers and field staff (http://moud.gov.in/nuis) (Fig. 13.12).

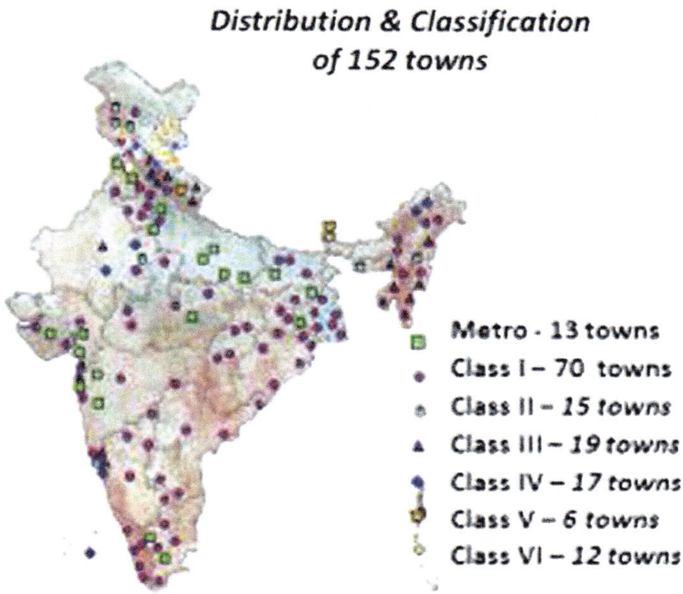

Fig. 13.12 Distribution and classification of 152 towns under NUIS scheme (http://www.sac.gov.in/SACSITE/SAC-Flyers/menu-links/society/6.1%20URBAN%20%20PLANNING.pdf)

13.13 Applications of e-Democracy Tools

13.13.1 Grievance Redressal and Management

Geo-enabled online systems are also the potential avenue for ULBs to deliver improved services to address the citizens' need. When location-specific information is available, it can lead to better management of services e.g., if the water leakage or pothole repair requests are received from a particular area, intervention can be taken up by the ULBs. Similarly, accidents are reported from a particular location, it will help in analysing and identifying the interventions such as road geometry correction or deployment of traffic personnel. These initiatives can also help in management of physical problems in neighbourhoods such as trash collection, loss of water services, maintenance of utilities, etc.

The Sarathi (System of Assisting Residents and Tourists through Helpline Information) initiative taken by the Pimpri Chinchiwad municipal corporation, Maharashtra, India is considered as one of the best practices under the right to information (http://www.cips.org.in/documents/2014/SPIPA/SARATHI.pdf). The information through SARATHI is made available using multiple channels—book, website, call center, mobile app, e-book and pdf book. As the call centre platform is interactive, SARATHI is also working towards resolving complaints of the citizens. The innovative use of technology to disseminate the information in the form of

Frequently Asked Questions (FAQs) as well as grievance redressal has become immensely popular with the citizens of Pimpri Chinchwad.

Similarly, the Virtual Charlotte (VC) system developed for the city of Charllotte, USA is one of the finest example of integration of Geospatial web 2.0 platform for citizen-centric services [26]. The system was developed to coordinate and manage service delivery seamlessly by city staff. It receives and records information regarding traffic accidents, construction projects, street maintenance services, vehicle location tracking, etc. The system provides an interface on Google maps to allow non-GIS specialists to use the system without training in a user friendly manner. All the city departments are integrated within this system that allows coordination among departments and enhances service delivery.

The initiative taken by the Hubli-Dharwad Municipal Corporation, Karnataka, India to set up 24-h control rooms for public grievance redressal is also listed as best practices by Ministry of Housing and Urban Poverty Alleviation (MoHUPA), Government of India (http://fincomindia.nic.in/writereaddata%5Chtml_en_files%5Coldcommission_html/fincom13/discussion/report08.pdf). A 24-h grievance redressal control room have been set up by the corporation for attending the public grievances like, road repairs, underground drainage cleaning, maintenance of street-lights, removal of garbage and carcass of dead animals. The status of the complaint can be checked online. However, if the system is made geo-enabled on Geoweb 2.0 platform, it can provide better results with location information and can assist in monitoring of complaints.

13.13.2 Location Based Services (LBS)

The Location Based Services (LBS) are software-level information service that uses coordinates of a location to serve specific service requirements of the users. Typically, the coordinates are acquired through GNSS and the LBS services are accessed through mobile devices which have embedded GPS location receiver. With the availability of high speed Internet connectivity in mobile devices, the location information can also be retrieved using geo-tagged IP address and triangulation network of Internet such as GPRS. LBS facilitates the services to identify a location of a person or object, such as finding out the nearest banking cash machine (ATM), nearest location of hospitals, parcel tracking, vehicle tracking service, etc. Under LBS, the personalized weather services and even location-based marketing, awareness and delivery of best services are possible.

LBS can be very effective tool for smart city planning where the citizen services can be delivered based on location of the citizen. The geo-enabled citizen-centric services will be the true implementation of smart city planning where the geo-location of available services and consumers will be automatically detected.

13.13.3 Real Time Mapping

With the advent of mobile based GIS, it is now possible to do real-time mapping for various online GIS applications. The availability of high speed internet connectivity in mobile devices are allowing to transfer large amount of geospatial data from server to mobile and vice versa. The interactive mobile GIS applications (also known as Mobile Apps) are allowing various online mapping tools to its users. Most of the time, these mobile apps are freely available to its users. Some of the popular Mobile GIS applications are listed in Table 13.10.

The mobile GIS applications also facilitate to auto-draw the geographic features as a map feature based on movement of persons or objects. The location of moving object is detected through GPS at certain interval and by connecting each recorded node, the feature is automatically drawn. This approach is very effective to collect authentic geospatial data based on ground survey.

Table 13.10 Popular mobile mapping applications

Mobile application	Description	Mobile platform	Status	URL
QField	A field data capture and management app fully compatible with QGIS	Android	Freeware/open source	www.qgis.org/en/
gvSIG Mobile Website	It a full GIS software for mobile devices. It runs on Windows Mobile and it is the ideal software for field data collection and updating projects	Windows, Android and other Java compatible mobile OS	Freeware/open source	www.gvsig.com/en
GeoODK	This software provide online mapping tools with real time transfer of this data server	Android	Freeware/open source	www.geoodk.com/
ArcGIS for Mobile	Online and offline mapping in mobile platform	Windows, Android and IoS	Commercial	www.esri.com
Mappt	Online and offline mapping in mobile platform	Android	Freeware/open source	www.mappt.com.au/
Supergeo Mobile Apps	Variety of mobile applications with SDK	Android, IOS and Windows	Commercial	http://www.supergeotek.com

13.13.4 Citizens as Information Providers

Volunteered Geographic information refers to "the explosion of interest in using the web to create, assemble and disseminate geographic information provided voluntarily by individuals or citizens" [27]. Earlier, traditional maps required skilled GIS specialists to generate GIS data but now with Geo web 2.0, citizens not trained in GIS can also generate information quickly over the web. This is also known as crowdsourcing in GIS. Smart phones/mobiles embedded with GPS devices are being used worldwide to geocode and document data by untrained users. Citizens are acting as a large collection of intelligent, mobile sensors which are ready to generate and share the information on near real-time basis.

Crowdsourcing is the act of outsourcing tasks to a large group of peoples or community also known as 'crowd' through an open call. Crowdsourcing has emerged as a powerful strength of 21st century in geospatial domain to generate user driven collective intelligence as part of geo-enabled planning and decision making. Through crowdsourcing the users generates their own dynamic contents and participate in data creation, validation, moderation, analysis and many other activities involved in planning. The platform for crowdsourcing could be personal computers connected with internet or portable devices such as smart mobile phones, PDA or tablets. The crowdsourced data get synchronized with central database with or without moderation and is accessible/sharable as a web-based databases, services and maps. Crowdsourcing is a participatory approach which makes possible the ways in which large groups of users come together to create data, serving as human sensors. Social networking websites such as Facebook, Twitter, Flickr and Instagram may be considered as potential example of crowdsourcing.

In geospatial domain, the crowdsourcing is considered as one of the successful mode of spatial data creation, field data collection and validation of ground truth data and information. OpenStreetMap (OSM) (www.openstreetmap.org) is one of the most successful project in internet where large scale geospatial data sets are created by "the public for the public" (Fig. 13.13). OSM is a collaborative project to create a free editable map of the world using public participation. The vector data on 1:10,000 scale or better is available freely for download. The project is still live which allows the users to create online spatial data of their own Area of Interest (AOI).

Crowdsourcing is getting very popular among GIS users and professionals using mobile platform. The field data collection is one of the arduous task in urban planning. The mobile based field data collection systems makes it easy, faster and interactive through user-centric platform. One of the most important benefit of this platform is that it can transfer a near real-time geo-tagged dataset to the central server systems. The ground data collected through these mobile apps can be easily linked with other geospatial data sets available at central server systems from

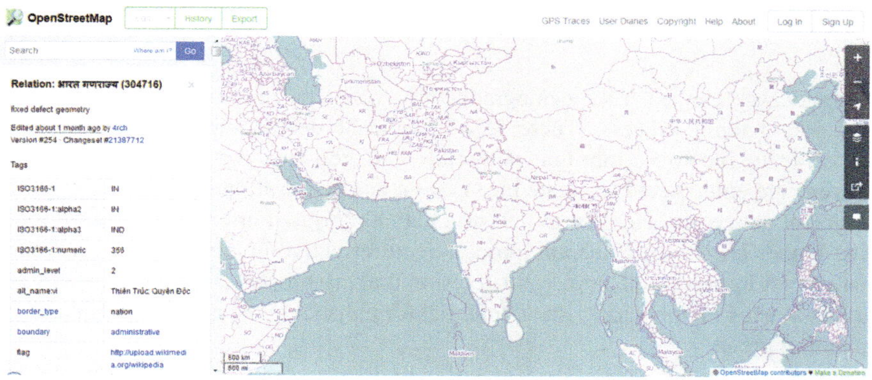

Fig. 13.13 Home page of OpenStreetMap. http://www.openstreetmap.org

Table 13.11 Mobile based real time geo-data collection systems

Software	Description	URL
Open Data Kit (ODK)	It is a free and open-source set of tools which help organizations to collect field data through mobiles	http://opendatakit.org/
EpiCollect	Developed by Imperial College London together with web based management tools. available as free and open source	http://www.epicollect.net/
GIS Cloud's Mobile Data Collection	It is developed for real-time field location, multimedia and attribute data collection. Available as commercial solution	http://www.giscloud.com/

various stakeholders. Some of the popular mobile application platform for online field data collection and reporting are listed in Table 13.11.

In smart city planning, the crowdsourcing and participatory GIS approach can be effectively used for better planning and decision making. These tools and technologies will allow direct participation and networking of the citizens for smooth execution of various citizen services. The citizen complain redressal system, geo-tagged reporting system for various services, location aware feedback system, citizen participation in planning, real-time monitoring of service vehicles, emergency management etc. are few smart applications of this technology. This mode of data collection has various implications for E-Democracy and public participation. It widens the domain of map making beyond professionals and facilitates the democratization of GIS [28]. This holds potential for collaborative map making by volunteers and generation of geo-tagged information with less resources spent by the government agencies. However, the accuracy and authenticity of data has to be ascertained before putting it for scientific uses.

13.13.5 Public Participation for Decision Making

E-democracy can be truly democratic only when citizen will participate in the decision making process. There is a significant developments in making data available to the users through geoportals and geo-data repositories. With VGI, citizens are empowered to create information and act on it, but still it is a long way to go for active participation in decision making through online geo-enabled tools. The higher levels of democratic process requires active involvement of citizens in making an impact on planning and decision making processes. The use of GIS and geo-enabled web tools have great potential for enhancing the public participation in planning process.

Using Geospatial web 2.0, it is possible to deliver complex geographical models of simulations of future growth, impact of development projects, alternative scenarios in simplified visual form that could be understood by the citizens. Until recently, most of the participatory approaches were either employed without the use of technology or if GIS technology was used then it was limited only to the feedback from participants. However, the technology has capabilities to provide information to the users in interactive manner in such a way that they could perform their own analysis, can interpret the results and make policy suggestions. Hence, the adoption of e-democracy tools for effective public participation is not a technological issue but more of institutional and political issue.

13.14 Geo-enabled E-Democracy: Challenges

The smart city programme is happening worldwide. The geo-enabling of database is the challenge for urban managers as they have to organize their infrastructural assets in such a manner that they are 'GIS ready'. It will create a single GIS data warehouse of available survey data, maps, images, tabular geo-tagged development data, cadastral data, etc. Another important component involves periodical updation of information. The biggest challenge faced today is the lack of technical capacity to plan, implement and monitor IT-driven or embedded projects. As the cities today are entering the realm of big data management, the urban managers need to address following major challenges:

- **Data integration**: The institutional labyrinth especially in developing world and coordinating information from different agencies often hinders the urban planning and reform process. The data required for urban planning lies with several urban local bodies. Therefore, the data integration holds the key towards the success of smart city mission. Large datasets are still in analogue format and need to be digitized whereas the data prepared through CAD software need to be aligned to GIS platform. The adoption of open standards for geospatial datasets has reduced the technical barriers, but political and organizational ones are hard to be addressed.

- **Privacy and security**: During the recent years, data availability especially from open sources has eased the process of urban planning. However, the datasets available at urban local bodies whether in analogue or digital format should be used by urban managers with precaution. Data security is even more of a challenge and should be adequately addressed under smart city programme.
- **Capacity building**: The viable utilisation of huge information by city managers for urban planning also calls for the need for capacity building. Lack of skill sets may well be the biggest barrier for the effective use of data by city managers. Managing and analysing large data sets and developing insights for effective policy making and overall improvement requires skills which are often deficient, particularly in the public sector. Cloud-based services, public-private partnerships, and open data strategies could help by providing access to a broader skills base. Cities can also build deeper relationships with academia. Many cities are working with local universities to establish smart city and urban informatics research programs.
- **Financing for Urban Infrastructure**: The smart city programme is investment intensive. It needs huge investment for realizing the goals of infrastructure creation and better planning. Basically, the investment need falls in two categories, firstly the capital and secondly the operational expenses. The urban managers should primarily think towards reducing/rationalizing the tax rates than raising the taxes. The increase in revenues after the realization of smart city mission may be expected while levying betterment tax, review of land use and floor area ratio (FAR), sale of recycled waste and user charges, etc.

13.15 Conclusions

Urbanisation is synonymous with growth and development as well as associated environmental deterioration. For sustainable growth of urban areas, smart city concept is supported, which aims to achieve overall and long term development of urban areas. This goal can only be achieved if the citizens are actively involved in decision making and planning process by the geo-enablement of e-democracy tools and services. Geo-enablement brings numerous advantages as it adds location to the existing information, which makes it easily comprehendible by the common citizen. The Google maps, Bing maps, *Bhuvan* Geoportals are some examples of the use of location based information by the masses. The latest developments in geospatial technologies and web GIS domain provides the necessary technologies and platforms for implementation of geo-enabled e-democracy tools. Now a variety of remote sensing datasets ranging from coarse to fine resolution as well as number of ground based sensors are providing multitude of datasets which enable generation of vast information for urban areas.

The advancements in GIS, web GIS and Geoweb 2.0 provides necessary platforms for integrating the remote sensing and ground based information on geo-tagged and interactive virtual environments. The availability of information on these platforms has brought paradigm shift in the usage of geographic information as now even an unexperienced, untrained user can understand, modify and analyse the geo-information. It enables to support three main components of any democratic process—information to collaboration and participation. At present, many geo-portals and geo-data repositories are available from where geographic datasets can be freely downloaded and can be used for the development of a range of smart city applications. These data sets are available at varied scales from global to local. However, still a lot need to be done for the creation of necessary datasets for multiple requirements of upcoming smart cities.

Bhuvan geoportal, developed and hosted by ISRO is one such portal which provides datasets, tools and applications for visualization, use and development of applications by urban managers in India. NUIS, municipal GIS, tourism GIS, Urban framework survey, etc. are few of the examples of the geo-applications available on *Bhuvan*. This portal also hosts very high resolution satellite data of Indian cities, which can be used for creation of a range of thematic information to be used for smart city applications. Similarly, the geo-enabled e-democracy tools is being used for grievance redressal, involving citizens in collection of information through volunteered interventions or crowd sourcing, provision of location based services, real-time mapping and participatory GIS. However, its role has been limited till now up to the dissemination or collection of information to and from the public. The collection and use of information in truly democratic way would be possible through the active involvement of citizens in the decision making and planning processes.

Smart cities focuses on integrated urban development through citizen participation by the use of technology, hence the geo-enabled e-Democracy tools definitely have potentials to enhance citizen participation and strengthen the democratic processes in local governance. Aspirations are high to achieve this goal through smart city mission, and geo-web tools and services have immense capabilities to provide information to the citizens in an interactive manner and easily comprehendible form so that they could participate in planning and decision making. The limitations is more of institutional and political rather than technological. There are various issues which need to be addressed such as data integration, privacy and security, skills gap and capacity building, and therefore, institutional setup and political will is necessary for the true implementation of these tools and services.

References

1. UN (2014) World urbanisation prospects, Department of Economic and Social Affairs, United Nations, https://esa.un.org/unpd/wup/Publications/Files/WUP2014-Highlights.pdf. Last accessed in Nov 2016
2. Mckinsey (2010) India's urbanisation: a closer look. http://www.mckinsey.com/global-themes/urbanisation/indias-urbanisation-a-closer-look. Last accessed in November 2016
3. Steudler D, Rajabifard A (2012) Spatially enabled society, the International Federation of Surveyors (FIG) and the Global Spatial Data Infrastructure Association (GSDI)

4. Nagy M, Legat R, Hřebíček J (2007) Electronic access to environmental information—an important fundament for eDemocracy and environmental protection
5. Anon (2016) India Smart City mission. http://smartcities.gov.in/writereaddata/winningcity/BHUBANESWAR_SCP.pdf. Last accessed in Nov 2016
6. Hai J, Ibrahim C (2007) Fundamental of development administration. Scholar press, Selangor. ISBN 978-967-5-04508-0, 2008, pp. 125–130
7. Gupta K, Roy A, Luthra K, Maithani S, Mahavir (2016) GIS based analysis for assessing the accessibility at hierarchical levels of urban green spaces. Urban Forestry and Urban Greening 18:198–211
8. Mahavir, Bedi P (2014) GIS for smart urbanisation. In: Vinod Kumar TM (ed) Geographic information system for smart cities. Copal publishing, Delhi
9. Jat MK, Garg PK, Khare D (2008) Monitoring and modelling of urban sprawl using remote sensing and GIS techniques. Int J Appl Earth Obs Geoinf 10(1):26–43
10. Gupta K (2013) Unprecedented growth of Dehradun Urban area: a spatio-temporal analysis. Int J Adv Remote Sens GIS Geogr 1(2):6–15
11. Bodhankar S, Gupta K, Kumar P (2015) Smart planning through geo-enabling of digital database for planning authorities: a case study of regional park zone (RPZ) pockets in Navi Mumbai notified area. In: ISG 2015 Conference at Jaipur, 16–18 Dec 2015
12. Gupta K, Kumar P, Pathan SK, Sharma KP (2012) Urban neighborhood green index—a measure of green spaces in urban areas. Landsc Urban Plann 105(3):325–335
13. Jensen JR (2007) Remote sensing of the environment: an earth resource perspective, 2nd edn. Upper Saddle River, NJ, USA, Pearson Prentice Hall
14. Bhaskaran S, Paramananda S, Ramnarayan M (2010) Per-pixel and object-oriented classification methods for mapping urban features using Ikonos satellite data. Appl Geogr 30:650–665
15. Jeffery C (2010) An introduction to GNSS, Novatel Inc., Canada. http://www.borealisprecision.com/pdf/An_Introduction_to_GNSS.pdf. Last accessed in Nov 2016
16. Rashid SM (1996) Economics of remote sensing. Manak Publications Pvt. Ltd. https://books.google.co.in/books?id=kpRNAQAAIAAJ
17. Steiniger S, Bocher E (2009) An overview on current free and open source desktop GIS developments. Int J Geogr Inf Sci 23(10):1345–1370
18. Peng ZR, Tsou MH (2003) Internet GIS: distributed geographic information services for the internet and wireless networks. ISBN: 0-471-35923-8, Mar 2003
19. Karnatak HC, Shukla R, Sharma VK, Murthy YVS, Bhanumurthy V (2012) Spatial mashup technology and real time data integration in geo-web application using open source GIS—a case study for disaster management. Geocarto Int 27(6):499–514
20. Karnatak HC, Saran S, Bhatia K, Roy PS (2007) Multicriteria spatial decision analysis in web GIS environment. Geoinformatica 11(4):407–429
21. Karnatak HC, Bhatia K, Saran S (2008) Multi-criteria decision analysis using spatial compromise programming. In: Proceedings of the 2nd national conference—INDIACom-2008, on computing for nation development, New Delhi, India. ISBN No-ISSN 0973-7529, ISBN 978-81-904526-2-5, pp 77–62, 08–09 Feb 2008
22. Karnatak HC, Saran S, Roy PS (2005) Spatial services, a click away. Geospatial Today 5:42–46 (2005)
23. Sayar A, Aktas M, Aydin G, Pierce M, Fox G (2005) Developing a web service-compatible map server for geophysical applications technical report. Indiana University, Indiana
24. ESRI (2004) Best practices: mobile GIS. http://www.esri.com/library/bestpractices/mobile-gis.pdf
25. Bhuvan User Hand Book (2015) National Remote Sensing Centre, Indian Space Research Organisation, Hyderabad
26. Ganapati S (2010) Using geographic information systems to increase citizen engagement. IBM Center for The Business of Government, Washington, DC, http://www.businessofgovernment.org/, Last Accessed in November 2016

27. Goodchild MF (2007) Citizens as sensors: the world of volunteered geography. GeoJournal 69:211. doi:10.1007/s10708-007-9111-y
28. Dunn CE (2007) Participatory GIS—a people's GIS? Prog Hum Geogr 31(5):616–637

URLs

29. http://activefiremaps.fs.fed.us/gisdata.php
30. http://aphousing.cgg.gov.in/Map.do
31. http://asterweb.jpl.nasa.gov/gdem.asp
32. http://bhuvan.nrsc.gov.in
33. http://bhuvan.nrsc.gov.in/bhuvan_links.php
34. http://bhuvan.nrsc.gov.in/data/download/index.php
35. http://bhuvan.nrsc.gov.in/governance/mospi_urbanframesurvey/#
36. http://bhuvan.nrsc.gov.in/governance/socialjustice_census
37. http://bhuvan.nrsc.gov.in/map/bhuvan/bhuvan2d.php
38. http://bhuvan.nrsc.gov.in/tourism/tourism.php
39. http://bhuvan.nrsc.gov.in/urban/sprawl/urbangrowth.php
40. http://bis.iirs.gov.in
41. http://cait.wri.org
42. http://cen.gisserver.nic.in
43. http://data.worldbank.org/topic/urban-development
44. http://database.espon.eu/db2/resource?idCat=43
45. http://dhte.puducherry.gov.in/download/rusa/1_Bhuvan_User_Handbook.pdf
46. http://download.geonames.org/export/dump/countryInfo.txt
47. http://due.esrin.esa.int/page_globcover.php
48. http://earthquake.usgs.gov/earthquakes/eqarchives/epic/kml
49. http://elearning.iirs.gov.in
50. http://en.beidou.gov.cn/
51. http://fincomindia.nic.in/writereaddata%5Chtml_en_files%5Coldcommission_html/fincom13/discussion/report08.pdf
52. http://freegisdata.rtwilson.com/
53. http://gecon.yale.edu
54. http://geodata.grid.unep.ch
55. http://geospatialworld.net/News/View.aspx?id=32320_Article#sthash.1dYjLUXc.dpuf
56. http://gis.nic.in
57. http://grass.osgeo.org
58. http://hydrosheds.cr.usgs.gov/index.php
59. http://ibin.gov.in
60. http://india-wris.nrsc.gov.in
61. http://Internetofthingsagenda.techtarget.com
62. http://live.osgeo.org
63. http://main.mcludhiana.gov.in/pdf/MCL_Blocks_gismap.pdf
64. http://mapguide.osgeo.org
65. http://maps.google.com
66. http://maps.mapmyindia.com
67. http://moud.gov.in/nuis
68. http://msi.nga.mil/NGAPortal/MSI.portal?_nfpb=true&_pageLabel=msi_portal_page_62&pubCode=0015
69. http://nelson.wisc.edu/sage/data-and-models/crop-calendar-dataset/index.php
70. http://nelson.wisc.edu/sage/data-and-models/global-land-use/grid.php
71. http://nelson.wisc.edu/sage/data-and-models/schneider.php
72. http://opendatakit.org/

73. http://openflights.org/data.html
74. http://openlayers.org
75. http://postgis.refractions.net
76. http://rgeostats.free.fr/
77. http://s3.amazonaws.com/simplegeo-public/places_dump_20110628.zip
78. http://sedac.ciesin.columbia.edu/data/collection/gpw-v4
79. http://sedac.ciesin.columbia.edu/data/collection/grump-v1
80. http://sedac.ciesin.columbia.edu/data/collection/ndh
81. http://sedac.ciesin.columbia.edu/data/collection/wildareas-v2
82. http://sedac.ciesin.columbia.edu/data/set/grump-v1-settlement-points
83. http://smartcities.gov.in/writereaddata/smartcityguidelines.pdf
84. http://srtm.csi.cgiar.org
85. http://themasites.pbl.nl/tridion/en/themasites/hyde/index.html
86. http://urban.jrc.ec.europa.eu/
87. http://urbandata.unhabitat.org/
88. http://vedas.sac.gov.in/vedas
89. http://visibleearth.nasa.gov/view.php?id=61004
90. http://wateriso.utah.edu/waterisotopes/pages/data_access/da_main.html
91. http://webarchive.iiasa.ac.at/Research/LUC/External-World-soil-database/HTML/index.html?sb=1
92. http://wikimapia.org
93. http://www.bhuvan.nrsc.gov.in
94. http://www.bing.com/maps/
95. http://www.cablemap.info
96. http://www.cgiar-csi.org/data/global-high-resolution-soil-water-balance
97. http://www.ciesin.columbia.edu/confluence/display/roads/Global+Roads+Data
98. http://www.cips.org.in/documents/2014/SPIPA/SARATHI.pdf
99. http://www.deegree.org
100. http://www.diva-gis.org/gdata
101. http://www.docs_wri.org/page_forest_gh_sc_final.zip
102. http://www.earthenv.org/landcover
103. http://www.eea.europa.eu/data-and-maps/data/urban-morphological-zones-2000-1
104. http://www.epicollect.net/
105. http://www.erdas.com
106. http://www.esri.com
107. http://www.exelisvis.co.uk/
108. http://www.fao.org/geonetwork/srv/en/main.home
109. http://www.fao.org/geonetwork/srv/en/metadata.show?id=30914
110. http://www.gaia-gis.it/spatialite
111. http://www.gdal.org
112. http://www.geobase.org
113. http://www.geoext.org
114. http://www.geofabrik.de/data/download.html
115. http://www.geohive.com
116. http://www.geomajas.org
117. http://www.geoodk.com/
118. http://www.geoserver.org
119. http://www.geotools.org
120. http://www.giscloud.com/
121. http://www.gvsig.com/en
122. http://www.hexagongeospatial.com
123. http://www.ibin.gov.in
124. http://www.ilwis.org/
125. http://www.intergraph.com

126. http://www.ipf.tuwien.ac.at/404/
127. http://www.isro.gov.in/3d-digital-model-of-gol-gumbaz-%E2%80%93-one-of-largest-dome-structure-world
128. http://www.isro.gov.in/irnss-programme
129. http://www.iucnredlist.org/technical-documents/spatial-data
130. http://www.ldeo.columbia.edu/chrr/research/hotspots/coredata.html
131. http://www.mappt.com.au/
132. http://www.maps.google.com
133. http://www.maps.mapmyindia.com
134. http://www.marine-geo.org/portals/gmrt/
135. http://www.mosdac.gov.in
136. http://www.mysql.com
137. http://www.naturalearthdata.com
138. http://www.navipedia.net/index.php/Galileo_Future_and_Evolutions
139. http://www.netdimes.org/new/?q=node/65
140. http://www.ngdc.noaa.gov/hazard/tsu_db.shtml
141. http://www.ngdc.noaa.gov/mgg/shorelines/gshhs.html
142. http://www.openstreetmap.org
143. http://www.opentopography.org/index.php
144. http://www.oracle.com/technology/products/spatial/index.html
145. http://www.osgeo.org
146. http://www.poi-factory.com
147. http://www.postgresql.org
148. http://www.qgis.org/en/
149. http://www.sac.gov.in/SACSITE/SAC-Flyers/menu-links/society/6.1%20URBAN%20%20PLANNING.pdf
150. http://www.seismo.ethz.ch/static/GSHAP
151. http://www.skylineglobe.com
152. http://www.start.umd.edu/gtd
153. http://www.supergeotek.com
154. http://www.terralib.org/
155. http://www.worldclim.org
156. http://www.worldpop.org.uk
157. http://www.worldwildlife.org/pages/conservation-science-data-and-tools
158. http://www.wudapt.org/
159. http://www-01.ibm.com/software/data/db2
160. https://en.wikipedia.org/wiki/E-democracy
161. https://fusiontables.google.com/DataSource?dsrcid=579353#rows:id=1
162. https://koordinates.com/about
163. https://mapzen.com/data/metro-extracts
164. https://nordpil.com/resources/world-database-of-large-cities
165. https://nsdiindia.gov.in
166. https://urban-tep.eo.esa.int/#!
167. https://velluminformation.com/2012/01/28/google-earth-world-bank-data-and-kml-files/
168. https://www.arcgis.com/
169. https://www.glonass-iac.ru/en/guide/
170. https://www.gsa.europa.eu/european-gnss/galileo/galileo-european-global-satellite-based-navigation-system
171. https://www.iscgm.org/gmd
172. https://www.nasa.gov/directorates/heo/scan/communications/policy/GPS_History.html
173. https://www.soilgrids.org/#/?layer=geonode:taxnwrb_250m

Chapter 14
Attaining E-Democracy Through Digital Platforms in Kenya

Romanus Opiyo, Baraka Mwau, Keziah Mwang'a and Dennis Mwaniki

Abstract The role of responsible citizens in any country—whether in a city or a village–is to actively participate in the development of their areas. This is a key pre-condition for the attainment of sustainable development, and is deeply entrenched in the sustainable development goals. While the citizens' right to the city and to actively contribute to a sustainable world cuts across the 17 goals, the provisions of goal 16, to "*promote peaceful and inclusive societies for sustainable development, provide access to justice for all and build effective, accountable and inclusive institutions at all levels*" clearly spells out the need for inclusiveness, participation, and access to information at all governance levels; which would promote a sustainable, inclusive and prosperous world (Kimani in Overview of SDGs in Kenya. Paper presented during the County Capacity Forum at Hill Park, Diani on 20 June 2016, [1]). These are central elements of democratic spaces, which are greatly being enhanced by today's advances in information and communication technologies (ICTs). Kenya has over the recent past made huge strides in the area of promoting e-government services, which has resulted in increased e-democracy alternatives. This has particularly been the case since the creation of the e-government strategy in 2004 (Government of Kenya in e-government strategy 2004. Government Press, Kenya, 2004, [2]), as well as the review of various national laws and policies including a change in the country's constitution in 2010.

R. Opiyo (✉) · B. Mwau
Urban Planning & Design Branch, United Nations Human Settlement Programme (UN Habitat), Nairobi, Kenya
e-mail: romanop2000@gmail.com

B. Mwau
e-mail: barakamwau@gmail.com

K. Mwang'a
Gran Sasso Science Institute, L'Aquila, Italy
e-mail: mwelu.keziah@gmail.com

D. Mwaniki
Urban Planning and GIS, Global Observatory Linking Research to Action (GORA) Corp, New York, USA
e-mail: denmwa02@gmail.com

T.M. Vinod Kumar (ed.), *E-Democracy for Smart Cities*, Advances in 21st Century Human Settlements, DOI 10.1007/978-981-10-4035-1_14

These reforms have expanded opportunities for citizens to participate in development processes, including governance. The advances in ICT in the country, which has seen mobile phone penetration rise from less than one percent to 88.1% in just 15 years (2000–2015), and broadband increase from a mere 28 Megabits per second in 2004 to 193.58 Gigabytes per second in 2015 (Export Processing Zones Authority in Kenya's information and communications technology sector 2005, 2005, [3]; International Telecommunication Union—ITU, 2016, [4]; Communications Authority of Kenya in First quarter sector statistics report for the financial year 2015/2016, 2015, [5]) has further expanded the options for citizen engagement in shaping the country's smart growth. As a result of these changes, more Kenyans today can easily access information on, and participate in various processes such as national and local budget preparation and plan formulation processes; demand for accountability from the political class on resource utilization; or even submit complaints on poor governance and service delivery through a diversity of platforms. Equally, citizens are able to easily and quickly access services through a few clicks on their phones or computers, services that traditionally took months to acquire through bureaucratic processes. This chapter outlines how Kenya has in the recent past adopted ICT tools and methods in a bid to attain inclusive, transparent, effective and efficient governance. It discusses how the country is utilizing various ICT avenues to promote access to information by the citizens and through which governance structures get feedback on issues. It further illustrates the existing avenues for active citizen engagement in democratic and governance processes and analyzes how effective the adopted ICT systems are in promoting inclusive and smart growth.

Keywords E-democracy · E-government/nce · Public engagement · ICT · Smart city · One-stop centres · Kenya · Nairobi

14.1 Introduction

Digital platforms are creating the next wave of participation, disruption, growth and breakthrough innovation. Information and communication technologies (ICTs) are often seen as catalysts for democratic processes, freedom of speech and fraud solution, while e-governance has been identified as having the potential to establish an environment for improved service delivery, increased transparency and accountability [6]. Responsive governments have the potential to translate e-government (implementing decisions with the help of ICT) to e-governance (using ICT to help make decisions), a situation that may lead to e-democracy if citizens are meaningfully engaged in the process.

While there is no single universally agreed definition of e-democracy, there are basic widely used concepts to describe the phenomenon. E-democracy has been viewed broadly as the use of ICTs to increase and enhance citizens' engagement in democratic processes [7]. Other writings have noted that prospects of e-government have been idealized as heralding in a new era of democratic involvement, with opportunities for unmediated discussions, direct participation and representation, and greater transparency and accountability through political openness [8]. Other scholars like Coleman and Norris [9] view e-democracy as a concept that includes all processes where ICT is used by those in power and the citizens they serve to interact with each other in order to inform and determine the way that power is utilized. Generally thus, e-democracy can simply be defined as the use of ICTs to facilitate democratic activities and processes such as citizen engagement and enhancing public awareness.

While in some instances e-democracy initiatives are people driven, many governments are now using ICTs to expand democratic spaces as well as enhance how they serve and relate with their citizens. The record of government-led ICT initiatives and experiments is as diverse as there are many governments. Projects range from one off single-issue activities to long-term and institutionalized initiatives at different levels of government/governance. An example of a government led e-democratic initiative can be drawn from the city of Santa Monica, California, which had adopted ICT to broaden citizen participation in public policy making in 1986 [10]. Though the platform used then, the Public Electronic Network (PEN) is no longer active, it provided an avenue for citizens to engage with the city, access information about city services, participate in city processes as well as bring to the attention of the city government community issues that needed to be addressed. One outcome of the citizen engagement was the SHWASHLOCK project which sought to provide homeless people with shower facilities. The SHWASHLOCK project was conceived by private citizens who then shared it with the city, though the PEN citizen forum.

Other examples of government-led e-democracy initiatives include the Finish Ministry of Finance '*Share your view with us*' forum started as a pilot project between 2000-03, which has now become institutionalized. The forum is intended to provide an open space where citizens could freely share their opinions and comment on government projects underway [11, 12]. A similar example is the REACH project by the government of Singapore which provides an online platform where citizens can give their views on national policies ranging from housing, education to annual national financial budget [13, 14].

While similar innovations and processes by governments are taking place in the developing world and particularly in Africa, there lacks extensive literature on the impact of these initiatives. However, part of the literature that exists demonstrates that there is increasing uptake of ICTs within the public sector, private sector, civil society and by the citizens themselves, largely expanding the arena for cross-cutting engagement. Indeed, digital platforms have been attributed to reversing a growing trend of disengagement between governments, political leadership and citizens [15]. This chapter seeks to fill the existing gap in literature by highlighting government-led or top-down e-democracy initiatives in the Kenyan context. The chapter uses case examples to highlight how ICT is being used to create democratic spaces in which citizens can easily and seamlessly enjoy and/or lobby for their rights to space and to engage in policy formulation processes. Case examples include the analysis of ICT based tools such as e-government platforms and one stop centres (virtual and physical spaces which are themselves highly ICT connected systems linking a multiplicity of government agencies) for enhanced service delivery and citizen engagement—and their associated legislative backings and standing.

14.2 E-Democracy in Kenya, a Conceptual Overview

E-democracy is a relatively new concept in Kenya, often appearing in the limelight in the last decade with the country's increased adoption of ICTs. Democracy, which is often related to the promotion of a people's right to "belong", to "place" and to influence decision making in their areas has deeply been entrenched in the constitution of Kenya since independence. The use of the term through the 1990s was however largely to push a political agenda, an agenda that would culminate in a shift from the then "oppressive" political regime. In respect, the term was broadly used to promote such aspects as the right to vote and freedom of expression. Since the turn of the century, the democratic setting in Kenya has been broadly shifting towards public participation, the right to space and place, the right to service provision, and citizen platforms pushing for accountability and action from those in power.

While the push for attainment of democracy through the 1990s was more physical (exhibited largely through mass protests and strikes), the modus operandi through the 2000s has been shifting towards more digital platforms and virtual campaigns. This can largely be attributed to an increase in the number of technology compliant citizens, and a growth in the national broadband and connectivity through mobile phones. These shifts, which are at tandem with global transitions towards use of digital platforms for efficient service delivery are already bearing fruits for various sectors in Kenya, and greatly improving citizen engagement, efficiency of service delivery and in turn enhancing the national productivity. For example, adoption of digital information sharing platforms such as websites and social media alternatives by various governmental and non-governmental institutions today form the core of mass information sharing in the country, and create a loop for getting feedback on efficiency of services, with the collective implication of enhanced performance by various institutions/levels of government. In addition, the high internet and mobile phone penetration in the country (estimated at 82.6 and 88.1% respectively in 2016 as noted by Communication Authority of Kenya), coupled with growth in social media platforms have expanded social, economic and political growth opportunities and also created unique avenues for citizen participation as well as platforms for mobilization and lobbying for various services by various arms of the government. This is notably achieved through various blogs and social media forums, which enable citizens to ventilate on various multi-sectoral issues, including discussing candidates for various positions, on-going or proposed activities and projects or opinion about a policy or any legal document. The collective outcome has been better visibility of complains on poor service delivery and in turn improved service delivery in some areas. There has also been a general improvement in citizen participation in decision-making. These are key pre-requisites for attainment of democracy.

In the subsequent sections, we discuss some key examples through which ICTs are being used to promote creation of democratic spaces in Kenya for citizens, and how this is in turn shaping the country's transition towards effective democracy.

14.3 Using ICTs to Attain E-Democracy and Smart Growth in Kenya, Some Emerging Approaches

In appreciation of the role and potentiality of ICT in attaining of e-democracy, it is important to understand the requisite for any country to reap the benefits associated with digital platforms. The basic requirement is to map the penetration and access to these tools.

Kenya is a country in transition from an analog to a digital nation. Over the last one and half decade, all the ICT indicators in the country have grown tremendously, making the country one of the fastest ICT adopting countries in the world, and putting it as an African leader in ICT adoption, innovation and use for creation of socio-economic opportunities. For example, the national mobile phone penetration has risen from less than one percent in 2000 to 88.1% in 2015, creating a highly connected and informed population. In just about 10 years, the country's broadband has increased from a mere 28 Megabits per second in 2004 to 193.58 GB per second in 2015 [3–5], creating a web of highly connected human settlements both locally and globally. In addition, the proportion of the Kenyan population accessing internet services reached 74.2 per 100 inhabitants in 2016 up from 69.0% per 100 inhabitants in 2015 [16].

This growth coupled with government openness to embrace the use of ICT has brought forth varied and positive changes to the country's democratic space, with new avenues for information dissemination, knowledge sharing, citizen engagement and enhanced service delivery. The bulk of ICT driven democratic channels in the country can be traced to the last decade, particularly after the creation of the E-government strategy in 2004. The major manifestations of the emerging spaces, particularly from the public sector (which is the major service delivering body and from which citizens demand the highest level of engagement in decision making) have included, but are not limited to;

- creation of rapid information dissemination platforms, largely through websites and social media applications
- creation of online and offline service delivery centres aimed at enhanced efficiency of the public sector
- consumer feedback systems for enhanced democracy and productivity

- creation of dedicated and exclusive online based channels for civic engagement and public participation
- creation of reporting mechanisms on public service achievement and for which the citizens have their say. Examples include state of the nation reports, citizen report cards, and various accounting systems promoted by the government

14.3.1 Websites and Social Media Platforms for Information Dissemination and Citizen Engagement

All government ministries and agencies have websites showing the various activities they are undertaking and a news segment which keeps the public informed about ministry specific current affairs. There are also provisions through which citizens can send emails, ask questions or comment in the website and wait for feedback from the agencies. This has enhanced interaction between citizens and their government in many ways including informing the public on various legal and policy frameworks guiding the agencies and the service charters related to provision of certain services. In addition to websites, social media platforms such as Twitter and Facebook form key tools used by government agencies and citizens to interact. Recent research shows that there are more than 5million and about 2 million Facebook and Twitter users in Kenya respectively. Kenya ranks fourth in Africa in the use of Twitter after Egypt, Nigeria and South Africa, with 10% of the tweets being politically oriented as compared to only 2% in the United Kingdom and the USA [17]. This shows the potential for social media use for political engagement. Significant and active social media channels include the president and deputy president's Twitter handles and Facebook pages; and the Inspector General of Police Twitter handle specifically dedicated to reporting crime and corruption related cases [18]. The current president has the highest Twitter followers in the country, making him accessible to majority of Twitter users.

Websites and social media platforms work in two ways. First they make the government visible and easily accessible to communities. The online presence of the government provides an opportunity for citizens to post their opinions or express distaste regarding diverse issues whether solicited or not. For responsible citizenry, social media in Kenya have been used as a tool for whistle-blowing and engaging government, non-government organizations officials and individuals on various issues of public interest, which some have ended up being acted upon by the respective offices. Notable examples include the online based petitioning and dismissal of public officials by the president, such as a 2014 dismissal of the then Cabinet Secretary for Interior based on online public petitioning and outcry against their handling of national security issues.

In another scenario, public outrage through social media and physical demonstrations led to the establishment of 'Anti Stripping Police squad' in 2014 to help curb rising cases of women stripping in Nairobi. The establishment of the unit was elicited by public pressure on the government to arrest the perpetrators of the crime that had gone unchecked. Using social media to organize a campaign under the hashtag #MyDressMyChoice, different groups came together, marched on the streets of Nairobi and pressured the government to take action- leading to the establishment of the police unit, and the arrest and charge of those involved in stripping cases. The movement and the cause drew support from other quarters such as the Law Society of Kenya that demanded they be accorded the permission to prosecute people caught harassing and stripping women.

Electronic participation, or e-Participation including web- streaming, social media (especially Facebook and Twitter), various blogs, discussion forums and decision support systems are gaining currency in Kenya. It is the use of ICT to facilitate political participation by enabling citizens to communicate with each other, civil society, their elected representatives and their government, which has changed the perception of the leaders and how they relate with the highly informed citizens. Previous studies show that stronger online participation through technology has the potential to improve the quality of political decisions and to increase the perceived legitimacy of the decisions taken. This aspect of e-Government is seen as ushering in a new era of democratic involvement, greater transparency and accountability. Other researchers note that governments have given little priority to technologies that enable citizens to contribute to decision-making, with some going so far as to call digital democracy and e-Participation the 'myths of e-government' [19].

The above examples show the potential for the use of social media and state online contacts by citizens to influence key decisions. However, there are complains that some of the websites and contact information provided by government officials are rarely active and in most cases even when active don't respond to public complaints [20]. On the other hand, public officials claim that citizens do not fully engage with the state in the platforms provided, and when they do, they result to unconstructive criticism or meaningless discussions [18, 20]. This highlights the need to not only provide avenues for engagement but to also ensure that the public and state officials can constructively use ICT tools for effective engagement. Indeed, the use of digital platforms to actively engage consumers of services amounts to e-participation rather than just using digital platforms to improve efficiency of operations and services. According to the OECD [21], there is need to pay more attention to quality and effectiveness of moderation of online consultations, and go further to integrate online and traditional methods of citizen engagement considering that there exits cultural and organizational barriers within the society; hence, attempts to reach every segment should be a priority.

14.3.2 Consumer Reporting and Feedback Systems for Enhanced Democracy and Productivity

In addition to the use of websites and social media platforms, some agencies have gone a step further to design flow-back chain systems specifically meant to promote interaction between users and service providers. This has mainly been in the form of short code services and other two-way communication channels through which consumers launch complains or queries and get assistance from the relevant authorities. Case examples include: *Maji*Voice by Nairobi water; Power Alert by Kenya Power, and the short message service (SMS) based voter verification by Independent Electoral and Boundary Commission. The *Maji*Voice, an initiative by the Water Services Regulatory Board (WASREB), the Water Sector Regulator in Kenya, provides a two-way communication between consumers and water providers through text messages (SMS) and/or through the internet. Using their mobile phones or computers, consumers use MajiVoice to communicate their concerns, provide feedback and complaints on service delivery, and where necessary, receive feedback on how issues raised are being addressed by the service provider. According to the water services board, the aim of the platform is to enhance efficiency, accountability, responsiveness and transparency of urban water service providers in Kenya, aspects that are critical in a democracy [22].

The Kenya power company on the other hand has created a special website referred to as PowerAlert that provides real-time power blackouts and status across the country. The PowerAlert amalgamates all planned outages in a central portal and users can access the PowerAlert portal by signing in Via Google, Twitter and Facebook. By providing location, consumers can also receive notifications via email, Twitter or SMS when an outage happens in their location. The website also provides a comments section where customers can report or check on the status. The power company also provides customers with an E-bill SMS service that allows clients to query their account balance by sending a short service via their mobile phones.

The Independent Electoral and Boundaries Commission (IEBC), the body in charge of all electoral processes and related boundary delineations, is moving towards digital voter registration and voting structures, which also allow instant transmission of electoral results. In built in the system is an SMS service that enables citizens to verify their voter registration details instantly using their mobile phones regardless of their location. This is an upgrade from the previous system that required citizens to walk to the IEBC county headquarters or wait until the commission specifically opens registration window period in local centres for voters to confirm their registration details. The registration period happens mostly once every 5 years, just before the general elections. This has largely enhanced the citizens right to vote, and in turn improved democracy in the country. The use of emerging technologies in these processes is rapidly creating many alternatives for e-democracy.

14.3.2.1 Service Delivery Centres and E-Democracy

In Kenya, various service delivery centres have emerged over the last decade, the key ones being the Huduma Kenya Programme, the E-citizen portal and various E-procurement systems. These platforms combine online and offline service delivery structures, and are all aimed at enhancing effective delivery of government systems to the citizens.

14.3.2.2 Huduma Kenya Programme—Online and Offline Service Delivery System

The Huduma Kenya programme is a government driven initiative that seeks to facilitate efficiency and transparency in the provision of Public Services by providing citizens access to various Public Services and information using an integrated service delivery mechanism. Established in 2013, the programme is built around 5 channels, which include [23]:

- **Huduma Centres**—which are one-Stop Shop physical locations offering over 20 national and county government services under one roof. There are now over 40 Huduma centres spread across the county.
- **Huduma Web portal**—which is an online platform where citizens can access government services by login in online regardless of their location.
- **m-Huduma**—a mobile Platform set to offer government services and information through mobile phones.
- **Huduma Call/Contact centre**—which allows citizens to enquire about services, offered by various government agencies using a single dialing prefix.
- **Huduma payment gateway facilitates**—a platform for payment for government services using a variety of both online and mobile money services.

Specific services offered under the Huduma Programme include among others, application of National Identity cards, birth certificates, renewal of drivers licences, assessment and payment of stamp duty, Search and registration of business names, National Health Insurance Fund (NHIF) registration and claims, Higher Education Loans Board (HELB) application and repayment, National Social Security Fund (NSSF) member statements and benefits, reporting corruption, police abstracts and public complains though the Commission on Administrative Justice. Tenders and public service vacancies can also be accessed through the Huduma Channels.

The establishment of the Huduma Programme was in response to the delays and red tape experienced in the delivery of public services that bred vices such as corruption and abuse of office by public officials. The initiative seems to have achieved its objectives of improving access to government services, by minimizing time taken to access service as well as elimination of bureaucracy and corruption loopholes in the delivery of public service. A 2015 study [24] at one of the Huduma Centers in Mombasa indicates that, overall, the one stop shop Huduma centres have greatly enhanced transparency and improved citizen satisfaction with public service delivery. Over 65% of those interviewed expressed satisfaction with the time taken to access services with 70% indicating that services offered met the expected standards. The initiative has also received various global awards and recognitions such as the United Nations Service Award for public service excellence; the African Gold Award on Innovative Management by the Association of African Public Administration and Management; the Best customer service award by the Institute of Customer Service Kenya and the Best Use of ICT award by the ICT association of Kenya [25].

The Huduma programme though more inclined to the improvement of public service, enhances e-democracy in several ways. One, in addition to bringing services closer to consumers, the program allows citizens to participate in the evaluation and monitoring of government agencies since all government agencies are now operating under performance contract and more importantly all of them have a service charter indicating how long someone should wait for a particular service and the cost for such services. This information is placed in strategic locations where those seeking services can easily access it. Access to information is a critical element in bolstering of democracy. Additionally, the Huduma programme provides platforms such as the call centres and citizen feedback mechanisms that allow citizens to post their comments and complaints concerning government services.

However the main challenges facing the programme is the limited human resource capacity in some of the centres and limited infrastructure development in the remote parts of the county, hence forcing citizen to travel long distances to access services in the available centres. Majority of the physical one-stop shops are located at the county headquarters while only a few people in the rural areas have access to internet connection that can allow them access the online web portal.

14.3.2.3 E-Citizen Portal—Online Based Service Delivery Platform

The E-citizen portal is an online portal launched in 2014 by the ICT Authority of Kenya to provide cross-agency, citizen-centric information and services, and to help citizens complete their transactions conveniently with ONE government regardless of their physical location [26]. The E-Citizen portal offers various services similar to those offered through the Huduma Programme which include: Provisional Driving License, Business Name Registration, Business Name Search, Notice of Marriage, Issuance of a registrar's certificate, Solemnization of marriage, Commissioning of affidavits, Special licenses for marriage, Driving Test Booking, Interim Driving License, Driving License Renewal, Driving Class Endorsement, Duplicate Driving License, Driving License Information Corrections, Official Search (Nairobi Blocks), Land Rent Clearance Certificate, Application for Passport and Application for Kenyan Visa. The portal works by allowing both international and local persons create an online account that they can use to access services from government agencies. For Kenyan citizens the portal is linked to the Integrated Population Registration Service.

The portal has not only made it easier for Kenyans to access services, but foreign nationals can apply for a Kenyan Visa online reducing the time taken going to Kenyan embassies abroad or waiting for clearance at customs.

14.3.2.4 E-Procurement Systems

In 2014, Kenya government launched an online system for submitting and evaluating government related procurement applications. The e-procurement system which is part of the Integrated Financial Management System (IFMS) under the National Treasury is set to reduce corruption in government tendering, reduce government spending and speed procurement processes. The initiative hopes to improve transparency and accountability in government procurement by ensuring that all applications submitted are recorded. The system also contains a in-built price referencing for all tenders that ensures that bids above a certain amount cannot be submitted, thus reducing the risk of consultants over quoting and over charging the government, a situation that has plagued government tenders [27].

The system has been successful but has nevertheless met several challenges. In 2015 the Kenya Council of Governors (CoG) requested for the suspension of the procurement system in County[1] related tenders. The reasons cited was the constant breakdown thus hampering county processes; limited technical infrastructure in some of the counties; limited access to internet for local contractors which the CoG claimed were being marginalized and the feeling the national government was recentralizing county services by requiring counties to use a national government hosted system [28, 29]. In the recent past, the IFMIS system is also being linked to major public funds mismanagement scandals, indicating its limitation to enhanced transparency and governance.

Although majority of the above initiatives can be broadly categorized as e-government and e-administration, they also provide avenues for citizens to easily access information, provide feedback and voice their opinions regarding varied issues thus contributing to decision-making processes in the respective government agencies.

14.4 The Future of E-Democracy in Kenya—A Legal, Political or Citizen Driven Agenda?

The Kenyan scenario shows both strong cases of bottom-up and top-down driven e-democracy initiatives. Government driven efforts have however been labeled as public relation stints and in some instances have been identified as been grossly

[1] A county is the second level of administration in Kenya, and is the highest level of devolution. There are 47 counties in the country and each county is headed by a governor.

wanting in meeting the purposes for which they were established, such as eliminating corruption, ensuring efficient delivery of services or providing a platform for communities to be heard. Going forward thus, it is imperative that the government puts into place mechanisms to ensure successful implementation of e-government that is built not only on the need to provide services and eliminate corruption but that seeks to loop in feedback from the citizens to improve service delivery as well as promote democracy by factoring in citizens voices in government practices and decision making, or simply translating e-government to e-governance.

Some of the key steps to consider include: the need to hinge e-government and e-administrative practices on law. For example, a reason cited by the CoG when they demanded the suspension of the requirement to use e-procurement was that such requirement was not envisaged in the Public Procurement and Disposal Act of 2005 which is the law governing procurement in Kenya [30].

Second, ensuring capacity and capability among staff and the public citizens/clients to understand and deal with ICT related systems. The setting up of electronic systems while necessary for the effective delivery and involvement of citizens is not sufficient. Such should be accompanied by investing in proper ICT infrastructure across the country, ensuring Internet access for all, and enhancing ICT literacy among government staff and the general public. Again using the case example of suspension of e-procurement system by the CoG, another reason cited was its inaccessibility by local consultants who did not have the required knowledge, equipment and resources such as Internet to access the system [29].

Third, there is need to consider stakeholder buy-in for services and systems that require multi-agency and multi-level systems of government to implement such programmes as the Huduma Programme, E-Citizen and e-procurement system.

Fourth, there needs to be genuine commitment to feedback and response to clients and public concerns. Although most government websites provide feedback forums, there is limited proof that such feedback is taken into consideration or even the feedback sites are monitored. A study monitoring the use of Twitter by government agencies and officials showed that while citizens posted comments and opinions, there was no feedback loop and most citizens felt that the government's online presence was only for public relations reasons and that they didn't care about reports, feedback and complaints from the public [18]. A similar study by ICT4D that randomly send emails to four government emails registered on government websites found that such emails weren't active as none of the emails received response [20].

While the government has a key responsibility in promoting e-governance, the Kenyan citizen too has a role to play in holding the government accountable, demanding representation in e-governance structures as well as utilizing the opportunities provided to participate and influence government decision making. A study by ICT4D [20] looking at the extent to which Kenya's ICT revolution has transformed e-government to e-governance found that ICT has not necessarily translated to better governance. The study carried out in Nairobi, the country's ICT hub found that State actors do not differentiate between e-government and e-governance, and that they were not attentive to citizens need for 'more public participation'. Similarly, despite the multiplicity of e-government tools and platforms provided by the government, only a handful of Nairobi residents utilize ICT to interact with government actors. Hence, this calls for concerted efforts both from the state and citizens to engage each other in the available channels and to create news ones to enhance government and citizens interaction.

Studies show that there is a growing number of citizens and civil society organizations using ICT platforms and social media channels such as Facebook and Twitter to raise critical issues and engage with the government. These include groups such as PAWA254, Ushahidi, several blogging platforms and Kenyans on Twitter (#KOT) online campaigns. These channels provide avenues for citizen engagement with the government and thus need to be fully harnessed and supported not only to bolster the voice of communities but also be used in a constructive way that reaches public leaders and state bureaucrats.

14.5 Review of Legal Setup for E-Democracy in Kenya

14.5.1 The Role of Political Setup, Legal and Policy Framework in Attaining E-Democracy in Kenya—A Historical Chronology

The evolution and use of ICT in Kenyan government related processes is highly entwined with the nations' political and legal policy framework, with various governments embracing technology driven democracy differently. Although the presence of internet and mobile phones started to be felt in late 1990's during President Moi's era,[2] Kenya's adoption of e-governance gained momentum in early 2000s under president Kibaki's regime.[3] The only notable implementation of an E-Government infrastructure project in Kenya under President Moi was in late 1990s under the Ministry of Local Government (MoLG) dubbed Local Authorities Information and Financial Operations Management System (LAIFOMS), which was part of the Kenya Local Government Reforms Programmes (KLGRP) partly funded by the World Bank [31]. LAIFOMS were supposed to improve accountability and service delivery in local authorities.

President Mwai Kibaki's regime is given credit for laying the foundation for e-governance in Kenya. In 2002 Kibaki's government established the E-government Secretariat to spearhead the computerization of government services [20]. Subsequently, the e-government strategy was created in 2004. In 2006, the government adopted the National Information and Communication Technology

[2]President Moi was Kenya's second president with a tenure running 24 years, from 1978 and 2002.

[3]President Kibaki was Kenya's third president with a tenure running for 10 years between 2002 and 2012.

(ICT) that spelt out the steps to be taken by the state to effectively implement the e-government strategy. Since then, the government embarked on the implementation of various e-government systems aimed at promoting efficiency, transparency and accountability within the public service administration [32]. The recent enactment of the 2010 constitution that requires citizen participation and access to public information has also bolstered the use of ICT in public engagement. It was also during Kibaki's era when Kenyan government launched the open data website in 2011 releasing several large datasets, including statistics on government spending at the national and county level, to enhance transparency in governance and access to information.

The current government headed by President Uhuru Kenyatta won the 2013 election on a campaign built around the promise to deliver a digital tech savvy government. The government has sought to make its promise true by building on the ICT gains achieved during Kibaki's regime as well as creating new ICT systems. Notable examples include the E-citizen initiative, the Huduma Kenya program and the efforts to enhance the penetration of digital platforms and ICT to all areas including plans to promote e-learning by equipping all public primary schools with ICT infrastructure. These reforms have expanded opportunities for citizens to participate in development processes, including governance.

On analysis of the political drivers of adoption of e-technology and provision of participation space to citizens, it is important to note that President Moi's regime was composed of mainly old and conservative leaders whose interest was oriented towards service delivery and not citizen participation, with most reforms being part of donor conditionality. It is also prudent to note that President Moi oversaw the last single party politics regime in Kenya, which did not give room for dissent voices and citizens choices were politically guided. President Mwai Kibaki's regime on the other hand had a hybrid composition of those in leadership, where young professionals left the private sector to join the Government to bring in new ways of doing things, hence fast tracking the process of e-governance. Kibaki's presidency was also buoyed by the promulgation of Kenya's constitution in 2010, where issues of participation and citizen rights are given prominence. Finally, Uhuru Kenyatta's government is composed of young leaders, and the presidency as a whole has been building on the momentum created by President Kibaki's regime to ensure that platforms such as the e-citizen digital platform operate optimally.

In the Kenyan context, like in many other countries, participatory development began with and was for a long time confined to community development projects [33]. A landmark event in the evolution of participatory development and law in Kenya was the enactment of the Physical Planning Act in 1996. The Statute does provide for community participation in the preparation and implementation of physical and development plans. Traditionally, government transactions have been provided face to face in office settings, but governments are increasingly adopting the use of ICTs to provide public information and services [34]. Policy-makers and researchers have argued that e-government has the potential to enhance government accountability and transparency, increase free flow of public information, and improve citizens' participation in the public policy process [35]. Public service oriented deliverables such as getting national identity cards, passports etc. for a long time was done in secrecy and it was very hard for citizens to make applications do a follow up. This has been solved by the e-Citizen, which is the first-stop portal for Government information and services, organized with citizen needs in mind.

14.5.2 Legal and Policy Framework Supporting E-Democracy Through Digital Platforms

(a) **National Information and Communication Technology (ICT) Policy (2006)**

The National ICT policy of 2006 is based on four guiding principles, namely;

(i) infrastructure development
(ii) human resource development
(iii) stakeholder participation and
(iv) appropriate policy and regulatory framework.

According to this policy document, the overall goal of e-Government is to make the Government more result oriented, efficient and citizen centered. The e-Government strategy focuses on redefining the relationship between Government and citizens with the objective of empowering them through increased and better access to government services. The broad objectives of e-Government is to: (a) Improve collaboration between Government agencies and enhance efficiency and effectiveness of resource utilization; 12 (b) Improve Kenya's competitiveness by providing timely information and delivery of Government services; (c) Reduce transaction costs for the Government, citizens and the private sector through the provision of products and services electronically; and (d) Provide a forum for citizens' participation in Government activities. The e-Government initiative aims to be a shared vision amongst the Government, private sector and civil society with its implementation process involving all stakeholders [36].

(b) **The Kenya Communication Act of 1998**

The Kenya Communication Act, which repealed the Kenya Posts and Telecommunications Act, provides the current framework for regulating the communications sector in Kenya. The Act unbundled the former Kenya Post and Telecommunication into five separate entities: Telkom, the fixed-line operator; the Postal Corporation of Kenya (Posta); the Communications Commission of Kenya (CCK), the sector regulator; the National Communications Secretariat (NCS) to advise the government on the adoption of a communication policy; and an Appeals Tribunal for the purposes of arbitration in cases where disputes arise between parties. It was the most influential document regarding ICT legislation and regulation in Kenya [37].

(c) **The Kenya Information and Communications Act Cap. 411A of 2011**

The Information and Communications Act 411A is an Act of Parliament to provide for the establishment of the Communications Commission of Kenya, to facilitate the development of the information and communications sector (including broadcasting, multimedia, telecommunications and postal services) and electronic commerce to provide for the transfer of the functions, powers, assets and liabilities of the Kenya Posts and Telecommunication Corporation to the Commission, the Telcom Kenya Limited and the Postal Corporation of Kenya, and for connected purposes. It seeks to address some of the challenges cited in the national ICT policy document. One of the key sections of the Information Communication Act is Part VII, on electronic transactions (e-transactions). In this part, the new Act, inter alia, gives legal recognition to electronic records; recognizes electronic messages as valid for the formation of contracts; and supports the use of electronic records and electronic signatures in government and its agencies.

(d) **The Constitution of Kenya, 2010 and related regulatory framework**

The promulgation of the Constitution of Kenya in August 2010 provided a strong legal foundation for the enhancement of participatory governance. The right to access information held by governmental authorities often referred to as Freedom of Information is a fundamental human right recognized in International Law. Most recently, Kenya has taken crucial steps towards recognition of this right in the New Constitution. Article 35(1) states that every citizen has the right to access information held by the State. Further, article 35(3) states that the State shall publish and publicize any important information affecting the nation [38].

In 2010, the government issued the following regulations specific to information and communications:

(i) Kenya Information and Communications (dispute resolution) Regulations, 2010, which stipulate the dispute resolution powers of the CCK and the processes of resolving disputes between consumers and service providers.
(ii) Kenya Information and Communications (tariff) Regulations, 2010, which provide a framework for the determination of tariffs and tariff structures, and seek: to ensure licensees maintain financial integrity and attract capital; to protect the interests of investors, consumers, and other stakeholders; to provide market incentives for licensees to operate efficiently; and to promote fair competition.
(iii) Kenya Information and Communications (compliance monitoring, inspections, and enforcement) Regulations, 2010, which stipulate the CCK's power in monitoring and enforcement of installations and maintenance of communication infrastructure.
(iv) Kenya Information and Communications (fair competition and equality of treatment) Regulations, 2010, whose purposes are: to provide a regulatory framework for the promotion of fair competition and equal treatment in the

communications sector; and protect against the abuse of market power or other anticompetitive practices within the communications sector.

(v) Kenya Information and Communications (interconnection and provision of fixed links, access and facilities) Regulations, 2010, which provide guidelines on interconnection.

(vi) Kenya Information and Communications (Universal Access and Service) Regulations, 2010.

(vii) Kenya Information and Communications (Radio Communications and Frequency) Regulations, 2010.

(viii) Kenya Information and Communications (Numbering) Regulations, 2010.

(ix) Kenya Information and Communications (Licensing and Quality of Service) Regulations, 2010.

(x) Kenya Information and Communications (Importation, Type approval, and Distribution of Communications Equipment) Regulations, 2010.

(xi) Kenya Information and Communications (Electronic Certification and Domain Name Administration) Regulations, 2010.

(xii) Kenya Information and Communications (Consumer Protection) Regulations, 2010.

14.6 Conclusion

Just like many developing countries, Kenya government is making good efforts in embracing digital platforms as a way and basis for sharing information, consultation and inducing citizen participation which is noted to be critical in creating and enhancing democratic space. The Kenya's constitution is likely to catalyze the role and adoption of ICT related tools, as it is a requirement that all policies and legal documents must be endorsed by citizens and experts before adoption. Today's advances in technology, and its rapid adoption and utilization in the country has made it a common platform for engagement among citizens and the government in Kenya. Various digital platforms are for example now commonly used by all government agencies in reaching out to citizens, and are also acting as a basis for stakeholder consultations in which citizen views on issues of public importance and interest are discussed. This has largely been empowering citizens through information and creating avenues for participation which is an asset in assessing Kenya's democratic gains. Thanks to the adoption of ICTs across various government sectors as well as rapid adoption and absorption of technology by Kenyans, major strides have already been made towards promoting citizen participation in governance, both through creating avenues for them to be heard and also promoting their democratic rights such as that to vote and participate in policy formulation. While the country is still struggling to attain optimal levels of e-governance and e-democracy which would positively contribute to enhanced socio-economic status, the next decade looks promising and is going to benefit from a consistently evolving policy framework and infrastructure development that favours electronic forms of interactions.

References

1. Kimani B (2016) Overview of SDGs in Kenya. Paper presented during the County Capacity Forum at Hill Park, Diani on 20 June 2016. http://www.devolutionplanning.go.ke/images/Overview%20of%20the%20SDGs%20%20in%20Kenya.%20Diani.pdf
2. Government of Kenya (2004) e-government strategy 2004. Government Press, Kenya
3. Export Processing Zones Authority (2005) Kenya's information and communications technology sector 2005
4. International Telecommunication Union—ITU. [Online]. Available: http://www.itu.int/net4/itu-d/icteye/. Accessed 27 Sept 2016. [Online]. Available: http://www.itu.int/net4/itu-d/icteye/. Accessed 15 Feb 2016
5. Communications Authority of Kenya (2015) First quarter sector statistics report for the financial year 2015/2016
6. Klopp J, Marcello E, Kirui G, Mwangi H (2013) Can the Internet Improve Local Governance? The Case of Ruiru, Kenya. Inf Polity 18(1): 21–42
7. Postnote (2009) E-democracy. Parliamentary Office of Science and Technology. www.parliament.uk/parliamentary_offices/post/pubs2009.cfm
8. Freeman J, Quirke S (2013) Understanding e-democracy: government-led initiatives for democratic reform. JeDem 5(2):141–154
9. Coleman S, Norris DF (2005) A new agenda for e-democracy. Oxford Internet Institute, Forum Discussion Paper No. 4, The University of Oxford for the Oxford Internet Institute, Jan 2005
10. Wittig M (1991) Electronic city hall
11. Macintosh A, Coleman S (2003) Promise and problems of e-democracy: challenges of online citizen engagement. Organisation for Economic Co-operation and Development
12. Khosrow-Pour M (2008) E-government diffusion, policy, and impact: advanced issues and practices: advanced issues and practices. IGI Global
13. Ke W, Wei KK (2004) Successful e-government in Singapore. Commun ACM 47(6):95–99
14. Singapore Government (2016) Reach, reaching evryone for Active Citinery@home
15. United Nations Development Programme and Panos Institute of West Africa (n.d) E-governance and citizen participation in West Africa: challenges and Opportunities. UNDP & PIWA
16. Communications Authority of Kenya (2016) Kenya's mobile penetration hits 88 per cent. http://www.ca.go.ke/index.php/what-we-do/94-news/366-kenya-s-mobile-penetration-hits-88-per-cent. Accessed on 10 Oct 2016
17. Portland (2016) How Africa tweets. Portland Communications. January 2012. http://www.portland-communications.com/publications/how-africa-tweets-2015/
18. Alfred Z (2014) Tweeting against corruption: fighting police bribery through online collective action, UPPSALA UNIVERSITET. Master of Social Science (Digital Media and Society)
19. Bekkers V, Homburg V (2007) The myths of e-government: looking beyond the assumptions of a new and better government. Inf Soc 23(5)
20. Salome N (2016) Has Kenya's ICT revolution triggered more citizen participation? ICT4Democracy. Naiorbi, Institute of Development Studies (IDS)

21. Organisation for Economic Co-operation and Development (OECD) (2003) Promise and problems of e-democracy. ISBN 92-64-01948-0. Accessed on 17 Nov 2016. http://www.oecd.org/gov/digital-government/35176328.pdf
22. Majivoice Website (2016) http://www.majivoice.com/?page=Introduction%20to%20MajiVoice. Accessed on 11 Oct 2016
23. Wanzala J (2016) There will be Huduma centres in all the 47 counties by end of year, says secretariat head. Standard Digital. Nairobi, Standard Group
24. Abdalla AG et al (2015) Effect of Huduma centers (one stop shops) in service delivery—a case study of Mombasa Huduma Centre. Int J Acad Res Bus Soc Sci 5(6)
25. Ochieng A (2015) Huduma centres win award. Daily Nation Nairobi, Nation Media Group
26. Mugendi E (2016) Kenya's ECitizen portal is saving the government a ton of money. Techcabal. Lagos, Techcabal
27. Matinde V (2014) Kenyan government launches e-procurement system. WebAfrica
28. Ndonga S (2015) Governors want e-procurement suspended in counties. Capital News, CapitalFM
29. Smith P (2015) Implementing eProcurement—Kenya runs into problems. Pubic Spend Forum from http://publicspendforumeurope.com/2015/09/25/implementing-eprocurement-kenya-runs-into-problems/
30. Sanga B (2015) Kenya: governors now drop e-procurement system. Standard Digital
31. Muganda-Ochara N, Jean-Paul VB (2008) Managing the e-government adoption process in Kenya's local authorities. Communications of the IBIMA 1
32. Wamoto FO (2015) E-government implementation in Kenya, an evaluation of Factors hindering or promoting e-government successful implementation. Int J Comput Appl Technol Res 4(12):906–915
33. Wakwabubi E, Shiverenje H (2003) Guidelines on participatory development in Kenya: critical reflection on training, policy & scaling up
34. Heeks R (2002) Information systems and developing countries: failure, success, and local improvisations. Inf Soc 18:101–112. http://www.elearning.jo/datapool/books/824/6705439.pdf
35. World Bank (2004) Building blocks of e-government: lessons from developing countries. August 2004. PremNotes No. 91. http://www1.worldbank.org/prem/PREMNotes/premnote91.pdf
36. Government of Kenya (2006) National information and communication technology policy. Government Printers, Nairobi
37. Waema T, Ndung'u M (2012) What is happening in ICT in Kenya. A Supply-and demand-side of the ICT sector
38. Omolo A (2011) Policy proposals on citizen participation in devolved governance in Kenya. The Institute for Social Accountability (TISA). http://www.tisa.or.ke/images/uploads/Policy_Recommendations_on_Citizen_Participation-TISA_2011.pdf

Chapter 15
Politics of Open Data in Russia: Regional and Municipal Perspectives

Yury Kabanov, Mikhail Karyagin and Viacheslav Romanov

Abstract The study is devoted to the development of open government data (OGD) policy in Russia, its regions and cities. We argue that although OGD are a potential driver of citizen participation and smart city innovations, these effects can be interfered with by political, institutional, socio-economic (structural) factors, as well as by agency. We use a mixed quantitative—qualitative methodology, and data gathered within the Infometer project, to unveil the determinants of OGD success and failure. Our findings suggest that although some general patterns can be traced, agency as a motivation to innovate plays a crucial role.

Keywords Open data · E-democracy · Online engagement · E-participation · Smart city · Russia · Regions of Russia · Comparative analysis · Neoinstitutionalism

15.1 Introduction

There is a growing number of studies that link the future of democracy and civic engagement to the development of the Internet-based tools [1, 2]. The recent innovation is the open government concept aiming at making public policy process more participatory, transparent and, eventually, democratic. Actively promoted by Obama's US administration, open government policies have diffused throughout the globe, creating different meanings in various countries and institutional settings [3].

Y. Kabanov (✉) · M. Karyagin
National Research University Higher School of Economics, St. Petersburg, Russia
e-mail: ykabanov@hse.ru

M. Karyagin
e-mail: karyaginm@gmail.com

V. Romanov
Infometer, St. Petersburg, Russia
e-mail: vromanov@infometer.org

T.M. Vinod Kumar (ed.), *E-Democracy for Smart Cities*, Advances in 21st Century Human Settlements, DOI 10.1007/978-981-10-4035-1_15

As argued by Meijer et al. [4] open government is a combination of two interconnected elements—*vision*, i.e. access to government information and opportunities to use it for engagement and control, and *voice*, i.e. public participation. In fact, both sides are crucial for effective open government performance: both the uninformed participation and information for the sake of informing cannot contribute to making governments more accountable. In this respect, open government data (OGD), as a means of informing citizens, are becoming an important part of public policy development [5]. OGD can be defined as data, produced by public agencies and free to be "used, re-used and distributed" [6: 6]. To be called open in that sense, the data need to meet several principles: be "complete, primary, timely, accessible, machine readable, non-discriminatory, non-proprietary, license free".[1] In these principles we see the *philosophy* and the *technology* of openness to be implemented.

OGD are now considered a cornerstone of *smart city* development, as big data approach helps to tackle urban problems and raise the quality of life [7], e.g. solving energy or traffic issues [8], boosting innovations in economy or civic education [9]. In terms of e-democracy discourse, OGD help to eliminate knowledge asymmetries, "facilitate joint fact finding" and maintain trust, hence provide a collaborative smart cities' governance [10]. As Davies puts it, "exploring OGD can give individuals greater understanding of the state, and direct access to facts and information can empower individuals in their interactions with the state" [11: 38]. In other words, OGD facilitate information provision, giving citizens a strong argument in favor of their citizen-expert position [12].

However, this optimistic view often confronts a more moderate reality. First, there are considerable doubts that OGD can resolve the problem of inequality in access to data between the rich and the marginalized. The technological elaborateness of OGD may in fact deepen the *digital divide* and worsen the asymmetry [13]. Secondly, there is ambiguous evidence on whether OGD have a real impact on socioeconomic and political processes, which is dependent upon the policy goals and strategy [14]. This makes the academia think of *why OGD policies fail and what the success factors are.*

Although this domain of research remains relatively new and undertheorized, several works have been published to address this issue. For instance, Susha et al. and Janssen et al. present comprehensive sets of success factors and barriers towards successful OGD policy implementation. The basic areas of study are regulations and legislation, political and managerial factors, technology, processes, stakeholders' communication etc. [5, 15]. As Susha et al. [15] have applied the framework to cases, they argue that the set of success factors depend largely on the context of OGD initiative implementation. It somehow resonates with the *understanding* perspective of e-government research, coined by Bekkers and Meijer [16], when we are trying to study the process within a context. Likewise, Conradie and

[1]Open Government Data Principles, https://public.resource.org/8_principles.html (2007).

Choenni [17] emphasize the importance of internal bureaucratic process in OGD release by local governments.

On the contrary, an outer perspective—*explaining* in Bekkers and Meiker's terms, aims at finding some general, one-fits-it-all patterns and factors. This approach is not so widely used in OGD research, but is quite popular in explaining e-government and e-participation performance. These are usually large-N studies using quantitative methods that underline primarily the importance of political variables (like regime), institutional factors (quality of government and bureaucracy), socioeconomic and technological development [18–20].

We argue that the synthesis of two approaches will be valuable in finding the answer to the question why OGD policy is a success or failure. In our view, OGD implementation is a process of actors' interaction that is run within a certain institutional and recourse constrains. The constellation of actors is definitely different across countries, and even across regions and cities within a state. And so vary the technological readiness, human capital or the level of democracy. Then we first need to find some basic factors (or *structures*) that influence OGD dynamic across polities and then look deeper to understand how *agency* moderate this effect.

15.1.1 What is Political about Open Government Data?

Unlike the view that OGD implementation is mainly about bureaucratic and managerial issues, we also take a political perspective. As OGD policy is normatively associated with the shifts in political sphere, it cannot escape the politics—that is, the competition and bargaining between politicians, bureaucracy and citizens.

There are several dimensions of the political in OGD release. First, it is the *input* meaning the publication of OGD. As Davies argues, "… datasets themselves are political objects, and policies to open up datasets are the product of politics. If you look beyond the binary fight over whether government data should be open or not, then you will find a far more subtle set of political questions over the what and the how of opening data" [21]. Second, it is the *output*, when we encounter the politics of OGD use and interpretation, as "[t]he political element enters when questions emerge about what the data tells us or means" [22].

But the *black box* of OGD release is also political in nature, at least for the following reasons:

- OGD release is a process of *resource allocation*. It implies, first, that OGD shift from government-owned to free resource, which deprives public bodies from an exclusive rent from it [23]. Secondly, OGD projects require budget funds or private investments, which need to be redistributed according to some criteria or as a result of political bargaining.
- OGD release implies the *politics of control* over bureaucracies. OGD policy is an intervention into the existing and stable behavior patterns of bureaucracy, provoking an institutional adaptation and resistance from the latter [24].

- OGD release is normatively a process of *empowering citizens* vis-à-vis the government. It might be a problem, especially in non-democratic political settings.
- OGD release, especially on the local and regional levels, encompasses the *political bargaining* between the national government and the subnational units. When the diffusion of OGD is not horizontal but pushed through the vertical influence patterns, there might be a sufficient resistance from below that hinder national initiatives. On the contrary, the subnational or municipal governments may face blocks from the center towards the further development of innovations.

In sum, from a political perspective, success or failure of OGD is mainly the result of the abovementioned interactions occurring in a certain institutional settings. To study OGD from this viewpoint, we should distinguish the key structural variables and the main actors involve and investigate how they shape OGD policy.

15.1.2 Research Question

To put simple, the research question of the paper is *what factors determine the success of OGD policy on the regional and municipal level?* We take OGD policies in Russian regions and cities as cases. Russia represents an interesting and understudied example of Open Government policy implementation. First, it is the case when normatively democratic innovation has diffused into a non-democratic polity. Secondly, Russia has a peculiar system of government, when the proclaimed federalism coexists with political centralization known as the *power vertical*, and it seems to influence the implementation of OGD. Thirdly, despite the overall stagnation of OGD policy fixed by international rankings, regions and municipalities represent different degrees of failure and success.

Taking our political perspective, we apply several analytical strategies to answer the research question. First, we use large-N analysis and quantitative methods of study to reveal some general factors that might influence the OGD dynamics on the regional level. Secondly, we take four representative cases and qualitatively analyze the politics of OGD that shape the outcomes of the policy. The data has been gathered by the *Infometer* project, an expert community analyzing Russian open government performance we cooperate with during our study.

The remainder of the paper is as follows. First, we outline the peculiarities of OGD policy in Russia in its national, regional and municipal dimensions. Secondly, we use Infometer data to reveal patterns of regional/municipal OGD performance, explaining the methodology of the project. Thirdly, we run quantitative study of factors determining OGD development on the regional level. Fourthly, we conduct case—studies to reveal other factors that may matter. Finally, we sum our findings up.

15.2 Authoritarian Modernization, Power Vertical and OGD Policy in Russia: A Review

The first elements of OGD policy in Russia started to be implemented in 2011, as a part of a broader Open Government concept. The latter moved from two directions and was related to the presidential and parliamentary elections period. The first line was associated with then President Dmitry Medvedev, who was to take the position of Prime Minister after the elections. He initiated the so-called *Big Government initiative*—an idea to create a kind of a consultative body operating along with the real government. Later on the idea transformed into the Open Government initiative, which encompassed measures to revive the existing consultative structures at public agencies, but also included some new technologies like OGD. The idea started to develop within a high-level Working Group, the deputy head of which—Mikhail Abyzov, became a minister in charge of the Open Government after the presidential elections. Another dimension is associated with Vladimir Putin, a nominal Prime Minister to be elected President. In his article *Democracy and Quality of Government* he pledged to transform government by means of new technologies. In particular, he initiated the e-petitions portal *Russian Public Initiative*, a cornerstone of the contemporary Russian e-participation.[2] After Putin was elected President, in May 2012 he signed a number of program decrees, known as *May Decrees*, to outline the key issues of his policy. One of them combined Medvedev's and Putin's initiatives related to open government and e-participation in one. Public consultations and OGD publication became then a priority task for Russian authorities.[3]

Soon the legal and institutional basis for OGD policy was created. The Governmental Commission led by Minister M. Abyzov coordinates the overall open government policy, including OGD through the Council on Open Data.[4] The conceptual document is the Standard of Openness, where OGD release as a measure to achieve public agencies' transparency and accountability.[5] A more practical document is the Guideline for Open Data Release. It is a legal document approved by the Russian Government that provides basic requirements, procedures and technicalities of OGD release, imposing obligations on federal, regional and municipal executive agencies. The current 3.0 version of the Guideline (2014)

---

[2]Putin, V.V.: Democracy and Quality of Government [in Russian]. Kommersant, 6.02.2012, http://www.kommersant.ru/doc/1866753.

[3]Decree No 601 of President of the Russian Federation On Basic Directions of Public Administration System Improvement. Adopted 7.05.2012, http://www.rg.ru/2012/05/09/gosupravlenie-dok.html [in Russian].

[4]What is Open Data//Open Government of Russia http://opendata.open.gov.ru/event/5598184/ [in Russian].

[5]Concept of Openness of the Federal Agencies. Adopted by the Government of the Russian Federation 30.01.2014, http://open.gov.ru/upload/iblock/37a/37a15dc7026ed84ae9a886bed7eab8a1.pdf [in Russian].

presumes that the publication of a certain dataset should be based on its demand, its readiness and cost-benefit analysis. OGD must be published in one of the three ways: on a special page of an agency's website, on a regional or the federal OGD portal, and voluntarily by alternative means (like FTP-server).[6]

In general, OGD policy in Russia organized somewhat in the middle between a hierarchy and a network. On the one hand, public agencies posses a certain portion of discretion in choosing the way and format they publish the data, although there are some obligatory datasets. As a result, we have a mosaic picture of different portals and webpages containing OGD of various formats and structures that are hard to link.[7] On the other hand, the federal government largely controls this process, with key decision-making concentrated in the governmental Commission.

The results of OGD policy on the federal level seem contradictory as well. First, from the governmental viewpoint, the policy is quite successfully implemented, and in fact, the costly and necessary infrastructure of OGD was created (like the federal Open Data portal).[8] As of the end of 2015 the government had released more than 12,000 datasets, 17 regional data portals and 56 regional OGD pages were launched.[9]

However, according to the international indices, the tempo and quality of OGD policy is declining. According to the *Open Data Index,* the country's rank declined from 32 in 2013 to 61 in 2015, as it still lacks some crucial datasets like on government budget, spending and geodata.[10] The *Open Data Barometer* indicates further a decline in 15 positions in 2015, as well as a low impact of OGD on political, social or economic transformations.[11] It means that OGD have little practical effect, do not contribute to democracy, institutional quality or economic growth.

Russia being an electoral authoritarianism [25], we may expect that OGD are becoming nothing more that a façade structure for building-up international or internal legitimacy. Plenty of works are devoted to this phenomenon, when no practical effects on democracy were expected at the moment of designing e-participation, e-government or open government in autocracies [26, 27]. And, in fact, we cannot expect democratization by means of OGD that are not used, in times of hardening Internet control and repression on the Web [28].

This is a part of the story, though. Legitimacy cannot be the only reason for innovations. At least, we can enumerate three, based on *democratic authoritarianism* concept. First, it might be rent seeking, when resources to create

[6]Methodological Recommendations on Open Data Publication, Vol. 3.0, http://data.gov.ru/metodicheskie-rekomendacii-po-publikacii-otkrytyh-dannyh-versiya-30 [in Russian].

[7]Unified Standard is Possible. Interview with A. Petrachenko, responsible for OGD release in the Sakhalin region//Infometer, http://read.infometer.org/sakhalin-interview [in Russian].

[8]http://data.gov.ru.

[9]Open Data//Administrative Reform in Russia, http://ar.gov.ru/ru/inform_otkritost_05_otkritii_dannie/index.html [in Russian].

[10]Global Open Data Index//Open Knowledge International, http://index.okfn.org/.

[11]Open Data Barometer//WWW Foundation, http://opendatabarometer.org/.

infrastructure and applications are re-allocated to certain groups. Secondly, requirements for OGD may be used to control elites and bureaucracies of different levels. Thirdly, *credible commitment* and a goal to make a *pocket of effectiveness* may also be the case [29].

We can think of whether e-democracy in Russia can be comparable to the so-called *authoritarian deliberation* practices (like in China), when online mechanisms are used for consultation and deliberation, making the voice of people heard, but no power shifts can be traced [30]. In fact, there are signs that the government to get new ideas and shape public policy agenda uses e-participation, but these attempts are rare and are more pertinent to regions and municipalities [31]. However, OGD play some role in fostering online engagement.

The key consumers of OGD in Russia can be divided into three main groups:

- *Software and applications developers.* OGD allow making an interesting product, both for commercial and non-commercial use. As a rule, commercial products are produced not solely on the basis of OGD. The clearest example is the number of applications developed to analyze public procurement. The *Public Procurement portal*[12] contains some open data, but it needs to be additionally re-arranged to use in apps. Some businesses use this situation to process the data and sell it to those working with governmental procurement system. We can also recall the non-commercial resource *GosZatraty* (*State Spending*)[13] that aims at raising public awareness in the sphere of state budgeting. Also there are several projects that use OGD for social welfare. For instance, there is the *Smooth.city*, a project that help to create routes for people with disabilities.
- *Civic activists.* Civil projects also use OGD. They are usually developed during *hackatons* and rarely lead to economic benefits. For instance, there is an anti-corruption *hackaton*, where developers look for the ways to use OGD to combat bribery, corruption and budget mismanagement.[14]
- *Researchers.* Social scientists usually use statistical data and information provided by the government. Machine—readable data largely facilitate the research.

International rankings rarely grasp the subnational dynamics of policies, and Russia demonstrates here a new mosaic of outcomes. That is why regional and municipal dynamics can be more useful to analyze the preliminary results of OGD policy in Russia. Before we turn to its analysis, several points regarding Russia's system of governance should be explained.

Although formally Russia is a federal state, its federalism is seriously diminished by the so-called *power vertical,* in fact a system of centralization of governance on the federal authorities [32]. It is an elaborate technique of redistributing budget and responsibilities that, put simply, impose constraints on regional policy choices and budgeting, making them comply with federal policies. As regions are unequal in

[12]http://zakupki.gov.ru/epz/main/public/home.html.

[13]https://clearspending.ru.

[14]http://hack.declarator.org/.

their economic, social and political development, they all need to meet at least some minimum federal standards of performance in various areas. Hence the gap between the most developed units (like Moscow or St. Petersburg) and laggards can be extremely broad. That is actually what happened in case of OGD. First, the regions and municipalities were just invited to join this innovation on a pilot basis as being a part of the *Open Region* and *Open Municipality* initiatives.[15] Also, some regions, like Moscow, had implemented some OGD policies before the national one came officially into force. These first regions and municipalities, regardless their genuine readiness, showed some motivation to innovate (be it sustaining legitimacy, boosting economy or other). Other regions and cities were then forced to comply with federal standards, and for many of them the need to report on the results to the higher levels of government has become the major, if not only, motivation to release OGD [33]. In other words, the power vertical, on the one hand, helps to force innovation by vertical coercion of regions and cities. The other side of the coin is the lack of motivation and growing disparities among entities.

15.2.1 Problems of Normative Regulation

The term *open data* was introduced to the legal field of Russia within the Presidential Decree demanding "to enable access in the Internet network to open data contained in informational systems of government bodies of the Russian Federation—up to July 15, 2013".[16] The idea of necessity to develop OGD segment in the Russian Federation fixed in the Presidential Decrees of May 2013 was reflected in the Federal Law #112-FZ dated June 7, 2013.[17] Thus, the obligation to publish governmental information in open data format was fixed by the Federal Law #8-FZ from February 9, 2009[18] while more detailed requirements were formulated in the Federal Law #149-FZ from July 27, 2006.[19]

In 2013, there was a problem of normative regulation since while necessity of information publication in open data format was declared, two questions of most importance were unanswered:

---

[15]Open Region//Open Government in Russia, http://open.gov.ru/openregion/ [in Russian].

[16]Decree No 601 of President of the Russian Federation On Basic Directions of Public Administration System Improvement. Adopted 7.05.2012, http://www.rg.ru/2012/05/09/gosupravlenie-dok.html [in Russian].

[17]Federal Law of the Russian Federation № 112, http://www.consultant.ru/law/hotdocs/26119.html [in Russian].

[18]Federal Law of the Russian Federation № 8 On Access to Public Information of State and Municipal Authorities, http://www.consultant.ru/document/cons_doc_LAW_84602/ [in Russian].

[19]Federal Law of the Russian Federation № 149 On Information, Information Technologies and Information Protection, http://www.consultant.ru/document/cons_doc_LAW_61798/ [in Russian].

- What information exactly should be published in open data format. One understands that it was impossible to transform the whole volume of governmental data to open data format in one moment (some of the data not even digitalized yet). The pool of topics was fixed by the Governmental Order #1187-r from July 10, 2013.[20]
- How to publish open data. Up to 2013, the community of developers and software programmers already formulated their requirements to OGD (formats, attributes, etc.) however there was no single correct way governmental officials could follow. The requirements to quality of open datasets should be formalized; that was made by the Methodical Guidelines[21] (actual version is now 3.0).

Such an extended large legal introduction is necessary in order to explain fundamental basic problems for regional OGD in the Russian Federation: legal regulation and technological divergence of regions.

While at federal level, the pool of OGD topics has been significantly extended by means of the Open Data Publication Roadmap, only datasets fixed by Order #1187-p remain obligatory for publication at regional level. Those 16 topics have no real practical significance or invest potential for software product and application development.

Thus, the first question remains unanswered at regional level since the list of topics is not much sound from its inception and has not been extended. For selection of OGD topics, regions are now left for their own resources and not all of them can soundly solve this problem by themselves.

Requirements defined in the Methodical Guidelines are comprehensive and take into account opinions of the professional community (the documents was developed by means of *Google Docs* and counted a large number of expert comments) but thanks to divergence in technological development level of the Russian Federation subjects, not all of them are able to follow those complex requirements due to either technological reasons (not all regions have resources enabling sound work with open data) or to lack of qualified personnel able to work with machine-readable data. When interviewing officials in charge of informatization in top executive government bodies of the Russian Federation subjects, the author sometimes could not find an official in charge of OGD development (Sevastopol); some other regional governments have no idea what OGD is (Dagestan, Ingushetia, Chechen Republics).

Thanks to such technological divergence, 18 regions of the Russian Federation have not published any open dataset up to now[22] while only six regions fulfill the actual OGD legislation fully.

[20]List of Publicly Available Information, Possessed by Federal Government and to be Released as Open Data, http://www.rg.ru/2013/07/29/info-site-dok.html [in Russian].

[21]Approved by Minutes of the Meeting of the Governmental Commission for Coordination of Open Government, #4, May 29, 2014.

[22]Results of OGD study in Russian regions with the author's direct involvement//*Infometer* system: http://system.infometer.org/ru/monitoring/371/rating/.

For municipalities, the situation is still more complicated: at municipal level, there is similar deficiency of normative regulation for dataset topics and still more technological divergence. Cities populated over a million have much more resources then smaller cities. This is why the study has fixed such difference in development of regions and municipal units.

15.3 Open Government Data Performance in Russian Regions and Cities: An Overview

15.3.1 Methodology of the Infometer Approach

To perform this study, we implemented a set of quantitative and qualitative methods and used software packages for processing large volumes of data. We performed qualitative content analysis of normative legal acts regulating the process of open governmental data publication. The methodology for monitoring government bodies' official websites of (hereinafter referred to as the Monitoring) involves studying and assessing the information content and technical features of the government bodies' official websites, calculating the coefficient that characterizes the information accessibility of each official websites, forming information accessibility rating of the official websites, and making recommendations to the government bodies with regard to maintaining their official websites.

To range formed analytical units (parameters) of quantitative content analysis (criterion: Availability) and of qualitative evaluation (Actuality, Comprehensiveness, Navigation, 3 format accessibility criterion) of datasets published, as well as to form a unified rating of open data development level in the cities, we used the Infometer project center's methodology.[23]

Final evaluation scores and rating were calculated by means of the Infometer automated informational system in accordance to the formula designed by the Infometer research team (1), where:

- IACoef—informational accessibility rate (in this case, for open datasets);
- i—parameter ordinal number (from 1 to n);
- j—criterion ordinal number (from 1 to 7);
- K_{ij}—Coefficient for the jth criterion, the ith parameter
- $CRCoef_{ij}$—Criterion Relevance Coefficient for the jth criterion, the ith parameter;
- $PRCoef_i$—Parameter Relevance Coefficient for the ith parameter;
- $SSCoef_i$—Social Significance Coefficient for the ith parameter.

[23]Methodology//Informeter, http://en.infometer.org/monitoring/Methodology-for-the-Monitoring-of-Government-Bodies%E2%80%99-Official-Websites/1.-opisanie-metodologii-monitoringa.

$$IACoef = \frac{\sum_{i=1}^{n}\left(\prod_{j=1}^{7}\left(K_{ij} \times CRCoef_{ij} + 1 - CRCoef_{ij}\right) \times PRCoef_{ij} \times SSCoef_{ij}\right)}{\sum_{i=1}^{n}\left(PRCoef_i \times SSCoef_i\right)} \times 100\% \tag{15.1}$$

Within the audit, evaluation parameters were based on normative acts regulating OGD publication procedures. Therefore, each evaluation parameters within the study reflects a normative requirement and the date calculated by means of the formula above reflects level of the published datasets compliance with actual legislative requirements.

To evaluate OGD publication, 72 Russian regions and 72 regional capitals/administrative centers populated over 100,000 were selected. Municipal administrations' OGD were evaluated by 57 parameters based on actual normative requirements. The parameters can be grouped into three key categories:

1. Usability of the OGD portal or section (existence, link, description of OGD usage conditions, statistical information, search tools, absence of UGD usage restrictions, etc.);
2. Dataset and metainformation handling (presence and correct filling of the register, passports, and dataset structures, versioning, contact data of responsible officials, possibility for visualization, etc.);
3. Organizational work with OGD (demand evaluation tools and results, list of application and services, data query and feedback modules, information on normative regulation in the OGD field, information on public events: conferences, hackathons, etc.)

Federal and regional normative acts do not establish mandatory publication of specific topical OGD at municipal level so that topical parameters were not included in the audit of large cities' OGD.

For evaluation of OGD publication by 72 Russian regions and their executive government bodies, 86 parameters were used, derived from actual normative requirements on OGD publication. The parameters can be grouped into five key categories:

1. Usability of the OGD portal or Sect.
2. Dataset and metainformation handling
3. Organizational work with OGD
4. Obligatory datasets following the list of publicly available information on activities of government bodies of the Russian Federation subjects (publication quality for 16 mandatory datasets defined by the Order #1187-p requirements)
5. Other datasets for various fields (e.g., public health, transport, etc.)

The first three evaluation categories are the same for regions and large cities. The fourth category and the fifth one evaluate filling of regional portals with obligatory and other datasets.

15.3.2 Evaluation Results

Final results of the audit show that cities fulfill OGD publication requirements by 22% at average[24] while regional authorities do that more than twice better, by 47.8%.[25] OGD development divergence is expressed at municipal level much more than at regional one. 36 city administrations (50%) publish no OGD and do not launch website sections for their publication while only two regional governments have got zero evaluation scores. Analysis of compliance (Table 15.1) with OGD publication requirements by separate categories of the parameters reveals more than double advantage of regional OGD portals and sections. At the same time, results for first three categories are similar for municipal and regional level: rate of technological requirement observance (usability and dataset handling) is much better than that for organizational work with OGD.

There is also divergence in numbers of datasets published by city and regional administrations: only two leading cities, Kazan and Krasnodar, publish more that 100 datasets (and five more cities, more than 20) while leading regions, Perm Territory and Tula Region, have published more than 800 datasets (and seven more regions, more than 200).

All datasets obligatory for publication are placed by 11 regions studied (Fig. 15.1). However, it is difficult to use the datasets published for developing interregional software application since there are no unified requirements for structure of those datasets. For instance, datasets named "Names of Civil Registry Offices" at OGD portals of the Bryansk Region[26] and of the Vologda Region[27] differ not only by number of structure elements but also by their filling (addresses, contact data, work schedules). We should point that the Tula city administration, the leader among municipal administrations, places its OGD at the portal of the Tula Region (Table 15.2).

Both regional and municipal authorities prefer to publish OGD at their official websites' sections. On the one hand, this requires no special resources, on the other hand, quality of such a technological site is poorer than that of OGD portals. Regional governments often combine various ways of OGD publication: they may use both regional and federal portals (5 cases), federal portal and regional website

---

[24]http://system.infometer.org/ru/monitoring/344/rating/ (for 72 regional administrative centers).

[25]http://system.infometer.org/ru/monitoring/371/rating/ (for 72 regions with administrative centers populated over 100,000).

[26]http://www.bryanskobl.ru/opendata/3234048116-civilreg/table.

[27]http://opendata.gov35.ru/datasets/3525082633-RegistryOfficesList.

Table 15.1 Number of government bodies for level of compliance with OGD publication requirements

Compliance level range (%)	0–10	10–20	20–30	30–40	40–50	50–60	60–70	70–80	80–90	90–10
Cities	40	4	3	3	4	4	10	2	1	1
Regions	5	0	7	16	13	10	8	6	1	6

72 regions and their administrative centers (capitals). *Source* Infometer

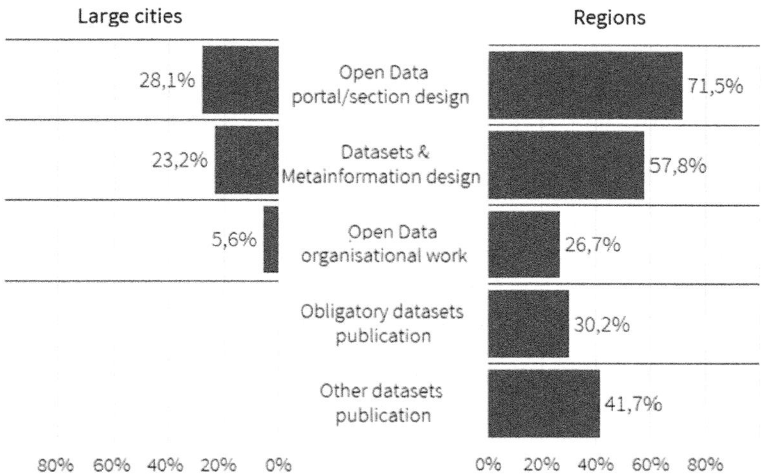

Fig. 15.1 Results of OGD monitoring in large cities and regions. *Source*: Infometer

Table 15.2 Ways of OGD publication by regions and city administrations (according to all sources found)

Publication way	Federal OGD portal	Regional OGD portal	Municipal OGD portal	A section at the body's official website
Regional administrative centers	2	2	5	28
Regions	28	21	–	50

Sects. (20 cases), or all possible ways (2 cases). City administrations usually neither implement such combinations nor synchronize datasets; only Belgorod city administration published OGD both at the federal portal and in section of its own website. For municipal units without sufficient funding to maintain OGD portals of their own, it looks logical to use possibilities of federal or regional portals but in fact they do not.

Poor participation of city authorities in OGD publication may also be caused by absence of clear normative regulation and of any federal, regional, or public control as well as by lack of interest to and trust in the OGD published from the business. While regional governmental officials attend largest all-Russian OGD events such as "Open Data—2015" summit, hold hackathons, and share their own best OGD practices, city administration officials lack attention of government entities in charge of project initiatives in the OGD field (such as the Open Government of the Analytical Center under the Government of the Russian Federation).

15.4 Explaining OGD Performance in Russian Regions: Quantitative Perspective

The first perspective we take to approach the development of OGD in Russia is quantitative. We attempt to reveal some structural factors that may influence the level of OGD performance, and here we continue a plethora of efforts conducted in many comparative works. One of the pioneering studies is one by Norris, who analyzed the level of e-government and e-democracy across nations, testing the importance of political, human and technological development. Her findings suggested that all three factors matter, although differently regarding various element of *electronic political systems* [34]. More recent studies expanded the number of hypotheses, e.g. including the quality of institutions, globalization, wealth, international contacts etc. [19, 35, 36]. They proved to be applicable and plausible even to the subnational level [37]. In case of Russian e-government, for instance, political regime, human capital, Internet access and the quality of bureaucracy were revealed as determiners of success [38].

Taking Norris' framework as a basis, we suggest that such structural factors might work in case of OGD. As a costly, technologically elaborate and politically sensitive issue, OGD release may be the function of more democratic governance, better technologies and the abundance of resources. However, we add some other factors. First, it is the level of bureaucratic efficiency, as public officials are those implementing OGD policy. Secondly, we hypothesize the link between OGD performance and the level of e-government development. We suppose that when a region demonstrates high commitment to e-services advancement, it provides institutional learning, as bureaucracies and politicians get accustomed to new technologies and acquire necessary competences. Moreover, better e-government creates demand for e-services among citizens, and the latter may be more interested in innovations, providing incentives for regional governments to provide better OGD policies.

There are several important limitations to the study we conducted. First, there is a problem of data access. That is why we use the average values for the periods available. Secondly, our analysis regards the regional level only, meaning we cannot grasp important peculiarities of region-city relations and distinctions addresses in the following section.

Our dependent variable (OPEN_DATA) presents a result of a region in 2016 Infometer's monitoring on open data in regions. We use the following independent variables:

- DEM—the level of democracy in the Russian regions, 2012–2014, surveys conducted by Moscow Carnegie Center and Higher School of Economics [39];
- HUM_CAP—the level of human capital in a region, calculated according to the UN's methodology, 2007–2011, multiple sources;
- TECH—the Internet access rate in a region, 2012–2014, data by Russian Statistical Service (Rosstat);[28]
- EGOV—the quality of e-government performance, Index of Russian Regions E-Readiness, 2007–2014;[29]
- BUR—the effectiveness of bureaucracy, measured as a quantity of civil servants per 1000 economically active citizens, 2007–2014, collected by Rosstat. We assume the more the share of bureaucracy, the less effective the apparatus is.

As control variables we use the number of citizens per region (POP) and the share of urban population (URBAN). Both are gathered from Rosstat and cover 2013–2014. The number of cases is 83 regions, as little data is on Sevastopol and the Crimea so far.

First, we ran the correlation analysis to look at whether the variables relate to each other at all. The results are presented in Table 15.3. We see that practically all variables significantly correlate with the level of OGD performance. The two exceptions are the political regime and the population. More important here is the

Table 15.3 Results of variables' correlation analysis with OPEN_DATA

Variable	Coefficient
DEM	183
HUM_CAP	339**
TECH	323**
EGOV	432**
BUR	−411**
POP	−098
URBAN	351**

Source Authors' calculations
**0.01 level (2-tailed)

[28]http://www.gks.ru [in Russian].

[29]Index of Russian Regions E-Readiness, http://eregion.ru/en/information-inequality-regions [in Russian].

Table 15.4 Results of regression analysis, dependent variable—OPEN_DATA models

Model no.	Variables included	Adjusted R-square	St. Error of estimation
1	BUR, POP, TECH, DEM, HUM_CAP, URBAN, EGOV	197	23.88
2	BUR, POP, TECH DEM, URBAN, EGOV	209	23.7
3	BUR, TECH, DEM, URBAN, EGOV	218	23.57
4	BUR, TECH, URBAN, EGOV	225	23.46
5	BUR, URBAN, EGOV	232	23.35
6	BUR, EGOV	230	23.39

Source Authors' calculations

Table 15.5 Coefficients of model 6

Variables	Standardized beta-coefficients	Significance
BUR	300	012
EGOV	−284	017

Source Authors' calculations

former, as it seems that OGD can develop in different institutional settings. Although it is hard to say that some regions are democracies in a generally authoritarian state, a relatively more democratic system is not an advantage. On the other hand, better OGD performance is associated with higher human capital, broader Internet access, e-government development, bureaucratic efficiency and urbanization. However, the coefficients are not so high to say that these factors heavily influence OGD development.

This is proved by the linear regression analysis, the results of which are shown at Tables 15.4 and 15.5. The best one is Model 6 including the level of e-government development and bureaucracy efficiency. All other factors proved insignificant when taken in interrelationship. The R-square remains low, and we actually can explain about 20% of the variance by the model (about 17 regions).

However, what the analysis shows is that in case of regional OGD the *traditional* variables, as democracy, human capital and technological development do not work well. Although the latter two accompany OGD advancement, forming a kind of a portrait of a successful region, they are not the drivers of innovation. What can matter are the factors that can be called policy-related ones. The development of e-government in a region may create a basis for further steps, and effective bureaucracy provides its success.

But again, explaining perspective seems to fall short to unveil the reasons for OGD disparities among regions, not mentioning the problem of city—region gaps. Structural explanations need to be supported by agency-based evidence from the case studies that are presenting in the next section.

Table 15.6 Matrix of region—municipality OGD performance

	Region	
Municipality	High/high	Low/low
	High/low	Low/high

15.5 Politics of Open Government Data: Four Cases

15.5.1 Approach

To describe specific cases within our study, we chose method of case studies allowing to reveal dominating factors for OGD development at regional and municipal governance levels. We formed four groups of development relations (Table 15.6).

Such a relation often appears to result from difference between policies implemented by regional and city administrations. General distribution of "region/municipality" relations is presented at the bar chart (Fig. 15.2). To describe each group in more details, we have analyzed four cases of different relations between the region and the municipality of its administrative center/capital (Table 15.7).

15.5.2 Tula city/Tula Region: An Example of Authoritarian Modernization

The first case is an illustrative example of "soft" authoritarian modernization when a regional government, having more administrative resource and being interested in the region branding, "pulls" the regional center up to its own high development level.

In Tula Region, OGD were initially developed at regional level. This region was the first to achieve 100% rate level in the Infometer rating.[30] Within a semi-formalized interview, Andrey Kolesnikov, Referent of the Mass Communication Department in the Tula Region Ministry for Informatization, Communications, and Open Governance Issues, stated that OGD appeared initially at regional level and then technologies, knowledge, and skills were transferred also to executive government bodies and municipalities of the region: "We started to develop an OGD portal in 2013; for us, it was a part of the open governance idea. It was an internal initiative of ours. Initially, we planned to launch our OGD portal in 2014 but realized in 2013 that it was possible to start it earlier. We were explaining to regional executive bodies and municipal officials why it was necessary to develop this direction so that the process began to establish. Today, in each body there are officials in charge of open datasets that body publishes. We have enabled

[30]Openness rating (spring 2015) http://system.infometer.org/ru/monitoring/259/rating/.

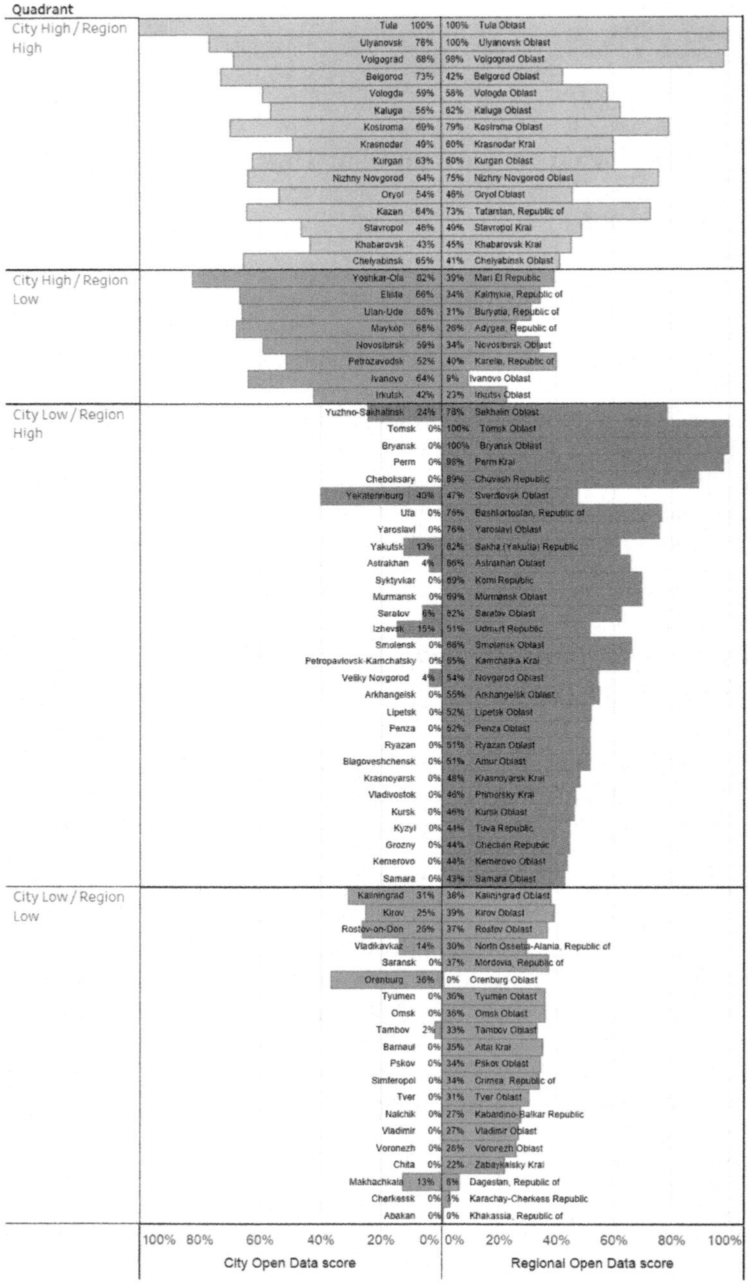

Fig. 15.2 General distributions of OGD performance in regions and municipalities. *Source* Infometer

Table 15.7 Cases selection

Relation	City/Region
City high/Region high	Tula/Tula Region 100/100
City high/Region low	Ivanovo/Ivanovo Region 64/9
City low/Region high	Bryansk/Bryansk Region 0/100
City low/Region low	Makhachkala/Republic of Dagestan 13/6

development of departmental normative acts assigning officials in charge and defining data actualization deadlines. Now, everything is much easier".

Top officials of the region support openness ideas actively so there are no conflicts within involvement of municipalities in OGD development; also, necessary funding is assigned. It is important to point that Tula Region implements sound product at relatively low expenses thanks to efficient work:

"If I remember correctly, the first version of the portal launched in 2013 cost us about 180,000 rubles," Kolesnikov commented. "We wrote a technical enquiry, announced an invitation to tender, and held all procurement procedures to get this amount as a result. The Center for Information Technologies, our subordinate organization, has its own professional developers. We had also a designer of ours own who helped us to create a layout for the new versions of the portal. We discussed some technological requirements and methodical guidelines among ourselves, wrote a small additional technical enquiry for upgrade of our portal and upgraded it. Costs are not large. Maintenance, upgrade, and introduction of new functionalities cost at average about 200,000–250,000 rubles per year".

Such an approach allows to introduce modern formats and to develop OGD-base software as well as to hold topical events.

15.5.3 Ivanovo city/Ivanovo Region

This case is possibly the most curious one. In all other cases, the region plays key role, impacting OGD level in its administrative center by means of influence, strong or weak, upon the city administration. In this case, the situation is opposite: at the city level, OGD are for whatever reason developed much better that at the regional level.

The Ivanovo Region portal has a section for open data[31] but only graphical PDF files are published there as "open data". If the publishers use CSV, a correct format, then a dataset contains a single string, causing questions whether such transformation of data into machine-readable format is feasible.

[31]http://www.ivzan.ru/opendata.

The city portal publishes not very much OGD (20 datasets) but they fully meet requirements of the methodical guidelines and can be used in services and applications. Larisa Kosterina, Head of the Department for Informational Resources in the Ivanovo City District, says that OGD started developing in the city of Ivanovo immediately upon issuance of the Presidential Decrees of May 2013 and subsequent normative amendments. However, besides the normative regulations, there was one more motivation factor: Ms. Kosterina's personal interest to OGD and her belief that OGD can being profit to the city. The city administration began to participate in all pilot informatization projects and to initiate development of new directions. Kosterina pointed, as one more positive factor, the fact that the city top officials are young (the new head of the city is 32) and that they support innovative solutions. Students of local universities also are involved in processes of data transformation into machine-readable formats and in development of OGD-based software products, therefore increasing practical significance of OGD. The city plans to launch on January 1, 2017, a new portal that will enable better data processing but there are not much resources for this: only two employees of the Department deal with open data.

The model for work with OGD in the Ivanovo city administration is similar in that of the Bryansk Region: structure units provide information to the Department for Informatization where it is transformed into machine-readable format.

In her interview, Larisa Kosterina several times called the region "depressive" and at the same time expressed belief that the OGD could provide possibilities for development. She did not answer why OGD are not developed at the regional level. In our opinion, the key reason if lack of motivation for officials of the Government of Ivanovo Region.

15.5.4 Bryansk City/Bryansk Region: Change of Management

The OGD section at the Bryansk Region portal[32] is an example how individual officials can promote openness policy within a government body when resources are limited. In the early 2015, when we studied regional OGD for the first time, Bryansk Region held the last position in the rating with the zero rate since they had no OGD web page, no datasets, and so on. Within our first remote online interaction with officials of regional government on OGD, the region improved its rate from 0 to 41% and held the 29th position (of 85) in the final rating.[33] A year later, Bryansk Region achieved the rate of 78% in the new similar audit and took the 8th position. For the latest audit, it leads the rating with 100%.[34]

---

[32]Open data portal http://www.bryanskobl.ru/opendata.

[33]http://system.infometer.org/ru/monitoring/259/rating/.

[34]http://system.infometer.org/ru/monitoring/371/rating/.

However, more detailed analysis of the Bryansk Region OGD web section shows that from the technological point of view, it lies behind OGD resources of other regions leading the rating. Datasets have rather simple structures and all data are presented in CSV and not in more complex form. Features of the datasets published makes one to conclude that they have been converted manually from Excel spreadsheets. The web section has neither any API (application programming interface) nor any innovative findings in the OGD field.

Alexey Podyapolsky, Head of the Section for Internet Projects and for Activities in Social Networks in the Department for Informational Supply of Activities of the Bryansk Region Governor and Government, is in charge of OGD in the region. In his interview, he pointed that only one more staff member helps him within OGD development and publication processes. Such is a small team is surely not enough for any serious development. He also told that they obtain raw data (in the best case, arranged in a table) from executive bodies and convert them into machine-readable format by themselves, also enriching them by addition new structural elements, e.g. geodata. According to him, officials from the Bryansk city administration had not asked them for any assistance and they had not resources to offer it by themselves.

Due to lack of resources, Bryansk Region is not able to follow the way of Tula Region and to "pull" its central city up to its OGD development level. They could develop only the website of their own up to certain quality level, having no resources to introduce serious software solutions and modern data formats.

Podyapolsky also pointed that in the work plan of the regional government up to 2017, there were no activities on OGD consulting for the Bryansk city administration.

15.5.5 Makhachkala City/Republic of Dagestan: What is Dead May Never Die?

Regions and municipalities fully or almost fully ignoring legislative requirements on OGD form a separate group in our rating. It is interesting that at the last time, public prosecution bodies actively initiate proceedings on administrative offense (Administrative Offense Code of Russia, Article 13.27 p. 2) for lack of online placement of information on activities of local self-government bodies if obligation to place such information is stated by federal law.[35] However, some regions, especially national republics, continue to ignore law requirements due to a number of reasons.

Low technological level. Paying attention to general level of informatization in national republic, one can notice that Internet implementation level is there much lower than in central regions of Russia; so is level of information knowledge of the

[35]Example: https://rospravosudie.com/court-sudebnyj-uchastok-po-buzdyakskomu-rajonu-respubliki-bashkortostan-s/act-200750130/.

public while republican informational resources of all power branches (legislative, executive, and judicial) usually hold last positions in openness ratings.

Lack of demand from civil society. For instance, no OGD event (e.g. hackathon) was held in the republic of the North Caucasus for 2016.[36]

Lack of republican authorities' interest in region branding on openness theme. For instance, the Republic of Tatarstan decided to make information and communications as a landmark for the region brand development; no republic of the North Caucasus make accents of that kind.

Lack of resources for building OGD portals and applications. Launch of an OGD portal and its actuality maintenance needs financial and human resources available not for all regions. Within growing economical crisis, some regional budgets have deficit not allowing to assign any additional funding for informatization. If there is a necessity to reduce budget, it is informatization expenses budget lines to be cut.

Therefore, a "perfect storm" appears, not allowing to develop OGD in the region in question. Regional authorities are unable to pull their subordinate municipalities to high OGD development level as they have done in Tula Region. Municipal authorities, meeting serious resource limitation and motivation vacuum, are not able to act as they do in Ivanovo. In such a case, only some changes at federal level or change of political climate in a region itself could make that region to work on OGD more actively.

15.6 Conclusions

This chapter is our effort to combine different levels (federal, regional and municipal) and various techniques of research to analyze factors affecting OGD success. Generally, our findings can be summarized as follows.

Although OGD normatively associate with democracy and can potentially lead to smart city development, reality shapes these causal links and may well interfere with them. Since OGD release is a complicated process involving political and managerial issues, plenty of factors need to be taken into account, and, what our study has also shown, there is always a variety of outcomes even in one and the same entity, be it country, region or city. Even if Russian OGD policy might seem a failure from the international perspective, the country is extremely disproportionate in policy performance, and even municipal and regional levels of the same region usually show different outcomes.

Our study has taken different perspectives to analyze OGD policy, its effectiveness and impact. From the quantitative, *explaining*, perspective, it seems that no single determinant can explain the patterns of OGD performance in Russian regions. The level of e-government development and bureaucracy effectiveness are the most significant factors in the regression models, and both generally relate to the

[36]Hackathon map of Russia: http://read.infometer.org/foiv2016_digest3.

ability of public agencies to adapt to new technologies of government. However, as correlation analysis shows, other factors, like human capital, urbanization and the Internet, are positively associated with OGD success.

But, generally, *explaining* perspective has many problems when applied to Russian regions. The qualitative, *understanding*, technique, seems to be the way to explain more about region—and city-specific issues. The interviews we have taken revealed that usually the success is not a result of good governance or democracy, but is agency—driven. The motivation of the low-level bureaucracy who are directly involved in OGD release, might be the only reason the policy is successful regardless the unfriendly environment he or she operates in. Another issue is the law implementation and enforcement, which is usually a problem for traditionalist regions. Thirdly, the deficit of financial resource can also be an obstacle. And, finally, it is an almost total lack of public involvement in OGD release. The demand on OGD is explicit only in the large cities—the absence of it may turn OGD process into a purely routine of meeting federal requirements.

Still a lot to be done in this area to get a solid understanding of political and bureaucratic processes, and in this regard our mixed quantitative—qualitative methodology seem a suitable and promising lens for the analysis.

Acknowledgements The paper was prepared within the framework of the Academic Fund Program at the National Research University Higher School of Economics (HSE) in 2016 (grant №16-05-0059) and supported within the framework of a subsidy granted to the HSE by the Government of the Russian Federation for the implementation of the Global Competitiveness Program.

References

1. Effing R, van Hillegersberg J, Huibers T (2011) Social media and political participation: are Facebook, Twitter and YouTube democratizing our political systems? In: Macintosh A, Tambouris E, de Bruijn H (eds) Electronic participation: third IFIP WG 8.5 international conference, EPart 2011, Delft, The Netherlands, August 29–September 1, 2011. Springer, Berlin, pp 25–35. doi:10.1007/978-3-642-23333-3_3
2. Gil de Zúñiga H, Jung N, Valenzuela S (2012) Social media use for news and individuals' social capital, civic engagement and political participation. J Comput Mediated Commun 3 (17):319–336. doi:10.1111/j.1083-6101.2012.01574.x
3. Coglianese C (2009) The transparency president? The Obama administration and open government. Governance 4(22):529–544. doi:10.1111/j.1468-0491.2009.01451.x
4. Meijer AJ, Curtin D, Hillebrandt M (2012) Open government: connecting vision and voice. Int Rev Adm Sci 1(78):10–29. doi:10.1177/0020852311429533
5. Janssen M, Charalabidis Y, Zuiderwijk A (2012) Benefits, adoption barriers and myths of open data and open government. Inf Syst Manage 4(29):258–268. doi:10.1080/10580530.2012.716740
6. Ubaldi B (2013) Open Government Data. OECD, 2013
7. Balch O (2013) Can open data power a smart city revolution? The guardian. https://www.theguardian.com/sustainable-business/open-data-power-smart-city
8. Hinssen P (Ed.) (2013) Open data power smart cities. Across Technology, Kankarbagh

9. Tauberer J (2009) Open data is civic capital: best practices for "open government data". https://razor.occams.info/pubdocs/opendataciviccapital.html
10. Bartenberger M, Grubmüller-Régent V (2014) The enabling effects of open government data on collaborative governance in smart city contexts. eJournal eDemocracy Open Gov 1(6):36–48
11. Davies T (2010) Open data, democracy and public sector reform. A look at open government data use from data.gov.uk. http://www.opendataimpacts.net/report/wp-content/uploads/2010/08/How-is-open-government-data-being-used-in-practice.pdf
12. Noveck B (2016) The rise of citizen expert. Policy Netw. http://www.policy-network.net/pno_detail.aspx?ID=5056&title=The+rise+of+the+citizen+expert
13. Gurstein MB (2011) Open data: empowering the empowered or effective data use for everyone? First Monday. 2(16). http://firstmonday.org/article/view/3316/2764
14. Davies T (2014) Open data policies and practice: an international comparison. Available at SSRN, https://ssrn.com/abstract=2492520 or doi:10.2139/ssrn.2492520
15. Susha I et al (2015) Critical factors for open data publication and use: a comparison of city-level, regional, and transnational cases. JeDEM-eJournal eDemocracy Open Gov 2(7):94–115
16. Meijer A, Bekkers V (2015) A metatheory of e-government: creating some order in a fragmented research field. Gov Inf Q 3(32):237–245. doi:10.1016/j.giq.2015.05.004
17. Conradie P, Choenni S (2014) On the barriers for local government releasing open data. Gov Inf Q 31:S10–S17. doi:10.1016/j.giq.2014.01.003
18. Jho W, Song KJ (2015) Institutional and technological determinants of civil e-Participation: solo or duet? Gov Inf Q 4(32):488–495. doi:10.1016/j.giq.2015.09.003
19. Lee C, Chang K, Berry FS (2011) Testing the development and diffusion of e-government and e-democracy: a global perspective. Public Adm Rev 3(71):444–544. doi:10.1111/j.1540-6210.2011.02228.x
20. Stier S (2015) Political determinants of e-government performance revisited: comparing democracies and autocracies. Gov Inf Q 3(32):270–278. doi:10.1016/j.giq.2015.05.004
21. Davies T (2013) The messy reality of open data and politics. The Guardian. https://www.theguardian.com/public-leaders-network/2013/apr/08/messy-reality-open-data-politics
22. Serewicz L (2011) The dark side of open data: politicized data. https://lawrenceserewicz.wordpress.com/2011/09/21/the-dark-side-of-open-data-politicized-data/
23. Peled A (2011) When transparency and collaboration collide: the USA open data program. J Am Soc Inf Sci Technol 11(62):2085–2094. doi:10.1002/asi.21622
24. Fountain JE (2004) Building the virtual state: information technology and institutional change. Brookings Institution Press, Washington, DC (2004)
25. Gel'man V (2014) The rise and decline of electoral authoritarianism in Russia. Demokratizatsiya 4(22):503–522
26. Katchanovski I, La Porte T (2005) Cyberdemocracy or Potemkin e-villages? Electronic governments in OECD and post-communist countries. Int J Public Adm 7–8(28):665–681. doi:10.1081/PAD-200064228
27. Linde J, Karlsson M (2013) The dictator's new clothes: the relationship between e-participation and quality of government in non-democratic regimes. Int J Public Adm 4(36):269–281. doi:10.1080/01900692.2012.757619
28. Karlsson M (2013) Carrots and sticks: internet governance in non–democratic regimes. Int J Electron Gov 3(6):179–186. doi:10.1504/IJEG.2013.058405
29. Brancati D (2014) Democratic authoritarianism: origins and effects. Annu Rev Polit Sci 17:313–326. doi:10.1146/annurev-polisci-052013-115248
30. He B, Warren ME (2011) Authoritarian deliberation: the deliberative turn in Chinese political development. Perspect polit 9(02):269–289. doi:10.1017/S1537592711000892
31. Chugunov AV, Kabanov Y, Zenchenkova K (2016) Russian e-Petitions portal: exploring regional variance in use. In international conference on electronic participation. Springer International Publishing, pp 109–122. doi:10.1007/978-3-319-45074-2_9

32. Gel'man V, Ryzhenkov S (2011) Local regimes, sub-national governance and the "power vertical" in contemporary Russia. Eur-Asia Stud 3(63):449–465. doi:10.1080/09668136.2011.557538
33. Kabanov Y, Karyagin M (2015) Deserted Islands of openness: problems of open data in Russia. Politex 4:38–51 [in Russian]
34. Norris P (2001) Digital divide: civic engagement, information poverty, and the Internet world-wide. Cambridge University Press, Cambridge
35. Gulati G, Williams C, Yates D (2014) Predictors of on-line services and e-participation: a cross-national comparison. Gov Inf Q 31(4):526–533. doi:10.1016/j.giq.2014.07.005
36. Zhang H, Xu X, Xiao J (2014) Diffusion of e-government: a literature review and directions for future directions. Gov Inf Q 31(4):631–636. doi:10.1016/j.giq.2013.10.013
37. Yun HJ, Opheim C (2010) Building on success: the diffusion of e-Government in the American States. Electron J E-Government 8(1):71–82
38. Kabanov Y, Sungurov A (2016) E-Government development factors: evidence from the russian regions. In: Chugunov AV, Bolgov R, Kabanov Y, Wimmer M, Kampis G (eds) Digital transformation and global society: first international conference, DTGS 2016, St. Petersburg, Russia, June 22–24, 2016, Revised Selected Papers, 85–95, Springer International Publishing (2016). doi:10.1007/978-3-319-49700-6_10
39. Petrov N, Titkov A (2014) Rating of democracy by Moscow Carnegie Center: 10 years in service. Moscow Carnegie Center, Moscow [in Russian]

Chapter 16
Urbane: Community Driven Architecture and Planning Through a Mobile Social Platform

Swapnil Shrivastav

Abstract A community, understands its needs the best; much better than a social scientist, architect or planner, politicians and various institutions such as municipal bodies and all stakeholders need to collaborate with the community to get the best results and fruitful outcome and impact. Most projects for the community are undertaken without input from the community dwellers. There are forums and municipal websites which formulate proposals for comments and but these proposals were not developed out of the community voice, needs and aspirations. With this background, this chapter proposes a mobile online social platform to connect different stakeholders to generate high levels of engagement in decision making, planning, idea generation and lead to resilient, smart and sustainable communities. People act as data donors by leveraging smartphone technology and the platform empowers the users to bring positive change.

Keywords Online social platform · Data-donor · Mobile · Community driven design

16.1 Introduction

Planning, urban design and architecture need to be disrupted in how they approach any design exercise. There is a need for a 'bottom-up' approach in how systems work and there is also a need to foster resilient, sustainable and strong local communities. Participatory design and design tools for the same [1] were explored in Seoul for many public projects. Public information, design workshop and feedback are the three major steps we need to deal with. With the advent of technology, we can create digital tools which can offer meaningful conversation, creative insight and deeper collaboration. Community participation needs to develop a totally new outlook.

S. Shrivastav (✉)
Uravu Labs, Hyderabad, India
e-mail: swapnil93ides@gmail.com

T.M. Vinod Kumar (ed.), *E-Democracy for Smart Cities*, Advances in 21st Century Human Settlements, DOI 10.1007/978-981-10-4035-1_16

The Indian context—In India internet penetration is about 300 million people out of which 220 million rely on smartphones and tablets and not desktops/laptops. Mobile devices are handier, better connected then ever and also very powerful.

With our Hon'ble, Prime Minister Mr. Narendra Modi proposing the creation of 100 smart cities, tools which take into account the ideas of the community are important, and a model which can be scaled to more cities and even villages remains a prime task.

Furthermore, getting people actually engage in discussion and collaboration is not really achieved through normal forums and websites. We need to exploit technology to provide the best of context and content for enhanced civic engagement and develop a sense of ownership.

A mobile social platform provides an incentive for people to remain engaged in discussion and collaboration. People get to see the ideas community members and can connect to it more easily. Also, when they have an idea or aspiration it can get backing from the community and local authorities can take actions to implement it.

If a new project is coming up, the community can give the development of such a project. People act as data donors.

The possibilities remain immense for architects and planners and local bodies when such data is validated, analyzed properly and used effectively. Such a platform empowers people.

16.2 Methodology

A rather simple methodology was adopted. A lot of learning based on current tools was done and efforts were made to thoroughly understand how it can be made better and different. More efforts were made to test the platform in the real world and get more feedback.

Case studies

A review of the project 'I wish this was' by artist Chang [2].

A case study of the platform 'OpenPlans' for community influenced maps [3].

A case study of the platform 'Neighborland' for collaborative planning [4].

Paper—Owning the city: New media and citizen engagement in Urban Design by de Lange and de Waal [5].

Developing the 'Urbane' App

A smartphone app based on Android mobile platform was developed that helped people take images and engage the community spaces, locations and areas where they aspire to see a change or have a great idea about how to enhance it. The data was stored on a web server and carefully consolidated and made available to the local community.

Data Collection

A simulated model for the process was carried out at NITC campus by drawing comparisons between city planning procedure and campus planning where we have the student body as the local community and hierarchical authorities who carry out the process.

Analysis

The collected geo-tagged data was laid on the location map and classified according to the type of data feed (planning, urban design, management etc.), was consolidated and was further validated so as to give a thorough understanding. The analyzed data was used in enhancing the ideas and wishes, for development efforts and predicting future trends.

16.3 Literature Case Studies

16.3.1 'I Wish This Was'

This was an interactive art project by artist Candy Chang conducted at New Orleans, USA in 2010. The aim was to strike a casual conversation in the city and using public art to garner interest and getting community involved in the design process. She had put up stickers with the line 'i wish this was' and a blank space, on various buildings and objects and citizens passing by filled in their ideas and wishes on the sticker. It provided a platform to engage community members in that neighborhood (Figs. 16.1, 16.2 and 16.3).

The whole project was amazing and simple enough to engage citizens in an effective manner.

What did the project miss?

- Data inputs from the community wasn't available for consolidation and further analysis and deliberation.
- As a public art project, it wasn't able to get into the mainstream process of urban design and planning.
- Scale of the project was limited.
- Feedback and dialogue between community and the experts was lacking.

Lessons learned

- Smart technologies need to be incorporated to improve community-expert communication.
- Public visualization techniques play a crucial role and has better effects than an online static website or discussion forum.
- Data gathered needs to be properly consolidated, filtered and analyzed.
- Social incentive is needed to engage people.

Fig. 16.1 'I wish this was' stickers

Fig. 16.2 Community members giving their inputs

Fig. 16.3 More examples

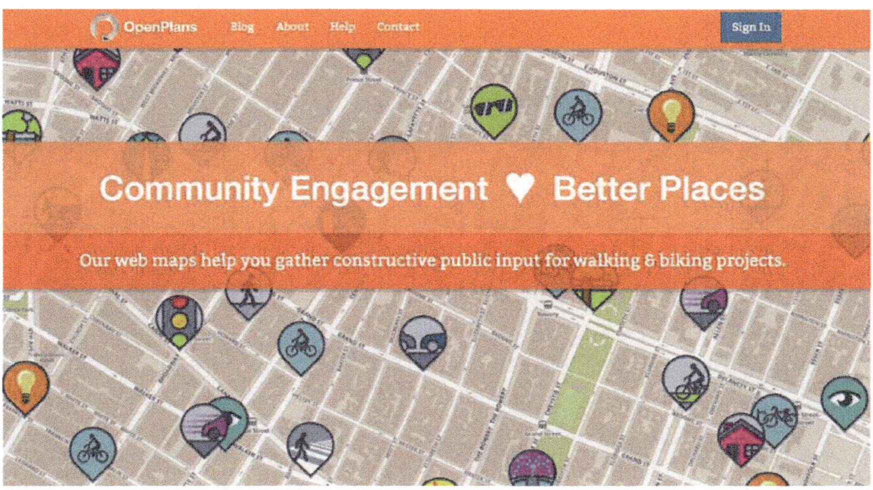

Fig. 16.4 Screenshot of the website

16.3.2 Open Plans

Open Plans is an online platform which lets users give constructive inputs on online maps and also provides access to the API's for enhanced customization (Fig. 16.4).

OpenPlans is all about maps and getting the right geo-tagged data based on various activities for which input is needed. It lets anyone build an online portal for getting the feedback and collecting data. One can customize the looks, the input tools, feel etc. and start the work (Figs. 16.5, 16.6 and 16.7).

Open Plans is pretty versatile and a good example of its use is the CDOT—Bike Parking Project [6].

The CDOT—Bike Parking project gathers inputs from the citizens via location based mapping. There is a feedback and approval process and also an online forum for users to comment. As a whole its very open and transparent.

Open Plans is also being used as an input tool for many other projects like NYC Dot Vision Zero [7] for street safety mapping and Philly Bike Share for bike share planning program.

Fig. 16.5 Open plans in a nutshell

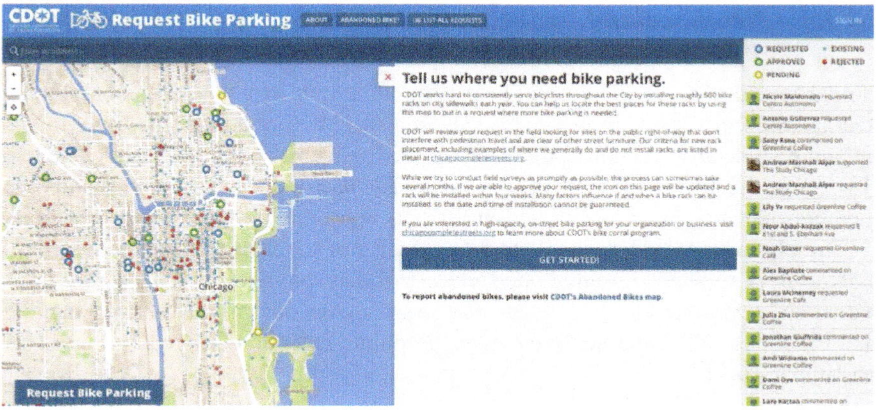

Fig. 16.6 Bike parking request

Fig. 16.7 Interactive map

What did the project miss?

- It's not mobile friendly and hence tagging location is a bit tough.
- Public visualization technique is limited to maps.

Lessons learned

- GPS based mobile mapping is required.
- Better public visualization and communication techniques are required.
- Social incentive and engagement needs to be in the backbone of the platform.

16.3.3 Neighborland

Neighborland is a website and tool for community participation, collaboration and idea sharing on various topics and needs in a neighborhood. It needs users to sign up and give details of the area they belong to. The website is pretty simple and uses 'I want' as the provoking tool for user input. The user thus creates an idea and other members can see, improve upon it and take action on a particular idea (Fig. 16.8).

The platform provides social incentive- sense of togetherness and belonging. It allows users to share resources and ideas and also promote it. It is highly transparent and open platform. Users can also share it over other social networking sites like Facebook or Twitter and have amazing public reception and reach (Figs. 16.9, 16.10 and 16.11).

What did the project miss?

- Its mobile friendly but merely adopted, and not built from scratch for mobile platform.
- No back-end data classification and analysis available for professionals, experts and agencies.
- Misses big on location-based mapping.

Lessons learned

- A complete re-thinking required for a mobile platform.
- Data classification, GIS based mapping and analysis options to be made available for deliberation.
- Better tools required for users to share ideas and donate data.

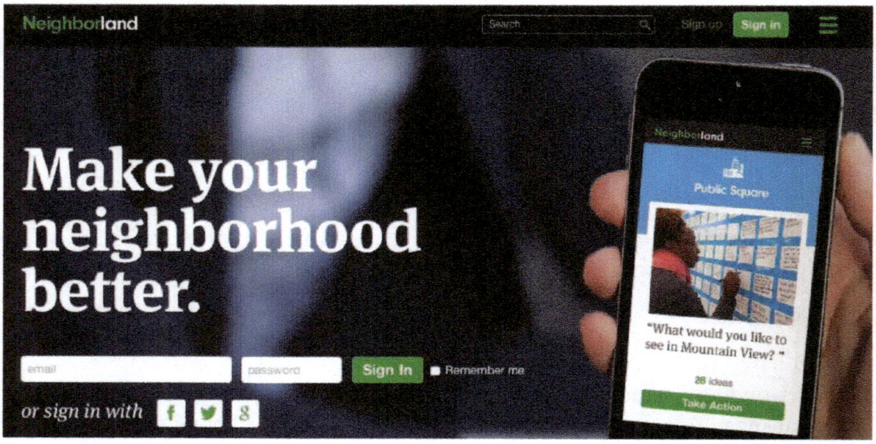

Fig. 16.8 Login page

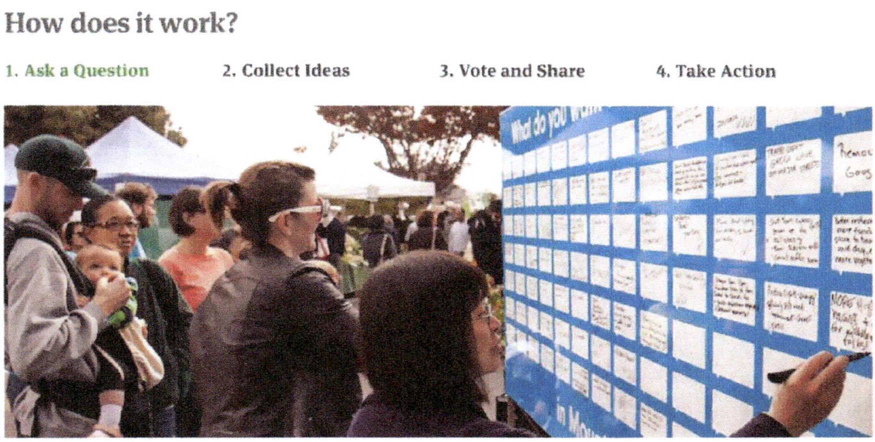

Fig. 16.9 How it works?

Fig. 16.10 An example

16.3.4 Owning the City

Owning the city: New media and citizen engagement in Urban Design by Michiel de Lange and Martijn de Waal proposes the notion of 'ownership' as a medium to take an alternative look at the role of urban new media in the city and in citizen engagement. It seeks to investigate how digital media and culture allows citizens to engage with, organize around and act upon collective issues and engage in co–creating the social fabric and built form of the city. The debate about the role of new

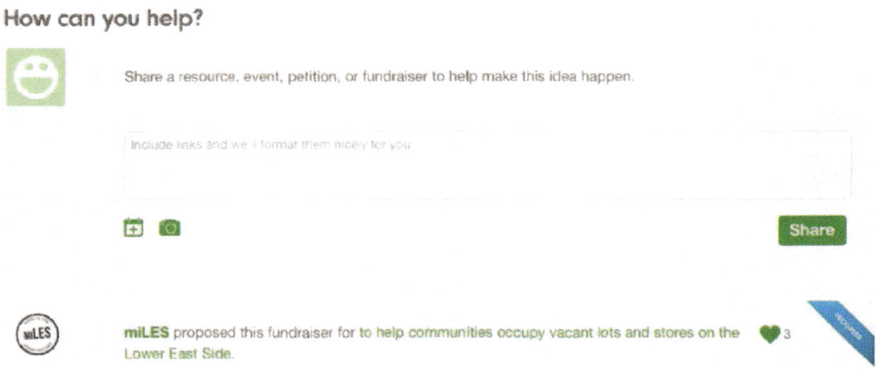

Fig. 16.11 Sharing resources and collaborating

media technologies in urban design needs to shift from an infrastructural to a social point of view, or from 'city management' to 'city making.'

Rapid urbanization has led to shifting of grounds and urban designers as well as citizens need to consider their role in city making.

A city has various stakeholders and complex networks. Engaging citizens with new media is pretty tough. The challenge is to balance these stories of personalization and efficiency on the one hand and of building collectives based on differences and mutualism on the other hand. Individuals need to cooperate more and devise avoidance strategies to address complex city life issues.

The various developments for improving citizen engagement and strengthening ownership are:

(a) Generating data commons

The city is considered as an information–generating system. A variety of technologies collect an enormous amount and range of data. 'Data Commons': a new resource containing valuable information for urban designers. Datasets can be used to bring out, visualize and manage collective issues. But for the establishment of a data commons and it to be wisely used the availability of and access to open data, and the skills citizens have to use the data in a meaningful way is of prime importance.

(b) Engagement: Sense of place

To engage people, it is essential that people envision themselves as part of the urban fabric, and understand that their individual actions make a difference to the common good. They also need to trust other citizens to act accordingly. Digital media can help in engaging citizens in many innovative ways and play an important part in this.

(c) Networked publics

'Networked publics' are groups of people that make use of social media and other digital technologies to organize themselves around collective goals or issues. In online culture, networks of 'professional amateurs' create 'user generated content' or take part in 'citizen science' projects. Think of open source software or Wikipedia as successful examples.

(d) Act: DIY urban design

Digital media have enabled mechanisms for managing collective action. With mobile and location–based media people can share more information more quickly and base adaptive decisions on it. With new participants having an active role, the terms 'co–creation' and 'crowdsourcing' are used for collective issues being tackled and managed collaboratively.

16.3.5 Inferences from the Case Studies

We need a tool for citizen engagement with some pretty amazing set of features and flavors which are meaningful and useful (Fig. 16.12).

- A mobile platform makes it natural to report and give out ideas quickly and easily, thus contributing to content generation. It also brings in the right context.
- All the geo-tagged data can be put into use to make informed decisions. Custom API's and a strong backend with the right set of visualization tools is required.
- More and more people should contribute to the network and sharing ideas and collaborating to make it come alive needs to happen.
- A strong social presence and elements of community feedback and response provide incentives to the users and keep them coming back to use the platform.

16.4 Urbane App—'The Urban Network'

The app revolves around generating relevant content in the right context, generating data commons which could be used for analysis, creating local communities which can take collective action and help in making city a better place. People act as data-donors.

> We don't need more civic apps, we need apps which are more civic!
>
> —MIT Media Lab

Fig. 16.12 Inferences

The beauty of such a mobile app is that all content gets geo-tagged and sophisticated location based mapping can be achieved.

16.4.1 Login/Signup

The various stakeholders are clearly identified and important details are asked. Users also have the ability to sign up using their existing Facebook, Twitter or Google accounts. It will make their inputs on the app more shareable and have wider reach. Similar stakeholders tend to be more comfortable with each other and unite easily for collective effort (Fig. 16.13).

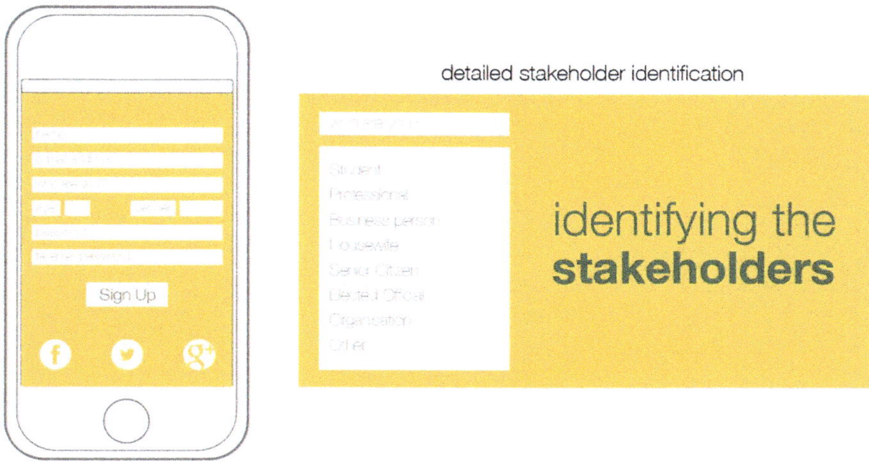

Fig. 16.13 Start screen

16.4.2 Input and Feed

The app takes advantage of the camera on your smartphone. The user can launch the camera from the app and give his/her input. For putting ideas and suggestions on the foreground 'I think' is used before their input and is a default.

For example—A user is walking down a street and sees an open space near a small garden in his college campus and thinks a canteen here would be a nice idea can launch the app, take a picture and share his input over the network. The user gets a choice to write a description and has to choose a relevant category for his input.

Any content shared by the user on the network goes to the feed and is seen by people following him. The content can also be discovered by the relevant category it was put to. Other users can upvote/downvote on the content, give their feedbacks, or share valuable resources.

Furthermore, local bodies and agencies can set up survey questions or ask for feedback and it can appear right in the feed. This makes it easier for the user to contribute (Fig. 16.14).

16.4.3 Public Channels

The various types of content users can put up may not be relevant and can lead to content clutter and rendering the app useless. The solution came in the form of hashtags (#)! The user when giving the input needs to categorize it into relevant categories by selecting the particular hashtag or a combination of hashtags. Such a

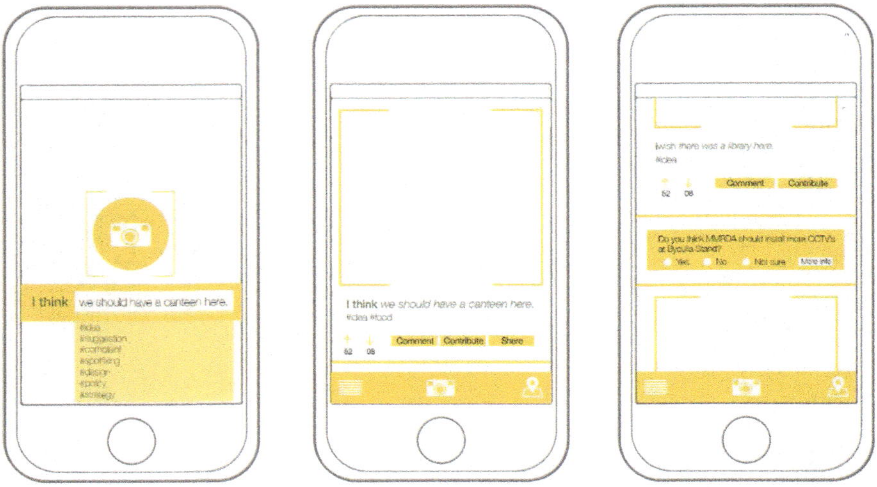

Fig. 16.14 User input and feed

small solution makes it very effective to organize such vast strings of data and generate 'data commons'.

Particular 'data commons' can be picked up by interested individuals or bodies and they can organize themselves around it.

The various categories identified count to 36 and were chosen to be simple to relate to by the everyday end user. These categories cover various aspects of urban and civic life ranging from bigger things like policy making and economy to small but important things like way-finding or trees. Also, a combination of hashtags can be put up to generate more relevant data sets (Fig. 16.15).

#idea
#spotfixing
#water
#crime
#policy
#energy
#mobility
#architecture
#publicspace
#safety
#heritage
#finance
#culture
#economy
#education
#health
#governance
#housing
#parking
#recreation
#streets
#trees
#environment
#tourism
#sanitation
#walking
#food
#women
#kids
#poverty
#seniors
#shopping
#tech
#wayfinding
#pollution
#waste

Fig. 16.15 List of public channels

16.4.4 Monitoring and Transparency

The variety of content on the network is open data and can be picked up by anyone. But for the citizen engagement to work more efficiently along with the government, a more filtered approach is a better solution.

In Kerala, various departments and agencies deal with various issues and concerns. We have a pretty long list and the most important ones which can handle all the proposed categories were identified. Most of the issues being local, the Municipal body always need to be in the loop along with departments like Public Works, Town and country Planning etc. These agencies will monitor the relevant hashtags concerning their purview and take actions. For categories like safety or crime, monitoring can be done in real-time and as everything is geo-tagged, quick actions can be taken and help can be provided (Fig. 16.16).

A detailed list of which hashtag will be monitored by which government department is as follows (Figs. 16.17 and 16.18).

The monitoring of data is an added functionality which is provided to these agencies via a backend dashboard along with added analysis tools to enhance planning procedures and take effective actions.

All this data is open, so other members of the community, or NGO's, or private bodies also have access and can contribute which remains the main aim of such a platform.

> The coolest thing to do with your data will be thought of by someone else.
>
> —Rufus Pollock, Director, Open Knowledge Foundation

Kerala Government Departments

Collegiate Education
Commercial Taxes
Commissionerate of Food Safety
Consumer Affairs
Drugs Control
Economics and Statistics
Election
Electrical Inspectorate
Environment
Finance
Food, Civil Supplies,Consumer Affairs
Health and Family Welfare
Information Technology
Industrial Training
Industries and Commerce
Kozhikode City Corporation
Kozhikode City Police
Labour
Law
Non-resident Keralites' Affairs
Police
Public Relations
Public Works
Revenue
Science and Technology
Social Welfare
Sports and Youth Affairs
Taxes
Technical Education
Tourism
Town and Country Planning
Vigilance and Anti-Corruption
Water Transport

Fig. 16.16 Government departments

#spotfixing	Public Works | Electrical Inspectorate | Kozhikode City Corporation
#water	Public Works | Water Transport | Kozhikode City Corporation
#crime	Drugs Control | Police | Kozhikode City Police
#policy	Collegiate Education | Commercial Taxes | Economics and Statistics | Environment Finance | Industries and Commerce | Kozhikode City Corporation | Labour | Law Revenue | Taxes | Town and Country Planning
#energy	Electrical Inspectorate | Environment | Public Works | Science and Technology
#mobility	Kozhikode City Corporation | Public Works | Town and Country Planning | Water Transport
#architecture	Kozhikode City Corporation | Public Works | Town and Country Planning
#publicspace	Environment | Kozhikode City Corporation | Public Works
#safety	Police | Kozhikode City Police | Public Works
#heritage	Kozhikode City Corporation | Tourism
#finance	Commercial Taxes | Economics and Statistics | Finance | Taxes
#culture	Kozhikode City Corporation | Tourism
#economy	Commercial Taxes | Economics and Statistics | Finance | Industries and Commerce |Labour
#education	Collegiate Education | Industrial Training | Social Welfare | Technical Education
#health	Commissionerate of Food Safety | Drugs Control | Environment | Health and Family Welfare | Social Welfare
#governance	Consumer Affairs | Election | Kozhikode City Corporation | Law | Public Relations
#housing	Health and Family Welfare | Public Works | Kozhikode City Corporation
#parking	Kozhikode City Corporation | Public Works
#recreation	Environment | Kozhikode City Corporation | Tourism
#streets	Kozhikode City Corporation | Public Works | Town and Country Planning
#trees	Environment | Kozhikode City Corporation
#environment	Environment | Kozhikode City Corporation | Public Works
#tourism	Kozhikode City Corporation | Tourism

Fig. 16.17 Monitoring I

#sanitation	Health and Family Welfare |Kozhikode City Corporation |Public Works
#walking	Environment | Kozhikode City Corporation | Public Works
#food	Commissionerate of Food Safety | Food, Civil Supplies,Consumer Affairs | Health and Family Welfare
#women	Health and Family Welfare | Social Welfare
#kids	Health and Family Welfare | Social Welfare
#poverty	Economics and Statistics | Finance | Kozhikode City Corporation | Social Welfare
#seniors	Health and Family Welfare | Non-resident Keralites' Affairs |Social Welfare
#shopping	Industries and Commerce | Kozhikode City Corporation | Tourism
#tech	Information Technology | Industrial Training | Science and Technology | Technical Ed.
#wayfinding	Environment | Kozhikode City Corporation | Public Works
#pollution	Environment | Kozhikode City Corporation | Public Works
#waste	Environment | Kozhikode City Corporation | Public Works

Fig. 16.18 Monitoring II

How all this content will be put to effective use, and maybe more is exciting. The potential is vast and can be a game-changer!

16.5 Experiment at Nitc Campus—Testing the App

For the experiment conducted at NITC campus, the basic prototype of the app was used which let users launch the camera, take images and upload contents along with putting hashtags. The public channels identified for the experiment were limited and also the monitoring bodies were carefully picked up.

The users were asked to go around the campus and give their inputs on any space they wished for a change or place they thought needed improvement. They were also encouraged to come up with new ideas for some vacant spaces within the campus premises (Fig. 16.19).

All the collected data was geo-tagged and was laid out on a map to show the various inputs. Also, users were given a chance to comment or give inputs on content uploaded by other users. The geo-tagged data was used to gain sufficient insights on improving the campus. The experiment was gracefully completed over a period of two days and the effectiveness of the app was tested out (Figs. 16.20, 16.21 and 16.22).

As seen, the content generated is simple and effective to understand and made useful. The relevant hashtags breathe a new life into data classification methods (Fig. 16.23).

A detailed look at one of the inputs and further discussion over it by other users shows how community members can come together, discuss, share ideas and resources and also act upon it.

#idea
#spotfixing
#architecture
#safety
#education
#environment
#recreation
#sanitation
#walking
#food
#waste

SAC
Director, NITC
Registrar
Dean, *Students Welfare*
Dean, *Plannning & Development*
HOD, *various departments*
Faculty
General Student body

Fig. 16.19 Public channels and monitoring for NITC

Fig. 16.20 Input map -shows the various locations where students gave their inputs for the experiment

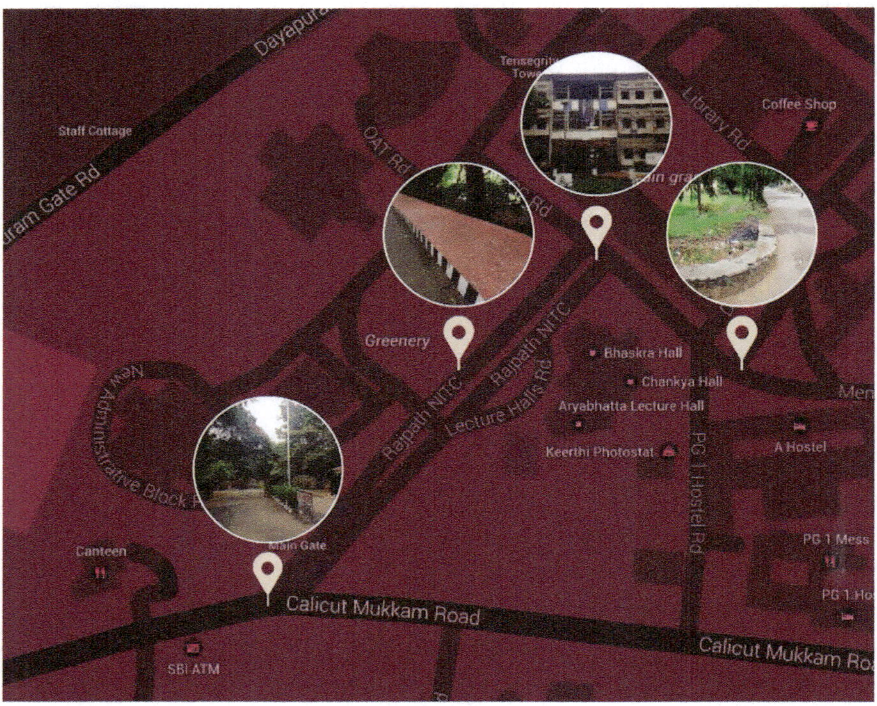

Fig. 16.21 Detailed input map—showing the content with the context. Part of the app

Fig. 16.22 User inputs

Such level of engagement at city scale will not only be amazing but also useful, will spark new levels of ownership and lead to stronger communities with effective governance and planning too.

16.6 Urbane Platform—Working and Scaling up

The app was found to be quite useful and it simply works. The process flow is pretty rational and easy to comprehend to. Users tend to enjoy it, it provides a social incentive. The idea sharing and content generation is made intuitive and profound. Users also get feedback and help from other members on the network. Monitoring by various agencies and open data for the city provides various opportunities and insights.

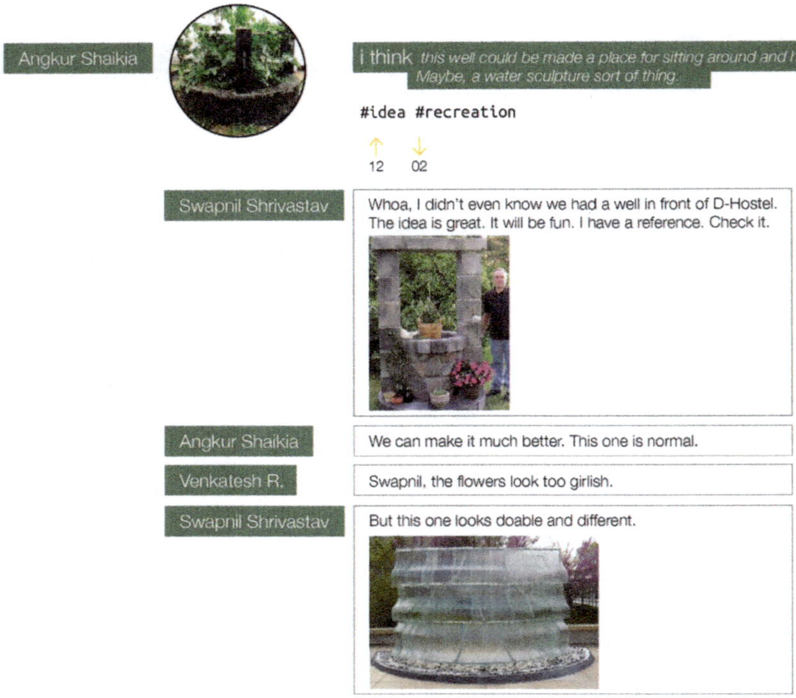

Fig. 16.23 Collaboration

The app when further improved from prototype stage can be scaled up to be made more effective in reach and use. There lies a potential for bringing positive change and improving the city (Figs. 16.24, 16.25 and 16.26).

16.7 Conclusions and Final Thoughts

Civic engagement and community participation needs awakening. We need smart and connected people to enhance civic life and also for making our cities better and more livable. We need to give citizens the tools to do so.

In totality, such a platform empowers the citizens and makes effective use of new media, mobile technology and social structure. It paves way for inspired living. Such a platform in future can improve governance, be an improved tool for democracy at all levels, be the next generation of social networks which gives meaningful conversation and deeper collaboration and also a platform for all of your urban life. It will help connect the digital to the physical.

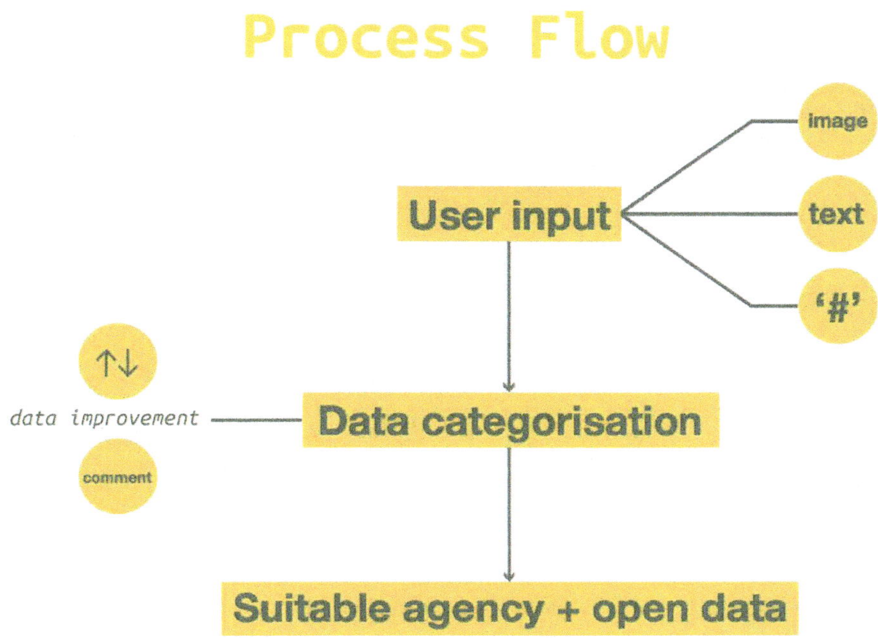

Fig. 16.24 Process flow

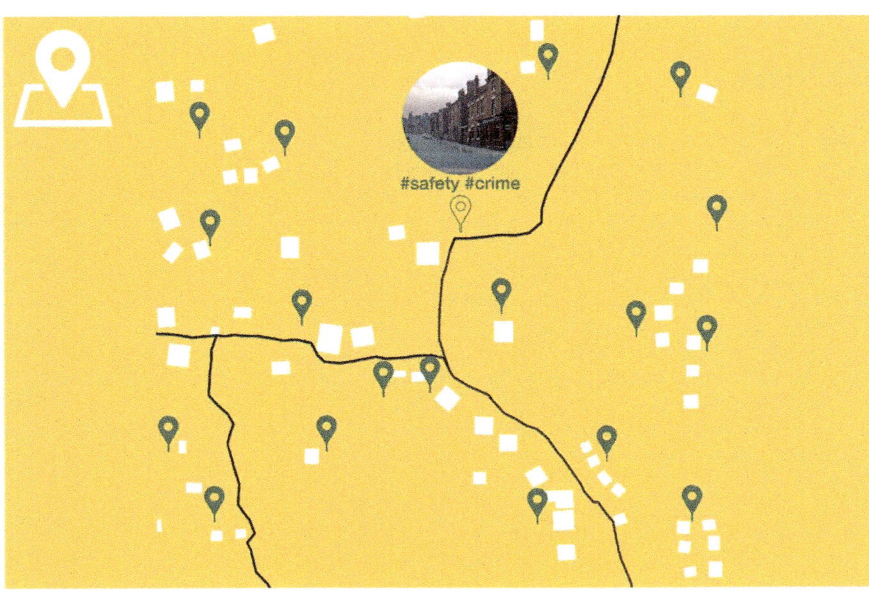

Fig. 16.25 Inputs at city scale

Fig. 16.26 Working at city scale

> You never change things by fighting the existing reality. To change something, build a new model that makes the existing model obsolete.
>
> —Buckminster Fuller

References

1. Ahn H-C (2007) Design tools and three steps in participatory design processes: a proposal for better communications among residents and experts, based on a case project of neighborhood park in Seoul, Korea. In: 6th conference of the Pacific Rim Community design network
2. Candy C (2016) I wish this was. Candychang.com. [Online]. Available http://www.candychang.com/i-wish-this-was/. Accessed 12 Feb 2015
3. Openplans.org (2016) [Online]. Available http://openplans.org/. Accessed 23 Feb 2015
4. Neighborland (2015) [Online]. Available https://neighborland.com/. Accessed 23 Feb 2015
5. de Lange M, de Waal M (2013) Owning the city: new media and citizen engagement in urban design. First Monday, 18 Nov 2013
6. C. Transportation (2015) Request bike parking. [Online]. Available http://bikeparking.chicagocompletestreets.org/. Accessed 23 Feb 2015
7. Vision Zero Community Safety Map (2015) [Online]. Available http://www.nyc.gov/html/visionzero/pages/maps-and-data/vz-input-map.shtml. Accessed 23 Feb 2015

Chapter 17
Blockchain-Powered Internet of Things, E-Governance and E-Democracy

Renming Qi, Chen Feng, Zheng Liu and Nezih Mrad

Abstract Digital technologies have dramatically changed people's daily life and made our life components much smarter. Nowadays, all users, including both human beings and devices, are connected to centralized servers. These servers act as the authorities, which are trusted by all users, making it possible to exchange critical information and money between untrusted users. However, maintaining large servers is costly and it's not affordable if such digital systems for cities' critical infrastructures are hacked. Blockchain, a technology revolution starting from 2014, offer the potential to solve these problems. It is essentially a tool that records every single transaction and digital event that happen in the virtual world. All the records are open to every user and the information asymmetries between two users are minimized. Thus, it's not possible for one user to cheat or hide information from another user. In other words, two strangers do not need to worry about being cheated by each other. They are allowed, for the first time in history, to do business without a centralized authority. Since a centralized authority is no longer a necessity, these two problems disappear naturally. This survey first explains how blockchain makes this magic happen and then introduces the blockchain's powerful applications in Internet of Things, E-governance, and E-democracy.

Keywords Blockchain · Internet of Things · E-governance · E-democracy

---

R. Qi
School of Engineering, EME1240-1137 Alumni Ave, Kelowna, BC V1V 1V7, Canada

C. Feng (✉)
School of Engineering, EME4285-1137 Alumni Ave, Kelowna, BC V1V 1V7, Canada
e-mail: chen.feng@ubc.ca

Z. Liu
School of Engineering, EME4205-1137 Alumni Ave, Kelowna, BC V1V 1V7, Canada

N. Mrad
Defence Research and Development Canada (DRDC), Department of National Defence, National Defence Headquarters, Major-General George R. Pearkes Building, 101 Colonel by Drive Ottawa, Ottawa, ON K1A 0K2, Canada

T.M. Vinod Kumar (ed.), *E-Democracy for Smart Cities*, Advances in 21st Century Human Settlements, DOI 10.1007/978-981-10-4035-1_17

17.1 Introduction

Our cities are getting smarter by utilizing digital technologies, such as online video meeting, self-driving cars, online medical insurance payment, and electronic voting etc. The application of digital technologies to smart cities can be, in general, divided into three categories: Internet of Things (IoT), E-governance, and E-democracy. The IoT refers to a network of connected devices (such as smartphones and sensors) where the network connectivity allows these devices to collect relevant information and take corresponding actions. For example, Apple Inc. can help its iPhone users to locate their missing phones by connecting every iPhone to iCloud and collecting GPS information from missing iPhones.

E-governance utilizes digital technologies to provide public services in an efficient and user-satisfactory way. For instance, in- stead of driving to a government building to file their taxes, nowadays citizens can choose to fill in and submit tax forms online, saving hours of unnecessary waiting time. E-democracy leverages digital technologies to facilitate citizens' participation in government's events and decision-making processes, holding political power running in an accountable way. For example, electronic voting provides voters a fast and convenient voting experience, attracting more citizens to vote.

Despite these great benefits, the application of digital technologies to smart cities still suffers from two major issues: high costs and insufficient security. For instance, with respect to IoT, it may incur a huge amount of operational expenditure (Opex) to build and maintain a large-scale *centralized* cloud platform (such as Apple's iCloud) to connect all devices. In addition, current cloud service providers are still not compatible with each other, leading to a high cost of labor to implement an information exchange hub working across different cloud providers. As for security, with an increasing number of connected devices embedded in cities' critical infrastructure (such as railways, tunnels and energy distributions), people cannot afford to under-estimate the risk of a cyber attack to a critical infrastructure that may cause severe damages to city properties and human life. Similarly, there is no proof or protection that our private information, such as bank statements and medical records, collected by E-governance will not be abused by incompetent management (perhaps due to corruption).

In this survey, we investigate a promising solution to all the existing issues: a technology revolution called "blockchain". The blockchain was first introduced by Nakamoto [1] as the ledger of Bitcoin, which is the first widely-deployed digital cash, very similar to the old American "gold-standard" currency. The blockchain can hold a record of every transaction made by Bitcoin users and provides a *decentralized* manner to process these transactions. Unlike traditional financial services hiring a bank to validate each transaction, there is no need for a centralized authority in blockchain. Many participants of blockchain volunteer to verify each transaction, leading to much lower Opex. To ensure the proper performance, these volunteers will be awarded for their correct work on validation and will, sometimes, be penalized for their incorrect work. In this way, blockchain participants only rely on trustable volunteers rather than a centralized authority like banks. On the other hand, if one participant wants to tamper with previous transactions, he has to persuade all the other users agreeing on him to do so, which is proven to be an extremely difficult task. Because of its low cost and high-level security, blockchain has emerged as an ideal technology to store records persistently, including contracts, diplomas, and certificates etc.

The application of blockchain is not limited to financial services. The industry extends the concept of transactions to smart contracts [2]. In order to distinguish blockchain of smart contracts from the previous blockchain of money, we call this stage blockchain 2.0 and the previous stage blockchain 1.0. A smart contract is a contract whose items are self-executing computer programs. Once two willing parties make a smart contract and publish it on blockchain, the associated computer programs will run without any human intervention. Since records in blockchain cannot be altered and the programs will run automatically, both parties have to obey rules in this contract and have no methods to break them. This enables, for the first time in history, untrusted parties to do business with each other without a centralized authority.

Blockchain is envisioned to transform the application of digital technologies to smart cities from two aspects: automation and security. For IoT applications, transactions and coordination between devices can be facilitated by blockchain. Instead of relying on a centralized cloud, devices are empowered to autonomously execute digital contracts, including buying electricity, selling collected information to peer devices and etc. For E-governance applications, the way of enforcing regulations is going to change. Traditionally, regulations are enforced by agencies such as food-safety agency and fiscal agency. With a smart contract in play, regulations can be translated into computer programs, which will be run by blockchain automatically. The human intervention will be taken out of the loop in the entire process, leaving no room for corruption. Security comes directly from the fact that there is no central point in blockchain. Hacking applications on blockchain requires hacking every single user, which is much harder than hacking a single central point. More details of applying blockchain to IoT, E-governance and E-democracy will be discussed in Sects. 17.3, 17.4 and 17.5.

17.2 Blockchain 1.0

Digital cash is infinitely copiable (like copying a file on computer for many times). If someone received a unit of digital cash, he/she can make several copies of it and pass them on to other people, which will definitely disturb the financial system. This problem is called double-spend problem. Previously people hire centralized intermediaries such as banks to solve this problem. When consumers do online payments, the bank will check whether the consumers have enough balance in their accounts and only after confirming that, it will send sellers the money and inform the seller to send consumers the goods. However, blockchain, as a breakthrough in digital cash research, solved the double-spend problem and avoids the centralized third party at the mean time. To explain how blockchain achieves this, we will describe the basic concept of blockchain, including transaction, block and distributed consensus algorithm over the peer-to-peer network.

A transaction in blockchain includes four components: (1) input: spender's address (or user account); (2) output: receiver's address; (3) amount: quantity of units transacted; and (4) metadata: additional information stored with each transaction [3]. Each transaction is signed by its creator with a digital signature, broadcasted to every node in the blockchain network and then recorded in public ledger after validation. Before recording any transaction, the verifying node must ensure two things: (1) the spender owns the money by verifying the digital signatures on the transaction; (2) the spender has enough money on his account by checking every transaction against spender's account in the ledger to make sure he has sufficient balance.

Transactions are passed node by node in the peer-to-peer network, so there is no guarantee that transactions will be received at a certain node or be received in order as they are generated. Here comes the problem of double-spend. Even though malicious user only owns ten dollars, he/she could broadcast two transactions containing 10 dollars to the network simultaneously. Nodes that only receive one transaction judge the received one as verifiable while nodes that receive both only admit the first one. This means that there is demand for a mechanism that enables the whole network to agree on the order of transactions. The mechanism is known as distributed consensus algorithm.

The distributed consensus on blockchain is a two-step process. At first, each node orders the transactions it received by placing them in a linear sequence of blocks. Each block consists of 4 components: (1) transactions collected in a certain period of time; (2) a reference to the block that comes immediately before it; (3) an answer to a hard-to-solve mathematical puzzle. The puzzle is unique for every block and it will be introduced in the following paragraph. The correctness of answer in a block is easy to verify; (4) a timestamp which indicates when the block is built [4]. The transactions in one block are regarded to be created at the same time. Every node can collect unconfirmed transactions into a block and broadcast the block to network as their suggestions on what should be added to the current sequences of blocks. In the second step, blockchain needs to decide which block among these suggestions should really be the next block and make every participant agree with this block. This turns out to be a hard problem since there is no centralized server which is responsible for collecting everyone's suggestions, making the final decision and forcing everyone to agree with the decision.

Blockchain solves this problem elegantly by using a race of solving a hard mathematical puzzle as mentioned above. The first one who solves the puzzle will make the next block. This is also known as "proof of work" [1], because by solving this puzzle, a node can prove that it does use lots of resources to win the race. For example, in Bitcoin, the puzzle is to find a number that makes the hash of the concatenation of this number, transactions contained in this block and the hash of previous block start with a certain number of zeros. The average efforts required to find the answer is exponential with the number of zeros. Bitcoin system will adjust the number of zeros according to the computing power of the network so that average time to find a required number is constant and nearly 10 min [5].

After receiving and verifying the block proposed by the node who first solved the puzzle, other nodes can show their agreement on this block by working on the new puzzle that contains the hash of this block. The block builder is usually called "miner". Miners receive financial awards for spending resources to make a block. In Bitcoin, after building a block, the builder will be awarded a certain number of bitcoins by Bitcoin protocol and each starter of transactions in the block have to leave him a tip. The awards can ensure miners work uprightly since if they conduct malicious behaviors, the blockchain is no longer trustable and then assets of miners on blockchain will become valueless. Besides, the network accepts only the longest sequences. As long as the majority of resources is controlled by miners who don't cooperate to attack blockchain, they will generate the longest sequence and the attacker will be outpaced. As a result, for the first time in history, untrusted parties to do business with each other without a centralized authority.

17.3 Blockchain 2.0

Blockchain 2.0 decentralizes more complicated agreements, i.e. contracts beyond financial systems. Techniques used to decentralize functionality of a ledger in blockchain 1.0 can be adopted to register, confirm and transfer all kinds of contract and assets in blockchain 2.0.

17.3.1 Smart Contract

Simply speaking, smart contracts are published computer programs agreed by both parties on blockchain [6]. A contract in a traditional sense is specific terms between two or more parties in which there is a promise to do something in return for some benefits. Each party must trust the other party to fulfill its obligations. If there is any disputation, they need to resort to courts for resolution. The contract on blockchain is smart in the sense that it allows contractors to solve common problems in a way that minimizes trust. Minimizing trust makes things easier by allowing human intervention to be taken out of the loop, thus allowing complete automation. The automation here means two things. One is after the contract is launched, it will be self-executing and there is no need for contact between contractors. The other is that the contract is able to arrange resources on its own behalf. This ability requires smart property which will be introduced later.

The main difficulty for smart contracts is that computer programs cannot easily and reliably tell what's happening in the physical world or who is telling the truth. Checking whether a financial payment is made is easy for computer programs, but real-world situations like whether one's work achieves the company's standard are hard for computer programs to evaluate. One solution is to use oracles—online service providers who broadcast data which can be used by users as input to smart contracts [7]. A good example is the inheritance gift. An old man leaves an inheritance gift to his grandson and sets the condition that the gift can be received after he passes away. The old man can make the contract and publishes it onto blockchain. An oracle can broadcast new entries in government registry of death. The contract receives this as an input and once the death is confirmed, the gift will be sent automatically.

17.3.2 Smart Property

Smart properties are properties whose ownership are controlled by blockchain using smart contracts [8]. The asset includes physical property (such as a house or a car) and intangible asset (such as votes, ideas, reputation, health data, rights, and shares in a company). The key idea of smart properties is that ownership of these properties is represented by a private key on blockchain. The owner can prove his ownership by showing having access to the certain private key. A good example is buying a car via blockchain. The car's computer requires authentication by using an owner key. Additionally, the car has a token from its manufacturer that has been published on blockchain to prove its existence. This token can also be used to identify its age, mileage and maintenance recording, giving enough information for the buyer to make the decision. In order to buy the car, the buyer makes a smart contract with the original owner with the following item, if the buyer pays the money, the token must be transferred to the buyer's account. The car's computer will be aware of this via blockchain and then the buyer can use its private key to open the car. No third party including government is necessary for this process.

17.4 Blockchain and IoT

Current IoT systems rely on centralized communication models. All devices are identified, authenticated and connected through various cloud servers that support huge processing and storage capacities. However, there is no single unified platform that connects all devices and no guarantee that cloud services offered by different manufacturers are interoperable and compatible. For instance, the field of logistics information support systems is fragmented and diverse. Every logistics company uses their own software products and very few of them can communicate meaningfully with each other. System integrations on any level are almost always built as one-on-one relationship between operational ERPs and their databases and are thus very much costly and tailored work. Human component as the information exchange hub is rather the norm than the exception. Actually, operational information is routinely passed over phone, email and even old-fashioned fax machine by human.

Blockchain works well for this problem. Blockchain achieves trustless connection without sacrificing data integrity, which is what supply chain business is virtually all about. Following this thought, a project named SmartLog is launched [9]. SmartLog is trying to introduce a scenario where all kinds and sizes of logistics companies are able to share and tap into a common blockchain, which will consist of all the relevant information pertaining to the movement of intermodal containers throughout the European Union transport corridors. Information will be gathered from companies' own information management system and then shared among all participants after being anonymized and filtered. There will also be a simple device attached to some of the containers, so companies can validate actual movements of containers in the corridors against the information which they see flowing into the blockchain. The data will serve several different purposes: the participating companies will have immediate access to it and can use it to greatly enhance their operations, resource management, and route optimization planning.

Another downside of current centralized communication model is that costs of installing and maintaining large centralized clouds are high. As the scale of IoT devices tends to grow explosively, these costs will grow substantially and cloud servers will become a bottleneck and a point of failure that can disrupt the entire network. Adopting a standardized peer-to-peer communication model between devices via blockchain will significantly reduce the costs associated with installing and maintaining large centralized data centers. It will also distribute computation and storage needs across the billions of devices from central servers to devices in IoT networks. This will prevent failure in any single node in a network from bringing the entire network to a halting collapse. However, the prospects of the decentralized model are beyond this. Without centralized management, each device manages its own roles and behaviors, resulting in an "Internet of Decentralized, Autonomous Things" [10]. We use the blockchain-powered self-driving car to show this idea. After taking a ride, people can pay directly to the car instead of the driver or the company it belongs to. The car can use these payments to maintain itself and the remaining money becomes the profit of the taxi company. If the car runs out of gas, it drives itself to a nearby gas station and pays the gas dump to get filled. When there are too many passengers for a car, the car can call its peer cars via blockchain to make every passenger served.

Here we present blockchain as "language" used by self-driving cars to communicate with the physical world (passengers) and peer devices (peer cars and the gas dump). We can be more aggressive. Blockchain can be used as general communication tools for artificial intelligence. The artificial intelligence found in Siri and Watson currently is a singular artificial intelligence, which is far from human beings. In contrast, human's society is a form of a multi-agent system, where each agent is independent and intelligent. The society is formed through communication and co-operation between human beings. Blockchain's mechanism of achieving persisting consensus in trustless peer environment makes a multi-agent system of intelligent devices possible. The consensus among devices makes each device get the same sense of a signal as others. These senses act like words in human's language, making device-to-device communication possible. Persistence means that the sense will not diminish with time. For humans, persistence means that we see further by standing on giant's shoulders. There is no necessity to invent calculus again when we use it. For devices, persistence means device-readable knowledge, making devices evolve their functionality automatically. There have already been some primitive discussions in this field, see details in [11, 12].

17.5 Blockchain and E-Governance

A key role of a democratic government is the appropriate distribution of resource among its citizens, both individual and corporate. This goes beyond the distribution of monetary resource and includes social intangibles such as security, the conditions for the maintenance of the rule of law. However, as we have witnessed, governments have become larger, more centralized and more remote from the individual citizens. Centralized model provides poor customer service and is no longer economic. Citizens' queries and needs cannot be responded in time. Governments have started to provide e-governance service, which uses information technology to improve service quality and efficiency. Though a prodigious amount of money has been invested to e-governance to improve service quality and governance efficiency, what is achieved by e-governance is trivial. Blockchain is about to change this by enabling governments to provide service and carry out governance in a detailed way.

A good example is the food safety issue. Consumers surprisingly know little about most of the products they use daily. A complex network of retailers, distributors, transporters, storage facilities and suppliers stand between consumers and the products they use. Governments set up food safety agencies to enforce every component of this chain to comply with standards but the recent Chipotle food contamination crisis [13], Chinese tainted milk scandal [14] show that governments failed.

Blockchain-powered food supply chain management may be a solution. Walmart, IBM and Tsinghua University just started to collaborate together to provide quality assurance for providers and consumers [15]. With blockchain, food products can be digitally tracked from an ecosystem of suppliers to store shelves and ultimately to consumers. Digital product information such as farm origination details, batch numbers, factory and processing data, expiration dates, storage temperatures and shipping detail are digitally connected to food items and the information is entered into blockchain along every step of the process. Governments can use each piece of information to detect critical data points that could potentially reveal food safety issues with the product. Since information on blockchain cannot be altered, this ensures all the information is accurate and that no involver can cheat.

Another example is the regulation in financial systems. Financial institutions need to submit digital reports to regulators to prove that they did not break the regulations. One of the main challenges is that they need to comply with existing regulations: EMIR in EU and Dodd-Frank in US [16]. Due to the myriad of regulatory obligations, the reporting process currently is rather complex. Moreover, regulations in EU and those in the US are not necessarily consistent with each other. This puts a heavy burden on industry and consumes substantial resources. For governments, validating these reports requires much human labor and time. The long delay of this process also enables bank managers to conceal their wrongdoings as they did in the crisis of 2008. And these reports often use different methodologies behind the calculations. This enables companies to take advantage of loopholes to avoid un-profitable regulation and sometimes leads to confusion. Blockchain can benefit both financial institutions and regulators. The regulations in legal files can be translated into smart contracts and then packaged to software. This software defines whether an action is permissible or not so that financial institutions can obey regulations by using API of this software instead of reporting to governments. Since transactions are stored in blockchain, access for regulators to them is easier and faster. Validation can be helped by the consensus mechanism of blockchain: impermissible behaviors cannot be admitted by the network and thus cannot be conducted.

The security of digital resources is also a headache of both governments and citizens. For governments, all information is stored on centralized servers. If one server is attacked, most of the confidential information will be leaked. For users, there is no proof or protection that their privacy collected by governments will not be abused. Even worse, the software used by governments and citizens may be built with backdoors, which can be used to conduct malicious behaviors without being detected. Blockchain provides elegant solutions to these security problems. We first introduce MIT's Enigma, a privacy-preserved blockchain platform [17]. Data is encrypted and stored in distributed nodes. Rather than hacking a centralized server to steal information, hackers need to hack the whole network. Moreover, any smart contract can directly run on encrypted data by using homomorphic encryption [18]. Thus, data on Enigma can be stored, shared and analyzed without revealed to the third party including the government itself, enabling trustless sharing of data and distributed computation without leaking users' privacy. If one wants to use back-doors to conduct malicious behavior, he will not succeed because other users on blockchain will reject his action. Moreover, every action is recorded on blockchain.

Big-data analysis on these records can be adopted to detect abnormal behaviors, prevent new frauds before they happen and even find the source of frauds. Block-Cypher has started to work on this idea [19]. It aims at building a blockchain-based platform on which every action will be given a risk score according to the user's history, behavior pattern, location and etc. The high-scored action which may be a fraud with high possibility will be rejected.

17.6 Blockchain and E-Democracy

Voting is the basic approach to democracy, but current voting system haves been plagued with voting frauds. According to [20], at least 7 billion Americans vote multiple times in federal elections. This duplicity can be used by any special interest groups who seek to gain advantage for the candidates they support. There is also no way for voters to assure that their votes will be accurately recorded and counted. It's not who votes but who counts the votes decides the winner. The voting fraud frustrates people from believing that their votes matter, so most people choose not to vote. In the American federal election of 2012, only 58% of people who are eligible to vote actually voted. Many governments became interested in the electronic voting machine, which could increase the transparency of voting to prevent frauds. However, this interest comes with security warnings. In [21], the authors revealed that one who gained physical access to the Diebold AccuVote-TS voting machine or its removable memory card could install malicious code capable of stealing votes without being detected. Many people will use the same voting machine so this single point failure has enormous impacts on results. Moreover, the voting machine is usually designed, produced and maintained by a single company. There is no proof that this company will not install malicious codes in the machine at the beginning.

Blockchain is an effective solution to problems in the voting system. The blockchain-powered voting runs as follows: (1) Before voting, a user sends his legal Identification and username on blockchain to an identity verifier. After being approved, the user will receive his unique ballot. The verifier is not necessarily the government. The task of verification can be undertaken by miners, who has the incentive to behave uprightly. And by using homomorphic encryption, the verification process can directly run on encrypted data, without leaking the user's identity; (2) During the election, the user completes his ballot and sends it to the blockchain-based ballot box. Each user uses a different device as the entrance point to voting so hacking single device has little effect on the voting results; (3) After the election, each user is allowed to audit the vote results because each user can access the open data on blockchain. In addition, a user's decision will not be known to other voters since users use their pseudo name on blockchain when voting. Many projects of blockchain-based voting have been launched already, such as BitCongress [22], AgoraVoting [23] and FollowMyVote [24].

The other impact of blockchain on democracy is the high-resolution information. Democracy government cannot avoid the wrong choices of public servants. Not every citizen in the city is aware of the city's political, economic and social circumstances. People may vote for a proposal based on other factors rather than its potential effect. IoT devices that are spread all over the city collect unbiased data about every aspect of the city, including the traffic situation, the road status, the water quality and etc. The collected data is published on blockchain and can reveal important factors about the circumstances of the city by adopting big-data analysis tools. When governments want to increase budget, say on municipal construction, citizens have extra facts to judge whether it's reasonable or not. During the election, citizens can turn to the quantized effect of their past work and choose the one with the critical capability, instead of listening to their debates and watching their immoral practices like tarnishing their opponent's reputation on the Internet. The intelligent devices themselves may also have a fair weight in voting. Since they monitor the runtime of city continuously, they are more knowledgeable of the city than citizens. They are also neutral inherently because they are controlled by smart contracts which are feasible only after being agreed by the majority of the network. As a result, their opinion provides great hints for making the right choice. Including them in the democratic system is not impossible in the future. At this point, E-democracy does include e-lives as its name implies.

17.7 Summary

In this survey, we investigate a novel technology named blockchain. We first explain the mechanism of blockchain which enables untrusted parties to do business with each other in a decentralized way. We then use concrete examples to explain the powerful applications of blockchain in Internet of Things, E-governance, and E-democracy. When introducing these examples, we focus on two aspects of blockchain:

1. *Automation*: by using smart contract, devices become self-serving and thus more intelligent. For governments, routine work can also be processed automatically, thus enable governments to provide services more efficiently. Less human intervention in the entire process also lowers the cost.
2. *Security*: one user being hacked is minimized. Besides, every action on blockchain is recorded and transparent to every user. Under such mass surveillance, conducting malicious behaviors without being detected is not possible.

References

1. Nakamoto S (2009) Bitcoin: a peer-to-peer electronic cash system. http://www.bitcoin.org/bitcoin.pdfhttp://www.bitcoin.org/bitcoin.pdf
2. Wood G (2016) Smart contract yellow paper
3. BitcoinWiki (2016) Transaction. https://en.bitcoin.it/wiki/Transaction
4. BitcoinWiki (2016) Block. https://en.bitcoin.it/wiki/Block
5. Croman K, Decker C, Eyal I, Gencer AE (2016) On scaling decentralized blockchains. Bitcoin and Blockchain
6. BitcoinWiki (2016) Contract. https://en.bitcoin.it/wiki/Contract
7. BitcoinWiki (2014) Oracle. https://en.bitcoin.it/wiki/Oracle
8. BitcoinWiki (2016) Property. https://en.bitcoin.it/wiki/Smart Property

9. SmartLog (2016) https://smartlog.kinno.fi/
10. Brody P, Pureswaran V (2014) Device democracy: saving the future of the internet of things. IBM
11. Maxim O (2016) How blockchain relates to artificial intelligence?—BICA Labs. https://medium.com/bica-labs/how-blockchain-relates-to-artificial-intelligence-f0111f39afc9
12. Worner D, Bomhard T (2014) When your sensor earns money: exchanging data for cash with Bitcoin. Exchanging data for cash with Bitcoin, ACM, New York, USA
13. Businessweek B (2016) Inside Chipotle's contamination crisis. http://www.bloomberg.com/features/2015-chipotle-food-safety-crisis/
14. Langman CB (2009) Melamine, powdered milk, and nephrolithiasis in Chinese infants. N Engl J Med 360(11):1139–1141
15. IBM (2016) How blockchain can help bring safer food to China. https://www-03.ibm.com/press/us/en/pressrelease/50816.wss
16. Carlo M (2016) Blockchain, financial regulatory reporting and challenges. https://www.finextra.com/blogposting/13102/blockchain-financial-regulatory-reporting-and-challenges
17. Zyskind G, Nathan O, Pentland A (2015) Enigma: decentralized computation platform with guaranteed privacy. arXiv preprint arXiv:150603471
18. Gentry C (2009) A fully homomorphic encryption scheme. Ph.D. thesis, Stanford University
19. BlockCypher (2016) blockcypher.com. https://www.blockcypher.com/
20. Watchdog (2014) http://watchdog.wpengine.netdna-cdn.com/wp-content/blogs.dir/1/files/2014/06/CrossStateCheckStatistics.pdf
21. Feldman AJ, Halderman JA, Felten EW (2006) Security analysis of the diebold Accuvote-Ts voting machine
22. BitCongress (2016) www.bitcongress.com. http://www.bitcongress.com/
23. AgoraVoting (2016) agoravoting.com. https://agoravoting.com/
24. FollowMyVote (2016) https://followmyvote.com/

Part IV
Conclusions

Chapter 18
E-Democracy for Smart Cities: Conclusion and Path Ahead

T.M. Vinod Kumar

Abstract The concluding Chapter attempts to put together various conclusions derived out of seventeen chapters of this book at one place. This work is accomplished in collaboration with the 42 authors editor of this book. This chapter then end with general conclusions and way ahead.

Keywords E-Democracy for smart cities · City studies · Domain studies · Issue and tools of E-Democracy

18.1 Introduction

This collaborative work of international scholars on "E-Democracy for Smart Cities" dwells more on the possibilities than based on current state of practice. No smart city today can boast that they have the best E-Democratic system in place. The reality is far away from that as can be seen in the city studies in Sect. 18.2. The objective of E-Democracy is accepted but miles to go to attain it. Therefore, the book title is "E-Democracy for Smart Cities" and not "E-Democracy in Smart Cities". The state of the art of democracy that was articulated in many republics has progressed considerably from ancient world to today but the state of E-Democracy is still in a nascent stage in most of the listed Smart Cities by leaned societies. Much work needs to done. The objective of this book is to show the way forward. Democracy in Republics is as old as the oldest written literature such as Veda but extending E-Democracy through wider applications into a smart city is a new opportunity which came into being with the explosive development of applications of Information and Communication Technologies in Smart Cities.

T.M. Vinod Kumar (✉)
School of Planning and Architecture, New Delhi, India
e-mail: tmvinod@gmail.com

T.M. Vinod Kumar (ed.), *E-Democracy for Smart Cities*, Advances in 21st Century Human Settlements, DOI 10.1007/978-981-10-4035-1_18

The introduction of digital ICT technologies and its diffusion in a massive and rapid scales such as ownership of smartphones, participation in Facebook, twitter and Instagram created a fertile ground for all democracies world over to transit to E-Democracy, but an achievement to this transition which is less costly and easy to implement has not been very impressive and widespread. Those who started using E-Democracy effectively in elections won the election leaving behind those who used less effectively in their election campaigns. Specialists in ICT related E-Democracy were selected by political parties to manage E-democracy in municipal and general elections. There was no Government policy to support E-Democracy in these countries and none was against the use of E-Democracy since it is an effective approach to further strengthen democracy. There are many sets of policy reforms to strengthen Governance through E-Governance in several countries but there is practically none for E-Democracy barring a few countries mentioned in the first chapter.

The introduction of ICT was welcomed by all countries and ICT is the foundation of smart cities. Smart City functions only in a city with a high level of deployment of ICT and its effective use in all walks of life. However, Smart Cities are emerging in the world urban-scape in very limited numbers. There is no scope that it will spread as fast as the spread of Smart Phones. The main reason is the creation of smart cities is only by design and takes all aspects of city functioning to a very efficient level, unknown today by people.

Smart cities creation happens only with the local people and no amount of public investment in smart technologies can create a smart city. To transform local people to smart cities they should embrace ICT in all walks of their life and leverage its strength for Smart Economy, Smart Living, Smart Governance, Smart Mobility and Smart Environment. This transformation is achieved by the active collaboration of Smart People. These aspects are covered extensively by Vinod Kumar [1–3]. Smart cities require an infusion of rapidly changing innovations in all these aspects discussed above. Smart people identifies such innovative possibilities and see that these innovations are implemented by collaborations of common people who live there. This process is achieved by E-Democracy while E-Governance only regulate these developments to see whether it is based on the principles postulated in the constitution. Constitutions postulates the rights and responsibilities of different sections of society such as Executives, Judiciary, Citizen, print and visual media, religious group, caste groups and so on. Although E-Democracy and E-Governance look like one and the same or at most inseparable twins but they are different in so many ways. E-Governance regulates but E-Democracy creates a plan, policy, strategies, tactics and so on within the framework of the constitution of that country. Smarter the city more innovative will be such creation. This establishes the role and importance of E-Democracy in Smart Cities.

This book is an attempt to throw more lights on E-Democracy for Smart Cities by scholars from many countries across the world both from East and the west and south and north. The book opens up in it' first chapter with "State of the Art of E-Democracy in Smart Cities". The book is divided into five sections. They are;

1. Introduction
2. E-Democracy State of the Art City studies
3. E-Democracy Domain Studies
4. E-Democracy Issues and Tools and
5. Conclusions.

There is no smart city in the world where E-Democracy is practised in its full potential in spite of the fact the driving force behind any smart city is the E-Democracy. Therefore, the underlying momentum of smart city functioning will be lost sooner unless E-Democracy is developed in a massive scale in Smart Cities. However, there are many instances of very rudimentary to the partial deployment of E-Democracy in many smart cities as well as non-smart cities. However, we do not have one smart city which is a smart user of E-Democracy in all six components of the smart city discussed earlier. The 100 Smart Cities Program of Government of India looks only at a very small part of the city and have no subprogram to develop E-Democracy but undoubtedly participation is highlighted as a prime requirement for selection of the candidate smart city for this project. Therefore, one can say smart cities are evolving the world over in limited numbers so also E-Democracy but at a much slower pace than smart cities than that is required.

The Introductory chapter deals with the State of the Art Survey E-Democracy for Smart Cities whose conclusion will be discussed later.

Section 18.2 deals with the state of the art studies of E-Democracy in selected cities. Selected cities are Lagos in Nigeria Africa, Feroke and New Delhi in India and Hong Kong in China. Each study looks at a different perspective of E-Democracy in practice or potential in these cities. Nigerian study is a study of status in all walks of life in Lagos. The Feroke study deals with the E-Democracy potential in the preparation of Outline Development Plan of Feroke. In Feroke a democratic way of preparing the Outline Development Plan was conducted by the study team of this book and then think about its potential transformation to E-Democracy. New Delhi study looks at E-Democracy as Smart City Planning Process. Finally, Hong Kong Study in China describe the existing E-Engagement in Hong Kong in urban development through case studies.

Section 18.3 deals with E-Democracy in Domain Studies. To starts with an Indian study links up the domain of E-Governance to E-Democracy and postulate an Indian theory of domain linkages to the city. Dubai study considers urban innovation as a domain and investigates how it leapfrog E-Democracy through the experiences of Dubai. Another study from India investigates smart water management domain in the E-Democracy. From a successful background of urban land management and city planning in Surat, the study extrapolates into the Preparation of Town Planning Scheme in an E-Democracy Framework for Citizen-Centric Planning. Another study from Surat looks at how E-Democracy can effectively exploit the domain of "Smart Project Selection, Prioritisation, Customisation and Implementation". Renewable energy backbone in a smart grid often assumes importance in smart cities. This aspect is explored from E-Democracy point of view in the chapter Smart Grid Evolution: e-Democracy and Citizens' Participation in e-Governance of Energy Backbone of Smart City.

Section 18.4 consists of six chapters on issues and tools. Each chapter identifies a dominant issue in the practice of E-Democracy and explores appropriate tools. In India, there are well designed urban information system some concentrate on the environment and yet another on urban planning but seems not integrated. The departmental nature of this information system is exhibited in this specialised emphasis. However, urban planning cannot ignore environment even though urban planning comes under Ministry of Urban development and Environmental planning under Ministry of Environment and Forest of Government of India. An Indian study looks at this issue and proposes an Integration of Existing Information Systems and integrate to a Spatial E-Democracy Tool for Smart Cities. Another Indian study makes an inventory of geospatial tools available and postulate with case studies how it can be of service to Smart Cities by further strengthening E-Democracy. In Kenya, Africa ICT penetration is taking place in a rapid phase in cities. Kenyan study investigates, how Kenyan cities can attain E-Democracy through a Digital Platform. Open Government Data which strengthen E-Democracy, has been practising in the city and regional governments in Russia recently. A Russian study looks at the Politics of Open Data in Russia in Regional and Municipal Perspectives. Another Indian study called "Urbane" investigate by an experiment how a community drove architecture and planning can be accomplished through a mobile social platform. A Canadian study investigate the tool Blockchain-Powered Internet of Things in E-Democracy and E-Government.

18.2 State of the Art of E-Democracy in Smart Cities

The theoretical foundation for "E-Democracy for Smart Cities" is presented in this chapter to form a backdrop for the book. This chapter describes first the state of the art of Democracy before discussing E-Democracy which is the ICT enrichment and transformation of Democracy. This chapter in detail studies the relationship of smart cities and e-democracy. The basic democratic principles are stated and then the culture of democracy that evolved through many centuries were studied. The present constitutional provision and ancient Greek and Indian models of Republican democracy were discussed and democracy in panchayat system was also studied. Advantages and disadvantages of democracy were discussed. E-democracy was defined, E-democratic models investigated and E-democracy typologies were postulated as well as ten models of E-Democracy. Finally, several tools used in E-democracy were enumerated. The chapter concluded with a vision of e-democracy in large urban agglomeration.

Although it is desirable to convert as many cities in India to smart cities, it is not simply possible. There should be potential for smart cities in states and union territories. Smart people and smart communities are primary requirements to trigger the development of smart cities. It is not that easy to convert all smart city citizen and communities to smart. This can only be achieved by instituting E-Democracy in smart cities. Smart cities function within the framework of E-democracy, E-governance, and E-government. They are there wherever there is high Human Development Index. It is only smart people who can create smart communities and therefore smart cities. Evidence of emerging urban economy is indicated by the high urban content (%) of the states geographically spread contiguously. The large impact of the smart economy can be there only in million-plus urban agglomeration. When one combines all these requirements, not all 53 million-plus cities in India that appeared in 2011 census in India can be potential candidates for smart cities. Even if few of this million-plus agglomeration or cities are shortlisted as candidates, they can never be smart cities unless they are having an appropriate institutional framework which can shape them to smart cities such as E-democracy, E-government, and E-governance. Seventy-fourth constitutional amendments of India provide for the metropolitan planning committees which need to be legislated by the respective states. Triple Helix Model is the only dynamic framework that can shape smart cities. It presupposes that members of this constitutional body, the metropolitan committee shall have representatives of democratically elected representatives such as mayor, chairman/president municipality, and panchayat in the metropolitan region. The legislative framework of many states in India do not follow this to play the Triple Helix Model in Metropolitan Development. The rigidity of the procedure driven legislations and lack of understanding of the dynamic of the Triple Helix Model are the main reasons. Members should be there from industrial sector or its representative bodies such as the chamber of commerce and industries. There shall be academicians such as vice chancellors, directors of higher educational institutions, directors of public research institutions, and eminent professors in metropolitan committee. This committee shall have an implementing authority with senior administrator heading implementing body whose main function is to implement the decisions of the metropolitan committee. The administrator shall be assisted by three wings namely metropolitan planning wing with highest level of capability for metropolitan regional development planning and capability to manage the modern centre for metropolitan geographic information system using E-democracy for E-government and E-governance; an engineering wing with senior engineer managing it and a finance wing whose main capability shall be the mobilisation of large-scale finances from national and international sources for implementing large-scale infrastructure and other development tasks in addition to traditional account keeping. All these wings shall have provision to employ the highest level of expert consultants to accomplish tasks which normally they cannot perform at market rate.

A global alliance of smart cities shall be formed where Indian smart cities shall be members to exchange experience and share software development as related to E-Democracy. Metropolitan planning committee shall have 20-year spatial and infrastructure development plan, 5-year metropolitan social and economic development plans in tune with state and national five-year plans and annual plans for implementation of the plan as per annual budget. A share of annual budget shall come from state government, and another central government and metropolitan planning committee shall have their own financial resources from entering commercial developments such as land development and leasing of land for industries, development of infrastructure to potential entrepreneurs or running major transport networks for smart mobility. Institute of Town Planners India can have a positive role in formulating model metropolitan planning committee legislation and model organisation framework for the smart city metropolitan committee and its implementing authority. This will considerably help the state government to formulate legislations for the creation of metropolitan planning committee. India's experience in the direction of the metropolitan planning committee is very meagre or one can say almost non-existent.

There are specialised hardware and software required for the functioning of E-Democracy In smart cities and smart communities. Many of these technologies are not implemented in Indian cities. Most of them aim at resource conserving metropolitan functioning of smart cities whether it is of energy or water use or on transportation. Outright purchase of technology and licence for local manufacture and commercial production is one approach for large-scale deployment of these technologies in emerging Indian smart cities. Another approach is collaborative research and joint production. Still another approach is to develop appropriate smart city technologies in Indian research centres based on specific needs. There is a need to appraise globally available smart city technologies and formulate appropriate Indian standards. Indian Standard Institution is competent to develop such standards, and Ministry of Science and Technology can conduct a global smart city urban technologies assessment.

There are no smart city policies for India or states of India. There is a need to conduct basic research leading towards the formulation of policies. This policy formulation can be from the study of other countries such as for example European community who is taking many steps in this direction. Another approach is to evaluate the existing smart cities to learn the experience and transfer it to Indian condition.

There are few experiences such as E-governance and E-commerce available in India. There are also resources conserving technologies being developed in India which can be used in smart cities. These needs to be studied and expanded in scope.

Smart cities open to an information or knowledge society. It is likely that information society may trespass in the privacy of a citizen. These needs to be studied and appropriate norms need to be applied to smart cities 1. E-Democracy for Smart Cities.

18.3 E-Democracy for Smart Lagos

Information and Communication Technologies (ICTs) have strengthened the democratic processes of most democracies in the world. The adoption of e-democracy is to encourage good governance and to properly monitor the democratic processes for the benefits of residents. Lagos has the fastest and most lucrative ICT market in Nigeria and Africa, yet despite this obvious and significant progress, little is known about its E-democracy adoption. Chapter 2 provides a comprehensive review of E-democracy to assess the levels of E-democracy implementation, evaluate critical success factors of E-democracy implementation and discusses the successes and failures of E-democracy. The analysis of the impact of E-democracy on service delivery in Nigeria is also provided. It notes that Nigeria is facing several challenges in the introduction of E-democracy. Given the importance of the successful implementation of electronic democratic services and from a practical perspective, the paper suggests that government should take a positive position towards the factors which will bring about effective and efficient E-democracy in Nigeria. The government should widen access to ICT services and guarantee an enabling environment for attracting the right level of investments. The chapter concludes that there is a lot of hope on the potential of E-democracy to transform the internal efficiency of government and the relationship of government with citizens.

From the foregoing, E-democracy is thriving in another aspect of the electioneering processes such as E-campaign and the screening of eligible voters. Another critical area where E-democracy is contributing positively is in E-monitoring of projects. This has largely been helpful in the provision of critical social infrastructure in Lagos metropolis. Therefore, the potential of E-democracy to transform the internal efficiency of government and the relationship of government with citizens is one of the ultimate panaceas that would place Nigeria's democracy on the global democratic map. However, E-democracy in Lagos is currently marked by some level of apathy arising from the constitutional framework and the history of elections in Nigeria. The history of elections in Nigeria has shown that Nigerians cannot rely on the technology as veritable means of installing the kind of leaders they want and by implication in changing the material conditions of their existence. The overarching implication is the need for government to sensitise Nigerians on the imperative of genuine transparency in the electoral process.

18.4 Outline Development Plan for Feroke Municipality in Execution Framework of Internet of People, Internet of Government, and Internet of Things

This chapter discusses an attempt made by National Institute of Technology Calicut, to explore possibilities of transforming the present democratic process to an E-democracy in the framework of a smart city. Feroke which is a newly formed municipality in the northern Kerala was selected for the case study and the preparation of Outline Development Plan (ODP) is selected as an example process to visualise the application of E-democracy. The ODP was intended to convert Feroke to a self-financing Smart City. The chapter discusses how the traditional approach to ODP or master plan preparation can get transformed to an E-Democratic process in a smart city.

The chapter explores the state of the art of participatory planning in Kerala, and also discuss the initiatives already made to modernise the planning process. Various forms of participatory planning exist in different parts of India, and adaptation of 73rd and 74th amendments by Federal State Government to the constitution institutionalised the process. Kerala has made advances in the participatory planning process in the past. However, when the pace of this process increased consequent to the adoption of globalisation in the 90 s, people were not able to spend much time for meetings and similar interactions. Hence the participation in such traditional methods of interaction has reduced progressively in Kerala. The ward committee meetings were consistently run with very low attendance in Feroke. Advances in digital communication can be of use in this case. All government organisations started publishing all the necessary information on the websites. Many of the governance processes have been automated. 'Sulekha' is an e-platform for facilitating planning and execution of projects in the three-tier local self-government. The chapter extensively discusses various processes facilitated by this platform.

The ODP preparation for Feroke was initiated with a meeting of all stakeholders of Feroke organised with the help of one of the most popular newspaper in Kerala. The meeting was intended to get a comprehensive view of all stakeholders about the major issues faced by them, and suggestions for a better future. The NIT Calicut team were given an opportunity to embark on this E-Democracy Project there. Secondly, the team conducted two sessions to identify major goals and objectives for preparing the ODP for Feroke. This was done by Delphi process, to systematically arrive at the most critical set of goals and objectives. Then a series of sample surveys were conducted to collect data for the ODP. These surveys were conducted with the help of public and people's representatives. The survey findings and various draft stages of ODP preparation were uploaded on a dedicated website for ODP. Suggestions from the public were invited through this website. The first draft of the ODP was presented to representatives of residential association and invited their suggestions on the proposal. These suggestions were incorporated into the second draft and presented to the municipal council. Their suggestions and concerns were recorded and were incorporated in the ODP. Final ODP was again presented to the councillors and was presented later to the public and the local Member of Legislative Assembly (MLA). The ODP was prepared with the maximum public participation possible with the existing infrastructure.

The chapter then looks at the significant proposals made by ODP to upgrade the democracy to E-Democracy. It discusses various Internet of Democracy (IOD), Internet of Government (IOG) and Internet of Things (IOT) tools proposed under six components of the smart city. These tools are intended to enhance public participation in every decision-making process that concerns the public in a day to day basis than occasionally. Many of such important decision can be systematically arrived at by the public through the proposed set of tools.

Once such tools are implemented and operational in the framework of a smart city, the democratic decision can become an easy process, and will be able to eliminate the need for elected representatives, and will allow each individual to take part in the decision-making process without sacrificing much of precious time to attend physical meetings. The chapter then discusses the change in the process of ODP preparation that will occur in three generations. The first generation of E-democracy will be an extension of available digital tools for ODP preparation. The stakeholders will be represented by the respective representatives, and the planners will be preparing the plan and stakeholders involvement will be limited to commenting on the proposals. The second generation will have dedicated tools developed for ODP preparation and will be able to assimilate the spirit of a large number of feedbacks received about the proposals through dedicated social media tools. Still participation of stakeholders will be through representatives, but their actions will be closely monitored by each member of the stakeholder community. The planners will prepare alternate scenarios for major decisions with the help of advanced tools and will leave the choice to the stakeholders. The third generation will have less obtrusive tools for the democratic process, where the system will rely more on the data generated by a plethora of sensors, IOT, IOD and IOG tools. There will be automated tools to gather public opinions from social media and activity pattern of individuals. Decision making will be done by individuals than the representatives. Planners will provide simulation tools to the public to try out combinations of alternatives solutions for various points of decision making in the planning. So E-Democracy in ODP is all about moving from exclusive and restrictive representative democracy to mass and totally inclusive participatory E-democracy using IOG, IOP and IOT.

18.5 E-Democracy in New Delhi Municipal Council: A Case of Smart City Planning Processes

A wave of democracy and more recently E-democracy has gripped South East Asia leading to decentralisation of political and financial authority to local governments. A similar movement is also being experienced in India since 1992 after the devolution of powers to urban local bodies. This chapter examines a unique case where various historical and current political forces have made a part of Delhi less rather than more democratic. New Delhi Municipal Council is one of those rare urban local bodies of the National Capital Territory of Delhi, which is governed by Government of India through an organisational arrangement led by senior civil servants who are assisted by professionals like engineers, architects, city planners and others. Normal democratic regimes and routines such as regular elections and re-elections are non-existent in the working of the New Delhi Municipal Council. As is well known, in the rest of the country, democratic processes are well and truly established since the early 1950s hailing India as the world's largest democracy.

With the declaration of the area under the jurisdiction of NDMC as the first smart city in Delhi, the highly planned and designed city by Edwin Lutyens aspires to become a benchmark for the global smart capitals. Covering the geographical area of 43.7 km^2 (2.95% of the total area of Delhi), and a population of only 0.3 million (1.79% of the entire city-state), NDMC expects to achieve the smart global capital status when one also considers that huge investments per person are being made year after year. The key issue that the answers in this chapter are: Can New Delhi Municipal Council un-leash E-Governance including E-Democracy, and pave the way for E-Democratic governance in the future for its citizens? Answers to this key question is arrived by looking at three cases apart from an analysis of the NDMC Smart City Proposal with a view to assessing how far the smart city processes have promoted democracy in general and E-democracy in particular (citizen engagement) in the NDMC area leading to transparency, accountability, and ease of doing business.

18.6 E-Engagement in Hong Kong

Citizen engagement has taken many centuries to evolve from its roots during the Greek period developing the concept of democracy to its current form as it is practised today. Participation in the planning process started in Great Britain in 1940s and began to be institutionalised in 60s, and became more popular during the 80s and 90s becoming known as participatory planning assumed to result in better decisions. Citizen involvement continues to evolve and varies with specific context and is generally aimed to influence the policies to shape better communities and cities that are more walkable, livable, sustainable and resilient. However, the tools of public engagement are transforming considerably with the use of ICTs during the digital age with very little differentiation between civic and political participation.

Planning in Hong Kong has evolved rapidly over the last two decades, transforming from top-down process to a more participatory process with active citizen engagement, more pronounced since 1997 after the handover from British Rule. Since then with a renewed sense of ownership, there is more involvement of Hong Kong people as there seems to be growing dissatisfaction with the planning and the political processes. However, through more engagement and participatory processes, there is a change for the better in various urban processes such as urban renewal, heritage conservation, harbourfront development in Hong Kong. In Hong Kong, information is readily available and accessible in Hong Kong both in English and Chinese both in printed and in on-line formats. In addition to participatory processes, e-engagement is becoming more prevalent for various projects in Hong Kong. However, it is not always clear that active participation is effective in leading to better results, but it is evident that citizen engagement is evolving in Hong Kong. The importance of community participation and involvement is recognised and with more e-engagement it will be possible to ensure better decision-making in the planning process.

The Kai Tak Development Case Study demonstrates that although the planning process has been underway for more than two decades, community involvement has been instrumental in its planning and development to date. There was extensive public and e-engagement starting with the South East Kowloon Development (SEKD) Statement Study in 1993, Feasibility Study for SEKD in 1998, Comprehensive Feasibility Study of the Revised Scheme of SEKD in 2001, Kai Tak Planning Review in 2004, to the more recent densification studies and the current Kai Tak Development Office and the Energizing Kowloon East Office. This public and e-engagement process have led to zero reclamation and the recent efforts for heritage conservation and environmental improvements of the area. Current planning of the Kai Tak area involves the transformation of the old airport and the surrounding area into the second CBD with office, hotels and mixed-use development including sports, leisure, recreation with transit and pedestrian-oriented development that is environmentally friendly.

In Hong Kong various Government Departments and Bureaus have websites where people in Hong Kong and around the world can get up to date information on various studies, under planning and development. The Town Planning Board transformed to digital Outline Zoning Plans and Outline Development Plans in the 1990s using GIS and other technologies and through its, Portal allows for easy access to information about draft plans for public inspection and consultation. Public input is secured through traditional means by mail but also through emails and online submissions providing more and easy access and also speeding up the process further. Also with the use of interactive 3D models and walk through animation allows the general public to visualise planned development much better than the printed plans before. More recently the use of Public Participation Geographic Information System (PPGIS) in district-wide studies is being used to enhance community participation through e-engagement. Even the private sector, NGOs, academia and community groups use ICT and social media to disseminate information and actively engage people. More informal forms of engagement such as the Magic Carpet and Public Space events in Hong Kong also use ICT and the social media to publicise the events but also encourage on-going e-engagement with the community.

With the emergence of the digital age and through the use of ICT e-engagement has become an integral part of citizen engagement. Through data analytics, data visualisation, simulations, open government, digital documentation, e-submissions, Internet of Things and social media more people can be engaged throughout the planning process. E-engagement has the potential to transform engagement through specific events such as workshops and forums to continuous engagement of the public. ICT also creates the opportunity to be more open, transparent and inclusive by creating a platform for on-going engagement and develop more trust between the community and the NGOs with the public and private sector. People in Hong Kong are very vocal and actively involved and engaged in getting their points across even though Hong Kong is not as democratic as they would like Hong Kong to be. Hong Kong needs to improve e-engagement and be more effective through the use of various tools and social media available to engage the wider community more effectively rather than hearing the voices of the vocal minority. This will further remove barriers to the formation of public, private and community partnerships to help develop Hong Kong into a more affordable, walkable, livable, sustainable and resilient city to become a model for other cities in Asia and the world.

18.7 The Indian Theory of E-Democracy and E-Governance and Their Linkage with the Smart Cities Mission

A theory is a simplified cognitive construct of a certain aspect of the reality built for the purposes of explaining that reality in a reasonably simplified form. Internal coherence is a central feature of a structure of any theory. One significant aspect of today's city and urban planning in India is the reality of the smart city. The government of India is steadfast in developing 100 smart cities. A large number of seminars, conferences, and panel discussions are held since the last two years in the major cities of India involving global and local policy makers, business people, and academics examining and presenting policy elaborations, assessment, analyses and critiques. So far a theory of E-Governance specifically in the context of the Indian Smart Cities Mission has not been built and this chapter is an attempt to fill this gap. One caveat, however, is necessary—a unique E-Governance theory for a smart city is necessary and possible for the Indian city given the special circumstances of Indian urbanisation and its unique historical and cultural settings. The E-Governance theory for a smart city can be constructed in the form of a number of prominent nodes including use of information and communication technologies for enabling economic, social and political transactions, selective commodification and intense development of places within cities, intensification of inequalities within the city as conceived and defined by census, and development to be carried out through projects disconnected from city plans. In the understanding of this author, E-Governance theory should prominently focus on issues such as fairness of decision-making processes in governments, businesses, NGOs, and most importantly decisions taken by the private sector and decisions taken by governments under the influence of the private sector. Second, E-Governance in Indian cities should focus on the nature of the technological agenda being pursued by governments and the private sector supported and promoted by big national and international think tanks. Third, the Indian E-Governance theory should also focus on reducing the material inequalities in the cities specifically related to access basic services such as clean drinking water, sanitation, education, health, and decent housing for all citizens. The fourth aspect of E-Governance theory is that it should encourage and promote economic activities that involve the city's urban poor. No smart city can have an adequate theory of E-Governance without simultaneously looking after its citizens because E-Governance theory in a smart city is essentially for citizens' welfare.

18.8 Smart Dubai: Accelerating Innovation and Leapfrogging E-Democracy

Electronic participation approaches in general and applications of the social web have played a crucial role in decreasing information asymmetry between the government and its citizenry. Dubai presents an in-depth view of the success factors, challenges and extracted lessons. The impact of public trust in the government and the societal readiness for utilising the social web to engage with residents has helped to make the public service design process more transparent and inclusive. This has led to greater citizens' satisfaction and participation in the co-production of public services, setting an example of good governance in the Arab Wold. The inclusive process has embraced as a modus operandi in the government, contribution to enhancing levels of trust in government, paving the way for a stronger partnership between the public and the government as well as smoother sustainable development efforts.

Since its inception in 1971, the United Arab Emirates has constantly been distinguished as an icon for innovation and creativity, enhancing its social and economic status and transforming into a primary destination for talents and businesses in record time. Believing that innovation is the future of human investment, the UAE Leadership emphasizes its importance across all sectors through the UAE Vision 2021: Innovation, research, science and technology will form the pillars of a knowledge-based, highly productive and competitive economy, driven by entrepreneurs in a business-friendly environment where public and private sectors form effective partnerships.

The emirate has quickly understood the value of becoming a smart city. Between 2003 and 2015 it saved AED 4.3 billion ($1.2 billion) due to the adoption of smart technology in government services. In December 2016, a social survey on the level of happiness on a scale of 0–10, happiness has increased from 7.9 in 2011 to 8.2 in 2016. The survey reflects 98% safety and security against 91% in Hong Kong and Singapore. Thus, the 2017 Dubai budget focus on innovation and increasing happiness levels of residents further.

From adopting an automated metro system in 2009 to monitoring solar panels in the desert to Hyperloop connection are the important aspects that are efficiently being implemented via smart technology. Dubai's goal to ensure 25% of all transport is driverless by 2030 will not only provide an economic boost but significantly reduce carbon emissions that is one of the world's biggest dilemmas going forward. Dubai's business friendly rating also will improve the field computing service for business licensing and registration process. Use of artificial intelligence software will also help in the analysis of many data will help the city escalate further. By keeping the doors open and inviting innovators from amateurs to entrepreneurs, Dubai is embracing one the key elements to any success. The emirate's experience and passion for the sector also are paying off in the form of relationships with other emerging cities wanting to take the knowledge path.

"The source of innovation and ideas, whether from citizens, employees or specialists, has not depleted and has formed one of the principles of action of Dubai government." In other words, engaging the citizens for better governance with strong adoption of digital governance and smart city initiatives with technical innovation enables a seamless inclusive engagement process. The UAE may not have a fully developed system of participatory governance in place yet, but engaging a tech-savvy society through digital means in an environment of trust, coupled with systematic 'big data' approaches may prove as inclusive as traditional participatory models.

The next 4 years of the run to Expo 2020 will take Dubai on an accelerated path of innovation and digitalization of government services as it prepares to host 30 million visitors, almost 10 times of its current population. Dubai's smart city revolution is in top gear and achieving its dream of becoming the smartest, healthiest and happiest city on earth is not very far!

18.9 Smart Water Management and E-Democracy in India

Water, the lifeline of human civilisation has become scarce for consumption due to overuse, misuse and mismanagement. The ideology of water management has evolved from provisioning water for use in an urban area to include sewer and drainage to integrating it like a water cycle to the present-day inclusion of democratic governance using technology. Water management is the activity of planning, developing, distributing and optimum use of water resources under defined water policies and regulations.

Water management practices have been prevalent since earliest times and have received patronage from the rulers in the form of construction of storage facilities for water supply to the capital cities and important towns. The citizens were also encouraged to build independent as well as community facilities to meet the domestic needs as in the form of wells and storage tanks. The citizens were responsible for construction and management of such facilities as the pre-British institutions were created on the principles of duties and obligations and the rights were overlapping in nature with a strong local fervour in the management of resources. Such decentralised traditional practices of managing water were practised all over the country. Traditional water management practices declined during the British period due to its imperialistic rule, lack of local ecological knowledge and systems introduced by the British that led to the disruption of the social structure in the country.

As per India's constitutional setup, the authority to formulate, legislate and implement policies on the water that entails water supplies, irrigation, canals, drainage, water storage and water power lies at the state level. There have been major efforts by the government to increase the accessibility of water for human consumption. Despite major programs and policies in India today access to drinking water in India remains a challenge even after six decades of planning and development and despite massive outlays for drinking water supply and management. The most recent initiative of Government of India, Smart Cities Mission, which calls for providing core infrastructure so as to provide a decent quality of life to its citizens, lays emphasis on water supply and infrastructure. Through Smart City solutions, the government intends to bring in smart water and waste management practices, encourage building of smart buildings, smart grid, smart environment and smart security and safety through smart urban planning using smart Information and Communication Technology.

Even though, water is the legislative responsibility of state governments, the process of water service delivery is marked by a top-down approach from the centre (Government in India), where the schemes are launched by the centre, to the states and from the state governments to the municipalities or the local government. All the programs and the policies are decided at the central or the state level and passed on to the local level agencies for implementation.

Presently, urban India is experiencing limited, irregular, unreliable, inequitable and polluted water supply. With growing urbanisation, the demand for water is increasing as is the generation of wastewater. Leakage and inefficiencies in the water supply system waste nearly 50% of usable water. The ground water level is declining at the rate of 10 cm per year. Over 70 per cent of surface water and groundwater resources are contaminated. In the absence of strong regulation, industrial and domestic wastewater is discharged into rivers, canals and underground water sources. About 70% of underground and surface water resources in India have been contaminated. All this is leading towards a water-scarce situation in most urban areas of the country.

In order to overcome the present water-scarce situation in the country, it is essential to shift from water management to water governance based on hydro-informatics in which information and communications technologies (ICTs) is applied in addressing the problems of the inequitable and inefficient use of water for different purposes. The smart water management practices that need to be adopted by the agencies will not only enable them in providing better services but will bring in transparency in the system. While E-Democracy will ensure accountability and transparency, hydro-informatics will ensure that decisions taken are well informed, thus providing smart water management solutions. Agencies like Kerala Water Authority, Bangalore Water Supply and Sewerage Board and Nagpur Municipal Corporation have made use of ICT in reducing non-revenue water losses, providing metered water to slums and outsourcing water supply for 24 h' water supply respectively.

Over and above community participation using ICT is an essential building block in water reforms. Only through a partnership between people who have suffered for want of water and water agencies who believe in democratic functioning can we ensure safe, equitable and adequate water and understand the need for conservation of resources and ensuring sustainable water systems.

18.10 Preparation of Town Planning Scheme and E-Democracy Framework for Citizen-Centric Planning

The chapter elaborates on the importance citizen participation framework in City Planning and details out the process of preparation of Town Planning schemes in Gujarat State in India. It focuses on the citizen participation mechanism provisioned in the legislation of Gujarat, the process followed and proposes ICT based framework.

Gujarat has been one of the progressive states of India so far as rate of urbanisation is concerned, more importantly, urban reforms are concerned, and Town Planning schemes which have been practised since early Twentieth Century in Urban Gujarat has been a model for other cities. The preparation of TP Scheme has provisions of public consultation at all the stages during the preparation of the TP Schemes and also stipulates timelines for the TP Schemes

Gujarat has pioneered in the field of Urban Planning, and have taken up more than 600 Town Planning schemes, which has been tested, modified and improved with a process re-engineering approach. Still, against stipulated time frame of 36–40 months, completion timeframe has been more than 70 months in the recent times. When literature states of it's a participative and democratic approach, processes mentioned fails to assess these attributes or even they are not a part of assessment either. This chapter has analysed the 128 TP Schemes implemented by Surat Municipal Corporation and Surat Urban Development Authority which depicts that while Schemes implemented prior to 1990 took more time to prepare draft TP Schemes, and the recent TP scheme preparation consumed less time probably due to use of new technology, but still the overall time consumed in approval of TP Schemes more or less remains unchanged as there has been no improvisation or re-engineering in the engagement of citizens including consultation and grievance handling.

As cities have now launched e-governance and m-governance mode to engage with citizens, a new ICT framework for TP scheme process has been proposed which could be integrated on the existing ICT frame of the Cities and connect with the citizens. Secondly, a backend analytical tool may be developed to assess the participation, suggestions, objections and guide decision making following the democratic methodology. The overarching GTPUD Act 1976 warrants for various statutory approvals including that of State Government, a suggestions is being made if such internal Government approvals and processes could be made online in a secured network. This would improve accountability and transparency in the process. A consultation framework compatible with the existing platform of Surat Municipal Corporation is proposed, and it is anticipated that the time required for preparation of TP Schemes can be reduced and would resource efficient as well so far as consumption of government manpower resource is concerned.

As the Urban population is growing at a faster pace, so are the challenges for City Administrators to match the supply of developed land and infrastructure.

Administrators are also finding it difficult to manage the growth of slums as wells the rising inequalities within the economic class of the Indian Cities. One of the reasons for this could be information asymmetry and regulatory capture. And these are outcomes of limited or poor citizen participations. A new approach to urban planning has emerged with the incorporation of the principles of urban planning.

Administrators of the central government are on the move to educate Indian citizens to practice the excellent platform of E-Governance. This will enable the large Indian mass to express their thoughts and willingness in a more free way. Therefore, the intention of this chapter is to throw light on ICT, specifically, the planning process adopted in a fast growing state of Gujarat. Apart from that, issues regarding E-Democracy initiatives are also addressed. Though the process of E-Governance is in its infant state from user's point of view, but this will surely gain momentum in the days to come. This practice will surely display high-Democratic values and plan implementation within desired time frame.

18.11 Smart City Project Selection, Prioritization, Customization and Implementation-Voice of Smart Citizens Case Study of Bhubaneswar and Surat

The selection process of Indian cities to be empanelled as smart cities under the Smart City Mission by Government of India trained the City Authorities to adopt a citizen participative approach for the first time in India a mega scale Citizen participation program were conducted. Considering, the threshold consultations requirement stipulated under the mission and the limited time for submission of proposals, cities adopted various innovative approaches to reach out to their citizens and get them engaged in the process and mostly were through online platforms. Moreover, the cities those scored high in the challenge were the ones which had adopted citizens identifying the problems and democratic process for selection of solutions.

Smart City Mission guided ULBs to increase participation of citizens and probably set the practice for city authorities to adopt citizen engagement for their endeavours be it in policy or in infrastructure development.

This chapter also assesses the stages identified for citizen engagement framework as suggested by the Department of Electronics and Information Technology under the National E-Government Program (NeGP) of Government of India and indicates about the stage where both Bhubaneswar and Surat stand. There are five distinct stages, Inform, consult, involve, collaborate and empower citizens in the citizen engagement framework. The chapter analyses the existing citizen engagement App by Bhubaneswar (My City My Pride-MCMP) and Surat (Citizen Connect, Surat) on the basis of principles for E-democracy, and suggests that how these tools could be utilised further in engaging citizens from identification of issues, project conceptualization, budgeting, and implementation. It is observed that while the Citizen Connect of Surat provisions for more number of functionalities and comprehensive in the e-mode the MCMP App by Bhubaneswar Municipal corporation engagement is more advanced so far as the engagement spectrum is concerned.

This paper also suggests that for all Smart Cities, such initiatives would be crucial and much needed to educate citizens to transform into Smart People. While the country has taken a strong step to promote cashless economy, the role of Cities being at the bottom of the pyramid shall be crucial. Different tiers of government obviously have different responsibilities and consequently different relationships with citizens. National governments, responsible for the welfare of a whole country, are unlikely to understand the needs of communities in the same way as local government. Therefore, many e-democracy researchers see the local government level as the most natural place for government to engage with citizens.

Democracy in the context of Project Selection or Policymaking when the projects or policies are conceptualised and are implemented, which are meant for the citizens ought to ensure the factor that these projects are "by" and "of" the citizens indicating proper consultation in the process. Most of the projects those are implemented without consultation, have failed due to many reasons one of them being ownership of Citizens. With this perspective, while cities have adopted e-governance and m governance to reach out to citizens for sharing information and receiving complaints, the systems could be made more comprehensive to transform these problems to create public welfare projects. Such an engagement would also educate citizens about the likely investment, source of funding, likely cost and pricing mechanism. This consensus building would help in decision making by the Authorities and also, more importantly, bring in social cohesion and ownership by the citizens. The success rate of such project from the utility perspective would also be high. The analysis of Apps put in place by Surat and Bhubaneswar gives an understanding about the degree of the citizen-centric approach adopted by the Authorities.

An analysis of a web-based citizen engagement activity undertaken by Surat Municipal Corporation for collaborating with citizens to assess the opinion on city beautification but the participation level is abysmally low which was below 300 in a city with more than 1 million households. One of the reasons for poor participation could be improper positioning and inadequate marketing or could be the interest of citizens to discuss or citizens are not habituated to participate and have accepted the representative democracy as a way of life.

When country like India strive to build Smart Cities, one of the crucial building blocks would be Smart People who adapt to the change and accept the new way of lifestyle using ICT, be it travelling in metros with cash cards, be it undertaking online classes, or be it surviving in a cashless economy, its people who would make it a success. Exposure of citizens to such a transformed set up shall require City Administrators to build up engage and expose citizens to such systems. Smart People shall make cities Smart.

18.12 Smart Grid Evolution: E-Democracy and Citizens' Participation in E-Governance of Energy Backbone of Smart City

India is transforming into a new democracy with novel and creative innovations in all sectors. The government is focusing on improving the quality of life of citizens by urban renewal and infrastructure development vide ambitious smart cities project. Energy, the electrical power has been the most crucial and the resource always in scarcity in India. Electricity has been proving itself as a major bottleneck for the growth of the country, because of its role in GDP, rapidly growing requirements and difficult to predict vibrant climatic situations. To fulfil the commitments of 24 × 7 availability of electricity, not only for smart cities but for everyone in India necessitates smart, inclusive and participatory management. Therefore, India has been transforming legacy conventional India is transforming into a new democracy with novel and creative innovations in all sectors. The government is focusing on improving the quality of life of citizens by urban renewal and infrastructure development vide ambitious smart cities project. Energy, the electrical power has been the most crucial and the resource always in scarcity in India. Electricity has been proving itself as a major bottleneck for the growth of the country, because of its role in GDP, rapidly growing requirements and difficult to predict vibrant climatic situations. To fulfil the commitments of 24 × 7 availability of electricity, not only for smart cities but for everyone in India necessitates smart, inclusive and participatory management. Therefore, India has been transforming legacy conventional non-smart non-intelligent unidirectional electrical power grids into modern smart grids which are bidirectional and intelligent in nature by leveraging ICT and e-Democracy. Smart grids shall serve as energy backbones of smart cities and involve the high interactive participation of citizens in energy management, based on humanitarian and customer centric approach. Different types of prosumers (producers + consumers), their different energy requirements at different timings, different types of energy resources and their switching feasibilities considering different aspects have been integrated.

Moving ahead from their earlier contributed chapter on E-Governance of the solar photovoltaic power grid, in an earlier book in this book series, in this chapter, presents an interesting description of smart grid development in India, Gujarat state and in upcoming smart city Ahmedabad. After presenting the historical perspectives and present situations of electrical power sector in India, interesting relationships of electricity, urbanisation and human settlement have been presented. An interesting case study of one of the most successful Indian smart grid pilot project by UGVCL at Naroda, Ahmedabad, Gujarat, India has been included with brief details regarding citizen participation tools based on internet and mobile. Experiences, issues, challenges along with vision and future roadmap have been shared to mark the end of the chapter.

18.13 Integration of Existing Information Systems as a Spatial E-Democracy Tool for Smart Cities in India

In the Indian context, Government of India focuses on 'sustainable and inclusive development in compact areas' basing the Smart City concept on core infrastructure elements. The smartness in the urban areas is to be brought about through smart solutions. The Smart Cities should essentially be developed based on citizen-friendly governance. The government has categorically stated the use of geospatial technologies for the building of Smart Cities to best capture real-time data.

Two important Information Systems for urban areas in India are National Urban Information System (NUIS) and Environmental Information System (ENVIS). National Urban Information System is a scheme for developing an urban database to build an information system to enable planning, management and decentralised governance in the context of the provision of scientific planning and implementation of the 74th CAA. While NUIS is based on standards and procedures defined at the national level, ENVIS is not so. ENVIS is a decentralised system with a web-based distributed network of environment specific databases. The basis of NUIS is Geographic Information System and it provides data towards master plan preparation among other planning and management activities of urban areas by urban authorities.

Over the last few years, the process of planning has been evolving into a democratic system with increasing public participation, hence, people being able to influence the decision-making process. Planning and its process have been much influenced by various policies and guidelines that have been formulated over the years since Independence, which have taken learning from best practices world over. With the advent and popularisation of technology, the shift has started towards E-Democracy, where Information Systems and communication technology is being used to support the democratic decision-making process.

These two important Information Systems of the urban areas, however, exist in isolation to each other. It is imperative that these be integrated to build smart cities. The framework proposed here is a conceptual structure to serve as a support for building an environmentally sensitive master plan. For the integration of the two systems, the existing design, methods and procedures of both the systems need to be assessed for conflict resolution and be built on the national standards. It would be essential for one organisation to take up the onus of integration. Since the area under consideration here is urban, it could be an organisation within the urban ministry that could take on the responsibility. With the integration of the two systems, the organisation can ensure the inclusion of not only environmental aspects in master planning and building smart cities but also e-participation of the people and hence spatial E-Democracy.

18.14 Geo-Enabled E-Democracy Tools and Services for Smart Cities

With growing urbanisation in different parts of the world, especially in Asian and African regions, there is a great need to plan, manage and monitor our urban areas efficiently. The most advocated concept of recent times for urban areas, the Smart City concept, aims to enhance the quality of life of its citizen, to improve the provision of infrastructure along with clean and sustainable environment through the applications of smart solutions. Gaining access to information and public participation in decision-making is one of the hallmarks of Smart City concept. It emphasises the importance of public participation through the use of technology or digital media to achieve this long-term goal and for identification of sustainable solutions. The E-democracy tools enable citizens to access the information, to collaborate and to actively participate in decision making. The participation of the citizen in decision making and planning process helps planners to address the concerns of target groups in advance which reduces later unrest and can speed up the implementation process.

The geo-enablement of e-Democracy tools has the advantage that it adds geo-location information to the data delivered to the public, which makes it easily comprehensible. The location information is used in everyday life directly or indirectly, be it identifying an address, location, facility or group of services. Geo-enablement is not only about developing Geographic Information System (GIS) data but it also emphasises the access to GIS-enabled data, which certainly carries geo-location information to the public. It empowers the public or citizen to comprehend the information in its locational setting as well as enables them to understand the impact of different components on decision making. For example, availability of geo-location information on a group of closely located unpacked bins can help planners to plan resources, which may be later targeted for finding a comprehensive solution. For geo-enablement of e-democracy tools, Geospatial technologies namely Remote Sensing, GIS and Global Navigation Satellite System (GNSS) can play a significant role as it is one of the major providers of geo-enabled data. The data acquired from remote sensors namely Satellite Images, Aerial Photographs, Airborne and Terrestrial LiDAR, UAV, smartphones and ground-based devices, today offer the data ranging from mm to km resolution pertaining to micro scales to global scales. It provides unique information on urban areas due to its capabilities of synoptic view, repetitive coverage and at present is one of the major sources of geo-enabled data. The advancements in GIS has taken a leap from single desktop user to the advent of geospatial web 2.0 to the cloud computing, Internet of Things and Service Oriented Architecture. These technological advancements provide the necessary platform for integrating the remote sensing and ground-based information on geo-tagged and interactive virtual environment. The availability of information on these platforms has brought a paradigm shift in the usage of geographic information as now even an inexperienced, untrained user can understand, modify and analyse the geo-information. It enables to support three main components of any democratic process—information to collaboration and participation. Similarly, the Global Navigation Satellite System or GNSS is also known as positioning infrastructure and provides high precision positioning services for a range of location-based applications search and rescue, vehicle tracking, fleet management, etc. With the availability of GPS-enabled Smartphones, this technology is emerging as one of the leading technology for the generation of geo-enabled data for use by planners and decision makers. The alliance of geospatial web 2.0 and GNSS technology has made it possible for any citizen whether trained in GIS or not, to gather locational data and share it with the world. The Google maps, Bing Maps and Open Street Maps are the biggest examples of the mass application of these technologies.

The latest advancements in geospatial technologies also made it possible to host Geo-enabled data to be made available in public domain, hence improving the access to information tremendously. A list of geo-portals and geo-data online repositories such as Google, Bhuvan, Wudapt, BIS, IBIN, etc. have been provided in this chapter from where geographic datasets can be freely downloaded and can be used for the development of a range of Smart City applications. These data sets are available at varied scales from global to local, however, still a lot need to be done for the creation of necessary datasets for multiple requirements of upcoming Smart Cities. The Bhuvan portal which is developed and hosted by the Indian Space Research Organisation (ISRO), provides multitude of datasets, tools, information on applications and very high resolution datasets of Indian cities, which are highly useful for developing base data as well as a range of applications for upcoming Smart City and Atal Mission for Rejuvenation and Urban Transformation (AMRUT) mission launched by Government of India. The portal also portrays few of the applications for urban managers such as municipal GIS, urban growth monitoring, urban framework survey, etc. which provide examples for developing similar systems for enhancing democratic participation of urban populace.

The geo-enabled e-democracy tools have been used in many aspects of urban governance and management from grievance redressal to involvement of citizens in the generation of information, tracking of vehicle fleet for service delivery enhancement, location-based services, real-time mapping and participatory GIS. However, still, its applications and use have been limited only to information dissemination to public and collection of data by involving citizen in data generation process. The collection and use of information in the truly democratic way would be possible through the active involvement of citizens in the decision-making and planning processes.

Smart cities focus on integrated urban development through citizen participation by the use of technology, hence the geo-enabled e-Democracy tools definitely have potentials to enhance citizen participation and strengthen the democratic processes in local governance. Aspirations are high to achieve this goal through Smart City mission, and Geo-web tools and services have immense capabilities to provide information to the citizens in an interactive manner and easily comprehensible form so that they could participate in planning and decision-making. The limitations are more of institutional and political rather than technological. There are various issues which need to be addressed such as data integration, privacy and security, the skills gap and capacity building, and therefore, institutional setup and political will are necessary for the true implementation of these tools and services.

18.15 Attaining E-Democracy Through Digital Platform in Kenya

The role of responsible citizens in any country—whether in a city or a village—is to actively participate in the development of their areas. This is a key pre-condition for the attainment of sustainable development, and is deeply entrenched in the sustainable development goals. While the citizens' right to the city and to actively contribute to a sustainable world cuts across the 17 goals, the provisions of goal 16, to "*promote peaceful and inclusive societies for sustainable development, provide access to justice for all and build effective, accountable and inclusive institutions at all levels*" clearly spells out the need for inclusiveness, participation, and access to information at all governance levels; which would promote a sustainable, inclusive and prosperous world. These are central elements of democratic spaces, which are greatly being enhanced by today's advances in information and communication technologies (ICTs).

Kenya has over the recent past made huge strides in the area of promoting e-government services which has resulted in increased e-democracy alternatives. This has particularly been the case since the creation of the e-government strategy in 2004, as well as the roll out of reforms such as the Kenya Local Government Reform Programme (KLGRP) in early 2000s and a review of the constitution in 2010. These reforms have expanded opportunities for citizens to participate in development processes, including governance. The advances in ICT in the country, which has seen mobile phone penetration rise from less than one percent to 88.1% in just 15 years (2000–2015), and broadband increase from a mere 28 Megabits per second in 2004 to 193.58 Gigabytes per second in 2015 has further expanded the options for citizen engagement in shaping the country's smart growth. As a result of these changes, more Kenyans today can easily access information on, and participate in various processes such as national and local budget preparation and plan formulation processes; demand for accountability from the political class on resource utilization; or even submit complaints on poor governance and service delivery through a diversity of platforms. Equally, citizens are able to easily and quickly access services through a few clicks on their phones or computers, services that traditionally took months to acquire through bureaucratic processes.

This chapter outlines how Kenya has in the recent past adopted ICT tools and methods in a bid to attain inclusive, transparent, effective and efficient governance.

It discusses how the country is utilizing various ICT avenues to promote access to information by the citizens and through which governance structures get feedback on issues. It further illustrates the existing avenues for active citizen engagement in democratic and governance processes and analyzes how effective the adopted ICT systems are in promoting inclusive and smart growth.

18.16 Politics of Open Data in Russia: Regional and Municipal Perspectives

A promising path to the development of electronic democracy in smart cities is based on the usage of the Open Government Data (OGD), i.e. data that is stored in machine-readable, non-proprietary formats that allow its processing without additional changes. OGD is becoming a part of a broader concept of Open Government, a philosophy of governance based on providing citizens with information and tools to control the government, as well as to set policy agendas. Despite the democratic connotations, OGD may advance economy as well as raise the quality of life. It is now widely used in urban planning and smart cities' development.

However, there are challenges governments need to tackle before OGD will strongly impact political and socio-economic life. There is evidence that OGD falls short of overcoming digital divides, and that the process of opening data implies politics, i.e. power relations and resources allocation. Hence, the existing political and institutional context strongly affects the outcomes of OGD policy on different levels.

The policy of OGD in Russia, taking into account the federal, regional and municipal level has been studied in this chapter. Although the federal policy has its drawbacks, and the authoritarian regime and the lack of political leadership generally negatively affect the results of the policy, other levels of government show various, if not polar, results. The authors have taken various research techniques to find out what determines the success in OGD development. Having applied quantitative analysis dealing with structural factors, they revealed that no single factor could explain the regional OGD performance. The most significant determinants are related to the ability of public agencies to adapt to new technologies, namely the effectiveness of bureaucracy and e-government development. However, agency, not structure, as argued by the authors, matters more. For instance, low level of motivation and skills of those public servants involved in OGD policy is usually the key obstacle to the policy success. The second problem is derived from the institutional environment and is related to the lack of legal and enforcement basis. Finally, it is the lack of public involvement in OGD release.

18.17 Urbane: Community Driven Architecture and Planning Through a Mobile Social Platform

Most projects for the community are undertaken without input from the community dwellers. There are forums and municipal websites which formulate proposals for comments, but these proposals were not developed out of the community voice, needs and aspirations. A community, understands its needs the best; much better than a social scientist, architect, planner, politicians and various institutions such as municipal bodies. All stakeholders must collaborate with the community to get the best results and fruitful outcome and impact. With this background, this chapter proposes a mobile online social platform to connect different stakeholders to generate high levels of engagement in decision making, planning, idea generation and lead to resilient, smart and sustainable communities. People act as data donors by leveraging smartphone technology and the platform empowers the users to bring positive change. An illustrative experiment is reported. Civic engagement and community participation need an awakening. We need smart and connected people to enhance civic life and for making our cities better and more liveable. We need to give citizens the tools to do so.

The reality is the world is highly-connected. One should not miss out on the opportunity to take into account the human element while planning or developing cities, especially smart ones. Mobile technologies, especially in developing countries like India, have a crucial role to play to aid in smarter planning. But for this to happen we need the right tools and the right approach.

With "Urbane" app the whole idea was to streamline the process of data collection and citizen engagement. The valuable data collected when fed into the system will be able to give insights and help in making key decisions about the planning and development of the city. The most important thing being, all this data is a log of actual desires and aspirations of people who reside in these cities and communities and who are the best judge of their needs and requirements. This bottom-up approach makes not only the process more inclusive but also responsible. With data-analytics taken into account, we can know how every department is being responsive in addressing the issues and suggestions. Such measures make sure there is accountability. Thus, we can create a robust system where true democracy in the digital age starts taking shape, a system which is responsive and responsible.

Furthermore, when datasets are made public and openly available, that is when a more evolved democratic platform starts taking shape. Not only transparency increases but third parties can also come in and start fulfilling the needs of the citizens. Governance can become more fluid and open. Other members of the community can start responding and collaborating in the true sense. The time taken to get things done will shorten. Feedback will be quick. New ideas will be appreciated and reach more people. Communities and citizens will start organising themselves around collective goals they want to achieve and will strive hard to achieve it by working with other members and government departments.

Democracy will not just be about elections every 4–5 years, it will be updating in real-time and moulding itself according to what citizens really desire and aspire for, what the majority wants. It definitely will not be a one-way street where only citizens' voice matters but even leaders can strike up a conversation with anyone and have a dialogue. Policies they want to implement or key decisions they want to take could be taken into voting before it's finalised. Amendments and suggestions on improving the policy or decision can be taken into account almost instantly and have a much faster roll-out.

The value of such a platform is having an opinion and acting towards an outcome. We can connect the right minds together like never before to solve our greatest challenges and build better communities and cities. The implementation will be tough and there will be a lot of learning to do but it will be for the better!

This open the floodgates to the human mind and spirit and let democracy work towards making our lives better in a way that it was always sought to be. This is not just limited to smart cities, any community or organisation can opt for such a manner.

"Urbane" platform empowers the citizens and makes effective use of new media, mobile technology and social structure. It paves way for inspired living. Such a platform in future can improve governance, be an improved tool for democracy at all levels, be the next generation of social networks which gives meaningful conversation and deeper collaboration and a platform for all of your urban life. It will help connect the digital to the physical.

18.18 Blockchain-Powered Internet of Things-Governance and E-Democracy

The goal of democracy is to offer every citizen freedom. Freedom is a stage where every citizen acts under his own will without any coercion from others. Currently, two dilemmas forbid us to attain this ideal stage: (1) the government is overpowered; (2) our limited knowledge stops us from making the most rational decisions. Our chapter has tried to demonstrate that blockchain's many concepts and features have been or will be extended to a wide variety of situations. These features will apply not only to the immediate context of currency and payments (blockchain 1.0) but also to all kinds of contracts in our contract-based society (blockchain 2.0). We paid special attentions to the current applications of blockchain which might affect the government and the democracy, including the use of blockchain to collect information; to make devices intelligent; to automate routine work and to enhance voting systems. We believe that, when blockchain is mature in the future, it will impact the government and the democracy as significantly as what the Internet has done.

In the famous book Denationalisation of Money, Hayek wrote, history is largely inflation engineered by governments. The individual's life quality subjects to the inflation rate and thus subjects to the government's behaviour. He calls for a completely free market in the production, distribution and management of money. When blockchain becomes common, this may be realistic soon. With a blockchain in place, Commercial banks can issue their own currencies, like Bitcoin and Lite coin, instead of a central bank issuing the national currency. The individual's economic life then is not subjected to the government's behaviours directly, so the power of government is greatly restricted without disrupting the social order. Blockchain also makes the democracy of the future society more "reliable". Most of the modern governments separate the three powers: the executive power, the legislative power, and judicial power, with a hope that each branch will check the behaviours of the other two branches. However, it gives us a false sense of safety. The three powers can collide if their interests happen to align. For instance, the President of United States can nominate the Chief Judges. If the political party which the president belongs to wins most seats in the congress, the nomination will be agreed without doubts. In this situation, the three powers are no longer separated but citizens still have a false sense that political powers are not controlled by any small group. The problem is that citizens cannot participant political affairs directly. For the consideration of efficiency, people are required to select representatives to represent them in most affairs. We have mentioned that the blockchain-based voting system is costless and convenient and that blockchain itself is a tool of consensus. So in the future, people can join political affairs more directly. Instead of watching news about whether a legislation is passed or not, they can take out their phones and vote for it. In the meantime, the legislation that citizens agree on can be enforced by smart contracts without human intervention. The executive power is controlled by all citizens instead of special agencies.

Despite limiting the power of government, blockchain also provides high-resolution information, which can change the way the government makes decisions. Every piece of information on blockchain is a consensus between intelligent agents including both human beings and intelligent devices, as we have mentioned in our chapter, so the information is of high resolution and provides the ground truth about the situations in the real world. Instead of sophisms and biased viewpoints, members of parliaments can directly use the facts on blockchain to reason about the legislations. Less time and efforts are devoted to meaningless debates so governments can react faster to fast-moving crises. Moreover, nowadays citizens are required to select representatives to represent them in most of the political affairs for the consideration of efficiency. We have mentioned in our chapter that adopting blockchain on a large scale is costless and that the consensus on blockchain is fast, usually every 10 min every user will achieve a consensus. Citizens might take out their phones to vote for a legislation instead of waiting for the results on TVs. The democratic government is no longer an elite group or a black box but a service provider that every citizen has access to.

However, we have to admit that blockchain currently is not ready to board millions of applications. The barriers come mainly from scandals and government regulations. Since blockchain was born, blockchain has started to be used for money laundering, drug-related, and other illicit activity—for example, the Silk

Road, which used to be the largest online platform for trading drugs. Blockchain is neutral as a technology and can be used for good or evil, but it is extremely powerful for malicious aims. Governments might impose strong regulations on the blockchain, which could make blockchain not comparable with other well-perceived technologies like cloud computing. Adopting blockchain on a large scale also hurts governments' interests. It's impossible for governments to carry out traditional taxation on transactions on the blockchain. Every transaction on blockchain is anonymous and paid by digital currencies. It' hard to trace the consumption of goods and services as before. Blockchain also threatens the powers of government officials since it can take over part of the government's duties without human intervention. The threatened officials might be invisible obstacles for adopting blockchain in the future.

Despite these potential limitations, there are no doubts that blockchain is still a disruptive force and that it will have significant impacts on the government and the democracy. Centralization is a good idea. By bringing banks and powerful governments, it greatly improved our civilisation. However, as human activities become more and more complex, the time of decentralisation has come. Blockchain is just the first large-scale implementation of decentralised models. Even if it fails, there would be something else that implements the decentralised the model and reshape the government and the democracy.

18.19 Conclusion

There is a widespread belief that E-Democracy is the driving force of Smart City functioning. While the models of republican democracy in ancient world was fashioned out of full people's participation in smaller settlement, the present reality is different. With increase rate of urbanization, the large size urban agglomerations are growing in a fester rat and very small size settlements are disappearing or merging with urban agglomeration. These larger settlements require democracy as much as the smaller settlements in yester years. This can be only achieved by E-Democracy. Representative democracy was only a solution to the increasing size of the country or settlements in an era of no or limited development of ICT. The ICT technologies accelerated in the 21st centaury and now it is possible for all citizen to actively involved in all democratic process of Governance using Smart phones. The access to smart phones is accelerating day by day and internet id is penetrating in a big way in democratic functioning of cities and nations. These development is making democracy move away from representative democracy to direct democracy of people.

This book shows that through few examples of city studies that E-Democracy in Smart Cities are showing only a beginning. There are miles to go but acceptance of E-Democracy is very much there. Whatever that is existing can influence the planning and implementing decisions based on the opinions and real needs of the people. However, policies that encourage rapid development of E-Democracy is not visible in many cities studied in this book.

While cities have been slower in adaptation of E-Democracies, many domains such as energy, and water for example showed rapid strides in adopting E-Democracy. This is well illustrated in many chapters of the book.

The major development of E-Democracy came from scientific and technological communities who build robust E-Democracy tools after carefully considering issues centered around the implementation of E-Democracy. This book discusses this in many chapters and there are more chanters in this section than other two aspects covered in the book. These technological innovations are awaiting to be used in many smart cities. This shows future of E-Democracy in Smart Cities is promising.

References

1. Vinod Kumar TM, Associates Editor (2014) Geographic information system for smart cities. Copal Publishing Group, New Delhi
2. Vinod Kumar TM (ed) (2015) E-governance for smart cities. Springer, Singapore
3. Vinod Kumar TM (ed) (2016) Smart economy in smart cities. Springer-Nature, Singapore